# Statistical Software
## A Comparative Review

### Ivor Francis
Cornell University

**North Holland**
New York · Amsterdam · Oxford

Elsevier North Holland, Inc.
52 Vanderbilt Avenue, New York, New York 10017

Sole Distributors Outside the United States and Canada:

Elsevier Science Publishers B.V.
P.O. Box 211, 1000 AE Amsterdam, The Netherlands

271,09

Library of Congress Cataloging in Publication Data

Francis, Ivor.
    Statistical software: a comparative review.

    Bibliography: p.
    Includes index.
    1. Statistics—Computer programs. 2. Mathematical
    statistics—Computer programs. I. Title.
QA76.4.F7        001.4'22'0285425        81-12568
ISBN 0-444-00658-3                       AACR2

B/19038

Manufactured in the United States of America

# PREFACE

At its business meeting in 1973 the Section on Statistical Computing of the American Statistical Association (ASA) expressed its interest in the quality of statistical software by appointing a Committee on Statistical Program Packages. At the 1974 meeting the Committee presented a proposal and was encouraged by the Section to evaluate the state of statistical software. (See Francis, Heiberger and Velleman, 1975.) Statistical societies in several other countries expressed a similar interest. Sessions of invited papers on the quality of statistical software became a regular feature at many statistical conferences, particularly those of the ASA and the Symposium on the Interface of Computer Science and Statistics. In 1977 an exhibition of statistical software was held at the first meeting of the International Association for Statistical Computing (IASC).

The proceedings of this exhibition were published in A Comparative Review of Statistical Software (CRSS) (Francis, 1979). This review contained descriptions, written by developers, of thirty-eight statistical packages. In addition, forty-six developers rated their packages on over fifty statistical capabilities, and an extensive bibliography was included.

The 1975 ASA Committee report proposed a series of reviews and evaluations to be carried out by users. Also, in the Foreword to the CRSS, the President of the IASC expressed the hope that future reviews would include evaluations by users. It was with this in mind, as well as the desire to increase the coverage of CRSS and bring the material up to date, that the present review was carried out.

The original impetus for the evaluation of statistical software in general and these reviews in particular came from the ASA Section, particularly its continuing Committee on Statistical Software which organizes invited-paper sessions at ASA meetings and Symposia on the Interface of Computer Science and Statistics. Thanks are due to the officers of the Section and to Committee members for these efforts. Thanks are also due to the members of the ASA's Subsection on Survey Research Methods, the International Association of Survey Statisticians, and several other organizations listed in Chapter 1, who participated in the early user surveys. We would like to thank the IASC and its officers for help in publishing the CRSS in 1979 and for encouragement to prepare this present book. The IASC maintains an interest in statistical software, particularly in its Committee on the Technical Information Exchange on Statistical Software. And we would like to thank Elsevier North-Holland and its editors for help in publishing this book.

The financial support for these reviews has come from a grant to Cornell University from the National Science Foundation, #MCS-13373-A02, and from the State of New York through the New York State School of Industrial and Labor Relations at Cornell University.

Several people deserve special thanks. Earlier collaboration with Richard Heiberger and Joseph Sedransk was essential in the evolution of the questionnare used here. Kelly Smurthwaite prepared all the graphical material in this book as well as most of the tables and assisted in the preparation and proof-reading of the material. Pat Mayo typed most of the manuscript. Francesca Verdier helped with the programming.

Lawrence Wood managed the surveys carried out in 1979-80 and contributed to the analysis of the data.  Evelyn Maybee not only assisted in the preparation of materials and their organization, but also provided the administrative glue that held the many pieces of this project together.  Without her thorough and dedicated work over many months in preparing this book for publication, this book, like the CRSS before it, would never have seen the light of day.

Finally, thanks are due to the many software developers, or their representatives, and users who have provided information for this review.  Without them there would be no review.  Their names are listed on the following pages.

We would ask any readers who know of statistical software not included here to send the names and addresses of the developers to us for inclusion in the next review.  Equally as important, any readers willing to provide users' opinions should send their names and addresses to us.

Ivor Francis
Department of Economic and Social Statistics
358 Ives Hall
Cornell University
Ithaca, New York 14853
June 1981

DEVELOPERS OR THEIR REPRESENTATIVES WHO

(a) COMPLETED Q2 QUESTIONNAIRE    (b) CONTRIBUTED MATERIAL FOR
                                      THE DESCRIPTIVE SUMMARIES

Program

| | | (a) | (b) |
|---|---|---|---|
| 2.01 | UPDATE | Jerome Weiland | Jerome Weiland |
| 2.02 | RAPID | Paul Cotton | Michael Jeays |
| 2.03 | ADABAS | Jane Jodeit | Catherine Vivona |
| 2.04 | SIR | Eli Cohen | Irene A. Navickas |
| 2.05 | MARK IV | Michael Amadio | Wesley Stukenberg |
| 2.06 | RIQS | Lorraine Borman | Ruth Kobbe |
| 2.07 | DATA 3 | Joel Achtenberg | Joel Achtenberg |
| 2.08 | FILEBOL | Donald H. Goldhammer | Donald H. Goldhammer |
| 2.09 | KPSIM/KPVER | Philip S. Sidel | Philip S. Sidel |
| | | | |
| 3.01 | GES | Richard Shively | Garry D. Kepley |
| 3.02 | CONCOR | Pedro Sust | Robert Bair |
| 3.03 | CHARO | William A. Gates | Anne Cooper |
| 3.04 | EDITCK | Theodore Karrison | Theodore Karrison |
| 3.05 | VCPLC | Jerome Weiland | Jerome Weiland |
| | | | |
| 4.01 | SYNTAX II | John Fitz | John Fitz |
| 4.02 | LEDA | Gerard Pougetoux | Gerard Pougetoux |
| 4.03 | CYBER GEN | G. Oakley | G. Oakley |
| 4.04 | ISIS | Pavel Kaiser | A. Rabenseifer, Pavel Kaiser & A. Scheber |
| 4.05 | VDBS | J. Grolleau | J.L. Grolleau & L. Gelabert |
| 4.06 | GTS | William T. Alsbrooks, Melroy D. Quasney, Charles King, Elbert Williams, Kathleen Chamberlain | Judy M. Bedell |
| 4.07 | CRESCAT | Kenneth Kaye | Kenneth Kaye |
| 4.08 | PERSEE | Jean Sousselier | Jean Sousselier |
| 4.09 | CENTS III | Robert Bair | Robert Bair |
| 4.10 | COCENTS | Robert Bair | Robert Bair |
| 4.11 | CENTS-AID | J. Beresford | Deirde Gaquin |
| 4.12 | FILE | Jerry G. Gentry | |
| 4.13 | OSIRIS 2.4 | M.C. Mesnage | J.C. Farget |
| 4.14 | GEN. SUMMARY | Richard Shively | Garry D. Kepley |
| 4.15 | TPL | Peter B. Stevens | Peter Stevens |
| 4.16 | RGSP | J.D. Beasley | F. Yates |
| 4.17 | NCHS | Gretchen K. Jones | |
| 4.18 | SURVEY | Frank James | |
| 4.19 | SAP | A. Barbetti | Annette Barbetti |
| | | | |
| 5.01 | HES VAR X-TAB | Gretchen K. Jones | |
| 5.02 | SPLITHALF | A. Barbetti | Annette Barbetti |
| 5.03 | MULTI-FRAME | Richard Shively | |
| 5.04 | MWD VAR | Don Young | Garry D. Kepley |
| 5.05 | GSS | Annette Barbetti | |
| 5.06 | CLFS VAR-COV | Rene Boyer | |
| 5.07 | CESVP | William D. McCarthy | |
| 5.08 | KEYFITZ | George Train | |
| 5.09 | CLUSTERS | M.C. Pearce | Elizabeth M. White |

| | | | |
|---|---|---|---|
| 5.10 | SUPER CARP | Wayne A. Fuller | Wayne A. Fuller |
| 5.11 | GENVAR | B. Causey | |
| | | | |
| 6.01 | BTFSS | M. Shanks & J. McCarthy | Margaret Baker |
| 6.02 | EXPRESS | Gerald Ruderman | Gerald Ruderman |
| 6.03 | EASYTRIEVE | David R. Malmstedt | Tom Thomason |
| 6.04 | FOCUS | Martin B. Slagowitz | Martin B. Slagowitz |
| 6.05 | SURVEYOR/SV | Charles Yarbrough | |
| 6.06 | DATAPLOT | James J. Filliben | James J. Filliben |
| 6.07 | PACKAGE X | Brian E. Cooper | Brian E. Cooper |
| 6.08 | SCSS | Joan Fee | Joan Fee |
| 6.09 | SPSS | Johnathan B. Fry | Joan Fee |
| 6.10 | SOUPAC | Joan A. Mills | Joan A. Mills |
| 6.11 | DATATEXT | David J. Armor | David J. Armor |
| 6.12 | P-STAT | Shirrell Buhler | Roald Buhler |
| 6.13 | DAS | Brian Cooper | Brian Cooper |
| 6.14 | SPMS | Toshiro Tango & Joji Kariya | Toshiro Tango |
| 6.15 | OSIRIS IV | Pauline Nagara | Ingrid Weisz |
| | | | |
| 7.01 | CS | John C. Kleinsen | John C. Kleinsen |
| 7.02 | SAS | Kathryn A. Council | Patti Reinhardt |
| 7.03 | OMNITAB | David Hogben | Sally T. Peavy |
| 7.04 | HP STAT PACKS | Thomas J. Boardman | Thomas J. Boardman |
| 7.05 | MINITAB II | T.A. Ryan, Jr. | T.A. Ryan, Jr. |
| 7.06 | BMDP | James W. Frane | James W. Frane |
| 7.07 | NISAN | Chooichrio Asano | Chooichrio Asano |
| 7.08 | GENSTAT | N. G. Alvey | N. G. Alvey |
| 7.09 | SPEAKEASY | Stanley Cohen | Stanley Cohen |
| 7.10 | TROLL | Peter Hollinger | Peter Hollinger |
| 7.11 | IDA | Faye E. Citron | Penelope Bingham |
| | | | |
| 8.01 | ISA | J.N.R. Jeffers | |
| 8.02 | GLIM | R.J. Baker | M.G. Richardson |
| 8.03 | ISP | Allan Wilks | |
| 8.04 | CMU-DAP | Samuel Leinhardt | Samuel Leinhardt |
| 8.05 | RUMMAGE | Del T. Scott | G. Rex Bryce |
| 8.06 | CADA | David L. Libby | David L. Libby |
| 8.07 | SURVO | Tuomo Myllynen | Tuomo Myllynen and staff |
| 8.08 | AUTOGRP+ | R.E. Mills | Henry Friedman |
| 8.09 | FORALL | John D. Kerr | John D. Kerr |
| 8.10 | AQD | Arthur Schleifer | Arthur Schleifer |
| 8.11 | STP | F. A. Sorensen | F.A. Sorenson |
| 8.12 | STATPAK | Russell R. Barr III | Russell R. Barr III |
| 8.13 | STATUTIL | J. B. Douglas | J. B. Douglas |
| 8.14 | MICROSTAT | J. Bordeane Orris | J. Bordeane Orris |
| 8.15 | A-STAT | Gary M. Grandon | Gary M. Grandon |
| | | | |
| 9.01 | AMANCE | C. Miller & J. Bachacou | J. Bachacou |
| 9.02 | MAC/STAT | David Rothman | David Rothman |
| 9.03 | REG | Arthur Gilmour | Arthur Gilmour |
| 9.04 | TPD | P. Dagnelie | P. Dagnelie |
| 9.05 | LSML 76 | Walter R. Harvey | Walter R. Harvey |
| 9.06 | LINWOOD/NONL | Fred S. Wood | Fred S. Wood |
| 9.07 | ACPBCTET | Hui S. Chang | Hui S. Chang |
| 9.08 | MULPRES | J. L. Aston | J. L. Aston |
| 9.09 | STATSPLINE | John Lambert | David Kendall |

| 9.10 | ALLOC | Jo Hermans | Jo Hermans |
|---|---|---|---|
| 9.11 | POPAN | C. Schwarz | A.N. Arnason |
| 9.12 | CAPTURE | Gary C. White | Gary C. White |
| | | | |
| 10.01 | GUHA | Thomas Havranek | Thomas Havranek |
| 10.02 | CATFIT | James Davis | |
| 10.03 | TAB-APL | John Fox | John Fox |
| 10.04 | MULTIQUAL | R. Darrell Bock | R. Darrell Bock |
| 10.05 | C-TAB | Shelby J. Haberman | |
| 10.06 | ECTA | Robert Fay | Robert Fay |
| 10.07 | MLLSA | Clifford C. Clogg | Clifford C. Clogg |
| | | | |
| 11.01 | TSP/DATATRAN | John Brode | John Brode |
| 11.02 | B34S | Houston H. Stokes | Houston H. Stokes |
| 11.03 | PACK | David P. Reilly | David P. Reilly |
| 11.04 | SHAZAM | Kenneth J. White | Kenneth J. White |
| 11.05 | TSP | Bronwyn H. Hall | Bronwyn H. Hall |
| 11.06 | QUAIL | David Brownstone | David Brownstone |
| 11.07 | KEIS/ORACLE | Charles G. Renfro | Charles G. Renfro |
| | | | |
| 12.01 | DATAPAC | James J. Filliben | James J. Filliben |
| 12.02 | IMSL | Walton C. Gregory | Jean D. Dickson |
| 12.03 | REPOMAT | N.P. Hummon | N.P. Hummon |
| 12.04 | NMGS2 | James W. Longley | James W. Longley |
| 12.05 | NAG LIBRARY | Trevor W. Lambert | Trevor W. Lambert |
| 12.06 | EISPACK | Burton S. Garbow | Burton S. Garbow |

## USERS WHO COMPLETED Q2U QUESTIONNAIRE

| | |
|---|---|
| John N. Adams | Robert Goldfarb |
| David Allen | Mark N. Greene |
| Gary Anderson | James E. Grunig |
| D.F. Andrews | Harold Gugel |
| James Ashurst | Craig S. Hakkio |
| O. Assereto | Judy Hallman |
| Murray Atkin | Martha S. Hansard |
| J. Barnard | Peter Harmanec |
| K.H. Barry | David Heilbron |
| M. Bayewitz | Henry K. Hess |
| Roger Belling | G.R. Hext |
| Harold Benenson | R. Hill |
| Jack Bergan | Ray Hinde |
| Anne S. Bourne | P. Hollister |
| Dave Brandt | Phyllis Hollyer |
| Nancy Brooks | R. Homel |
| Lawrence A. Bruckner | John Hon |
| Rocco L. Brunelle | Linda Hutton |
| Barry V. Bye | J. Ivanek |
| Ginger Caldwell | Teyssier Jacque |
| Gordon R. Caldwell | K.J. Jones |
| Rick Carr | T.C. Jones |
| Mai Carey | Daniel A. Kaberun |
| D.L. Chapman | Barbara Kanki |
| William R. Clark | Jorg Kaufmann |
| Clifford C. Clogg | R. Keating |
| Elliot M. Cramer | James E. Keith |
| D.L. Crawford | Paul Keller |
| P.M. Cullagh | Margaret E. Kepner |
| Brian Curran | Bradford W. Knapp |
| Allen L. Davis | Stephen L. Koffler |
| Michael Decker | Andrew Kolstad |
| Beatriz de Salazar | A.J. Kooter |
| Alain Desrochers | Neal Koss |
| Marita Di Lorenzi | James A. Krupp |
| D.A. Dodd | Sherman Landau |
| Ira Dobrow | Alan J. Laub |
| John Donnelly | Marit Leringe |
| Patrick Doreian | John G. Lewis |
| Jozsef Dornyei | Susan Linacre |
| David Duncan | Jeffrey H. Loesch |
| Thomas Dwyer | William Lorenz |
| Peter Endebrock | B.A. Maaskant |
| Rita Englehardt | Harold Mack |
| Bonnie Erickson | Jay Magidson |
| J.S. Fenlon | Kim Malafant |
| H. Fishkind | Walter Maling |
| Allen I. Fleishman | Bryan F.J. Manly |
| Jimenez Flores | S.R. Matheson |
| Bruce E. Foster | R. Mays |
| Leslie J. Fyans, Jr. | Ruth McDougall |
| George E. Gantner III | Hank McLeod |
| Vincent P. Giorgi | H.W. Merrill |

Bryan E. Melton
Michael Meyer
J. Philip Miller
K.H. Mills
William Morrison
Joru Motokura
Kathy Mulder
B.P. Murphy
John Murphy
Kathryn J. Myers
David Nasatir
Ronald Nations
Hugh Neuburger
T. Nosanchull
Vilem Novak
Nancy L. Oshier
R. Eugene Parta
Douglas K. Pearce
R. Gerard Pence
Robert Petersen
P. Phillips
John C. Pickett
Janace Pierce
W.A. Powers
John I. Quebedeaux
D. Ratcliff
J. Rattenbury
M. Bourgeolet Remy
D. Renaud
John R. Rice
Jarvis Rich
Donald Richter
James Robbins
Richard D. Roistacher
M.D. Rowe
Keith Rust
L. James Savage
D. Savin
Peter Schmitz
David A. Seaman
Barry F. Sexton
Douglas P. Sharp
Richard G. Sheehant
Patrick Shrout
John H. Simpson
Gary Skoog
Ronald D. Snee
B.A. Sobel
Gene Solomon
Frank W. Stitt
Houston H. Stokes
Kim Streitburger
G. Thackray
David Thissein
Olga Towstopiat
Kenneth Train
R.L. Trosper

Ronald J. Usauskas
Richard Valente
Nicholas Vasilatos
J.P. Vila
Debby Vivari
John C. Volz
Peter Ward
Jack Wasil
K.L. Wearne
Stephen Weiss
Charles J. Wilcox
Michael E. Williams
Jean Wood
H.R. Wright
Sara Wyant
Fritz Yurck
Suzanne Zimmer

# TABLE OF CONTENTS

## LIST OF TABLES

## LIST OF FIGURES

CHAPTER 1


A SURVEY OF STATISTICAL SOFTWARE


## 1.1  THE AIM OF THIS BOOK

Packaged computer software has become an important and permanent feature of
statistical practice as an increasing number of researchers rely on packaged
programs for their analyses.  Beginning some twenty years ago, computer programs
for statistical analyses began to invade the statistical environment.  This
invasion was seen at first by some statisticians as a plague that threatened
their accustomed habitat, but by now, not only have most statisticians learned
to live with this invader, but many have experienced a very productive synergism
with it.  Furthermore, many researchers who have not had training in statistics
can now perform their own statistical analyses, requiring only the guidance of
a statistician and not his computational assistance.

Statistical programs have thrived in this environment, some growing
together to form "packages", others evolving into "languages" or "systems"; some
have interbred with programs for non-statistical uses; some take root every-
where, while others grow only under very special conditions.  The name used for
all these varieties of statistical programs, the summum genus, is "statistical
software".

Over the years, this software has germinated and grown without control or
management.  Little has been known of the number of these programs, their
detailed characteristics, and which ones were dangerous.  Under these circum-
stances it is not surprising that no standards exist for either the development
or use of such programs:  some program developers have given insufficient attention to accuracy and to methods of protecting the user against his misusing the
programs; users, on the other hand, in publishing the results of analyses in
which computers have been used, typically fail to identify precisely the program
and computer used.  In addition, with minimal pruning and management some programs could have been transplanted to similar habitats, thereby eliminating the
need for developing new programs which essentially duplicated others already in
existence.

This book attempts to bring some order to this self-seeded disorder by the
traditional scientific and statistical means, namely, describing and classifying:  the discipline of Statistics has been called "a method of research
which tends to clarify the systematic collection of quantitative observations by
the comparison of numerous typical groups." (Kaufmann, 1913).

The first thing a reader will discover is that this book is not an in-depth
evaluation of statistical software.  Its scope is broad but shallow.  Very simply, we have collected a great deal of information from the developers of this
software, we have checked some of it with experienced users, and we have organized it and presented it in a form which we hope will make this information
comprehensible to a potential user.  It will enable this user to quickly
identify those programs that are potentially useful in terms of statistical and
portability characteristics, and to compare these potentially useful systems
using the developers' and users' ratings on a number of coarsely defined
capabilities.

Figure 1.1 outlines the methodological steps which lead to the articulation of standards (box 2.3) and ultimately to the improvement of statistical software (box 5.1). The categories listed in this figure were used in Francis (1977 and 1979) to review and classify the literature in statistical software up to 1977. In terms of these categories, the surveys and reviews in this book fall into categories 1.1, 1.2, 1.3, and 1.4.

To reiterate, the broad comparisons in this book do not constitute in-depth evaluations. Moreover, just because a package is popular with users does not mean that it is the most accurate. Ideally, programs should be evaluated by examination of the algorithms or by comparative experiments. Some evaluations which compare certain aspects of a few programs have been conducted: Chambers (1973) -- least squares algorithms; Francis and Sedransk (1979) -- analyzing survey data; Heiberger (1976) -- analysis of variance; Ling (1974) -- algorithms for means and variances; Searle (1979) -- unbalanced analysis of variance; Velleman, Seaman and Allen (1979) -- multiple regression; and many others listed in the bibliography of Francis (1979). Unfortunately these evaluations are difficult and often expensive, and so they can examine only a few features of the software; furthermore, the software is constantly changing and so reviews are soon out of date. One of the reasons for changes in the software is that evaluations and reviews frequently disclose errors or shortcomings.

We believe there is a need for some standards of accuracy. How these standards should be formulated is debatable, but meanwhile, in the absence of articulated standards, the results of published comparative experiments do provide some de facto standards, since the best performances give users some indication of what they can expect from new programs. These de facto standards will evolve over time as the state of the software art improves. However, they may not be sufficient for evaluating major innovations such as new statistical languages, new methods of analysis, or new software development tools.

One advantage of such de facto standards is that they are clearly attainable, which would not necessarily be the case for standards promulgated by some visionary committee. A number of papers in recent years have proposed desiderata for statistical systems, but some of these standards may be impossible to attain. For example, Wexelblat (1978) proposes that one criterion for a good statistical computing language is "naturalness", and that "a system is natural to use when you talk to it in the same vocabulary and notation that you use when solving problems without the system." Brode (1978), in proposing language standards for statistical computing, sees on the horizon "Artificial intelligence. This covers automatic error correction...and the determination of how to solve a problem." Taken literally, these two criteria assume the automatic translation of human language, which has eluded a generation of researchers. Dreyfus (1972), in his book entitled, "What Computers Can't Do", argues that there are real limits to what computers can do: "the recent difficulties in Artificial Intelligence, rather than reflecting technological limitations, may reveal the limitations of technology." Statisticians and other users cannot expect statistical software systems to replace the problem-solving capability of the human mind, and our criteria for these systems must reflect these limitations.

We hope that this book, by presenting this information and a taxonomy of statistical software, will stimulate the in-depth evaluations that are needed, and thereby contribute to the improvement of statistical computing and of the practice of Statistics itself.

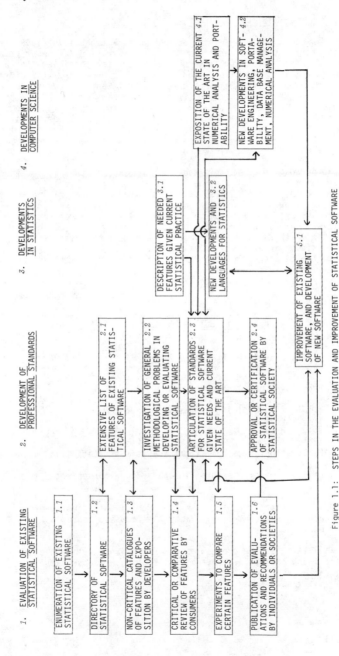

Figure 1.1:   STEPS IN THE EVALUATION AND IMPROVEMENT OF STATISTICAL SOFTWARE

## 1.2  A TAXONOMY OF STATISTICAL SOFTWARE

The world becomes comprehensible, according to Wittgenstein (1958), through its linguistic representation, and "naming is a preparation for description." Hartigan (1975) points out that "It is not necessary (or possible) that a naming scheme be best, but for effective communication it is necessary that different people give the same name to the same object".  "A taxonomist chooses carefully the variables to be used in clustering, rejects many as irrelevant or too variable, and gives more or less importance to others.  These decisions are often subjective ones, disagreed on among the experts, subject to later revision.  The weighting decisions for variables are made interactively with the establishing of clusters.  Thus a variable which does not distinguish well between established clusters will be reduced in weight." (Hartigan, 1972).

The basic paradigm used in this book to organize and present the information about a program is built upon the four questions in Table 1.1 that a potential user would have; the first part of the paradigm describing "Capabilities" follows the sequence of procedures which would be employed in a complete statistical analysis from beginning to end, the "life cycle" of a statistical analysis.  This provides the organization for the fifty-six items on which each program is rated in the survey of statistical software described in the next section.  These items are the variables and the ratings are their values which will assist in the clustering of programs into a taxonomy of statistical software.

---

### TABLE 1.1:  A USER INVESTIGATES A PROGRAM'S USEFULNESS

1. <u>Capabilities</u>:  Was the program designed to help solve problems like mine?

   1.1 Processing and displaying data

       1.1.1  file building and manipulation
       1.1.2  editing
       1.1.3  data display

   1.2 Exploration and mathematical analysis of statistical data

2. <u>Portability</u>:  Can the program be transported conveniently to my computer?

3. <u>Ease of learning and using</u>:  Is the program sufficiently easy to learn and use that it will actually be useful in solving problems?

4. <u>Reliability</u>:  Is the program maintained by some reliable organization, and has it been extensively tested for accuracy?

The paradigm has evolved over several years beginning with a report to the American Statistical Association summarizing the opinions of over a hundred contributors (Francis, Heiberger, and Velleman, 1975.) The views of many participants in a workshop at the 1976 Symposium on the Interface of Computer Science and Statistics were incorporated into a paper by Francis and Sedransk (1976) which led to the paradigm. It was used first by Francis (1979) in a clustering of ratings and programs which produced a taxonomy similar to the one described here.

Table 1.2 displays our Taxonomy of Statistical Software, a two-way clustering of twenty variables (items) and one hundred and seventeen cases (programs) from the survey. The clustering of the programs provides the organization for this book. Each of the programs was rated by its own developer on each of the twenty items on a four-point scale reflecting the (developer's opinion of the) program's level of coverage of that statistical capability: "full" coverage scored 3, "moderate" scored 2, "low" scored 1, and "none" scored 0. A definition of what was to be considered "full" coverage was given with each item in the questionnaire. In Table 1.2, the scores 0 to 3 are represented by the thickness of a bar.

This taxonomy employs only twenty items from the statistical section of the questionnaire, and so the clustering of the programs reflects only these capabilities and not a program's portability, ease of use, or maintenance. As Hartigan suggested, the choice of variables was done carefully, redundant or non-informative ones being omitted, and they were selected and clustered (ordered) simultaneously with the clustering (ordering) of the programs. The original order of the variables -- that is, the questions on the questionnaire-- was generally in accordance with the paradigm, but minor reordering improved the clustering of ratings.

Generally speaking, the clustering in Table 1.2 was done so as to cluster the "3" ratings down the main diagonal. To the extent that this was possible, therefore, the ordering of the programs, and therefore of the chapters, reflects the "life cycle" ordering of the capabilities; in other words, the programs are roughly in the order that they would be used in a complete statistical analysis: file management, editing, data description, mathematical analysis.

In a few cases additional information was used in the ordering of the programs. For example, IMSL (in Chapter 12) apparently could have fitted into Chapter 8; POPAN and CAPTURE in Chapter 9 have a particular statistical analysis capability (estimating animal populations) which is not reflected in any of the chosen items; the programs in Chapter 8 and Chapter 9 were separated according as they were interactive or batch (item 54). But as much as possible, the grouping of programs was done according to what the developers claimed to be the strengths of their respective programs. (The extent to which users agreed will be seen in the individual chapters.)

Although the clustering down the main diagonal is far from perfect, this visual presentation allows the imperfections to be easily seen. Also, this "top left, bottom right" clustering was used locally as well: for example the programs in Chapter 7 are uniformly highly rated on statistical analysis items, while those in Chapter 6 are less highly rated on statistical analysis but are generally strong on file management and tabulation, two areas of strength that are needed in handling data from large surveys.

## TABLE 1.2: A TAXONOMY OF STATISTICAL SOFTWARE - RATINGS BY DEVELOPERS

CAPABILITIES

RATING KEY

LEVEL OF CAPABILITY

- =3: FULL
- =2: MODERATE
- =1: LOW
- =0: NONE

| Chapter | Package Name | DATA MANAGEMENT | | EDITING | | TABULATION | | ESTIMATION FOR SURVEYS | | STATISTICAL ANALYSIS | | | | | | | | ECONOMETRICS | | MATHEMATICS | |
|---|---|---|---|---|---|---|---|---|---|---|---|---|---|---|---|---|---|---|---|---|---|
| | | FILE MANAGEMENT | COMPLEX STRUCTURES | CONSISTENCY CHECKS | PROBABILISTIC CHECKS | COMPUTE TABLES | PRINT TABLES | SURVEY ESTIMATES | SURVEY VARIANCES | MULTIPLE REGRESSION | ANOVA/LINEAR MODEL | LINEAR MULTIVARIATE | NON-LINEAR | NONPARAMETRIC | EXPLORATORY | ROBUST | MULTI-WAY TABLES | TIME SERIES | ECONOMETRICS | OPERATIONS RESEARCH | MATH FUNCTIONS |
| | | 14 | 11 | 18 | 19 | 24 | 25 | 43 | 44 | 30 | 31 | 32 | 39 | 36 | 37 | 38 | 33 | 35 | 41 | 48 | 47 |
| 2 DATA MANAGEMENT | UPDATE RAPID ADABAS SIR MARK IV RIQS DATA 3 FILEBOL KPSIM/KPVER | | | | | | | | | | | | | | | | | | | | |
| 3 EDITING | GES CONCOR CHARO EDITCK VCP-LCP | | | | | | | | | | | | | | | | | | | | |
| 4(i) TABULATION ONLY | SYNTAX II LEDA CYBER GEN. ISIS VDBS GTS CRESCAT PERSEE CENTS III COCENTS CENTS-AID FILE 2.0 OSIRIS 2.4 GEN. SUMMARY | | | | | | | | | | | | | | | | | | | | |
| 4(ii) TABULATION AND ARITHMETIC | TPL RGSP NCHS-XTAB SURVEY SAP | | | | | | | | | | | | | | | | | | | | |
| 5 SURVEY VARIANCE ESTIMATION | HES VAR X-TB SPLITHALF MULTI-FRAME MWD VARIANCE GSS EST. CLFS VAR COV CESVP KEYFITZ CLUSTERS SUPERCARP GENVAR | | | | | | | | | | | | | | | | | | | | |

*A Survey of Statistical Software*

| 6 SURVEY ANALYSIS | 7 GENERAL STATISTICAL ANALYSIS | 8 SPECIFIC INTERACTIVE STATISTICAL ANALYSIS | 9 SPECIFIC BATCH STATISTICAL ANALYSIS | 10 ANALYSIS OF MULTI-WAY TABLES | 11 ECONOMETRIC ANALYSIS | 12 MATHEMATICAL SUBROUTINE LIBRARIES |
|---|---|---|---|---|---|---|
| BTFSS | CS | LSA | AMANCE | GUHA | TSP/DATATRAN | DATAPAC |
| EXPRESS | SAS | GLIM | MNCA/STAT | CATFIT | B34S | IMSL LIBRARY |
| EASYTRIEVE | OMNITAB | ISP | REG | TAB-APL | PACK | REPOMAT |
| FOCUS | HP STATPACK | CMU-DAP | TPD-3 | MULTIQUAL | SHAZAM' | TSO |
| SURVEYOR/SV | MINITAB II | RUMMAGE | LSML76 | C-TAB | TSP | NAG-52 |
| DATAPLOT | BMDP | CADA | LINWOOD/MONIL | ECTA | QUAIL | NAG LIBOARY |
| PACKAGE X | NISAN | SURVO | ADDETECT | MLLSA | KEIS/ORACLE | EISPACK |
| SPSS | GENSTAT | AUTOGRP+ | MRLPRES | | | |
| SCSS | SPEAKEASY III | FORALL | STATSPLINE | | | |
| SCDPAC | TROLL | AQD | ALLOC | | | |
| DATATEXT | | STP | POPAN | | | |
| P-STAT 78 | | STATPAK | CAPTURE | | | |
| DAS | | STATUTIL | | | | |
| SIPMS | | MICROSTAT | | | | |
| OSIRIS IV | | A-STAT 79 | | | | |

## 1.3  THE SURVEY OF STATISTICAL SOFTWARE

Most people would agree that, ideally, statistical software should be evaluated by some "consumer organization" doing in-depth evaluations totally independent of the developers.  However, given the dynamic nature of software this
is at present impossible or at least enormously expensive.  An alternative
would be to require a developer to prove a product's safety before it could be
distributed.  This is unrealistic at present, and would also be very expensive.

For this review we chose a much less expensive method, namely mail questionnaires, to obtain information from developers, followed by as much validation as possible by users, given our time and financial constraints.  Fortun-
ately we do not (yet) have an adversarial relationship between manufacturers
and consumers, as sometimes seems to be the case in the automobile marketplace
for example, where the fierce nature of competition leads to some misleading
claims by salesmen.  With very few exceptions, the developers of the programs
in this review have been cooperative, and indeed some have been unduly modest, in
our opinion, in rating their packages.

The following sections describe in detail the three steps taken in
gathering the information:

1) developing a frame, a list of programs which perform one or more of the
tasks of processing or analyzing statistical data;

2) obtaining (a) ratings, and (b) technical information from developers on
their own programs; and

3) obtaining ratings from users on these programs.

### 1.3.1  THE FRAME DEVELOPMENT SURVEY

To initiate the frame development process, a questionnaire (Q1a) was sent
in 1977 to all members of the Subsection on Survey Research Methods of the
American Statistical Association and to all North American members of the International Association of Survey Statisticians, a total of approximately 2500
questionnaires in all.  The 375 respondents named 194 programs used in their
statistical computing.

In 1978 an improved version of this questionnaire (Q1b) was sent to all 123
institutional members of AAPOR (American Association of Public Opinion
Research), all 36 Survey Research Centers listed in Survey Research, and 115
statisticians selected purposively from the Federal Statistical Directory, as
well as to most of the respondents to Q1a.  The objectives were to identify
appropriate computer programs and to obtain very general information about their
purported emphases.  Thus, if the contacted organization used computer programs
in the processing or analysis of data from sample surveys or other sources of
large or complex data sets, the respondent was asked to provide the names of the
programs and the names and addresses of the developers, as well as users in his
organization who were familiar with the programs.  These programs were to be
broadly classified according to the following five tasks:  (1) data management
and file building, (2) editing:  error detection, correction and imputation,
(3) data description, tabulation and plotting, (4) estimation of finite population parameters and associated variances for complex sample surveys, and
(5) statistical analyses and model building.

Also in 1978 additional programs were identified from responses to notices placed in journals such as <u>Amstat News</u>, <u>International Statistical Institute Newsletter</u>, <u>SIGSOC Bulletin</u>, the Royal Statistical Society's <u>News and Notes</u>, <u>Newsletter of the Institute of Statisticians</u>, <u>Software World</u>, <u>Computing</u>, <u>Bulletin of the V.V.S.</u> (Vereniging voor Statistiek, the Netherlands Society for Statistics), <u>Communications in Statistics B</u>, and the <u>American Economic Review</u>.

Developers of programs identified in the frame development phase were requested to complete a lengthy questionnaire (Q2) about program capabilities. This self-evaluation questionniare contained a small number of questions pertaining to each of the five tasks listed above. The objectives were to provide basic summary information about package capabilities over a broad range of tasks, and to identify packages of purported excellence in carrying out specific tasks.

The first Q2 self-evaluation was carried out in 1977, when a purposive sample of developers of 75 widely-used statistical programs were requested to complete questionnaire Q2a. These 75 programs were selected from the 194 packages identified by Q1a, plus those known from personal knowledge, from the Conference of European Statisticians of the United Nations Economic Commission for Europe, and from lists appearing in Schucany, Shannon and Minton (1972), Rowe and Scheer (1976), and Kohm, Ryan and Velleman (1977). The 44 responses were presented at an Exhibition of Statistical Software at the 1977 meeting of the International Association for Statistical Computing. These and the exhibits prepared by 38 of the respondents appear in Francis (1979).

## 1.3.2  THE DEVELOPER SURVEY, 1979-80

The information in this book was collected in 1979 and 1980 when a revised questionnaire (Q2b) was sent to the developers of over 200 packages identified in the frame development process and to newer programs described in the <u>American Statistician</u> and elsewhere. This questionnaire appears in the Appendix of this book, and the self-ratings by the one hundred and seventeen responding developers, or their representatives, appear in Table A.1. (For a description of other numbers in Table A.1, see Section 1.3.4).

The questions on Q2b addressed, in order, a potential user's interests listed in Table 1.1. The structure of the questionnaire is displayed in Table 1.3.

In order that the ratings could be easily reported and comprehended in a form such as in Table 1.2 and Table A.1, all of the questions numbered 6 through 61 were chosen to have a four-point, ordinal scale. The two coding schemes, described on the second page of the questionnaire, are listed in Table 1.4.

The precise meaning of the ratings for each characteristic must be obtained from the questionnaire. For example, on question 30 regarding a package's capabilities in multiple regression analysis, a "0" rating indicated "no capabilities"; a "1" indicated that this was a minor purpose; a "2" indicated "moderate" capabilities or a "significant" purpose; and a "3" indicated complete coverage of all multiple regression procedures including stepwise regression, all possible regressions, ridge regression, wide variety of residual plots available, summary statistics (Durbin-Watson, etc.), and regression through origin. For questions with more qualitative answers, the meaning of the four codes were spelled out: for example, on question 55 concerning the amount of training in statistics needed by the typical user to be an effective

TABLE 1.3:  STRUCTURE OF THE QUESTIONNAIRE

I.  Names and Addresses:  questions 1 to 5

II.  Program Capabilities:  questions 6 to 48

    A.  Architecture and Generality of Purpose

    B.  Processing and Displaying Data

        B1  File building and manipulation
        B2  Editing - Error detection and reporting
        B3  Handling of erroneous or missing data
        B4  Tabulation, display, and graphics

    C.  Mathematical Analysis of Statistical Data

        C1  User language and extensibility
        C2  Capabilities of procedures for statistical analysis
        C3  Selection and analysis of complex sample surveys
        C4  Capabilities for other mathematical procedures

III.  Portability:  questions 49-54

IV.  Ease of Learning and Using:  questions 55 to 59

V.  Reliability and Maintenance:  questions 60 to 61

---

TABLE 1.4:  QUESTIONNAIRE CODING SCHEMES

0 to 3 Ratings:

0  No facilities in this area, or not an intended purpose.

1  A few functions in this area are present, or a minor purpose or byproduct.

2  Moderate capabilities, or a significant purpose.

3  Complete coverage in all aspects of the area, or one of the principal purposes.  (In the statement of all of the questions with a 0-3 rating scheme, we have defined what we consider to be complete coverage.)

a to d Ratings:

    These describe a program's generality of use to people other than the developer, ranging from "a" which is "low", to "d" which is "high".  For precise meaning, reference must be made to the specific question.

user, a rating of "a" indicated a bachelor's degree in statistics, or equivalent
professional training and experience; "b" indicated a year's study or experience
in applied statistics; "c" indicated an elementary, college-level course in statistics, or equivalent; and "d" indicated virtually no training in statistics.

The responses to questions with the "a" through "d" coding scheme were
recoded to a "0" through "3" scale for reporting in this book. Thus, in Tables
1.2, A.1, and elsewhere, ratings for all characteristics are on a four-point
ordinal scale representing general usefulness of the program as far as each particular characteristic is concerned, where "3" is high and "0" is low. (For a
few of the qualitative questions, such as questions 54 and 55, it might be
argued that the scales are not strictly ordinal.)

## 1.3.3 DEVELOPERS' TECHNICAL DESCRIPTIONS

In addition to being asked to complete this questionnaire, each developer
was asked to provide information of a more detailed and technical nature covering each of the topics in Table 1.5. This technical information is presented
in Chapters 2 to 12 of this book. For each program the information is structured in a common format in the order of the topics listed in Table 1.5, so
that the reader will very easily be able to make comparisons across programs.
The majority of the developers provided all the requested information and we
were able to report it with only organizational and editorial changes. However
for some programs we had to prepare this technical description from other
sources, but in all cases the developer was given the opportunity to check our
descriptions.

Finally, each developer was asked to supply a list of three users who
might be willing to give their assessment of the package's capabilities. These
users should (1) not be at the institution that supports the package, (2) not
have been connected with the development of the package, and (3) be familiar
with at least one other statistical package so that they would have some basis
for comparison. This was the primary source of users for the user survey
described in Section 1.3.4 below.

## 1.3.4 THE USER SURVEY

The developer of a program can surely be excused for being a little
biassed in rating the usefulness of his own program. Firstly, even though the
program may indeed be perfect when used by the developer for the purposes for
which it was designed, nevertheless when the program is transplanted to a different environment under different conditions -- and for the most part in this
book we are looking at programs which are being used by people other than the
developer -- the program may not be as useful to others as it is to the developer. Secondly, because of his complete familiarity with his program and his
knowledge of its undocumented eccentricities, he may not be aware of its shortcomings as others see it and therefore unintentionally over-rate its usefulness
to others. Thirdly, a developer, possibly with some commercial motives, may
intentionally over-rate his program.

To control for these possible biases, we proposed to undertake a survey
of users: each respondent would rate a program's capabilities on a questionnaire similar to that used for the developers. Our difficulty was to obtain a
sufficient number of users who would be at once (a) sufficiently knowledgeable
of particular programs and of statistical software in general, and (b) sufficiently independent of the developer's influence or enthusiasm to be able to

offer an unbiassed judgement of the program's usefulness. We wanted users who
would know, or who could readily find out, whether a program had a particular
facility.

The names of statistical software users who would probably be willing to
complete a long questionnaire about a program were obtained from two sources
mentioned above: we call them the "Independent Users" (I-users) and the
"Developer-Suggested Users" (DS-users). The I-users came from the frame development survey (Section 1.3.1): the hundreds of respondents were asked to provide the names of users in their organizations who were familiar with their
statistical software. The DS-users came from the developers who were all asked
(see Section 1.3.3) to supply the names of three independent, knowledgeable
users.

A DS-user is, by definition, known to the developer, and is familiar with
more than one program. He is, therefore, most likely to be very experienced in
the use of the particular program. An I-user may be a less frequent user, and
may even use the program for limited purposes. On the other hand an I-user may
have had to struggle with the documentation and other aids in learning to use

---

TABLE 1.5:   TOPICS COVERED IN THE TECHNICAL DESCRIPTIONS

1.  General Purpose and Origins

2.  Capabilities (i): Processing and Displaying Data

3.  Capabilities (ii): Statistical Analysis

4.  Extensibility

5.  Interfaces with other systems

6.  Proposed additions in next year

7.  References to the program's use in the literature

8.  Sample Job (i) Input

9.  Sample Job (ii) Output

10.  Developer's name and address

11.  Distributor's name and address

12.  Computer makes on which the program has been installed

13.  Operating Systems

14.  Source Languages

15.  Cost of obtaining the program

16.  Documentation

the program without the help of the developer, and might therefore be in a better position to assess the documentation or level of training required.

In the ideal factorial experiment which would take account of all factors affecting an assessment of the quality of software, including experience of users, statistical training of users, interactive versus batch use, etc., a controlled variety of users would be desirable. In this review, however, the practical constraints prevented this: we did not have enough I-users and, in any event, their familiarity with statistical software extended over only a small fraction of all the programs in our review. Therefore, for several reasons, some good, some not so good, we used the DS-users. But as a check on the differences between DS-users and I-users we conducted a pilot survey in which both types of users rated five well-known packages. The comparison of their responses, as well as a profile of the background and experience of the users, is reported in Section 1.6.

Table A.1 displays the average DS-user rating for each item for each program for which we had a DS-user. For each average, the number of users responding to that item "with confidence" is reported immediately under the average. Thus for those programs for which we had users, for each item three numbers appear stacked vertically: the top number is the developer's rating (see Section 1.3.2), the middle number is the average of the users' ratings, and the third number is the number of users. In those cases where the average user rating is not an integer, for example 1.7, this number appears as 1.

These ratings, the developer's and, where they exist, the users', are also plotted graphically in the figures at the beginning of each chapter. In each of these figures, for example Figure 6.1, for each program the developer's rating is plotted, the height of each point from the lower edge of each box representing a number from 0 to 3, and each of these developer's points is connected by a continuous line to assist the reader's eye. For example, for SPSS the developer's first two ratings on the sixth and seventh questions are 3 and 3, followed by ratings of 2, 3, 2, 1, 3, 2, 2, 3, 0, etc. In addition, for those programs with responses from users (which excludes FOCUS and SCSS, and so these two are placed at the bottom of Figure 6.1), the users' average rating is plotted on the same grid. If the users' rating differs from the corresponding developer's rating, a vertical line is drawn connecting the two points: if the users' rating is higher than the developer's rating, the line is thin; if the users' rating is lower than the developer's rating, the line is thick. Thus a thick line indicates that the developer has over-rated his package compared with the users. It is therefore easy to see, for example, that developers and users of SPSS agree rather well, except that the users over-rated the editing capabilities and the developer over-rated on question 55: the developer feels that no training in statistics is needed for a user to be an effective user, but all users agree that a bachelor's degree in statistics is needed.

On the other hand it is also easy to see, again on Figure 6.1, that the developer of EASYTRIEVE has a higher opinion of this program than do the users. Recall that all users were suggested by the developers. Even so, one is unable to tell whether this discrepancy is due to the users' (two of them) being totally unaware of any statistical analysis capabilities, which may possibly exist as an unpublished option, or whether the developer is simply over-stating these capabilities. A clue might be given in question 29, a "filter" question for that group of items: a rating of 3 on that question indicates that the program offers complete coverage of statistical analysis features, and yet for none of the specific statistical analysis procedures in questions 30 through 41 is

the coverage complete.  Indeed, on all similar filter questions, numbers 8, 16,
21, 23, 29, and 42, the developer rates EASYTRIEVE as "3", and yet the specific
questions reveal something less than complete coverage, except for "Filing".

It is of course entirely possible that the wording of some questions was
not entirely clear or, even if it was clear, that it was not understood.  For
example, judging by the differences between developers' and users' ratings on
questions pertaining to editing and missing data, some confusion exists concerning editing and imputation.  These two areas have been relatively undevel-
oped in the literature in the past, but recently have attracted increasing
attention.

There is one other possible source of discrepancy:  the user survey had to
be conducted after the developer survey, and in some cases several months separated the two.  In all cases respondents were asked to identify the version of
the program they were rating, but some users did not know.  For example, the
MINITAB developer rated a new release which included time series (question 35,
Figure 7.1) but the users probably were not aware of this new addition.

It can be seen from these figures and from Table A.1 that users' responses
to questions on portability were not recorded.  The reason was that few users
would have this information.

Only 74 of the 117 developers supplied users who were willing to return a
questionnaire.  Many of the 43 programs without users did not supply any users,
even after reminders:  some apparently did not have any that met the conditions
(i.e. not associated with the development); others said the program was only
used internally, for example, in a particular government department; some wished
to protect the privacy of their customers; a few, in particular SCSS, AQD, and
SPEAKEASY, did send names of users but too late for us to contact them.

Even for the programs where we have one, two, or even three users, any con-
clusion that a reader may draw based on these users' ratings must be qualified
on account of the small sample size.  However, we repeat our one strong
defense:  all of these users were suggested by the respective developers.  In
addition, users were asked to attach to each individual rating a "confidence
rating" on a four-point scale (0=don't know, 1=low, 2=medium, 3=high) reflecting
that user's confidence that his rating was accurate.  With the exceptions noted
in the next paragraph, only those ratings made with a confidence of at least
"2" were recorded; all others were recorded as "don't know", and were ignored
when computing the users' averages.  For this reason, the number of users reported on Table A.1 may vary from item to item on a particular program.

The exceptions mentioned in the previous paragraph were made for three
users, two for BMDP and one for OSIRIS IV.  Before our main user mailing took
place, we tested the questionniare on the five well-known packages listed in
Table 1.9.  Some of the respondents misunderstood the request for a confidence
rating for each capability rating.  This aspect of the questionnaire was rede-
signed and a complete remailing was done.  These three early respondents did
not respond to this re-mailing, and so we had no confidence ratings for them,
but they were included in Table A.1.

## 1.4  PRESENTATION OF RESULTS BY CHAPTER

The results of the developer and user surveys are organized into eleven
chapters (2 through 12) as outlined in Table 1.2.  They are presented in figures and charts designed to maximize a reader's ability to comprehend the

information and make comparisons between similar software.  In addition, the numerical results from which all charts and figures were derived are recorded in Table A.1 in the Appendix.  Also Chapter 13 provides some overall assessments of the strengths of the software, and makes a review of software portability by machine type.  (See Section 1.5).

For the eleven Chapters x, where x = 2,3,...,12, the structure of each chapter is as follows:

1)  Table x.1:  Developers' ratings on all relevant items,

2)  Figure x.1:  Developer/user comparative rating chart on all items,

3)  Figure x.2:  Summary D-scores and U-scores for appropriate attributes (groups of items),

4)  Technical descriptions of all programs in the chapter.

## 1.4.1  TABLE x.1:  DEVELOPERS' RATINGS ON ALL RELEVANT ITEMS

Table x.1 reproduces from Table A.1 the developers' ratings on all items deemed relevant for the programs in Chapter x.  Specifically these items include (i) all items in the taxonomy of Table 1.2, plus (ii) all capability items related to the principal purposes of the programs in that chapter which had not been used in the taxonomy, plus (iii) all items numbered 49 through 61 related to Portability, Ease of Learning and Using, and Reliability and Maintenance.  This table gives the developers' views of their programs.  It is an expansion of Table 1.2, broken down by chapter.

## 1.4.2  FIGURE x.1:  DEVELOPER/USER COMPARATIVE RATING CHART ON ALL ITEMS

Figure x.1 was described in Section 1.3.4.  Briefly it is a plot of developers' ratings on all items, 6 through 61, joined by a continuous line to assist the reader's eye.  The height of each point from the lower edge of each box represents the developer's rating, from 0 to 3.  In addition, for those programs with responses from one or more users, the users' average rating is plotted on the same grid.  If the users' rating is higher than the developer's, a thin vertical line is drawn connecting the two; if the users' rating is lower than the developer's, a thick vertical line is drawn connecting the two.  Thus a thick line indicates that a developer has over-rated his program relative to the users' rating.

The order of the programs in Figure x.1 is the same as in Table x.1 (and Table 1.2), except that those programs without user responses are placed at the bottom.

## 1.4.3  FIGURE x.2:  SUMMARY D-SCORES AND U-SCORES

The Figures x.1 contain all the results from the developers' and users' rating surveys.  We suspect that for many readers this presentation will be sufficiently succinct that they will require no further analysis.  However, for those who would like a simpler picture at the expense of some loss of information, we have computed ten pairs of weighted averages of ratings for the ten summary attributes listed in Table 1.6.  The particular items in Table 1.6 were chosen because they appeared to be the most unambiguous and informative items for these attributes.

For each of these ten attributes we have computed a D-score and a U-score, except that $U_a$ does not exist. The values for all scores appear in Table A.1, and the relevent ones are plotted in Figures x.2 in the various chapters.

A D-score is simply the average of the developer's ratings for a particular attribute:  for example, for file management,

$$D_f = \frac{1}{7} \sum_{i=9}^{15} d_i$$

where $d_i$ is the developer's rating on the $i^{th}$ item.  On the other hand, U-scores are the averages of the smaller of the developer's and users' ratings:  for example:

$$U_f = \frac{1}{7} \sum_{i=9}^{15} \min(u_i, d_i)$$

where $u_i$ is the average of the users' ratings on the $i^{th}$ item, as defined in Section 1.3.4.

---

TABLE 1.6:  SCOPE OF D-SCORES AND U-SCORES

| D-Score | U-Score | Attribute | Item Numbers | Number of Items |
|---------|---------|-----------|--------------|-----------------|
| $D_f$ | $U_f$ | file management | 9-15 | 7 |
| $D_d$ | $U_d$ | detecting and editing errors | 17-20 | 4 |
| $D_t$ | $U_t$ | tabulation | 24,25 | 2 |
| $D_v$ | $U_v$ | variance estimation for surveys | 43,44 | 2 |
| $D_s$ | $U_s$ | statistical analysis | 30-32,34,36,39 | 6 |
| $D_c$ | $U_c$ | contingency table (multi-way) analysis | 33 | 1 |
| $D_e$ | $U_e$ | econometrics and time series analysis | 35,41 | 2 |
| $D_m$ | $U_m$ | mathematical functions | 47 | 1 |
| $D_a$ |  | availability | 49-54,60 | 7 |
| $D_u$ | $U_u$ | usability | 55-59 | 5 |
|  |  |  |  | 37 |

Each D-score and U-score, in common with individual ratings, has a maximum value of 3 and a minimum of 0. A value of 3 for a U-score would indicate full capability for that attribute in the unanimous opinion of the developer and all the users.

A U-score can be viewed as the usefulness of this attribute of a program from the user's perspective. We are assuming that the users who rated a package were well qualified (they were suggested by the developers and were confident in their ratings) and were therefore more familiar than most users with the program's capabilities. Thus if a developer gave a rating of "3" on some item but the users averaged less than "3", even a "0", then either the developer was over-rating the program or its "complete coverage" of this item was partially or perhaps completely unknown to these users, perhaps through lack of documentation. In any event, to these surveyed users the usefulness of the program was not what the developer claimed it to be.

On the other hand, when a developer gave a rating of "0" and the users averaged "3", or indeed any time the developer's rating was less than the users', for computing the U-score we took the developer's word since he ought to know if his own package has some capability, and concluded that the users misunderstood the question or perhaps were not familiar with the state of the art in this area.

It should be recognized that the U-scores penalize the program that does just one thing even though it does it very well. Thus, for example, $U_s$ is an indication of the totality of a package's capabilities for statistical analysis. (In Chapter 13, we compute a $\Omega$-score, another weighted average of users' ratings, but in this case the Capability items used are only those for which the developer claimed some strength. Thus on the $\Omega$-score, a special-purpose program can rank with the most general. See Section 1.5.)

If all the responding users for a particular program responded "don't know" or were not confident about their rating on a particular item, a "-" appears in Table A.1. When this "missing value" was encountered in the computation of a U-score, the developer's rating was used to impute this value.

In addition to displaying the relevant scores for the programs in the chapter, Figure x.2 also lists those programs from other chapters which rank highly on the most relevant D-score for that chapter. Thus, for example, Figure 4.2 includes D-scores not only for those specifically tabulation packages grouped into Chapter 4, but also points to packages from other chapters which claim major tabulation strengths.

## 1.4.4 TECHNICAL DESCRIPTIONS OF ALL PROGRAMS

The Technical Descriptions of all programs in Chapter x are included in the order of Table 1.2 or Table x.1. Almost all of these Descriptions have a similar structure and include information on most topics listed in Table 1.5. Most of this information was supplied by developers and has not been verified independently.

## 1.5 AN OVERVIEW IN CHAPTER 13

Chapter 13 contains two overviews of all the software in the book:

1. a two-way chart (Figure 13.1) which relates all programs to all the hardware on which versions of the programs have been installed, and

   2.  a scoring system, called an $\Omega$-score, and a related $\Delta$-score, which provides an overall ranking of all programs which have users:  on this scoring system a good program will score high whether it be a (good) general-purpose package or a (good) special-purpose package.

   The items on which a program is to be rated in the $\Omega$-score are chosen by the developer.  This is in contrast to the choice of items in Table 1.6 for the U-scores and D-scores described in Section 1.4 and used in Figures x.2 in Chapters 2 to 12:  those items were chosen a) to cover the range of capabilities which make up the attributes in Table 1.6, and b) because, from a study of the questionnaire results, they seemed to be unambiguous, informative, and reliable.  Thus, the U-scores and D-scores are computed from our choice of items and therefore are measures of our definition of a good program.  On the other hand, the $\Omega$-score in Chapter 13 measures how well a program stands up, in the users' eyes, against its developer's definition of a good program.

   The specific items used in the $\Omega$-scores and $\Delta$-scores are as follows: beginning with the 37 items in Table 1.6, exclude items numbered 49 to 57 and 60.  This leaves 27 items.  The subset (S) of these 27 used in computing $\Omega$ and $\Delta$ for a particular program consists of items 58 and 59, plus those items from the remaining 25 on which the developer rated his program as "2" or "3".

Define
$$\delta_i = \begin{cases} 1 & \text{if } i^{th} \text{ item is in S} \\ 0 & \text{otherwise,} \end{cases}$$

then
$$\Delta = \frac{\sum\limits_{S} \delta_i\, d_i}{\sum\limits_{S} \delta_i},$$

and
$$\Omega = \frac{\sum\limits_{S} \delta_i\, \min(u_i, d_i)}{\sum\limits_{S} \delta_i}$$

Thus $\Omega$ is an overall average user rating on items 58 and 59, which rate documentation, plus those items which the developer claims to be "significant" or "principal" purposes of his program.  A good small program can score as well as a good large program on this $\Omega$-score.

   All programs with users are ranked by their $\Omega$-scores in Table 13.1.  Those programs with the highest $\Omega$-scores might be considered to be on our "honor roll".

   All programs with or without users are ranked in Table 13.2 by their overall D-scores which are defined like the individual D-scores in Table 1.6 except that the overall D-score is defined on the 27 items defined above, that is, on the 37 items in Table 1.6 excluding items numbered 49 to 57 and 60.  The overall U-score is defined on the same 27 items.

   Tables 13.1 and 13.2 list the values of $\Omega$, $\Delta$, U, and D as appropriate.  In addition, these two tables provide a summary profile of each program according to the developers' claims:  for the first eight attributes in Table 1.6, the initial letter of the attribute is printed if the D-score for that attribute is equal to 1.5 or greater.

## 1.6  PROFILE OF USERS

In Section 1.3.4 we distinguished between the DS-users -- the "Developer-Suggested" users -- and I-users -- "Independent" users whom we learned about by other means.  In this section we give a brief profile of the background of both groups of users, and describe a survey designed to examine differences between these two groups.

Tables 1.7 and 1.8 display the distributions of formal training and occupation respectively for the two groups of users.  The questions from which these tables were compiled were asked of all users in a section of the user questionnaire that sought information on the user's background.  Not all users answered all of these questions.

The I-users in these two tables, and in Tables 1.9 and 1.10, were a subset of the entire pool of available I-users.  All available I-users had been surveyed to determine which statistical programs they were familiar with.  As mentioned earlier, the experience of these I-users was largely confined to a few well-known packages, which is one reason why we eventually used DS-users.  However there were a sufficient number of I-users who were familiar with the following six packages:  SAS, BMDP, OSIRIS IV, SPSS, TPL, and DATATEXT.  It was with this subset of I-users, twenty-five in number, and the DS-users suggested by the developers of these six packages, that we conducted the comparison described later in this section.  And it is these I-users whose background is described in Tables 1.7 and 1.8, along with the background of all one hundred and fifty-one DS-users.

Because of this method by which the I-users were selected, the two parts of Tables 1.7 and 1.8 are not directly comparable:  the twenty-five I-users were selected in part because they were experienced in the six chosen packages, five of which are large statistical analysis systems, well-known in academic computing circles.  The DS-users in Tables 1.7 and 1.8 include all DS-users, including those with specialized experience, for example in file management, who would have different training and professional backgrounds from statistical users.

This said, we will simply observe in Table 1.7 that, of the 151 responding DS-users, 97 had some statistical training, 75 having received a bachelor's degree or more in statistics.  Of the same 151 DS-users, 55 had formal training in computing, 39 having a bachelor's degree or more in computing.  Of the 25 I-users surveyed, all 25 had some formal statistical training, and 23 some formal training in computing.

The distribution of occupations appears in Table 1.8 for those users who answered this question:  142 (or 126 plus 16) of the 151 DS-users, and 19 of the 25 I-users.  The spread of the distribution for North American DS-users indicates the breadth of application of the software surveyed in this book.

The pilot experiment, mentioned earlier in this section, compared the ratings of I-users with those of DS-users.  Table 1.9 displays the averages of the users' ratings on six of the attributes listed in Table 1.6 (except that "statistical analysis" also included items 33 and 35), plus a weighted average over these attributes, for the five packages SAS, BMDP, OSIRIS IV, SPSS, and TPL.  (Our DATATEXT I-users did not respond.)  Recall that each of the I-users had claimed to be experienced in the respective programs.  (Each user rated

just one program, although all users were familiar with at least two statistical
programs.)  Also each of these DS-users had been recommended by the developers
of their respective programs.

TABLE 1.7:   DISTRIBUTION OF USERS' FORMAL TRAINING

| Training | DS-USERS | | I-USERS | |
|---|---|---|---|---|
|  | Statistics | Computing | Statistics | Computing |
| Ph.D. | 22 | 6 | 7 | 0 |
| M.S. | 23 | 16 | 8 | 2 |
| Graduate courses | 6 | 8 | 0 | 3 |
| B.S. | 24 | 9 | 2 | 2 |
| Courses | 16 | 16 | 7 | 11 |
| Other | 6 | 0 | 1 | 5 |
| Total | 97 | 55 | 25 | 23 |

TABLE 1.8:   DISTRIBUTION OF USERS' OCCUPATIONS

| OCCUPATION | DS-USERS | | I-USERS |
|---|---|---|---|
|  | NORTH AMERICA | ELSEWHERE |  |
| Academic Faculty | 24 | 6 | 7 |
| Academic Computing | 35 | 2 | 2 |
| Government | 34 | 8 | 8 |
| Business | 25 | 0 | 2 |
| Medicine | 8 | 0 | 0 |
| Total | 126 | 16 | 19 |

By and large the two sets of users agree very well in Table 1.9.  The only consistent differences are in the "File Manipulation" and "Detecting Errors/ Editing" columns where, in all cases except for TPL's File management, the DS-users rated the packages higher than the I-users.  Since this pattern of differ-ences is not maintained across attributes we may infer that the differences in these two columns are not the result of bias but rather a difference in famili-arity.  No doubt the developers recommended only those users whom they thought would be familiar with <u>all</u> aspects of their programs.

Further evidence of this difference in familiarity can be gleaned from the confidence ratings.  All users were asked to assign a "confidence" rating-- high, medium, or low -- to every single rating.  Of the total of all confidence ratings for the DS-users in this pilot experiment, 76.5% were "high", 19.1% were "medium", and only 4.4% were "low".  On the other hand, for the I-users, only 59.1% were "high", 29.2% were "medium", while 11.7% were "low".

We stated earlier in Section 1.5 that we used only DS-users for the compar-ative results presented in the rest of this book.  The primary reason was that the familiarity with the range of statistical software in the pool of I-users extended mainly only to a few well-known packages, and not over the wide range of software in our survey.  In the light of the results of the pilot experiment, however, we feel this was a happy choice we were forced to make since it appears that there is no consistent "pro-package" bias among DS-users, and furthermore they tend to be more familiar with all aspects of the packages.  Moreover, 80% of the DS-users stated that their work involved "heavy use" of the program they were rating.

TABLE 1.9:  COMPARISON OF DS-USERS' AND I-USERS' AVERAGE RATINGS

| | DS-USERS | | | | | | | I-USERS | | | | | | |
| | File Management | Detecting errors/editing | Tabulation | Statistical Analysis | Variances for Surveys | Usability | Average | File Management | Detecting errors/editing | Tabulation | Statistical Analysis | Variances for Surveys | Usability | Average |
|---|---|---|---|---|---|---|---|---|---|---|---|---|---|---|
| SAS | 2.8 | 2.5 | 2.8 | 2.5 | 1.5 | 2.8 | 2.6 | 2.5 | 1.8 | 2.4 | 2.2 | 1.4 | 2.5 | 2.2 |
| BMDP | 2.1 | 1.1 | 0.8 | 2.1 | 0.0 | 2.1 | 1.7 | 1.1 | 0.7 | 1.1 | 2.0 | 0.3 | 2.0 | 1.4 |
| OSIRIS IV | 2.4 | 2.4 | 1.7 | 1.6 | 1.6 | 1.9 | 2.0 | 2.3 | 2.2 | 1.7 | 1.4 | 1.2 | 1.7 | 1.8 |
| SPSS | 2.1 | 1.2 | 1.7 | 1.2 | 0.4 | 1.7 | 1.5 | 2.0 | 0.5 | 1.9 | 1.3 | 0.8 | 2.1 | 1.5 |
| TPL | 1.5 | 0.8 | 2.7 | - | - | 2.1 | 1.6 | 1.8 | 0.7 | 2.7 | - | - | 2.2 | 1.8 |

We present one final tabulation in Table 1.10 concerning users and what they find important in the software they use.  Each DS-user was asked to write down a list of "What features, capabilities, etc., of this package are especially important to you as a software user?"  The 151 responding DS-users wrote down a total of 379 items which we coded and then grouped to form Table 1.10.

TABLE 1.10:  FREQUENCY WITH WHICH USERS MENTIONED VARIOUS

TOPICS AS BEING "ESPECIALLY IMPORTANT".

| STATISTICAL CAPABILITIES | FREQUENCY | TOTALS |
|---|---|---|
| Wide Range of Capabilities | 38 | |
| Specific Capabilities | 88 | |
| | | 126 |
| **DATA HANDLING** | | |
| File Management or Complex Data Structures | 37 | |
| Error Checking | 5 | |
| Data Documentation | 4 | 46 |
| **CONVENIENCE OF USER LANGUAGE** | | |
| Ease of Specifying Model or Convenience of Use | 46 | |
| Flexible Input Format | 11 | |
| Language Design | 27 | |
| Ease of Data Transformation | 7 | |
| Extensibility, Modularity, or Macro Facility | 16 | |
| Interactive | 15 | 122 |
| **OUTPUT** | | |
| Useful Output | 25 | |
| Plots or Graphics | 9 | 34 |
| **USER CONVENIENCES** | | |
| Documentation | 9 | |
| Interface with Other Systems | 10 | |
| Cost or Efficiency | 10 | |
| Portability | 4 | 33 |
| **RELIABILITY** | | |
| Reliability or Accuracy | 18 | 18 |
| | | 379 |

The first group of items, "Statistical Capabilities" is no surprise: users will understandably choose a program because it can do what they want it to do, either a specific analysis or a broad range of statistical operations.  The notable feature of this table is the group entitled "Convenience of User Language."  Expressions such as "ease of use", "convenience" or "language" were repeated time and again.  In contrast, topics related to output, documentation, cost, portability, and surprisingly even accuracy, were seldom mentioned.  We believe that Table 1.10 contains a message to developers that major improvements in statistical software from the user's point of view will come from the improvement of the user languages, and underscores the need for developing statistical computing languages.  But it also contains a message for the statistical profession, that users appear to be unaware of questions of accuracy of results.  Much more needs to be done in evaluating and developing standards for statistical software.

CHAPTER 2


DATA MANAGEMENT PROGRAMS


CONTENTS

INTRODUCTION

Data management is playing an increasingly important role in statistical analyses, in part because the advent of the computer has made possible the exploration and analysis of large and complex data sets.  One reason for the early popularity of SPSS was that it permitted the user to easily create, name, store, and retrieve files of data.  More recently SIR, a program in this chapter, was designed as a file management system to interface with existing statistical packages.  The degree to which it successfully fills a need is demonstrated by its being ranked first in Table 13.1:  among all programs in this book its users rated it highest on those capabilities which its developer also claimed were its strengths.

The major strengths of the programs in this chapter lie in some aspect of data management.  These programs range from small programs designed for simple data entry or manipulation, such as UPDATE and KPVER/KPSIM, to very large Data Base Management Systems, such as ADABAS and MARK IV.  This can be seen from Table 2.1, Figure 2.1, and Figure 2.2, which are explained in detail in Section 1.4 of Chapter 1.  Table 2.1 gives the developers' ratings on selected items, Figure 2.1 compares developers' and users' views on all items, and Figure 2.2 summarizes developers' ratings (D-scores) and users' ratings (U-scores) on several relevant attributes.  As indicated in Figure 2.2, additional programs with strong file management capabilities can be found in other chapters.

It is impressive to see in Figure 2.1 the agreement between the developers and users of RAPID and SIR.  More noticeable, however, is the disagreement between the developer and users of RIQS.  But the reader should remember that for the three programs at the bottom of Figure 2.1 we do not know how well the users would have agreed with the developers.

The remainder of this chapter consists of the individual technical descriptions of these data management programs.  Each description follows, more or less, the format described in Table 1.5.

## TABLE 2.1:  RATINGS BY DEVELOPERS ON SELECTED ITEMS

### (i) Capabilities

Usefulness
Rating Key:
3 - high
2 - moderate
1 - modest
. - low

| | Data Set Size | Flexible Data Input | Complex Structures | Filing Missing Data | Storage/Retrieval | File Manipulation | Flexible Output | Consistency Checks | Probability | Compute Tables | Print Tables | Multiple Regression | Anova/Linear Model | Linear Multivariate | Multi-way Tables | Nonparametric | Exploratory | Robust | Non-linear | Time Series | Econometric |
|---|---|---|---|---|---|---|---|---|---|---|---|---|---|---|---|---|---|---|---|---|---|
| | 9 | 10 | 11 | 12 | 13 | 14 | 15 | 18 | 19 | 24 | 25 | 30 | 31 | 32 | 33 | 36 | 37 | 38 | 39 | 35 | 41 |
| 2.01 UPDATE | 3 | 2 | . | . | . | 3 | . | . | . | . | . | . | . | . | . | . | . | . | . | . | . |
| 2.02 RAPID | 3 | 2 | 2 | 1 | 3 | 2 | 2 | . | . | . | . | . | . | . | . | . | . | . | . | . | . |
| 2.03 ADABAS | 3 | 3 | 2 | . | 2 | 2 | 2 | . | . | . | . | . | 2 | . | . | . | . | . | . | . | . |
| 2.04 SIR | 3 | 3 | 3 | 3 | 3 | 3 | 3 | 3 | . | 1 | . | . | . | . | . | . | . | . | . | . | . |
| 2.05 MARK IV | 3 | 3 | 3 | 1 | 3 | 3 | 2 | 2 | . | 1 | 1 | . | . | . | . | . | . | . | . | . | . |
| 2.06 RIQS | 2 | 3 | 1 | 1 | 3 | 3 | 3 | 1 | . | . | . | . | . | . | . | . | . | . | . | . | . |
| 2.07 DATA3 | 1 | 3 | 2 | 2 | 1 | 2 | 2 | 1 | 1 | . | . | . | . | . | . | . | . | . | . | . | . |
| 2.08 FILEBOL | 3 | 3 | . | . | . | 1 | . | 1 | . | . | . | . | . | . | . | . | . | . | . | . | . |
| 2.09 KPSIM/KPVER | 2 | 1 | 1 | 2 | 1 | . | . | 2 | . | . | . | . | . | . | . | . | . | . | . | . | . |

### FOR DATA MANAGEMENT PROGRAMS

#### (ii) User Interface

| | Survey Estimates | Survey Variances | Math Functions | Operations Research | Availability | Installations | Computer Makes | Mini Version | Core Requirements | Batch/Interactive | Stat. Training | Computer Training | Language Simplicity | Documentation | User Convenience | Maintenance | Tested for Accuracy |
|---|---|---|---|---|---|---|---|---|---|---|---|---|---|---|---|---|---|
| | 43 | 44 | 47 | 48 | 49 | 50 | 51 | 52 | 53 | 54 | 55 | 56 | 57 | 58 | 59 | 60 | 61 |
| UPDATE | . | . | . | . | . | . | 1 | . | 2 | . | 3 | 3 | . | 1 | 3 | 2 | . |
| RAPID | . | . | . | . | 2 | 1 | 1 | . | . | 1 | 3 | 2 | 2 | 2 | 3 | 3 | . |
| ADABAS | . | . | . | . | 3 | 3 | 2 | 3 | . | 3 | 3 | 3 | 2 | 3 | 3 | 3 | . |
| SIR | . | . | . | . | 3 | 2 | 2 | 2 | 1 | 3 | 3 | 2 | 3 | 2 | 3 | 3 | 1 |
| MARK IV | . | . | . | . | 3 | 3 | 2 | . | . | . | 3 | 3 | . | 3 | 3 | 3 | 1 |
| RIQS | . | . | . | . | 2 | 2 | . | . | . | 3 | 3 | 3 | 3 | 2 | 3 | 3 | . |
| DATA3 | . | . | . | . | 1 | . | 1 | 3 | 3 | 2 | 2 | 3 | 3 | 1 | 3 | 3 | . |
| FILEBOL | . | . | . | . | 2 | 1 | . | 1 | 2 | . | 3 | 3 | 2 | 2 | 3 | 2 | . |
| KPSIM/KPVER | . | . | . | . | 2 | . | . | . | 3 | 2 | 3 | 3 | 2 | 1 | 1 | 2 | 1 |

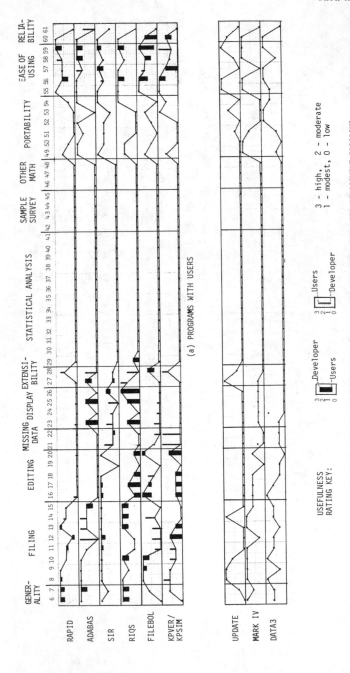

(a) PROGRAMS WITH USERS

FIGURE 2.1:  RATINGS BY DEVELOPERS AND USERS ON ALL ITEMS FOR DATA MANAGEMENT PROGRAMS

RATING KEY:    3 - high,    2 - moderate
               1 - modest,  0 - low

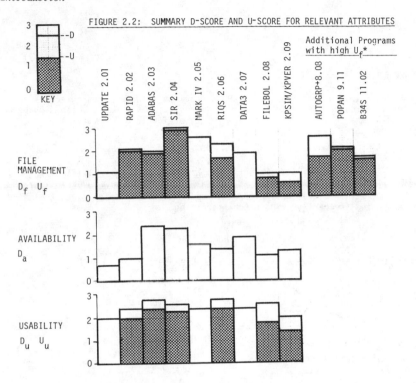

FIGURE 2.2:  SUMMARY D-SCORE AND U-SCORE FOR RELEVANT ATTRIBUTES

Other chapters generally containing programs with strong Filing
Capabilities are Chapters 4, 6, and 7 (see Figures 4.2, 6.2 and 7.2`

*In addition, for these specific programs, $U_f > 1.5$.

## 2.01  UPDATE

### INTRODUCTION

UPDATE is a program used to make corrections to data.  Its principal application is in correcting address data or survey data.  It was developed by Jerome
Weiland at the Institute for Survey Research of Temple University.  It is a
batch program.

### CAPABILITIES

Program provides the ability to delete or insert data to a file, change a
set of columns that are not contiguous and change a set of columns that are
contiguous.  The program was written for use on survey-related files that have
to be updated frequently using batch processing.

## UPDATE

Developed by:

    ISR
    Temple University
    1601 N. Broad Street
    Philadelphia, Pa.  19122

Distributed by:

    Temple University
    1601 N. Broad St.
    Philadelphia, Pa.  19122

Computer Makes:

    CDC CYBER series
      6000 series

Operating Systems:

    CDC NOS
    SCOPE 3-4

Source Language:

    FORTRAN

Cost:

    Not currently set.

Documentation:

Jenne, Carolyn., Use of Update Program.  Internal Memo.

### SAMPLE JOB

Samples of input cards and partial output are shown.

Partial Input of UPDATE Program

```
030185G  0462
030187G  0462
030188G  0465
030203H0104704806
030205H0104704854
030223H0104704840
030224H0104704840
030226H0104704840
030236G  0463
030265H0104704846
030284G  0461
030286G  0462
030301H0104704854
030307G  0464
030382G  0463
030406G  0463
030493G  0463
030515G  0463
030636G  0464
030652G  0461
030659G  0456
030677G  0464
030687G  0461
030691G  0463
030695G  0461
030709G  0461

NO. OF CARDS READ    36
```

Partial Output of UPDATE Program

## UPDATE SPECIFICATIONS

```
TAPES BLOCKED    8 * 40
SORT KEY ORDER  2   1   0   0   0
TAPE4
      COLUMNS   1   4   0   0   0
      RANGES    2   3   0   0   0
TAPE8
      COLUMNS   1   4   0   0   0
      RANGES    2   3   0   0   0
```

## VARIABLE FORMATS

```
IFMT4 (    I  2  ,1  X,I3   )
IFMT8 (          I   2  ,  1    X,I 3     ,A1)
IFMTD (80  A1)
IFMTD1(9   X,  25(I3,A1))
IFMTD2(7    X,I2,2I3,  65  A1)
IFMTD3(1X,     I   2   ,   1    X,I 3   )
IFMTD4(6   X,A1,2X,2I3,65  A1)
```

```
0300165 4 4 1 3 3 2 1 1 09 04 00 09 311014621197101457 9021
0300165 4 4 1 3 3 2 1 1 09 04 00 09 311014621125101457 9021
0300315 4 4 3 1 1 1 3 1 00 09 06 08 2060550211 907077610575
0300315 4 4 3 1 1 1 3 1 00 09 06 08 2060550212 907077610575
0300454 4 6 3 2 2 2 2 3 42 20 20 08 810114521197050652 7461
0300454 4 6 3 2 2 2 2 3 42 20 20 08 810114521128050652 7461
0300575 1 3 1 3 3 3 3 1 04 01 00 14 604234711197030358 9107
0300575 1 3 1 3 3 3 3 1 04 01 00 14 604234711136030358 9107
0300845 4 4 1 2 2 2 1 1 22 02 00 09 211032421297041034 4153
0300845 4 4 1 2 2 2 1 1 22 02 00 09 211032421254041034 4153
0300975 1 1 1 3 3 3 3 1 00 01 03 14 608234121197100548 6443
0300975 1 1 1 3 3 3 3 1 00 01 03 14 608234121106100548 6443
0301343 3 4 4 1 3 3 2 3 35 24 20 09 805294811197082462 9590
0301343 3 4 4 1 3 3 2 3 35 24 20 09 805294811122082462 9590
0301655 2 4 4 3 3 3 3 3 10 08 12 14 606115721297061873103 44
0301655 2 4 4 3 3 3 3 3 10 08 12 14 606115721223061873103 44
0301784 3 4 2 3 3 3 3 3 31 11 07 14 808214911112305157510499
0301784 3 4 2 3 3 3 3 3 31 11 07 14 808214911112305157510499
0301845 4 6 2 2 3 2 2 1 32 14 07 10 408144821123052476105 69
0301845 4 6 2 2 3 2 2 1 32 14 07 10 408144821123052476105 69
0301855 3 4 3 3 3 3 3 3 30 12 13 14 811162411151081076105 79
0301855 3 4 3 3 3 3 3 3 30 12 13 14 811162411125081076105 79
0301875 3 4 3 2 3 3 1 1 00 07 09 11 401294911123011377105 91
```

## 2.02  RAPID

### INTRODUCTION

RAPID (Relational Access Processor for Integrated Databases) is a generalized database management system which is typically used to process census and
survey data.  It was developed by Statistics Canada initially to process the
Canadian 1976 Housing and Population Census (total of 30 million records) but
now is used extensively by Statistics Canada for other applications.  RAPID is
actively maintained by Statistics Canada and will be used to process the 1981
Census.  RAPID is installed in 6 different statistical or government offices.

### CAPABILITIES:  Processing and Displaying Data

RAPID is based on the "Relational" model as first popularized by E.F. Codd
of IBM.  All RAPID relations are viewed as simple matrices which contain rows
(records) and columns (variables).  The system's primary function is to act as
the "Access Processor" to users of RAPID utilities as well as the application
programmer and the designer of customized software.  RAPID processes "Integrated" databases in that RAPID processes and manages both data descriptions
(metadata) and the application data itself.  The system provides access to this
information by a consistent set of facilities which ensure the integrity
between the data and its description.  By "Database" we mean any collection of
RAPID files which are seen by a user as being related in some way.  A RAPID
database evolves as new application functions are added or as separate databases are integrated (connected) into a common application.

RAPID stores each relation in an IBM BDAM file as a fully transposed file
which provides fast access for statistical retrievals (many records and few
variables).  Key access is also supported for informational queries (few
records and many variables).  A single RAPID file can contain up to 536,870,911
variables, and 2,147,483,647 records.  Each individual RAPID file is limited in
physical size only by the limitation on the IBM BDAM access method.

RAPID uses its own data dictionary which has both an online and batch
interface.  Generalized loading utilities are available to load metadata and
data into RAPID files.  A full set of database administrator utilities are
provided including utilities to create, expand and shrink a RAPID file, backup
and recovery programs, RAPID file analysis programs and a single relation query
facility.

Application programs can be written in PL/1, COBOL, FORTRAN or ASSEMBLER.

The DBM'S memory requirements vary depending on the physical characteristics of the RAPID files being processed.  Most applications at Statistics
Canada run in 200K to 512K.

RAPID's performance has been compared against such commercial packages as
TOTAL and ADABAS.  In every facet of database manipulation except random update
(rare in statistical applications) RAPID performs as well or better than these
packages.

## INTERFACES WITH OTHER SYSTEMS

A RAPID-SPSS interface allows SPSS users to read RAPID files directly
with no intervening 'copy' step.  New variables created during the SPSS job can
be saved back on the original input RAPID file.

An interface exists between the EXTRACTO report writing system and RAPID
(this software is supported by the commercial vendor of EXTRACTO).

## PROPOSED ADDITIONS IN NEXT YEAR

Interfaces between RAPID and both SAS and TPL are being considered.  New
data conversion facilities will provide a more independent view of data on a
RAPID file to application programs.

## REFERENCES

Cotton, P., PRODUCE-Using Database Technology to Provide Efficient Access to
    Population DATA, Notas de Pobacion, Revista.Latinoamericana de Demografia,
    CELADE, Santiago, Chile, April, 1980.
Davies, B., Data Base Management Systems in National Statistical Services, 42nd
    Biennial Congress of the International Statistical Institute, Manila,
    Phillipines, December, 1979.
Verm, J.C., Sundgren, E., Data Base Techniques on Statistical Data Processing,
    42nd Biennial Congress of the International Statistical Institute, Manila,
    Phillipines, December 1979.

Output from RAPID is dependent on application program's use of the system.

RAPID, Version 2.01

Developed by:                                    Distributed by:

    Special Resources Sub-Division     Special Resources Sub-Division
    Systems Development Division       Systems Development Division
    R.H. Coats Bldg.                 R.H. Coats Bldg.
    Statistics Canada               Statistics Canada
    Tunney's Pasture               Tunney's Pasture
    Ottawa, Ontario, Canada       Ottawa, Ontario, Canada
    K1A 0T6                        K1A 0T6

Computer Makes:                                  Operating Systems:

    IBM 370 Compatible            IBM MVT, MVS, OS/VSI

Source Languages:                                Interfaced Systems:

    PL/1, Assembler              SPSS, EXTRACTO

Cost:

    RAPID is available to non-profit, government and academic institutions for a one-time charge of $100 (Canadian). The system is not available to commercial users. The fee includes program source modules, load modules, installation instructions and manuals all on magnetic tape.

Documentation:

Turner, M.J., Hammond, R., Cotton, P., A DBMS for Large Statistical Databases,
    Very Large Database Conference, Rio de Janeiro, Brazil, October 1979.
Cotton, P., Turner, M.J., Hammond, R., RAPID: A Database Management System
    for Statistical Applications, 42nd Biennial Congress of the International
    Statistical Institute, Manila, Phillipines, December 1979.
Full set of user manuals available from Statistics Canada.

## 2.03  ADABAS

### INTRODUCTION

ADABAS is a flexible, general purpose data base management system which
features ease of implementation and use, efficient data retrieval, data independence and reliability.  ADABAS' general capabilities make it suitable for all
industries; this fact is verified by our diverse customer base.

### CAPABILITIES:  Database Management and Report Generation

ADABAS has been designed to accomodate large applications environments; it
is able to handle 255 files each with a capacity of 16 million records.  It is
multithreaded.

Any existing file can be integrated into the ADABAS structure.  ADABAS
fields can be fixed or variable length, contain multiple values, be associated
with groups of repeating fields and contain data of any type (fixed point,
alpha-meric).  Compression eliminates storage space requirements.

ADABAS uses fully inverted files.  It selects and fetches data rapidly by
searching for descriptors (key fields) in core instead of accessing each record.
It provides a multi-data base file-coupling capability giving the user unlimited
file access.

The ADABAS macro interface, ADAMINT, provides the application programmer
with a high-level call interface to ADABAS.  The ADAMINT interface contains a
module which permits access to and manipulation of a select subset of the data
base for each "user view".  Therefore, only the data needed or "user view"
requested is accessed.  ADAMINT provides outstanding data base independence and
facilitates easier application programming and maintenance.

The interactive data dictionary describes the data base, files, fields, and
the relationships in the dictionary.  Twenty standard reports are provided.
These reports are designed to meet the data entry cross-reference needs of the
DBA.

A new add-on "instant online applications" product called NATURAL was
released in January 1980.  NATURAL, when used with ADABAS, reduces program
development time by up to 90%.  It is a truly interactive all-in-one language
that includes facilities for online data entry and update, online query, report
generation and program development.

NATURAL relieves the programmer of the mechanics of conventional languages
such as file definitions, OPENs, etc.  NATURAL also serves as an interactive
compiler, and a map generator independent of the TP monitor and the terminal.
NATURAL facilitates the data base environment by allowing users to form logical relationships from information in several different files.

Random access into the data base is most efficiently done through ADAM
(ADABAS direct access method).  ADAM is a logical randomizer that facilitates
direct access as opposed to indexed access.  This facility saves about 75% of
the CPU and I/O time normally used for random access operations.

Two unique aspects of ADABAS are 1) a phonetic retrieval capability and 2) a data entry encryption or cypher option for ensuring data base security.

ADABAS provides multiple security schemes.  Fifteen (15) levels of security for the protection of data from unauthorized accessing and updating is one. Security-by-Value is the lastest extension to the ADABAS security sytem.  Where current procedures provide for the control of access or update functions against files and/or fields, Security-by-Value extends this control to the value content of specific fields.  Thus, the user can further limit access to specific sets of records in the data base.  Additional security can be provided through the use of ADAMINT, the ADABAS Macro feature.

ADABAS is distributed with a full range of data base modification and maintenance utilities.

INTERFACES WITH OTHER SYSTEMS:

ADABAS interfaces with COM-PLETE (Software AG's TP monitor), CICS, TSO, INTERCOMM, SHADOW II and TASKMASTER.

ADDITIONS IN NEXT YEAR

Major enhancements to ADABAS were incorporated in Version 4.1 (released in 1980).  Software AG will continue to enhance product features and internal "efficiencies".

## ADABAS   Version 4.1.1

Developed by:                              Distributed by:

    Software AG                                Same, and international
    Darmstadt, Germany                           affiliates in 6 continents

    Software AG
    Reston, Virginia
    North America

Computer Makes:

    IBM 360, 370, 43XXseries
    all OS, MVS, DOS systems
    (Itel, AMDAHL and other IBM
     compatibles)
    Siemens 4004

Source Languages:

    Assembler

Cost:

    The permanent license fee for ADABAS begins at $99,000.  Lease prices are
available starting at $1,875.

Documentation:

    A full range of manuals are available:  introductory manuals, planning
guides, reference manuals, and installation manuals.

## 2.04  SIR


### INTRODUCTION

SIR is an integrated, research-oriented data base management system which supports hierarchical and network file structures and interfaces directly with SPSS and BMDP. It can be run in either batch or interactive mode. SIR Version 1.0 was released in January 1977 and the program has been under continuous and active development since then. The program has been installed at over 70 sites, worldwide.

### CAPABILITIES

A SIR data base is defined using SPSS-like data definition (schema) commands. These commands allow for multiple record types, the definition of hierarchical and network relationships, data editing and checking (valid values, ranges and data consistency), data security at the item and record levels and multiple data types (integer, real, alpha, date, time, scaled, categorical, etc.).

SIR provides a wide range of batch data entry and update options including new data only, replacement data only and selected variable update.

The SIR retrieval language is structured and fully integrated with the rest of SIR. There is no need to learn a host language such as FORTRAN, COBOL or PL/I. The retrieval language has full arithmetic and logical operations. Retrieved information can be subjected to simple statistical analysis (descriptive statistics, frequency distributions, printer plots), used in reports, used to create SPSS or BMDP SAVE FILEs or a new SIR data base or written to a formatted, external data set.

The interactive subsystem includes a text editor, storage of user-written procedures and an interactive retrieval processor. The CALL (macro) facility enables the creation of generalized procedures.

Other features include various utilities for restructing, subsetting, merging and listing, as well as automatic creation of journal files when data is added to or changed in the data base.

### INTERFACES WITH OTHER SYSTEMS

As noted above, SIR creates SPSS and BMDP SAVE FILEs. It thus interfaces with any program that can read these files (such as SAS and P-STAT). SIR can also produce a formatted, raw data file.

### PROPOSED ADDITIONS IN NEXT YEAR

The major additions proposed are a forms-input processor, a FORTRAN host-language interface and a new report generating capability.

## REFERENCES

Cohen, E., Navickas, I.A., and Robinson, B.N., SIR - A Data Base Management
    System for the Medical Researcher, presented at the 4th Annual Symposium
    on Computer Applications in Medical Care, Washington D.C., November 1980.
Harasymiw, S., and Stahl, P., "Computer-Based Medical/Psychosocial Information
    Systems in Rehabilitation," presented at the International Association for
    Social Sciences Information Services and Technology, Washington D.C., May
    1980.
Navickas, I.A., Miller, A.H., Cohen, E., Dunn, J.K., Phillips, L.S. and Dunne,
    J.P., "Information Management in the Study of Diabetes in Pregnancy," pre-
    sented at the 9th Annual Meeting of the Society for Computer Medicine,
    Atlanta, GA, November 1979.
Sjoreen, Andrea, "An Application of the SIR System to a Large Geologic Data
    Base," presented at the 7th Annual Indiana University Computing Network
    Conference on Academic Computing Applications, Bloomington IN, April 1980.

## SAMPLE JOB

The following sample job uses the SIR Editor to create a retrieval program
which retrieves information from a clinical trials data base. Two variables
are retrieved from record type 1: treatment group and menopausal status;
recurrence information is retrieved from record type 2. An SPSS SAVE FILE is
created with two subfiles corresponding to pre- and post-menopausal status.
(The line numbers are generated by SIR.)

```
   SIR EDITOR READY
 > add
10 retrieval
20 .    process cases all
30 .        process rec 1
40 .            move vars treatment, menopaus
50 .            compute recur = 2
60 .            process rec 2
70 .                if (exists(recdate) = 1) recur = 1
80 .            end process rec 2
90 .        end process rec 1
100 .       perform procs
110 .    end process cases
120 var labels      recur, recurrence of cancer/
130 value labels    recur (1) recurrence  (2) no recurrence/
140 spss save file  filename = xfile/ subfiles = menopaus (1)pre (2)post/
150 end retrieval
160 =
   150 WAS LAST LINE
   run
```

    After entering the retrieval text, we execute it from within the Editor using the RUN command.  The output is a brief description of the SPSS SAVE FILE created by the retrieval.

```
TRANSLATE RETRIEVAL
ENTER RETRIEVAL
FILENAME        = XFILE
VARIABLES       = 5
SUBFILES        = 2
RECORDS ON FILE = 194

SUBFILE NAME      RECORD COUNT

   PRE                  58
   POST                136
   ELSE                  0 *** DELETED ***
RETRIEVAL FINISHED
SIR EDITOR READY
>
```

XFILE will contain subfile structure information, together with all variable labels, value labels and missing value indicators.

    The following input deck would then use the SPSS SAVE FILE created by the above job.

```
RUN NAME          CLINICAL TRIAL STUDY - CANCER TREATMENTS
GET FILE          XFILE
RUN SUBFILES      EACH
CROSSTABS         TABLES = RECUR BY TREATMENT/
STATISTICS        1
FINISH
```

### SIR (Versions 1.1 and 2.0) - Scientific Information Retrieval

Developed by:                              Distributed by:

    SIR, Inc.                              Same
    P.O. Box 1404
    Evanston, IL 60204

Computer Makes:                            Operating Systems:

    Version 1.1:
    CDC 6000 and CYBER                     all

    Version 2.0
    IBM 370                                OS/VS (TSO and CMS)
    PRIME 450 through 750                  PRIMOS
    SIEMENS                                BS 2000
    UNIVAC 90 series                       VS/9
    VAX 11/780                             VMS

Source Languages:   SIRTRAN - A macro preprocessor that generates FORTRAN and Assembler.

Cost:

    The initial license fee for SIR is $2000 for universities; $3500 for governments and nonprofit organizations; and $5000 for commercial use.  The cost in subsequent years is 80% of the current year license.

Documentation:

Robinson, B.N., Anderson, G.D., Cohen, E., Gazdzik, W.F., Karpel, L.C., Miller, A.H., and Stein, J.R., SIR User's Manual Version 2, 1980.

## 2.05  MARK IV


### INTRODUCTION

Mark IV is a full programming language designed as a COBOL replacement. It is fundamentally nonprocedural, which allows a programmer to address the problem and not the computer. It is an application development tool or system.

Mark IV evolved in the mid 1960's and was initially released in 1968. The original developer was John Postley. The current system is Version 8 which is installed at over 2,000 computer sites domestically and internationally. Version 9 will be released in the fall of 1981.

### CAPABILITIES:  Processing and Displaying Data

Mark IV has facilities for creating and maintaining files and data bases, as well as for generating reports and graphs. It has many automatic features including automatic report formatting capabilities, although a user can optionally format a report exactly as he wants. It can process arrays of data. It can automatically handle multiple files, such as bringing in a master file, up to nine coordinated files and one transaction file, which can contain virtually unlimited transaction types, in one simultaneous processing.

### CAPABILITIES:  Statistical Analysis

Mark IV has somewhat limited statistical analysis capabilities although it has standard arithmetic functions. It can compute averages, ratios and percentages and in the array processor can work with the usual arithmetic commands on arrays of data. For more complicated computations the program can exit to specific routines for the calculations and then return to Mark IV for a report generation. It has transparent interfaces with IMS or DL-1 and a generalized interface to other data bases.

At present all processing is done in batch mode although programs can be prepared interactively with syntax checking under VMCMS or TSO.

### PROPOSED ADDITIONS IN NEXT YEAR

The next generation Mark V will be a fully on-line interactive processor. The IMS version is expected to be released in September 1981 with CICS, TSO and VMCMS versions to follow.

### REFERENCE

"From COBOL to Mark IV", Jim Flynn and Dick Kimber, Datamation, Jan. 1977. Other references available from the Distributor.

MARK IV

Developed by:                                Distributed by:

  Informatics, Inc.                            Same
  21050 Van Owen Street
  Canoga Park, CA   91304
  (213)-887-9121

Computer Makes:                              Operating Systems:

  IBM 360/370/303X, 4300 Series                All
  Manufacturers plug compatible with
    IBM Mainframe's such as AMDAHL and
    ITEL

Language:

  IBM Assembler

Cost:

  Mark IV is highly modular and the cost of acquiring it is therefore very
dependent on the requirements.  It runs from $15,000 for the DOS Reporter to
$100,000 for the full OS version.

Documentation:

  Extensive-available from the Distributor.

## 2.06  RIQS

### INTRODUCTION

RIQS/RIQSONL is an information storage and retrieval system applicable for small data bases (less than 90,000 records) and for small to moderately-sized records.  RIQSONL is used interactively; RIQS non-interactively.  Data base records may contain up to 500 items, each of which may consist of a text strings, a numeric value, or a date, or  a list of text strings, numeric values or dates.  The number of items and the type of each is defined for the data base by the user.  Records are stored sequentially.  No file inversion or key-word linkage is done.  All record information is available to the user for searching and sorting.  Records may be grouped according to criteria based on record content.  Unnecessary record reads may be eliminated in RIQS operations by specifying the record groups or record numbers to be used in the operations.

### CAPABILITIES - RIQS

New data base files may be created.  Records for existing files may be added, deleted, or updated.

The file may be searched.  Arithmetic expressions containing numeric/date data base items, constants, trigonometric functions, numeric operations (+, -, /, *, **), and numeric variables may be used.  String expressions containing literal text strings or text strings from data base items may be used. RIQS commands may be conditionally executed.  The conditionals may include text and/or arithmetic expressions.  Variables, data base items, and literal strings may be output in a user-defined format via a report generator.  Commands may be executed for each record to be searched, and, optionally, before and after records are searched.  Up to 10 searches may be executed in one pass through the data base file.

Item information may be used to produce sort output.  Key Word In Context (KWIC) sorts list key words with following (and optionally preceding) context. INVERT sorts list key words with record numbers in which the words appear. Key Word Out of Context (KWOC) sorts list key words or phrases along with data base record information output in a user defined format via a report generator. Key words may be restricted by listing words to be used/not to be used in sorting.

### CAPABILITIES - RIQSONL

RIQSONL contains only the search capability described above for RIQS.  One search is done per pass through the data base file.  In addition to the above, RIQSONL allows the user to define, save, and retrieve search procedures and to obtain plot output.

### INTERFACES WITH OTHER SYSTEMS:

Variables and item information from a RIQS search can be written to a file in SPSS format in both RIQS and RIQSONL.  In addition, RIQSONL can suspend RIQSONL operations, execute SPSS, and return to RIQSONL when the SPSS run is completed.

## PROPOSED ADDITIONS

RIQS has been rewritten to (1) Alter sort operations to allow multi-level sorting, allow additional sort options (e.g. word stem sorts, ascending/descending sorts), and include more user options in the sort output, (2) Allow existing data base files to be restructured to add space for new items, delete items, and alter types of existing items, (3) Add additional features to search, e.g., allow string expressions to contain word stems, arithmetic expressions to be truncated to integer, searching for null item information to be done, and include more user options in search report output. The new version is in its final testing stage.

## REFERENCES

Dominick, W. and Borman, L., "User/System Interfacing in an Interactive
       Retrieval, Statistical and Graphical Analysis Environment" Proceedings,
       5th ASIS Mid-Year Meeting, May 20-22, 1976, pp. 34-49.
Mittman, B. and Borman, L., "An overview of Personalized Data Base Systems"
       Proceedings: INFOPOL-76 Conference, Warsaw, Poland, March 1976.
Mittman, B. and Borman, L. Personalized Data Base Systems, John Wiley and
       Sons, Inc., 1975.
Borman, L., Hay, R., and Mittman, B. "Information Retrieval, Statistical Analy-
       sis, and Graphics: an Integrated Approach" Information Storage and
       Retrieval, (9), pp. 309-319, 1973.

## SAMPLE JOB

The data base used contains information on US presidents. Item number 28 of the file contains comments about the presidents. In the first sample, a key word in context sort is done using the key words DEMOCRATS and REPUBLICANS. The context following the key words is printed. The test in item #28 is used. Each time a keyword is encountered in the item, a sort record is constructed with the key word and its context. The sort records are sorted and listed. In the second sample, the item is searched for the word ASSASSINATED or the word DIED. If either word is found, the president's name and the years in which the president was elected are listed. Item number 7, election year, was defined to contain a list of entries. In addition, a count is accumulated and printed at the end of the search.

Output from sample 1:

```
DEMOCRATS AS THE PARTY OF SECESSION AND TREASON.  COMMENTS -
DEMOCRATS AT ST.  LOUIS IN JUNE.  THOMAS A.  HENDRICKS OF IN
DEMOCRATS CHOSE THE FIRST DARK HORSE NOMINEE FOR THE PRESIDE
DEMOCRATS CONVENING AT BALTIMORE, MAY 5-7, UNANIMOUSLY RENOM
DEMOCRATS FOUND IT MORE AND MORE DIFFICULT TO RECONCILE THEI
DEMOCRATS IN CONVENTION AT BALTIMORE IN JULY ACCEPTED GREELE
DEMOCRATS MEETING AT CHARLESTON, APRIL 23-7AY 3, 1860) FAILE
DEMOCRATS MEETING AT CHICAGO IN AUGUST NOMINATED A TICKET OF
DEMOCRATS MEETING IN TAMMANY HALL, NEW YORK CITY, IN JULY OR
DEMOCRATS UNANIMOUSLY ENDORSED JACKSON AND REPLACED CALHOUN
DEMOCRATS - VOTED ALONG STRICT PARTY LINES TO AWARD THE 19 V
DEMOCRATS, WHO LIKE THE WHIGS MET AT BALTIMORE IN JUNE, HAD
DEMOCRATS.  GREELEY HAD CONSISTENTLY CRITICIZED THEIR PARTY,
REPUBLICANS AND SEVEN DEMOCRATS - VOTED ALONG STRICT PARTY L
REPUBLICANS AT PHILADELPHIA IN JUNE HAD UNANIMOUSLY RENOMINA
REPUBLICANS ENDORSED A PROTECTIVE TARIFF AND CONDEMNED THE E
REPUBLICANS IN NEW YORK PUT UP DEWITT CLINTON, WHO THEREUPON
REPUBLICANS MET AT CINCINNATI IN JUNE TO SELECT RUTHERFORD B
REPUBLICANS MET, AS THE UNION REPUBLICAN PARTY, AT CHICAGO I
REPUBLICANS NOMINATED JOHN C.  FREMONT OF CALIFORNIA AND WIL
REPUBLICANS PROMISED CONGRESSIONAL ACTION TO PRESERVE FREEDO
REPUBLICANS RELIED ON GRANT S WAR RECORD TO OVERCOME THE SCA
REPUBLICANS TO OPPOSE GRANT S RENOMINATION.  CONVENTIONS AND
REPUBLICANS TO OPPOSE LINCOLN S RENOMINATION.  CONVENTIONS A
REPUBLICANS USED THE BLOODY SHIRT TACTIC, ATTEMPTING TO LABE
REPUBLICANS WAS HAMILTON.
REPUBLICANS, BUT HAMILTON WAS LESS THAN HAPPY ABOUT THE CAND
REPUBLICANS, STYLING THEMSELVES THE NATIONAL UNION PARTY, ME
REPUBLICANS, WHO TENDED TO SUPPORT FRANCE.  THE APPARENT DES
REPUBLICANS.  NOMINATIONS - THE CONGRESSIONAL CAUCUS CHOSE J
REPUBLICANS.  THE CAMPAIGN - THE FEDERALISTS GAVE THEIR OPPO
```

Output from sample 2:

```
TAYLOR, ZACHARY          ELECTION YEARS 1848
LINCOLN, ABRAHAM         ELECTION YEARS 1860 * 1864
GRANT, ULYSSES S.        ELECTION YEARS 1868 * 1872
GARFIELD, JAMES A.       ELECTION YEARS 1880
MCKINLEY, WILLIAM        ELECTION YEARS 1896 * 1900
HARDING, WARREN G.       ELECTION YEARS 1920
ROOSEVELT, FRANKLIN D.   ELECTION YEARS 1932 * 1936 * 1940 * 1944
KENNEDY, JOHN F.         ELECTION YEARS 1960
```

8 PRESIDENTS DIED IN OFFICE OR WERE ASSASSINATED.

RIQS/RIQSONL

Developed by:                                    Distributed by:

  Vogelback Computing Center                       Same
  Northwestern University
  Evanston, Illinois

Computer Makes:                                  Operating Systems:

  CDC 6000,                                        SCOPE, NOS, NOS/BE
   70 Series

Sources Languages:  Primarily FORTRAN, Input/output, sort, and utilities in
COMPASS.

Cost:

  The one-time fee for RIQS/RIQSONL is $3000 for universities; $3000 for
governments and nonprofit organizations; and $9000 for commercial use.
Commercial installations receive a 50% discount for secondary sites.  Prices
are subject to change.

## 2.07  DATA3

### INTRODUCTION

DATA3 is a general-purpose data entry and management system which is oriented to applications in which the case record consists of one or more forms
or reports accumulated over an indefinite period of time.  The system is
designed to reproduce the content and chronological organization of such a
record so as to facilitate data entry and feedback in the clinical setting
while at the same time providing full file search and data summarization capabilities for research purposes.  DATA3 was developed by Joel Achtenberg at the
Washington University School of Medicine and has been in use there in about a
dozen research studies, departments, and centers.

### CAPABILITIES:  Data Entry, Management, and Display

1)  Chart-Like Storage of Data  Just as a hospital chart consists of lab
reports, nurses notes, and physician orders, the record for each case in DATA3
consists of an unlimited number of DOCUMENTS maintained in chronological order.
Each document contains values for a series of variables.  Each case is identified by a unique ID number, and by a user-defined alpha-numeric "name" which
need not be unique, so that references to a case by name may return several
cases with the same name, from which the proper selection can be made.

2)  Two-Level Dictionary.  DATA3 is a table-driven system with the complete
definition of each variable contained in a dictionary.  DATA3 contains an
additional dictionary which defines the organization of items into documents,
which specifies any skip patterning or special processing options, and which
serves as an interface between a) data entry and processing programs and b) the
dictionary of items.

3)  Flexible, User-oriented Data Entry.  Data entry is designed to accomodate
a number of real-world complications that might require pre-coding or editing
prior to data entry in other systems.  For example, numeric values can be
entered in units other than those called for by the definition of the item
(e.g. feet rather than inches), and conversion takes place automatically; or
the user can enter an arithmetic expression whose result is to be used as the
response value.  A response entered as DITTO tells DATA3 that no change in the
value of the variable has taken place since the last data entry for that
patient.

4)  Precise Input Editing.  Provisions have been made for defining (in the dictionary) precise editing specifications.  Seventeen predefined format types are
available, including dates, times, names, phone numbers, yes/no, abnormal/
normal, alpha-codes, integers, decimal numbers, and others.  These are combined with absolute and "normal" range limits, and lists of "legal" codes.  In
addition, the dictionary can call for cross-checking of one response with
others previously entered (in the same, or earlier, documents).  The result is
to largely eliminate the possibility of illegal responses being entered and to
greatly reduce the occurrence of legal, but erroneous data.

5) Data Access Support. A number of facilities permit flexible access to the
information in a DATA3 database. Access to individual cases is provided by
QUERY and PRINT. Other facilities (SEARCH, REPORT, SORT, STATS, XTABS) access
a range of cases which may be defined by user-specified selection criteria, by
predefined sub-groupings, or the entire population may be accessed. In making
retrievals, users may refer to variables by keyword rather than item number.
In addition, the user may specify a response modifier to indicate which response
among potentially multiple responses the user wishes to access. Response modi-
fiers include EARLIEST, LATEST, ALL, HIGHEST, LOWEST AVERAGE, and MOST FREQUENT.

6) Report Generation. Case by case reports can be prepared interactively by
specifying the items to be included and formatting options. Such reports can
be columnar or completely free-formatted, and can include titles, labelling,
and standard text segments; and format specifications can be retained for
repeated use.

7) Interactive Inquiries. Inquiries requesting specific information about a
specific patient/case are handled flexibly using the QUERY facility, an example
of which is included on the following page.

8) Full-File Searching. The SEARCH facility permits searching the entire file
or a subset to locate cases meeting user-specified criteria. For example, one
might retrieve patients having a history of heart disease, and cholesterol at
last hospitalization under 200. The results include a count of the "hits", and
an IDLIST containing all the hits which can be saved for use in subsequent
searches, reports, or statistical analyses.

## CAPABILITIES: Statistical Analysis

DATA3 itself provides only limited statistical processing. Univariate
statistical summaries (mean, range, standard deviation, high & low outlying
values, and frequency distribution) are provided by the STATS facility. Non-
statistical cross-tabluations (N-way) are provided by XTABS. All other statistical analysis is done by invoking a semi-automatic interface facility which
creates an input stream to SAS (SAS Institute, Cary, NC, 1979) including the
data and all SAS statements to create a fully-documented SAS dataset and perform desired statistics. This interface is described fully in Achtenberg &
Miller, 1978.

## CURRENT STATUS AND PROPOSED ADDITIONS WITHIN NEXT YEAR:

DATA3 is currently fully operational and in use at Washington University
Medical School in several departments. Major revisions to improve the efficiency throughout the system are projected for completion within 6 months. A
completely rewritten SEARCH component with much greater specification of search
criteria is also due to be completed by mid-1981.

A TYPICAL INQUIRY INTO A PATIENT'S RECORD*

WHAT FACILITY DO YOU WISH TO USE?   QUERY

WHAT PATIENT?   3-TEST GROUP
PATIENT'S NAME IS JONES, JOHN X
    IS THIS THE CORRECT PATIENT?   YES

WAHT VARIABLE?   SEX --- SEX (8)
THE VALUE 'MALE'
    WAS NOTED IN AN ENROLLMENT FORM DATE 5-MAY-79

WHAT VARIABLE?   272:   FASTING PLASMA GLUCOSE
WHAT MODIFIER?   ALL
THE VALUE '154 MG/DL'
    WAS NOTED IN AN INPATIENT EVALUATION DATED 6-SEP-77.
    THIS VALUE FALLS ABOVE THE NORMAL RANGE OF VALUES.
THE VALUE '176 MG/DL'
    WAS NOTED IN AN INPATIENT EVALUATION DATED 6-SEP-77.
    THIS VALUE FALLS ABOVE THE NORMAL RANGE OF VALUES.
THE VALUE'119 MG/DL'
    WAS NOTED IN AN INPATIENT EVALUATION DATED 7-AUG-78.
THE VALUE '110 MG/DL'
    WAS NOTED IN AN INPATIENT EVALUATION DATED 8-MAY-79.

WHAT VARIABLE?   272:   FASTING PLASMA GLUCOSE
WHAT MODIFIER?   AVERAGE
    THE AVERAGE IS 139.75 MG/DL BASED UPON N=4 VALUES.

WHAT VARIABLE?   *NDOCS
ENTER NAME OF DOCUMENT TO BE LOOKED AT,
OR * TO LOOK AT ALL OF THEM:   INPATIENT EVALUATION
NO. OF ENTRIES FILED IS 2.

WHAT VARIABLE?   *LASTDOC
ENTER NAME OF DOCUMENT TO BE LOOKED AT,
OR * TO LOOK AT ALL OF THEM:   INPATIENT EVALUATION
DATE OF LATEST ENTRY FILED IS 6-SEP-77.

WHAT VARIABLE?   AGE
ENTER REFERENCE DATE (OR 'CURRENT'):   CURRENT
PATIENT'S AGE IS 59.37.

WHAT VARIABLE?   RETURN

WHAT FACILITY DO YOU WISH TO USE?   QUIT
HAVE A GOOD DAY!!
EXIT

* User responses are underlined.

## DATA3

Developed and Distributed by:

    Joel Achtenberg
    Washington University Med. School
    Box 8019
    660 S. Euclid Avenue.
    St. Louis, MO 63110

Computer Makes:

    Any system with an ANSI MUMPS oper-
    ting system and interpreter, such as:
    DIGITAL      Burroughs
    Data General  HARRIS
    IBM 360/370   PRIME
    TANDEM

Operating Systems:

    MUMPS

Interfaced Systems:

    SAS

Source Languages:  ANSI MUMPS

Cost:

    Distribution, support, and financial details by arrangement.

Documentation and References:

    A preliminary DATA3 User's Guide, now available for review, provides a
basic introduction to the use of the system. Particular attention is given to
the initial creation of variable and document dictionaries, to data entry, and
to the basics of retrieval and analysis runs. Formal documentation will be
released mid-1980.

Achtenberg, J., Miller, J.P., Cryer, P.E., & Weldon, V.V., "A diabetes center
    patient registry". In Zimmerman, J.(ed.). Proc. of the 1975 MUMPS User's
    Group Meeting, pp. 1-7, MUMPS User's Group, St. Louis, MO 1975.
Achtenberg, J., Miller, J.P., Santiago, J.V., and Cryer, P.E., "DATA3 - A forms-
    oriented data management system". In Rothmeier, J.(ed.). Proc. of the 1976
    MUMPS User's Group Meeting, pp. 1-8, MUMPS User's Group, St. Louis, MO
    1976.
Miller, J.P., Moore, P., Achtenberg, J., & Thomas, L.J. Jr., "Computerized
    information handling for long term clinical studies". Proc. of the 1977
    IEEE Conference on Computers in Cardiology, pp. 151-159, Rotterdam, 1977
    (IEEE #77, CH-1254-2C).
Achtenberg, J., & Miller, J.P., "Interfacing a MUMPS-based data entry system to
    SAS". In Strand, R.H.(ed.). Proc. of the Third Annual Conference of the
    SAS User's Group International, pp. 161-167, SAS Institute, Raleigh, NC
    1978.
Achtenberg, J.,"DATA3: A Current Report". In Zimmerman, P.J.(ed.) Proc. of the
    1978 MUMPS User's Group Meeting, pp. 1-5, MUMPS User's Group, Bedford, MA
    1978.

    This project has been supported by Grants from the National Institutes of
Arthritis, Metabolism, and Digestive Diseases, (Grant #P17 AM 17904-01A1) to
the Diabetes and Endocrinology Center, and from the Division of Research
Resources, National Institute of Health (Grant RR 00036) to the Washington
University School of Medicine Clinical Research Center.

## 2.08  FILEBOL

### INTRODUCTION

FILEBOL is a data manipulation utility language and processor.  It is
typically used to update and generate reports from sequential files.  It is
designed to process large files efficiently.  Its principal applications have
been in administrative data processing and utility file handling (copying,
listing,...).  It was developed originally by Jarvis Rich of the National
Opinion Research Center and Donald Goldhamer of the University of Chicago Computation Center in 1972-5, and has undergone one minor revision since then.  It
is installed at 9 sites and is currently the most heavily used batch program at
the University of Chicago.

### CAPABILITIES:  Processing and Displaying Data

The FILEBOL language is free format and English-like.  All processor parameters have default values which reduce the number and size of statements
needed to perform a task.  The language provides for conditional processing,
for creation and execution of non-recursive internal procedures, for flexible
addressing of data fields, for simple access to execution and record parameters
(eg. the current date, the current record length and position in its file,...),
and for unrestricted abbreviation of frequently used phrases or statement
fragments (which are automatically cross-referenced).

FILEBOL can process any number of sequential input, output, and report
files simultaneously,  limited only by operating system restrictions.  FILEBOL's
general file processing facility will, by default, copy one sequential file.
It contains options for reformatting, scanning, selecting, sampling, and
sequencing records, for comparing fields, for converting data formats, and for
limited numeric computations (four arithmetic operators).

FILEBOL contains an automatic file updating facility which allows most
master/transaction processes to be specified in two or three statements.
Options in this facility provide for non-numeric keys, for the construction or
modification of keys from master and/or transaction records, for the implementation of user collating sequences, and for user specification of the update
process at seven logical points.

FILEBOL's report-generating facility will produce a simply formatted file
listing with one command (two words).  Options allow for complete specification of page format and size, for headings of any size which may contain record
data, and for annotated file totaling and counting.

## SAMPLE JOB

Update the master file using a file of transactions.  Treat all non-matching transactions as errors and ignore them.  For each matching transaction, obtain a data field origin from columns 10-12 and data field contents from columns 20-29 of the transaction record.  Replace the contents of the designated field in the master record with a specified content.  Place the entire updated record into a new master file.

For each updated master record, format the contents of columns 10-15 in standard monitary form, and generate a report with these figures and a total for this field across the entire file.

```
HEADING 1 1Ø  WIDGIT VALUES
AT OUTPUT.
COMPUTE 1Ø1-1Ø9 '$ ,    Ø.ØØ' = 1Ø-15.  PRINT FROM 1Ø1-1Ø9.
TOTAL 1Ø-15,
EDIT 1 23 '$     ,     Ø.ØØ',
NOTE 1 1 'TOTAL WIDGIT VALUE IS'.
ID=1-9.  UID=1-9.  AT ADD.  ERROR.
AT CHANGE.
MOVE TRANSACTION 2Ø-29 TO OUTPUT ADDRESS TRANSACTION 1Ø-12.
* COULD HAVE BEEN ABBREVIATED AS:   M T 2Ø-29 TO O A T 1Ø-12.
//SYSUTI DD *
111111111 45678
222222222 9Ø123
333333333123456
//UPDATES DD *
2222222222Ø1Ø         98765
555555555Ø33         98765
```

Partial Output from Sample Program:

```
CONTROL CARD LIST       JOB=FILBRPT STEP=GO 8/05/80  11:42:06.520 VERSION 2.0  PAGE 1
     ...5...10...15...20...25...30...35...40...45...50...55...60...65...70...75
00001  HEADING 1 10 WIDGIT VALUES
00002  AT OUTPUT.
00003  COMPUTE 101-109 '$ ,  0.00' = 10-15.  PRINT FROM 101-109.
00004  TOTAL 10-15,
00005    EDIT 1 23 '$ ,  0.00',
00006    NOTE 1 1 'TOTAL WIDGIT VALUE IS'.
00007  ID=1-9. UID=1-9.  AT ADD. ERROR.
00008  AT CHANGE.
00009  MOVE TRANSACTION 20-29 TO OUTPUT ADDRESS TRANSACTION 10-12.
00010  * COULD HAVE BEEN ABBREVIATED AS:  M T 20-29 TO O A T 10-12.
```

```
FILEBOL INPUT-OUTPUT TOTALS      8/05/80  11:42:06.520 VERSION 2.0  PAGE 2
     3 INPUT RECORDS FROM SYSUT1        0 ADD TRANSACTIONS
   + 0 RECORDS ADDED FROM UPDATES       1 CHANGE TRANSACTIONS
   - 0 RECORDS DELETED FROM SYSUT1      0 DELETE TRANSACTIONS
   ------                               1 TRANSACTION RECORDS REJECTED
     3 NET INPUT RECORDS                ------
     1 INPUT RECORDS CHANGED            2 TRANSACTION RECORDS
```

| RECD CNT | DDNAME | DATASET NAME | VOLSER | ORG | FM, | LRECL, | BLKSZ |
|---|---|---|---|---|---|---|---|
| 3 | OUTPUT RECORDS ON SYSPRINT: | ........... | ...... | PS | F A, | 133, | 133 |
| 3 | OUTPUT RECORDS ON SYSUT2 : | ........... | ...... | PS | FB , | 80, | 80 |
| 3 | INPUT RECORDS FROM SYSUT1 : | ........... | ...... | PS | FB , | 80, | 80 |
| 2 | INPUT RECORDS FROM UPDATES : | ........... | ...... | PS | FB , | 80, | 80 |
| 10 | INPUT RECORDS FROM SYSIN : | ........... | ...... | PS | FB , | 80, | 80 |
| 11 | OUTPUT RECORDS ON CNTLLIST: | ........... | ...... | PS | F A, | 133, | 133 |

ALL DONE  AT 11:42:07.730

```
        WIDGIT VALUES                                              PAGE 1
$456.78
$987.65
$1,234.56

RUN TOTALS                                                         PAGE 2
TOTAL WIDGIT VALUE IS      $2,678.99
```

## FILEBOL, Version 2.0

Developed by:                          Distributed by:

  University of Chicago                Same
   Computation Center
  5737 South University
  Chicago IL  60637
  (312)753-8439

Computer Makes:                        Operating Systems:

  IBM 360, 370 or equivalent          IBM - OS/VS, SVS, MVS

Source Language:  IBM Basic Assembly Language (BAL)

Cost:

    The one time permanent license fee for FILEBOL is $300.  The fee includes program source and load modules, a sample catalogued procedure, and the FILEBOL manual on magnetic tape.

## 2.09 KPSIM/KPVER

### CAPABILITIES: KPSIM

KPSIM.FOR is an on-line, general-purpose data entry program. It is a create-oriented editor, designed to simulate keypunch operations. A format-specifications file is entered at the beginning of a run (in some ways similar to the drum card on a keypunch). This file is set up by the user to define fields, indicating: an identifier for each field (to be typed at the terminal as a prompt to the operator), the width of each field, whether the entry is to be right- or left-adjusted, and any restrictions on the data (such as whether the field is alpha, integer or decimal and what characters or values are allowed). Default missing-value codes may also be set. "Duplicate" or constant fields are also available, as are copy fields. There is no limit on the length of a record, and since record marks can be defined as a constant in a given field, multi-record data layout can be used.

Once the specifications file has been input, the operator simply identi-fies the output file, then enters data field-by-field in response to prompts. A limited repertoire of editing commands ($TO, $TYpe, $REplace, $DUplicate, $ENd) enables the operator to skip fields, to go back and correct fields, to duplicate fields from the previous case, and to exit from the program.

### CAPABILITIES: KPVER

KPVER.FOR is an on-line, generalized program for entering data in a for-matted, fixed-field data file, and verifying the entries variable-by-variable against corresponding entries in a previously created file. Whenever there is a discrepancy between the current entry and the previous entry, the program will display both versions at the terminal and allow the operator to enter the correct form of the variable into the final data record. As with KPSIM, a for-mat-specifications file is entered at the beginning of a run to define fields.

Once the specifications file has been input, the operator simply identi-fies the field to be verified and an output file, then enters data field-by-field in response to prompts. The same editing commands are available as in KPSIM. Default vaules are entered in skipped fields, and are verified against corresponding values in the previous file, just as explicit entries are veri-fied. Verification can be interrupted and restarted with ease.

### PROPOSED ADDITIONS

None anticipated in next 12 months. Planned modification for indefinite future is use of multiple format specifications files to allow for entry of hierarchical or other complex data sets.

Part of a Sample Session (Input only):

```
>SAMPLE.ONE


CURRENT PROGRAM LIMITS:
   NTAGS  =   230
   MAXOUT = 1250
   VXMAX  =   400
   NCMDS  =     7
   MAXLIN =   200
IID           -- 5>12054
FID           -- 4>3660
FAMIN         -- 1>2
FE            -- 8>
HFE           -- 1>
FLV           -- 8>
HFLV          -- 1>
ININ          -- 1>2
FNHM          --10>NASUKAWA
ALTFN         --10>
PERN          --10>HARUE
1ALTP         --10>$TOSEX


SKIPPING AHEAD TO:SEX

SEX           -- 1>D
FA            -- 5>$TOMTO


SKIPPING AHEAD TO:MTO

MTO           -- 5>12053
MPL           --13>$TO4MPL


SKIPPING AHEAD TO:4MPL

4MPL          --13>MAKAENO-62
4DIV          -- 8>$TO4MPL


GOING BACK TO TAG: 4MPL
CURRENT ENTRY IS: MAKAENO-62
4MPL          --13>MUKA^U
MUKAENO-5\5\62
MUKAENO-62
4DIV          -- 8>$TO/
```

### KPSIM:  Keypunch Simulator
### KPVER:  Keypunch Verification Simulator

Developed by:                                    Distributed by:

    Philip S. Sidel & K.Y. Chang              Same
    Social Science Computer
    Research Institute
    232 Lawrence Hall
    University of Pittsburg
    Pittsburg, PA 15260

Machine:                                          Operating Systems:

    DEC-10/20                                 TOPS-10

Source Language:  FORTRAN-10

Cost:

    The one time charge for KPSIM.FOR is $20 plus materials for universities, governments and nonprofit organizations, and commercial use.  KPVER is included.

Documentation:

Machine-readable HELP file is distributed with program.

# CHAPTER 3

## EDITING PROGRAMS

## CONTENTS

## INTRODUCTION

Editing of data, in the sense of detecting and correcting errors, poses philosophical as well as technical problems.  It includes not only the recognition and handling of extreme observations, but also the treatment of missing values:  the appropriate treatment may depend on the purpose for which the data was collected.

Most editing is done with programs custom-written for a particular survey. There has been little success in producing general-purpose editing programs. The editing requirements for processing data from large surveys or censuses were listed in Francis and Sedransk (1976) under five headings:

i) Structural Checks

   a.  file structure checks
   b.  skip checks

ii) Consistency Checks

   c.  "wild code" or range checks
   d.  logical checks between items
   e.  internal arithmetic checks
   f.  external checks with previous surveys

iii) Probabilistic Checks

   g.  univariate tests, probability plots
   h.  multivariate techniques; e.g. principal component analysis

iv) Error Display

   i.  error messages
   j.  user controlled criteria for listing a record
   k.  ability to remember errors

TABLE 3.1:   RATINGS BY DEVELOPERS ON SELECTED ITEMS

(i) Capabilities

Usefulness
Rating Key:
3 - high
2 - moderate
1 - modest
. - low

| | Complex Structures | File Management | Editing Language | Consistency Checks | Probability Checks | Error Handling | Compute Tables | Print Tables | Multiple Regression | Anova/Linear Model | Linear Multivariate | Multi-way Tables | Nonparametric | Exploratory | Robust | Non-linear | Time Series | Econometric |
|---|---|---|---|---|---|---|---|---|---|---|---|---|---|---|---|---|---|---|
| | 11 | 14 | 17 | 18 | 19 | 20 | 24 | 25 | 30 | 31 | 32 | 33 | 36 | 37 | 38 | 39 | 35 | 41 |
| 3.01 GES | 2 | 2 | 1 | 3 | 1 | 3 | . | . | . | . | . | . | . | . | . | . | . | . |
| 3.02 CONCOR | 1 | 2 | 3 | 3 | . | 3 | . | . | . | . | . | . | . | . | . | . | . | . |
| 3.03 CHARO | . | . | 2 | 3 | . | . | . | . | . | . | . | . | . | . | . | . | . | . |
| 3.04 EDITCK | . | . | 2 | 3 | . | 3 | . | . | . | . | . | . | . | . | . | . | . | . |
| 3.05 VCP-LCP | . | . | . | 3 | . | 2 | . | . | . | . | . | . | . | . | . | . | . | . |

## FOR EDITING PROGRAMS

### (ii) User Interface

| | Survey Estimates | Survey Variances | Math Functions | Operations Research | Availability | Installations | Computer Makes | Mini Version | Core Requirements | Batch/Interactive | Stat. Training | Computer Training | Language Simplicity | Documentation | User Convenience | Maintenance | Tested for Accuracy |
|---|---|---|---|---|---|---|---|---|---|---|---|---|---|---|---|---|---|
| | 43 | 44 | 47 | 48 | 49 | 50 | 51 | 52 | 53 | 54 | 55 | 56 | 57 | 58 | 59 | 60 | 61 |
| GES | . | . | . | . | 2 | . | 2 | 1 | 1 | 3 | 2 | 1 | . | 1 | 2 | 3 | 1 |
| CONCOR | . | . | . | . | 2 | 2 | . | . | 2 | . | 3 | 1 | 2 | 1 | 3 | 3 | 1 |
| CHARO | . | . | . | . | 2 | 1 | 2 | 3 | 2 | . | 3 | 3 | . | 1 | 3 | 2 | 1 |
| EDITCK | . | . | . | . | . | . | 1 | 3 | 2 | . | 3 | 1 | . | 1 | 2 | 3 | 1 |
| VCP-LCP | . | . | . | . | . | . | 1 | . | 1 | . | 3 | 3 | 1 | 1 | 1 | 3 | . |

v)  Error Correction

   l.  delete faulty records
   m.  delete faulty items and treat as non-sampled elements
   n.  automatic correction
   o.  user's flexibility in correcting records
   p.  imputation methods
   q.  creating new file

   In the present survey, many developers and users thought that a program
was good at editing (item 16 on the questionnaire) if it could provide consistency checks of types (c) or (e).  Few programs provide probabilitistic checks
(item 19) and even fewer attempt any error correction or imputation (item 22).

   Of the programs in this chapter, and indeed in this book, CONCOR is the
only serious attempt to include most of the above features in a general-purpose
program.  But even this program must undergo major enhancements to facilitate
general sample-survey editing.  We hope that the recent interest in editing and
in the quality of data will continue in the literature and among software developers.  It might be noted that no users were found for any of the five programs
in this chapter.

   The ratings by the respective developers on selected items are displayed in
Table 3.1.  Figure 3.1 compares developers' and users' ratings on all items,
while Figure 3.2 summarizes developers' ratings (D-scores) and users' ratings
(U-scores) on several relevant attributes.  These are explained in detail in
Section 1.4 of Chapter 1.  Figure 3.2 also points to other chapters which
describe programs with strong editing capabilities.

   The remainder of this chapter contains descriptions of these editing
programs in the format described in Table 1.5.

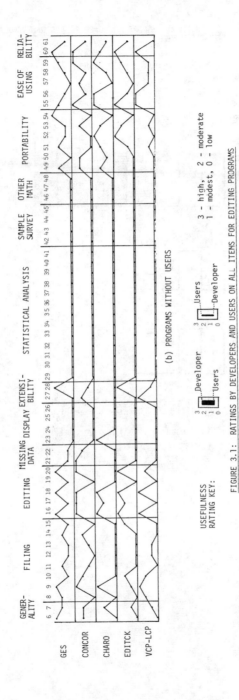

(b) PROGRAMS WITHOUT USERS

USEFULNESS
RATING KEY:

3 - high,  2 - moderate
1 - modest,  0 - low

FIGURE 3.1: RATINGS BY DEVELOPERS AND USERS ON ALL ITEMS FOR EDITING PROGRAMS

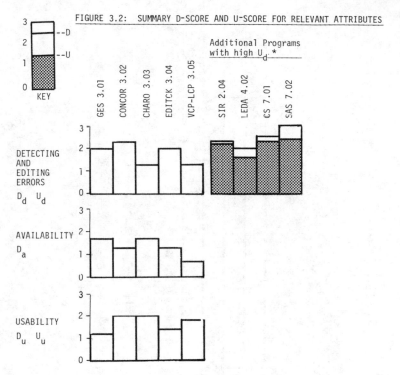

FIGURE 3.2:  SUMMARY D-SCORE AND U-SCORE FOR RELEVANT ATTRIBUTES

Another chapter generally containing programs with strong
Editing Capabilities is Chapter 6 (see Figure 6.2).

*In addition, for these specific programs, $U_d > 1.5$.

## 3.01 <u>GES</u>

### INTRODUCTION

During the past decade, increased utilization of computer programs for processing survey data in the United States Department of Agriculture brought with it a considerable duplication of programming effort. Because of the variety of questionnaires and methods of processing survey data in Washington, D.C., and in the field offices, the number of computer programs to perform such functions as data editing grew rapidly. Programming was repeated for basically the same function - editing survey data. Furthermore, changes in questionnaires and procedures resulted in frequent computer program modifications.

The dynamic characteristics of the surveys being conducted and the increasing commitment of resources to data processing, caused the Agency to decide on the development of generalized systems for federal programs.

In September 1971, a task force was established to develop a generalized system to edit raw data from Statistical Reporting Service surveys. This Generalized Edit System is the result of work by that task force.

### CAPABILITIES: Processing Statistical Data

The system is designed to accept raw data from most surveys now conducted by the Agency and prepare it for summary following minimum manual review.

Some of the benefits from a computer edit system for general use include the following:

1) Its flexibility and external control permit its use for various surveys without the need for changes in the source programs. New surveys or revised questionnaires can be more easily accommodated.

2) Greater uniformity and consistency of editing data are possible.

3) The time required to automate a survey is reduced.

4) Data Processing capabilities can be extended to additional surveys.

5) The productivity of personnel is increased. Training for a multiple number of systems is not needed.

6) Commodity personnel can maintain more control over the specifications for data manipulation through the use of parameters.

7) The system provides documented information such as error counts by type and record counts that might not otherwise be available.

The Generalized Edit System can interface with the Generalized Summary System (see Chapter 4.14) and with Multiframe Summary (see Chapter 5.03). This system is not applicable to general public use at present.

## GES:   Generalized Edit System

Developed by:

    Systems Branch
    ESCS, Statistics
    U.S. Department of Agriculture
    Washington, D.C.  20250

Computer Makes:

    IBM 360/370
    UNIVAC 1100

Language:

    COBOL

Cost:

    Materials

Additions next 12 months:

    None anticipated.

## 3.02 CONCOR

### INTRODUCTION

COBOL CONCOR is a special-purpose statistical software package which is used to identify data items that are invalid or inconsistent, automatically correct data items by "hot-deck" or "cold-deck" imputation, create an edited data file in original or recoded or reformatted form, create an auxiliary data file, produce an edit diary summarizing errors detected and corrections made, and perform error tolerance analysis. The original CONCOR edit and imputation system was developed in 1975 by the United Nations Latin American Demographic Center (CELADE) at Santiago, Chile and was written in IBM Assembler Language Code (ALC). The latest version of CONCOR, COBOL CONCOR Version 2.1 was developed in 1979 by the International Statistical Programs Center (ISPC) of the U.S. Bureau of the Census and contains significant improvements over the original version developed by CELADE. COBOL CONCOR Version 2.1 was written in the COBOL language and is designed to run on virtually all computer systems that have an ANSI COBOL compiler, random access disk storage, and a minimum amount of core storage.

### CAPABILITIES: Processing and Displaying Data

CONCOR can be used to inspect the structure of an interview or questionnaire, the validity of individual data items, and the consistency among data items both within a logical record and across logical records within the same observation (interview or questionnaire).

The system will automatically generate error messages as the data are being inspected based on the validity and consistency rules set forth in the user's program. The user can supplement CONCOR's message system by supplying any specifically desired messages which will be displayed as errors are detected. Error messages can either be generated on a case-by-case basis or summarized over any desired area.

Corrections to data can be imputed using "hot-deck" arrays, "cold-deck" arrays, or through simple arbitrary allocations. The system maintains counts of the frequency with which the original values of data items are "changed."

The system will automatically produce an "edited" output file identical in format to the unedited input file which allows the output to be "test read" back through the system to determine if changes made during the editing process have introduced other invalidities or inconsistencies.

A derivative output file of any format can be produced concurrent with the edit run.

Data are automatically converted to binary before editing for faster processing.

The command language is completely free-format in design and the system provides a comprehensive syntax analysis function which protects the user against coding errors and execution time errors.

## CAPABILITIES:  Statistical Analysis

CONCOR displays the frequency with which data items have been tested, the frequency of errors found, and the error rate.  These statistics can be displayed for the total run or specific disaggregate levels.  CONCOR allows the user to set tolerance limits above which the system will "reject" a defined work unit as being unacceptable.

## INTERFACE WITH OTHER SYSTEMS

CONCOR can read files produced by many other packages and the outputs of CONCOR are compatible as inputs to CENTS III and COCENTS.

## PROPOSED ADDITIONS IN NEXT YEAR

Major enhancements to facilitate general sample survey editing will include:  a manual corrections sub-system to provide interface for adding, deleting or modifying specific records on the file;  an additional input file to allow check-in of sample cases and verification of geographic ID codes;  providing frequency distributions of values used in "hot-deck" imputation;  providing weighting scheme to show estimates in edit diary for sample cases; and expanding the maximum size of data fields allowed from 9 to 18 digits.

## SAMPLE JOB (Partial listing)

```
        RANGE P001 VALUE 1 TO 3,9

            NOMATCH 99

            MESSAGE 7'RANGE 1 TO 3 AND 9',P001;

        ASSERT P011-MARITAL-STATUS EQUAL TO
        EVER-MARRIED

          EQUIVALENT P012-YEARS-MARRIED <> 0

          MESSAGE 6, 'MARITAL STATUS/YEARS MARRIED

            INCONSISTENT', P011-MARITAL-STATUS,

              P012-YEARS-MARRIED;

        ALLOCATE P005-AGE=

          A001-AGE-FROM-RELATIONSHIP-AND-SEX (P006-SEX,
          P002-RELATIONSHIP);

        UPDATE A001-AGE-FROM-RELATIONSHIP-AND-SEX
              (N006-SEX,N002-RELATIONSHIP)

                  =P005-AGE(1);
```

Sample Output:

```
                                    C O N C O R                                        PAGE    68
                         ( C O N S I S T E N C Y   A N D   C O R R E C T I O N )       RUN DATE  12/31/79
                         E D I T   A N D   I M P U T A T I O N   S Y S T E M
                                   EDIT STATISTICS REPORT
                                 BY EDIT SPECIFICATION COMMAND

CONTROL AREA: 10:17:201:1:0

xxxxxxxxxxxxxxxxxxxxxxxxxxxxxxxxxxxxxxxxxxxxxxxxxxxxxxxxx
x   U S E R - S P E C I F I E D   M E S S A G E S   x
xxxxxxxxxxxxxxxxxxxxxxxxxxxxxxxxxxxxxxxxxxxxxxxxxxxxxxxxx
-----------------------------------------------------------------------------------------------------------
SOURCE                                                                           NUMBER     NUMBER      PER
 LINE   ERROR                                                                    CASES      CASES      CENT
NUMBER  NUMBER  MESSAGE                                                          TESTED     FAILED    FAILED
-----------------------------------------------------------------------------------------------------------
    26      1   STRUCTURE ERROR - INVALID RECORD TYPE EXISTS                       12          0       0.00
    45     10   STRUCTURE ERROR - NO HOUSING RECORD                                2           1      50.00
    64     20   STRUCTURE ERROR - MORE THAN 1 HOUSING RECORD                       2           1      50.00
    91     25   STRUCTURE ERROR - EXCESS HOUSING RECORDS                           0           0       0.00
   107     30   STRUCTURE ERROR - WRONG NUMBER OF POPULATION RECORDS              12           3      25.00
   120     40   STRUCTURE ERROR - NO PERSONS IN OCCUPIED UNIT                     12           0       0.00
   125     50   STRUCTURE ERROR - PERSONS PRESENT IN VACANT UNIT                   0           0       0.00
   188     60   GROUP QUARTERS NOT SHOWN IN TYPE OF UNIT                           3           0       0.00
   209     70   RELATIONSHIP CODE SHOULD BE GROUP QUARTERS                        14          14     100.00
   229    100   STRUCTURE ERROR - NO HEAD OF HOUSEHOLD                             8           2      25.00
   386    110   STRUCTURE ERROR - EXCESS HEAD RECORD                              12           0       0.00
   509    120   STRUCTURE ERROR - EXCESS SPOUSE RECORDS                            7           0       0.00

                    TOTAL NUMBER OF USER-DEFINED MESSAGES = 12
                         TOTAL NUMBER OF CASES TESTED = 84
                         TOTAL NUMBER OF CASES FAILED = 21
                         PERCENTAGE OF CASES FAILED = 25.00


xxxxxxxxxxxxxxxxxxxxxxxxxxxxxxxxxxxxxxxxxxxxxxxxxxxxxxxxxxxx
x S Y S T E M - G E N E R A T E D   M E S S A G E S  x
xxxxxxxxxxxxxxxxxxxxxxxxxxxxxxxxxxxxxxxxxxxxxxxxxxxxxxxxxxxx
-----------------------------------------------------------------------------------------------------------
SOURCE                                                                           NUMBER     NUMBER      PER
 LINE                                                                            CASES      CASES      CENT
NUMBER          MESSAGE                                                          TESTED     FAILED    FAILED
-----------------------------------------------------------------------------------------------------------
   116          ALLOCATE H15 (1) = H03                                            3           3     100.00
   197          ALLOCATE H01 (1) = 98                                             0           0       0.00
   214          ALLOCATE P02 (N05) = 8                                           14          14     100.00
   282          ALLOCATE P02 (N22) = 0                                            1           1     100.00
   390          ALLOCATE P02 (N05) = 5                                            0           0       0.00
   513          ALLOCATE P02 (N05) = 5                                            0           0       0.00
   531          RANGE H01 VALUE 0 TO 9 NOMATCH 0                                  11          0       0.00



                                    C O N C O R                                        PAGE    11
                                 EDIT STATISTICS REPORT                                 RUN DATE  12/31/79
                                    BY QUESTIONNAIRE

CONTROL AREA:        10:17:201:1:0

QUESTIONNAIRE IDENTIFICATION:    1017:201:01:000:012
-----------------------------------------------------------------------------------------------------------
SOURCE
 LINE   MESSAGE
NUMBER  NUMBER   TEXT                                                          IDENTIFIER(S) WITH CURRENT VALUE(S)
-----------------------------------------------------------------------------------------------------------
   229    100    STRUCTURE ERROR - NO HEAD OF HOUSEHOLD                        P02 (1) = 3
   555   SYSTEM  RANGE H07A, H07B VALUE 0 TO 19, 99 NOMATCH 99                 H07B (1) = 33
   583   SYSTEM  RANGE H14A, H14B, H14C, H14D VALUE 1, 2, 9 NOMATCH 9          H14A (1) = 5
   583   SYSTEM  RANGE H14A, H14B, H14C, H14D VALUE 1, 2, 9 NOMATCH 9          H14B (1) = 3
   583   SYSTEM  RANGE H14A, H14B, H14C, H14D VALUE 1, 2, 9 NOMATCH 9          H14C (1) = 6
   583   SYSTEM  RANGE H14A, H14B, H14C, H14D VALUE 1, 2, 9 NOMATCH 9          H14D (1) = 0
   598   SYSTEM  RANGE P04 VALUE 0 TO 99 NOMATCH 0                             P04 (1) = 100
   602   SYSTEM  RANGE P03 VALUE 1,2 NOMATCH 0                                 P03 (1) = 10
   643   SYSTEM  RANGE P04 VALUE 0 TO 98 NOMATCH 0                             P04 (1) = 8
   646   SYSTEM  ASSERT P02 = 0 IMPLIES P04 >= 12  . IF THE RELATIONSHIP IS HEAD,   P02 (1) = 0   P04 (1) = 8
  1069   SYSTEM  RANGE P06 VALUE 1,2,9 NOMATCH 9                               P06 (1) = 0
  1078   SYSTEM  RANGE P07 VALUE 0101 TO 1927, 2000 TO 8300, 9999 NOMATCH 9999  P07 (1) = 10
  1098   SYSTEM  ASSERT P08B = 1           . MUST BE POPSTAN CITIZEN           P08B (1) = 8
  1141   SYSTEM  RANGE P11 VALUE 0, 11 TO 16, 21 TO 26, 31 TO 39, 99 NOMATCH 99   P11 (1) = 95
  1236   SYSTEM  RANGE P12C VALUE 1, 2, 9 NOMATCH 9                            P12C (1) = 3
  1248   SYSTEM  RANGE P12D VALUE 1, 2, 9 NOMATCH 9                            P12D (1) = 0
  1272   SYSTEM  RANGE P14 VALUE 1 TO 9 NOMATCH 9                              P14 (1) = 0

           TOTAL ERROR FREQUENCY (FAILED) FOR THIS QUESTIONNAIRE =     17
```

### COBOL CONCOR, Version 2.1:  COBOL Consistency and Correction 2.1

Developed by:

    U.S. Bureau of the Census
    International Statistical
      Programs Center
    Washington, D.C.  20233

Computer Makes:                          Operating Systems:

    IBM System 360/370                       IBM - OS, VS, DOS/VS
    IBM 303X series
    (planned next 12 months)
    ICL 2900 series
    NEAC 500                                 Interfaced Systems:
    HONEYWELL 66
    FACOM M-160                                  CENTS III, COCENTS
    UNIVAC 1100 series

Source Language:  ANSI COBOL

Cost:

    There is a one-time charge of $350 for COBOL CONCOR.  This charge covers
expenses for providing an installation tape containing source modules, installation guide, and computer readable documentation.  Hard copies of the system
reference manual, error diagnostics guide, systems guide, installation guide,
and procedures library are also included.

Documentation:

CONCOR System Reference Manual, International Statistical Programs Center,
    U.S. Bureau of the Census, Washington, D.C., 1980.
CONCOR Diagnostic Message Guide, International Statistical Programs Center,
    U.S. Bureau of the Census, Washington, D.C., 1980.
CONCOR Systems Guide, International Statistical Programs Center, U.S. Bureau
    of the Census, Washington, D.C., 1980.

3.03  CHARO

## INTRODUCTION

This program was developed by William A. Gates at the University of Wiscon-
cin, Madison.  The first version written in COBOL, has been installed on IBM 370
and UNIVAC machines.  The second version, written by Michael L. Stouffer in
FORTRAN 77 and DEC Command language, is operating on a VAX.

The IBM and UNIVAC version is not actively maintained.  The VAX version is
still undergoing some development and documentation, and will probably be available for distribution by the end of 1981.

## CAPABILITIES:  Processing Data

CHARO is a character-overview program which displays specific alphabetic,
numeric or special characters which occur in each location of a file on tape.
It is not a column frequency program, but prints all the different characters or
codes found in a specific location of the record after reading the whole file.
Blocked records can be processed.  A maximum of 750 characters of a record could
be examined (although the record length itself could be much bigger).  Up to
five character-location groups can be checked (e.g. locations 1-10, 15-35,
9-200, 210-215, 250-301).  The total of 750 characters allowable could be
enlarged by a simple system modification.  In addition, the VAX version will
examine up to 4096 columns in one pass.

The VAX version is accompanied by a program FREQO, which computes column
frequencies on the ten digits, blanks, -, +, alphabetic, and other characters.

## CHARO - Character Overview

**Developed by:**

    William A. Gates (IBM-UNIVAC)
    University of Wisconsin
    Madison, Wisconsin  53706

    Michael L. Stouffer (VAX)

**Distributed by:**

    Center for Demography and Ecology
    3224 Social Science Building
    University of Wisconsin
    Madison, Wisconsin  53706

**Machine:**

    IBM 370
    UNIVAC
    VAX

**Language:**

    COBOL - IBM, UNIVAC
    FORTRAN - VAX

**Cost:**

    One time charge for materials.

## 3.04   EDITCK

### INTRODUCTION

The program EDITCK is a general program for the detection of errors on data forms. Its primary use has been in the editing of multi-center clinical research data. The program was written by Theodore Karrison in 1978 and has been further developed since that time.

### CAPABILITIES

The program identifies four basic error types: unanswered items (invalid blanks), illegibilities, values out of range, and inconsistencies. The detection of all four types of errors can be accomplished by the use of an edit table. The fourth type, inconsistencies, does not lend itself to as simple a programmable solution as the first three. This flexibility makes EDITCK useful for screening a variety of data forms and reduces the time and effort spent in this area.

The program also includes a subroutine which calculates the exact number of days falling between two dates and allows this value to be compared with an arbitrary constant. This procedure can be useful if, for example, the interval between two dates on a form is not permitted to exceed a specified number of days. An additional feature is the optional use of variable flags (if maintained in the data base) to suppress specific error messages if so desired. (An example would be the suppression of error messages for previously confirmed out-of-range values.)

Error messages are printed in a convenient format (called an edit statement) for future correction (see sample output). Error counts by variable and error type are also printed to help identify problems in the design and completion of data forms.

### EXTENSIBILITY

The program is written in standard Fortran IV with the exception of a subroutine which uses in-core I/O facilities available, in one form or another, at most installations. It is therefore easily added to or modified, and can be run on most systems.

### SAMPLE JOB

The program is run in batch mode. The user supplies a title and date for the edit statements and the necessary file specifications for the data file and associated edit table. An edit table for a form containing 19 variables and 23 consistency checks is given below for illustration. Details on the format of edit tables can be found in the program documentation. A sample edit message appears on the following page.

An Edit Table:

051    19 2        -1   6

```
**************************************
++++++++++++++++++++++++++++++++++++++
(5X,A8,14X,A1,4A2,10A1,A2,2A1,A6,6I1)
NO 2ND FORMAT
NO 3RD FORMAT
  1 051 101 2            1 1   0        1        3 0
  2 051 102 2 FV NO      2 0   0        1       10 0
  3 051 103 3 MO         2 1   1        1       12 0
  4 051 104 3 DAY        2 1   2        1       31 0
  5 051 105 3 YR         2 1   3       75       79 0
  6 051 106 4-A          1 1   0        1        2 0
  7 051 107 4-B          1 1   0        1        2 0
  8 051 108 4-C          1 1   0        1        2 0
  9 051 109 4-D          1 1   0        1        2 0
 10 051 110 4-E          1 1   0        1        2 0
 11 051 111 4-F          1 1   0        1        2 0
 12 051 112 4-G          1 1   0        1        2 0
 13 051 113 4-H          1 1   0        1        2 0
 14 051 114 4-I          1 1   0        1        2 0
 15 051 115 4-J          1 1   0        1        2 0
 16 051 116 5-NUMB       2 0   4        1       20 0
 17 051 117 5-NONE       1 0   5        1        1 0
 18 051 118 6            1 1   0        1        4 0
 19 051 119 7            6 1   6    10000    20000 2
  1    1=       1    2*
  2    1=       2    2*
  3    1=       3    2=
  4    6=       1   16=
  5    6=       1   17*
  6    8=       1   16=
  7    8=       1   17*
  8   11=       1   16=
  9   11=       1   17*
 10   12=       1   16=
 11   12=       1   17*
 12    6=       2    8=      2  11=      2  12=      2  16*
 13    6=       2    8=      2  11=      2  12=      2  17=
 14   16*              17*
 15   16=              17=
 16    6=       1   18=      4
 17    8=       1   18=      4
 18   10=       1   18=      4
 19   11=       1   18=      4
 20   12=       1   18=      4
0
 21      5   6      5   8      5  10      5  11      5  12      1  16 = 51
 22      5   6      5   8      5  10      5  11      5  12      1  18 = 52
 23      5   6      5   8      5  10      5  11      5  12      1  18 = 53
```

Sample Edit Message:

## SAMPLE EDIT

### EDIT 06/01/80

| PATIENT ID.NO. | FORM | ITEM NO. | ERROR TYPE | CURRENT VALUE | CORRECT VALUE | VAR NO. |
|---|---|---|---|---|---|---|
| 3106231 | SU0 | 3 YR | UNANSWERED | | | 5 |
| 3106231 | SU0 | 4-I | UNANSWERED | -- | -- | 14 |
| 3106231 | SU0 | 2 | INCONS- 3 | 3 | - | 1 |
| 3106231 | SU0 | 2 FV NO | INCONS- 3 | - | - | 2 |
| 4302952 | SU0 | 3 YR | OUT OF RANGE | 67 | -- | 5 |
| 4302952 | SU0 | 2 | INCONS- 1 | 1 | - | 1 |
| 4302952 | SU0 | 2 FV NO | INCONS- 1 | 03 | - | 2 |
| 4302952 | SU0 | 4-A | INCONS- 4 | 1 | -- | 6 |
| 4302952 | SU0 | 5-NUMB | INCONS- 4 | - | | 16 |
| 4302952 | SU0 | 4-A | INCONS- 5 | 1 | -- | 6 ** |
| 4302952 | SU0 | 5-NONE | INCONS- 5 | 1 | - | 17 |
| 4302952 | SU0 | 4-A | INCONS- 16 | 1 | - | 6 ** |
| 4302952 | SU0 | 6 | INCONS- 16 | 4 | - | 18 |
| 4410611 | SU0 | 4-A | INCONS- 12 | 2 | - | 6 |
| 4410611 | SU0 | 4-C | INCONS- 12 | 2 | - | 8 |
| 4410611 | SU0 | 4-F | INCONS- 12 | 2 | - | 11 |
| 4410611 | SU0 | 4-G | INCONS- 12 | 2 | - | 12 |
| 4410611 | SU0 | 5-NUMB | INCONS- 12 | 2 | -- | 16 |
| 4410611 | SU0 | 5-NUMB | INCONS- 14 | 2 | -- | 16 ** |
| 4410611 | SU0 | 5-NONE | INCONS- 14 | 1 | - | 17 |

## EDITCK:  Edit-Check

Developed by:                                              Distributed by:

    Theodore Karrison                                          Same
    University of Chicago
    Abbott Hall - Room ˙336
    947 E. 58th St.
    Chicago, Illinois 60637

Source Language:  FORTRAN

Cost:

    A one-time charge is estimated at $25 plus postage for a punched deck and $35 plus postage for a magnetic tape.  If a magnetic tape is provided, the fee will be $20 plus postage.

Documentation:

Karrison, Theodore;  EDITCK Program Documentation, CCE Program Library, University of Chicago, August, 1980.
Karrison, Theodore, "Data Editing in a Clinical Trial," Controlled Clinical Trials, to be published.

## 3.05  VCP-LCP

### INTRODUCTION

VCP-LCP is a set of programs used to clean data.  They have been used in various types of survey research data.  They were developed by Jerome Weiland at the Institute for Survey Research of Temple University.  They are set up as batch programs and are presently being used at Temple University on a CDC CYBER 174.

### CAPABILITIES

The Valid Code Program checks the data for card sequence (assuming more than one card per case) and detects validity of data in each column or set of columns of the data.

The Logical Check Program makes a set of tests among different columns of the data which can be on different cards or records.  The program has 20 subroutines each of which performs a specific consistency check in the data.  The programs assume files of data in Fortran or Cobol Format.

The Logical Check Program has a limit of forty cards per case due to memory restrictions of the computer at Temple University.  The Valid Code Program has a limit which is a combination of cards and tests per card.  This makes the program more flexible and easy to use.

### EXTENSIBILITY

The Logical Check Program can easily add more subroutines if general rules are set up.  At present the program requires recompilation to add additional subroutines.

### SAMPLE JOB

* Find all cases that have invalid codes for race in card 2 column 18.
* Find all cases that do not satisfy the following consistency check: If Card 2 col. 18 equals 8 THEN Card 2 col. 19 equals 9.

See Output on following page:

Sample Output:

NO. OF CARDS PER CASE   2   MAXIMUM NO. OF TESTS PER CARD   25

FORMAT FOR CARD NO. AND CASE NO.   (I1,I4,75X)

VALID CODE FOR  MANPOWER STUDY

2  1 18 181    0    7    9    18    18  999999999    0

PARAMETER CARDS READ =  2

```
         1         2         3         4         5         6         7         8
1234567890123456789012345678901234567890123456789012345678901234567890123456789012345678901234567890
COLUMNS THAT HAVE AN ERROR   18
   ONE OR MORE ERRORS IN THIS CARD  20309   888888888888888088888888023211
```

686 TESTS MADE ON  686 CARDS.   685 WERE GOOD    1 WERE BAD

OUTPUT FROM LOGICAL CHECKS

1 218 219 1 1 TEST= 1    8    8    9    9    9999

PARAMETER CARDS READ =  2

```
         1         2         3         4         5         6         7         8         9        10        11        12
1234567890123456789012345678901234567890123456789012345678901234567890123456789012345678901234567890123456789012345678901234567890123
TEST NO. AND COLUMNS THAT HAVE AN ERROR   1 2 18- 2 19---
   ONE OR MORE ERRORS IN THIS CARD   1 2 18- 2 19---
20309   88888888888A888088888888023211
```

343 TESTS MADE ON  343CARDS.   342 WERE GOOD    1 WERE BAD

## VCP-LCP:  Valid Code Program-Logical Check Program

Developed by:                                    Distributed by:

   Institute for Survey Research          Same
   Temple University
   1601 N. Broad Street
   Philadelphia PA 19122

Computer Makes:                                  Operating Systems:

   CDC CYBER Series                       CDC NOS, SCOPE 3.4

Source Language:

   Fortran easily converted to any machine with Fortran.

Cost:

    Not Currently Set.

Documentation:

Weiland, Jerome., Use of Valid Code and Logical Check Programs.   Internal memo.

CHAPTER 4


TABULATION PROGRAMS


CONTENTS

INTRODUCTION

     The first stage in a statistical analysis after the data has been filed is
a descriptive presentation of the data, often in the form of tables.  Indeed for
most people who are employed in "statistical" activities, and for most censuses
and many surveys, tabulation is also the last stage in a statistical analysis.
It can be seen from the taxonomy (Table 1.2) that, along with an ability for
file management, the ability to compute tables is the most common capability.

     Chapter 4 has been divided into Parts (i) and (ii) based on whether the
developers claimed some ability to compute population estimates in addition to
simply tabulating the data:  see Table 1.2.

     The two parts of Chapter 4 taken together include several tabulation pro-
grams of government agencies which frequently conduct large censuses or surveys.
It should be noted that for the most part these heavily-used programs are
poorly equipped to compute variances of population estimates.

TABLE 4.1:   RATINGS BY DEVELOPERS ON SELECTED ITEMS

(i) Capabilities

Usefulness Rating Key:
3 - high
? - moderate
1 - modest
. - low

| | Complex Structures | File Management | Consistency Checks | Probabilistic Checks | Compute Tables | Print Tables | Multiple Regression | Anova/Linear Model | Linear Multivariate | Multi-way Tables | Nonparametric | Exploratory | Robust | Non-linear | Time Series | Econometric |
|---|---|---|---|---|---|---|---|---|---|---|---|---|---|---|---|---|
| | 11 | 14 | 18 | 19 | 24 | 25 | 30 | 31 | 32 | 33 | 36 | 37 | 38 | 39 | 35 | 41 |
| 4.01 SYNTAX II | 3 | 2 | 3 | . | 3 | 3 | . | . | . | . | . | . | . | . | . | . |
| 4.02 LEDA | 2 | 3 | 2 | . | 3 | 2 | . | . | . | . | . | . | . | . | . | . |
| 4.03 CYBER GENINT | 2 | 2 | 2 | . | 3 | 3 | . | . | . | . | . | . | . | . | . | . |
| 4.04 ISIS | 2 | 2 | 3 | . | 2 | 3 | . | . | . | . | . | . | . | . | . | . |
| 4.05 VDBS | 3 | 3 | . | . | 3 | 3 | . | . | . | . | . | . | . | . | . | . |
| 4.06 GTS | 3 | 3 | . | . | 3 | 3 | . | . | . | . | . | . | . | . | . | . |
| 4.07 CRESCAT | 3 | 3 | . | . | 3 | 2 | . | . | . | . | . | . | . | . | . | . |
| 4.08 PERSEE | 3 | 2 | 1 | . | 3 | 2 | . | . | . | . | . | . | . | . | . | . |
| 4.09 CENTS III | 3 | 1 | 1 | . | 3 | 3 | . | . | . | . | . | . | . | . | . | . |
| 4.10 COCENTS | 3 | 1 | . | . | 3 | 3 | . | . | . | . | . | . | . | . | . | . |
| 4.11 CENTS-AID | 2 | 1 | . | . | 2 | 2 | . | . | . | 1 | . | . | . | . | . | . |
| 4.12 FILE  2.0 | . | 1 | 1 | . | 3 | 3 | . | . | . | . | . | . | . | . | . | . |
| 4.13 OSIRIS 2.4 | . | . | 1 | . | 3 | 3 | . | . | . | . | . | . | . | . | . | . |
| 4.14 GEN. SUMMARY | 1 | . | . | . | 1 | 2 | . | . | . | . | . | . | . | . | . | . |
| 4.15 TPL | 3 | 2 | 1 | . | 3 | 3 | . | . | . | . | 1 | . | 2 | . | . | . |
| 4.16 RGSP | 2 | . | 3 | 2 | 3 | 3 | . | . | . | . | . | . | . | . | . | . |
| 4.17 NCHS-XTAB | 1 | . | . | . | 2 | 2 | . | . | . | . | . | . | . | . | . | . |
| 4.18 SURVEY | . | 1 | 3 | 1 | 2 | 2 | 2 | . | 1 | . | . | . | . | . | . | . |
| 4.19 SAP | 1 | . | 1 | . | 1 | . | . | . | . | . | . | . | . | . | . | . |

## FOR TABULATION PROGRAMS

| | Survey Estimates | Survey Variances | Math Functions | Operations Research | (ii) User Interface | | | | | | | | | | | | |
|---|---|---|---|---|---|---|---|---|---|---|---|---|---|---|---|---|---|
| | | | | | Availability | Installations | Computer Makes | Mini Version | Core Requirements | Batch/Interactive | Stat. Training | Computer Training | Language Simplicity | Documentation | User Convenience | Maintenance | Tested for Accuracy |
| | 43 | 44 | 47 | 48 | 49 | 50 | 51 | 52 | 53 | 54 | 55 | 56 | 57 | 58 | 59 | 60 | 61 |
| SYNTAX II | 1 | . | 1 | 1 | 2 | 1 | . | 2 | 1 | . | 3 | 3 | 3 | 2 | 2 | 1 | 2 |
| LEDA | . | . | . | . | 2 | 2 | 2 | . | . | 1 | 3 | 2 | 3 | 3 | 3 | 3 | 1 |
| CYBER GENINT | . | . | . | . | . | . | . | . | . | . | 3 | 2 | 2 | 2 | 2 | 2 | 1 |
| ISIS | . | . | 1 | . | 2 | 1 | 2 | . | 1 | . | 3 | 1 | 2 | 2 | 3 | 2 | 2 |
| VDBS | . | . | . | . | . | . | . | 3 | 1 | 3 | 3 | 2 | 2 | 3 | 3 | 3 | 2 |
| GTS | . | . | . | . | . | . | 1 | 1 | . | 1 | 3 | 3 | 3 | 3 | 3 | 3 | 2 |
| CRESCAT | . | . | 1 | 1 | 2 | 1 | . | . | . | . | 3 | 3 | 2 | 2 | 2 | 1 | 1 |
| PERSEE | . | . | . | . | 3 | 2 | 3 | 2 | 1 | . | 3 | 3 | 3 | 2 | 3 | 3 | 1 |
| CENTS III | . | . | . | . | 2 | 3 | . | 3 | 3 | . | 3 | 1 | 2 | 2 | 3 | 3 | 2 |
| COCENTS | . | . | . | . | 2 | 3 | 3 | 3 | 2 | . | 3 | 1 | 2 | 2 | 3 | 2 | 2 |
| CENTS-AID | . | . | . | . | 3 | 3 | 2 | 1 | 2 | 0 | 3 | 3 | 3 | 2 | 3 | 3 | 2 |
| FILE 2.0 | . | . | . | . | . | . | . | . | . | . | 3 | 3 | 3 | 1 | 2 | 1 | 1 |
| OSIRIS 2.4 | . | . | . | . | 2 | 1 | 2 | . | . | . | 3 | 2 | 2 | 2 | 2 | 2 | 1 |
| GEN. SUMMARY | . | . | . | . | . | . | 2 | 1 | . | 3 | 2 | 2 | . | 1 | 1 | 2 | 1 |
| TPL | 2 | 2 | . | . | 3 | 3 | 1 | . | . | . | 3 | 3 | 3 | 3 | 3 | 3 | 2 |
| RGSP | 2 | 2 | . | . | 2 | 2 | 2 | 2 | 1 | . | 2 | 3 | 2 | 2 | 3 | 3 | 2 |
| NCHS-XTAB | 2 | . | . | . | . | . | 1 | 1 | . | . | 3 | 3 | . | 1 | 3 | 2 | 2 |
| SURVEY | 1 | . | . | . | 1 | 3 | . | . | . | 2 | 3 | 3 | 3 | 3 | 2 | 1 | 2 |
| SAP | 1 | 1 | . | . | 1 | 1 | 2 | 1 | 2 | . | 3 | 2 | . | 1 | 1 | . | 1 |

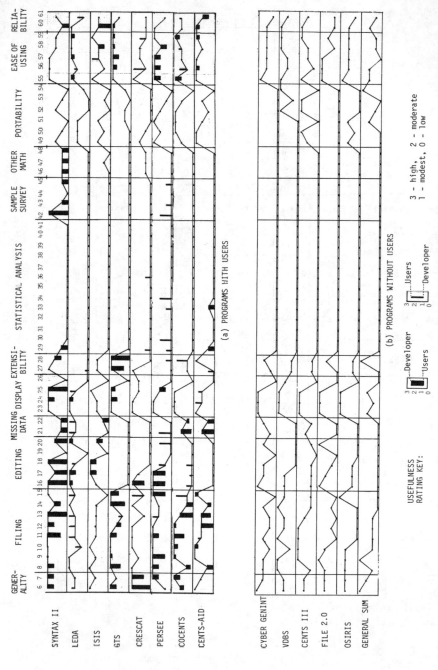

(a) PROGRAMS WITH USERS

(b) PROGRAMS WITHOUT USERS

USEFULNESS
RATING KEY:                    3 - high,   2 - moderate
                               1 - modest, 0 - low

FIGURE 4.1(i):  RATINGS BY DEVELOPERS AND USERS ON ALL ITEMS FOR TABULATION PROGRAMS (i)

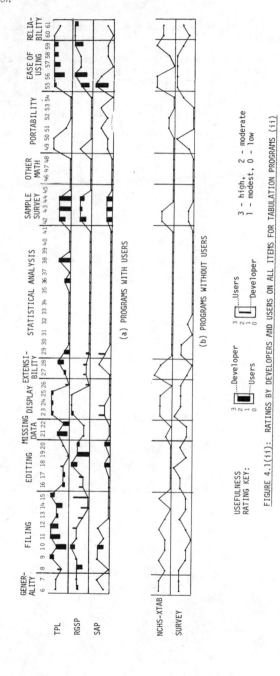

FIGURE 4.1(ii):  RATINGS BY DEVELOPERS AND USERS ON ALL ITEMS FOR TABULATION PROGRAMS (ii)

The tabulating abilities of several of these programs are compared in Francis and Sedransk (1979).

The ratings by the respective developers on selected items are displayed in Table 4.1.  Parts (i) and (ii) of Figure 4.1 compare developers' and users' ratings on all items, while Figure 4.2 summarizes developers' ratings (D-scores) and users' ratings (U-scores) on several relevant attributes.  These are explained in detail in Section 1.4 of Chapter 1.  Figure 4.2 also points to other chapters which describe programs with strong tabulation capabilities.

The  remainder of this chapter contains descriptions of these tabulation programs in the format described in Table 1.5.

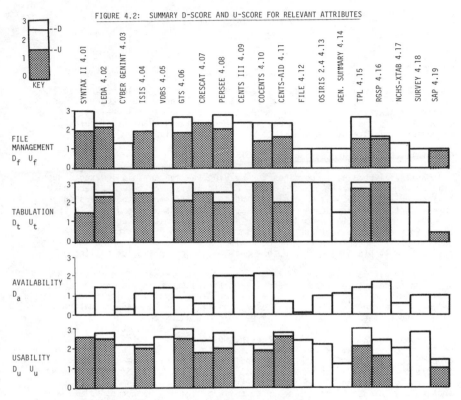

FIGURE 4.2:   SUMMARY D-SCORE AND U-SCORE FOR RELEVANT ATTRIBUTES

Other chapters generally containing programs with strong Tabulation Capabilities are
Chapters 6 and 7 (see Figures 6.2 and 7.2)

## 4.01  SYNTAX II

### INTRODUCTION

SYNTAX II is a generalized report generator, cross-tabulator, and utility program.  It was developed by John Fitz at the California Department of Public Health for statisticians and non-programming personnel and is presently in use in most California government departments, the University of California, and several other universities and state agencies nationwide.

### CAPABILITIES:  Data Processing and Display

SYNTAX II recognizes commands for reading any sequential, indexed sequential, or IMS database file, and copying or listing all or selected records or fields from those records or producing up to 15 different output files in any format sorted into any sequence.  The listing commands may be used to prepare reports on any selected records and fields with reformatting and sorting of lines into any order, with or without summarization of fields on any number of control breaks on sort keys, line numbering, and computations on summarized results.  Multiple line sorted output enables the production of mailing labels.

Procedural commands are available in order to generate or calculate new data and to select and process records and fields according to any logical criteria, and thus to edit and detect errors or other conditions in the data and to make corrections or changes.  Two or more files may be read and matched or merged and tables of information may be loaded and searched.

Cross-tabulations may be defined on two or more variables and the columns of the tabulation may be specified according to any format, including selection of different records for each column and computation of columns based on tabulated results in other columns.  Tabulations may have up to 31 printed columns (but any number of unprinted intermediate columns) and any number of rows showing the intersections or one or more fields for any number of values with labelling of those values.  Percentages may be calculated and results may be printed in histogram format.  Any number of titles and footnotes may be included on tabulations.

The completeness of the SYNTAX II language enables reports to be prepared according to any specifications by the writing of two or more steps if necessary.  Graphic displays are possible for the sophisticated user using matrix definitions.

### CAPABILITIES:  Statistical Analysis

SYNTAX II does not offer any statistical measures directly.  The computational instructions, however, provide all operations from which such measures may be determined in the production of reports, including arithmetic operations, square roots and exponentiation, trigonometric functions, and logarithmic and exponential functions.  These operations may be used in procedural instructions, listing commands, and cross-tabulation definitions.  Means and standard deviations, for example, can be produced by using division and square root extraction on summarized columns in a tabulation.

## EXTENSIBILITY

SYNTAX does not interface with other languages, since it is a compile and go system.  However, it provides a LIBRARY feature for accessing routines written in SYNTAX and a subroutine and remote statement execution capability so that complicated processes may be developed by users and stored in library files.  The data handling and control functions in SYNTAX are sufficient to execute any process which could be written in any other language.

## INTERFACES WITH OTHER SYSTEMS

SYNTAX can read any file in any format.  Special procedures are required to access IMS databases; for other database structures, the file handling routine would have to be modified.  Care needs to be taken in reading FORTRAN output files since FORTRAN inserts a decimal point into floating-point fields.

## PROPOSED ADDITIONS IN THE NEXT YEAR

No changes are contemplated in SYNTAX II during the next twelve months. Two new table definition functions (ROW ACCUMULATE and ROW COMPUTE) are projected but there is no scheduled date for completion.

SYNTAX II will be available on the UNIVAC 1100 by January, 1981.  Two modified versions are available on the HP-3000 (REX/3000) and the 8080 microcomputer and related hardware (ACCESS/80).

## SAMPLE JOB

Read a file containing survey data, produce a table on variable SEX against SCHOOLING with labels, showing missing data, and computing the average age at marriage for each cell.

```
READ FILE INPUT BLOCKED 39*80
SCHOOLING//47 CODES 1-5,OTHER; LABELS GRADE SCHOOL, HIGH SCHOOL,
      SOME COLLEGE, COLLEGE 4 YEARS,GRADUATE SCHOOL,MISSING DATA
SEX//33 CODES 1,2,OTHER; LABELS MALE,FEMALE, MISSING DATA
AGE//34-35; AVE//F6.1
TABLE 001. SCHOOLING BY SEX WITH AGE AT MARRIAGE
STUB SCHOOLING; SPREAD SEX
COLUMN COMPUTE AVE/AGE MARRIED/= (C7+C8)/(C1+C2)
COLUMN COMPUTE AVE/AGE MARRIED MALES/ = C7/C1
COLUMN COMPUTE AVE/AGE MARRIED FEMALES/ = C8/C2
```

Also produce a listing of missing data records by survey number:

```
NUMBER//1-6
NAME//7-31
REPORT 001. LISTING OF RECORDS WITH MISSING DATA
SELECT SEX NE 1,2 AND SCHOOLING NE 1-5
LIST NUMBER,10X,NAME,10X, SEX,10X, SCHOOLING SORTED BY NUMBER
```

Output

TABLE 001. SCHOOLING BY SEX WITH AGE AT MARRIAGE

| SCHOOLING | TOTAL | MALE | FEMALE | SEX MISSING DATA | AGE MARRIED | AGE MARRIED MALE | AGE MARRIED FEMALE |
|---|---|---|---|---|---|---|---|
| TOTAL | 90 | 74 | 16 | 0 | 34.3 | 36.6 | 23.6 |
| GRADE SCHOOL | 1 | 0 | 1 | 0 | 15.0 | .0 | 15.0 |
| HIGH SCHOOL | 5 | 4 | 1 | 0 | 31.6 | 35.7 | 15.0 |
| SOME COLLEGE | 82 | 62 | 13 | 7 | 34.8 | 37.0 | 24.2 |
| COLLEGE 4 YEARS | 2 | 1 | 0 | 1 | 37.0 | 37.0 | .0 |
| GRADUATE SCHOOL | 4 | 3 | 1 | 0 | 38.0 | 40.5 | 33.0 |
| MISSING DATA | 4 | 3 | 0 | 1 | 26.7 | 26.7 | .0 |

REPORT 001. LISTING OF RECORDS WITH MISSING DATA

| NUMBER | NAME | SEX | SCHOOLING |
|---|---|---|---|
| 075750 | HENRY, MALCOLM | 1 | 3 |
| 101731 | DAVIDSON, LAWRENCE | 1 | 0 |
| 123451 | HERNANDEZ, DOLORES | | 3 |
| 123718 | PETRUCCHIO, CHARLES | | 4 |
| 124121 | GRANT, CARRIE | | 3 |
| 187453 | LEWIS, ALICE | | 4 |
| 227475 | APPLEBAUM, DARIUS | | 4 |
| 291564 | KRIEGER, BRIAN | | 3 |
| 391132 | VAN LEDEN SELS, JANICE | | 9 |
| 400000 | ELPHINSTONE, DANA | 1 | 9 |
| 306204 | DIXON, BRUCE | | 3 |
| 850020 | SUGAR, DENISE | | 3 |

## SYNTAX II

**Developed by:**

    University of California
    2200 University Avenue
    Berkeley, California  94720

**Distributed by:**

    For information write:
    Health and Welfare Agency
    Attn. SYNTAX Consultant
    112 J Street
    Sacramento, California  95814

**Computer Makes:**

    IBM 360/370
    UNIVAC 1100 (by 1981)

**Operating Systems:**

    OS/MFT, OS/MVT, MVS

**Source Languages:**  BAL through-written; Assembly language (Univac 1100), (IBM)

**Cost:**

    SYNTAX II is in the public domain.  Nominal fees are charged for providing copies of the program and manuals, and for updated versions.  No maintenance contract is available.  Routine consulting is available at nominal cost from Q.E.D. Computing, 3109 Lewiston Street, Berkeley, California, 94705.

**Documentation:**

John Fitz, SYNTAX II Manual, University of California, July 1977; obtainable
    from Q.E.D. Computing, address above.
John Fitz, SYNTAX II Documentation, unpublished, April, 1977, also available
    from Q.E.D. Computing, address above.

## 4.2  LEDA

### INTRODUCTION

LEDA is a package for data handling, error detection, editing, recoding and mainly tabulation.  It is composed of three independent modules:  CASTOR, POLLUX and HELENE.  Each one is a COBOL program generator.  The three phases use a free-format language, specific for each one but with the same French-like or English-like syntax.

It was designed by the French Institute of Statistics and Economic Studies (I.N.S.E.E.) between 1971 and 1973.  It has been distributed out of INSEE since 1975.  LEDA is installed in statistical offices of French administrations and in national statistical offices of some countries.

### CAPABILITIES:  Processing and Displaying Data

The CASTOR phase checks the structure of the input data set resulting from data entry, and also the validity of each variable.  It allows some automatic editing such as HOTDECK.  Then it creates a "standard file" which consists of a dictionary and a hierarchical data set.

The POLLUX phase is optional.  It processes the standard file and creates a new standard file.  It allows detection of inconsistencies, rectification of the sample, creation of new variables and makes it possible to obtain a sub-file.

The main recoding tool in POLLUX is the decision-tables.  The user may also insert COBOL statements within the POLLUX language and it is possible to call any external sub-program.

The HELENE phase creates and prints cross tables.  The number of criteria it is possible to cross tabulate is unlimited.  However each table will appear physically as a three dimensional one (maximum).  It is not usually necessary to sort the data set before cross tabulation; however in a few cases (i.e. if one table contains more than 32767 cells) a sort on one or two criteria may be necessary.

It is possible to store files of tables and to compute tables from one or two previously computed tables without re-reading the initial data set.

### CAPABILITIES:  Statistical Analysis

The HELENE phase allows to compute percentages, means, standard deviations.  The file of tables may be converted into a FORTRAN readable file in order to use statistical analysis packages.

### PROPOSED ADDITIONS IN NEXT YEAR

Enrichment of the dictionary, in order to introduce new printing informations such as titles, labels, etc...  These informations will be easily handled through the HELENE language.

REFERENCE

R. Mandel, "Les Logiciels Généraux à l'INSEE" in "Le Courrier des Statistiques"
    (No. 4, 1977) Paris.
"Produits de Depouillement. D'Enquetes ou D'Aide a L'Exploitation des Donnees
    Statistiques", CdF Informatique , Paris (1979).
"Le Traitement Informatique des Enquetes Statistiques" by "Groupe de Démogra-
    phie Africaine IDP-INED-INSEE-MICOOP-ORSTOM".   INSEE, Paris, 1978.

SAMPLE JOB

The table represents the distribution of the average income according to
the age of the head of household and the income type.

The request is:

```
DEFINE;
LAB-AGE : TITLE-LIST LENGTH 14 VALUES
'0 TO 24 YEARS','25 TO 39 YEARS','40 TO 59 YEARS',
'60 TO 64 YEARS','65 YEARS AND +';
LAB-INC : TITLE-LIST LENGTH 15 VALUES 'SALARIES',
'FAM ALLOWANCES','PENSIONS','AGRIC INCOMES',
'COMERC INCOMES','PROFITS','INC INDEP PROF',
'PERS ESTATE','REAL ESTATE','OTHER INCOMES','TOTAL.';

TABULATION;
PAGE-TITLE 'DISTRIBUTION OF AVERAGE INCOME, BY INCOME TYPE ',
'ACCORDING TO AGE CATEGORY OF HEADS OF HOUSEHOLD';
TABLE-1 ; DISTRIBUTION FLOATING;
CRITERIA  /TYPE-INC. TITLE LAB-INC#
          /AGE SECTION 0,25,40,60,65 TITLE LAB-AGE
SUM NUMB-INCOME ✶ RATE # ANNUAL-INCOME ✶ RATE;
COMPUTE SE(1) # SE(2)/SE(1);
HEADINGS 'NUM-INC' # 'AVERAGE INC';
FORMATS'-----9' # '-----9V.99';
```

Output:

TABLE-1                                      NULLFILE.T        800804162I*    I*FRAC 1/ 1    PAGE   1

PAGE CRITERIA     :  ( INC-CAT )
LINE CRITERIA     :  ( INC-CAT )
COLUMN CRITERIA   :  ( AGEC )

DISTRIBUTION OF AVERAGE INCOMES, BY INCOME TYPE ACCORDING TO AGE CATEGORY OF HEADS OF HOUSEHOLD

|                 | 0 TO 24 YEARS | | 25 TO 39 YEARS | | 40 TO 59 YEARS | | 60 TO 64 YEARS | | 65 YEARS AND + | |
|-----------------|----------|-------------|----------|-------------|----------|-------------|----------|-------------|----------|-------------|
|                 | NUMB INC | AVERAGE INC | NUMB INC | AVERAGE INC | NUMB INC | AVERAGE INC | NUMB INC | AVERAGE INC | NUMB INC | AVERAGE INC |
| SALARIES        | 28005    | 60265.98    | 72005    | 99506.66    | 74785    | 119875.34   | 27070    | 98215.77    | 17425    | 69292.39    |
| FAM ALLOWANCES  | 3915     | 52127.24    | 61870    | 91108.27    | 38170    | 108773.97   | 1040     | 82950.00    | 520      | 114979.20   |
| PENSIONS        | 0        | 0.00        | 5240     | 74195.20    | 25590    | 80280.10    | 17965    | 54444.29    | 3640     | 40376.60    |
| AGRIC INCOMES   | 0        | 0.00        | 2080     | 66744.82    | 0        | 0.00        | 0        | 0.00        | 1040     | 86520.00    |
| COMMERC INCOMES | 0        | 0.00        | 4170     | 105531.10   | 5965     | 75195.23    | 1040     | 39726.75    | 2090     | 58662.99    |
| PROFITS         | 0        | 0.00        | 1825     | 100682.63   | 4170     | 91733.06    | 1560     | 80172.40    | 785      | 38474.14    |
| INC INDEP PROF  | 0        | 0.00        | 1305     | 162683.80   | 3650     | 409172.75   | 1040     | 103950.00   | 0        | 0.00        |
| PERS ESTATE     | 0        | 0.00        | 0        | 0.00        | 1570     | 133084.07   | 1040     | 107658.60   | 3120     | 91166.60    |
| REAL ESTATE     | 0        | 0.00        | 520      | 113001.00   | 7290     | 298384.16   | 3640     | 71176.50    | 6250     | 60089.66    |
| OTHER INCOMES   | 3425     | 53656.79    | 5475     | 112996.18   | 38975    | 147771.38   | 4680     | 65701.53    | 4425     | 62368.64    |
| TOTAL.          | 35345    | 56724.05    | 3760     | 99947.79    | 185      | 125353.45   | 59075    | 79195.04    | 39295    | 47601.21    |

## LEDA Version 4.4

Developed by:                               Distributed by:

Institut National de la Statis-            Same
tique et des Etudes Economiques
18 Boulevard Adolphe Pinard
75675 Paris CEDEX 14 (France)

Computer Makes:                            Operating Systems:

IBM 360,370,303X, AMDAHL V7                 OS (MFT,MVT,VSI,MVS)
CII-HB IRIS 80                              SIRIS 8
CII-HB IRIS 45-60                           SIRIS 3

User Language:  Two versions : French and English

Source Languages:  Primarily CPL1, assembler for input/output.  The sources
modules are not supplied.

Cost:

One time charge:  200,000 French Francs including a 10-days training, documentation, one year maintenance.  After one year, the annual cost of maintenance is 20,000 French Francs.

Documentation:

In French and English:  Presentation, statistician's manual, CASTOR-POLLUX user
manual, Tabulation requester manual, Reference manual, Programmer's guide.

All booklets edited by INSEE-FRANCE.

## 4.03  CYBER GENINT

### INTRODUCTION

CYBER GENINT is an applications software package for processing a statistical collection from data entry to final tabulations.  It was developed by the Australian Bureau of Statistics (ABS) during 1975-78 and has been used for processing the annual Integrated Economic Censuses.  The predominant mode of usage is batch jobs but some interactive functions are availabe.  The package has a Data Dictionary System based on the commerical data base management system SYSTEM 2000.  There has been no development of the package since 1978.  It is only installed on the ABS CYBER computer.

### CAPABILITIES:  Processing and Displaying Data

CYBER GENINT has a user language which is used to program the required edits, tabulations and data manipulations for the collection being processed. It has its own file handling facilities which include many of the features of a generalised data management system including retrieval of unit records through secondary indexes, processing of hierarchical data structures, privacy checking and backup and recovery.

Table matrices are stored in the data base and are accessible to user programs by simple procedural language statements.  These matrices may also be accessed by an interactive query facility which allows simple conditional statements, creation of subfile from the matrix, some functions and the saving of a procedure for future execution.

CYBER GENINT has an amendment system allowing the user to adjust, insert or delete records and fields within records.

The language would allow users to implement most edits the application system would require and there are facilities available to assist in formatting the error report.  The ability to enter FORTRAN coded routines through a CALL statement should enable implementation of those requirements not possible in GENINT.

The package can cross tabulate and summarise data and put it in a report format suitable for direct publication.  The Data Dictionary is used to hold much of the report specification and descriptive text of the classifications.

Data files can be created for input to other packages.  The external file interface is fixed length records, fixed length blocks and character data fields.

### CAPABILITIES:  Statistical Analysis

CYBER GENINT has no specific facilities for statistical analysis. Features of the system allow either extraction of data onto a subfile to go into another package or interface through the CALL statement to statistical subroutines.

## FUTURE DEVELOPMENT

No further development is planned on this package and maintenance is restricted to fixing major problems that would affect production running of the Integrated Economic Censuses.

The ABS is reequipping with FACOM M-200 machines and a package called PLEAT is being developed for that machine.  It has similar functionality to CYBER GENINT but uses ADABAS data base management system for file handling.

## REFERENCES

Australian Bureau of Statistics,  CYBER 72 Systems Manual, Manual No 12, GENINT
      Version 4-0, 1978.
Australian Bureau of Statistics,  CYBER 72 Systems Manual, Manual No 13, QUERY
      GENINT, 1978.

## SAMPLE JOB

This is an example of a GENINT tabulation assuming the appropriate parts of the Data Dictionary have been established.

Produce a summary table of major variables by industry and area.

```
SECTION TABLE EXECUTION, SCAN = ACCESS PATH
REQ     R1,     GET(COMPUTER SERIAL NUMBER) PERFORM (TAB 1)
FINISH TABLE EXECUTION
SECTION TAB1
REQ     TABS 1,    PTABLE((STATE*AREA CODE) (INDUSTRY)
                  (MALES, FEMALES, TOTAL EMPLOYMENT, WAGES, VALUE ADDED,
                  SALES))
FINISH TAB1
REPORT TABS1
TOGGLES     OMIT STANDARD HEADING, REPLACE NULL = "NA"
TITLE (LINE = 1, AT = "C", TEXT = "SUMMARY OF OPERATIONS BY INDUSTRY")
FOOTNOTE (LINE = 59, AT = 10, TEXT = "FOR DEFINITION SEE CHAPTER 5)
WAFER (STATE, NOTOTAL) (AREACODE, SUBFIELDS = S1, S2)
STUB (INDUSTRY, SUBFIELDS = S2, S1)
COLUMN (1, COVER = 3, TEXT = "EMPLOYMENT AT 30 JUNE 1980")
```

POSTSET (TOTAL EMPLOYMENT = MALES + FEMALES)
CONCLUDE TABS1

PARTIAL OUTPUT FROM EXAMPLE

SUMMARY OF OPERATIONS BY INDUSTRY

NEW SOUTH WALES

101    SYNDEY STATISTICAL AREA

| INDUSTRY | | EMPLOYMENT AT 30 JUNE 1980 | | | WAGES | VALUE ADDED | SALES |
|---|---|---|---|---|---|---|---|
| | | MALE | FEMALE | TOTAL | | | |
| | | NUMBER | NUMBER | NUMBER | $'000 | $'000 | $'000 |
| RETAIL | | | | | | | |
| DEPARTMENT VARIETY AND GENERAL STORES | | | | | | | |
| DEPARTMENT STORES | 4811 | 119 | 219 | 338 | 611 | 1 092 | 5 114 |
| VARIETY STORES | 4812 | 17 | 37 | 54 | NA | 268 | 1 288 |
| GENERAL STORES | 4813 | 14 | 28 | 42 | NA | 79 | 542 |
| TOTAL | 481 | 150 | 284 | 434 | 611 | 1 439 | 6 944 |
| FOOD STORES | | | | | | | |
| SUPERMARKETS | 4821 | 21 | 23 | 44 | 108 | 257 | 1 845 |

.
.
.
.
.

FOR DEFINITIONS SEE CHAPTER 5.

### CYBER GENINT Version 4.0

Developed by:                          Distributed by:

   Australian Bureau of Statistics        Same
   PO Box 10
   BELCONNEN  ACT 2616
   AUSTRALIA

Computer Makes:                        Operating Systems:

   CDC CYBER 72                           KRONOS 2.1

Source Languages:

   Primarily FORTRAN and COBOL
   Same assembly language (COMPASS) to improve efficiency.

Cost:

   The Australian Bureau of Statistics will consider requests for the provision of software individually and will take into account the intrinsic importance of the request, the resources required to meet it and existing exchange
arrangements.  ABS will not be able to undertake any support activities in
relation to software.

4.04  <u>ISIS</u>


## INTRODUCTION

The main objective of the ISIS (later SOFIS) project was to design and to implement a software system which would automate selected areas of statistical data processing in particular, but that could also be used in other data processing systems with similar characteristics.

The ISIS/SOFIS software was developed in two stages. The first stage covered the period from 1972 to 1975 and produced the first version of the ISIS software. This version consisted of four user products:
. the input module user language - INLAN
. the data base management module - DBMM
. the output module user language - OUTLAN
. COBOL host language extensions - ISLE

New features in SOFIS include:

1.  A data management system with data directory and directory module, a single file access module (sequential index/sequential and direct organization) and a relational interface (with multiple access module).

2.  A graphical output module (including a dialogue and report generator).

## CAPABILITIES:  Processing and Displaying Data

INLAN is a problem-oriented user language designed for the programming of input data checks, corrections and basic conversions. A program written in INLAN manipulates external data files and makes it possible to describe external file structures, to process one or several external files, to check input data and signal errors, and to correct data.

OUTLAN is a problem-oriented, user language designed to process and present data within the ISIS system. A program written in OUTLAN makes it possible to perform the following functions: the selection criterion, input file sorting prior to processing, input file processing, the output document printing.

An OUTLAN program consists of two parts - declaration part and algorithmic part. Basic elements of OUTLAN syntax that create respective parts of a program are declarations and statements. Both declarations and statements fall into either commonly used types of declarations or statements (e.g. declarations of integers, reals, arrays or assignment, loop and IF statements) or specialized OUTLAN declarations and statements.

ISIS/SOFIS - Integrated Statistical Information System

Developed by:                              Distributed by:

    Computing Research Centre          same
    Dept. of Programming System
    Dubravska 3, 885 31 Bratislava
    Czechoslovakia

Computer Makes:                            Operating Systems:

    IBM 370                            OS, VS
    CDC 3300                           MASTER
    RJAD/EC/                           RJAD/OS, DOS

Source Language:

    Pascal

Cost:

    By individual agreement with each installation.

## 4.05  VDBS

### INTRODUCTION

V.D.B.S. is a package for data management, processing, tabulating and editing.

Its principal applications have been in areas such as financial statistics. and socio-economic indicators.

It was developed by J.L. Grolleau and L. Gelabert in 1978 to meet the requirements of the Development and Cooperation Directorate and has been under continuous development since.

It is used interactively or in batch by people without any knowledge of programming at the O.E.C.D. headquarters in Paris.

### CAPABILITIES

V.D.B.S. accepts data, physically stores them, without any need for the user to bother about the concept of files; given the identification of the datum, the system can store, update, access, process, and retrieve it. Given a preliminary description of the data, the system maintains a documentation of the contents of the data base.

The system can perform arithmetic calculations and simple statistical functions on rows and columns of user specified matrices, these matrices being documented, tabulated, added, ordered, and then printed or displayed on V.D.U. at the user's convenience. In the same run, it is possible to create new data, produced by calculations performed from basic data, and update the data base with them. Primary and derived data can be plotted. Tables can be stored for further processing in order to produce tapes, directly suitable for photocomposition. For external requests, the system can extract subsets of the data base and dump them on tape.

V.D.B.S. provides a conversational routine, giving a step by step guide to the user, in order to create, store, and update sets of parameters; subsequently, given the name of such a set, the system will answer the request either in batch or in conversational mode.

Calculations include: arithmetic operations, logarithms, antilogarithms, growth rates, means, weighted means, standard deviation, marginal distributions, simple regressions, comparisons, flagging of missing data.

Security facilities: the access to the system can be restricted by usercodes and passwords; subsets of data can be attributed different usercodes and security levels. The system provides automatic data enciphering.

## INTERFACES WITH OTHER SYSTEMS

V.D.B.S. can accept data using user-defined translating tables.  The system currently processes:

*    World Bank Atlas tapes, World tables, Debtor Reporting System.
*    International Monetary Fund tapes:  International Financial Statistics, Balance of payments, Direction of Trade and Government Financial Statistics.
*    O.E.C.D. Foreign Trade, Creditor Reporting System.

## PROPOSED ADDITIONS IN NEXT YEAR

New statistical routing at user's request, improvement of photocomposition routines, new interfaces.

## REFERENCES:  Publications produced by the system

O.E.C.D. DAC Chairman's Report:  Statistical Annex.
Geographical Distribution of Financial Flows to Developing Countries.

## SAMPLE JOB

This sample job shows the marginal distribution of a financial flow, e.g. each cell represents the percentage of the flow from a donor to a recipient vis-a-vis the total flow of this donor to all his recipients.

This job can be executed by these statements:

```
RUN VDBS;   FILE FILE99 = PAR/MARGIN/SAMPLE;
FILE FILE8 = RECIPIENTS/BY/INCØME;
```

where PAR/MARGIN/SAMPLE is the name of the set of parameters created in conversational mode under the control of V.D.B.S. itself: RECIPIENTS/BY/INCOME is the name of a file containing the names of the countries grouped by income.

Partial Output
10/1./80

PERCENTAGE DISTRIBUTION OF INDIVIDUAL DAC COUNTRIES NET ODA 1978.                                                   VDES

UPPER MIDDLE INCOME

| | UNITED STATES | FRANCE | GERMANY | JAPAN | UNITED-KINGDOM | NETHER-LANDS | SWEDEN | CANADA | AUSTRALIA | BELGIUM |
|---|---|---|---|---|---|---|---|---|---|---|
| ALGERIA | -0.10 | 5.75 | 7.69 | 9.91 | 0.29 | 3.00 | 17.23 | 4.57 | 1.11 | 16.18 |
| ARGENTINA | -0.52 | 2.72 | 2.72 | 0.94 | 0.95 | 0.48 | 0.00 | -0.08 | 0.00 | 2.76 |
| BARBADOS | -0.10 | 0.00 | 0.00 | 0.01 | 0.16 | 0.05 | 0.00 | 23.38 | 4.44 | 0.00 |
| BRAZIL | -2.28 | 1.44 | 9.61 | 17.53 | 0.82 | 2.17 | 0.00 | 8.42 | 0.56 | 6.46 |
| CHILE | -1.24 | 1.43 | -1.20 | 0.54 | 2.76 | 1.65 | 0.00 | -1.05 | 0.56 | 4.64 |
| CYPRUS | 1.97 | 0.07 | 1.16 | 0.01 | 1.04 | 0.10 | 0.00 | 0.00 | 0.56 | 0.00 |
| DJIBOUTI | 0.10 | 3.27 | 0.05 | 0.01 | 0.04 | 0.00 | 0.00 | 0.00 | 0.00 | 0.36 |
| GUADELOUPE | 0.00 | 15.32 | 0.00 | 0.00 | 0.00 | 0.00 | 0.00 | 0.00 | 0.00 | 0.00 |
| GUIANA, FRENCH | 0.00 | 6.30 | 0.00 | 0.00 | 0.00 | 0.00 | 0.00 | 0.00 | 0.00 | 0.00 |
| HONG KONG | 0.00 | 0.00 | 0.20 | 0.38 | 0.83 | 0.00 | 0.00 | 0.00 | 2.22 | 0.15 |
| IRAN | -0.33 | 0.26 | 1.49 | 48.36 | -0.54 | 0.05 | 0.00 | 0.00 | 1.67 | 0.80 |
| IRAQ | -3.10 | 0.00 | 0.11 | 19.96 | 0.10 | 0.13 | 3.94 | 0.00 | 0.56 | 0.73 |
| JAMAICA | 2.59 | 0.00 | 0.79 | 0.03 | 63.64 | 14.73 | 3.94 | 44.35 | 6.67 | 0.07 |
| LEBANON | 1.35 | 0.02 | 0.67 | 0.23 | 1.11 | 1.55 | 1.79 | 8.40 | 17.22 | 2.83 |
| MALTA | 0.93 | 0.07 | 2.46 | 0.01 | 12.07 | 0.09 | 0.00 | 0.40 | 1.11 | 0.00 |
| MAYOTTE | 0.10 | 1.05 | 0.00 | 0.00 | 0.00 | 0.00 | 0.00 | 0.00 | 0.00 | 0.00 |
| MEXICO | -0.41 | 0.24 | 2.03 | 2.37 | 2.76 | 0.46 | 3.94 | 0.00 | 1.11 | 2.25 |
| NETHERLANDS ANTILLES | 0.00 | 0.00 | 0.00 | 0.00 | 0.00 | 29.77 | 0.00 | 0.00 | 0.00 | 0.00 |
| PACIFIC ISLANDS (U.S.) | 10.37 | 0.00 | 0.00 | 0.27 | 0.00 | 0.00 | 0.00 | 0.00 | 1.67 | 0.00 |
| PANAMA | 1.76 | 0.23 | 0.23 | 0.17 | 0.26 | 0.15 | 0.00 | 0.00 | 0.00 | 0.51 |
| PORTUGAL | 4.67 | 0.00 | 1.54 | 0.03 | 0.35 | 0.99 | 59.86 | 15.94 | 0.00 | 0.58 |
| REUNION | 0.00 | 33.15 | 0.03 | 0.09 | 0.00 | 0.00 | 0.00 | 0.00 | 0.00 | 0.00 |
| SURINAM | 0.00 | 0.00 | 0.13 | 0.00 | 0.00 | 41.31 | 0.00 | 0.18 | 0.00 | 0.29 |
| TAIWAN | -0.93 | 0.00 | 1.57 | -5.30 | 0.00 | 0.00 | 0.00 | 0.00 | 0.00 | 0.22 |
| TRINIDAD & TOBAGO | 0.00 | 0.00 | 0.02 | 0.06 | -0.16 | 0.11 | 0.00 | 1.34 | 2.22 | 0.36 |
| TURKEY | -3.11 | 0.19 | 41.58 | 2.53 | -1.41 | 4.27 | 9.50 | 0.00 | 0.56 | 53.05 |
| URUGUAY | -0.21 | 0.06 | 0.78 | 0.12 | 0.00 | 0.77 | 0.00 | 0.00 | 0.00 | 1.23 |
| YUGOSLAVIA | -3.53 | 2.28 | 1.09 | -1.78 | 0.00 | 0.07 | 0.00 | 0.00 | 0.00 | 0.07 |
| TOTAL | 10.58 | 65.95 | 74.79 | 96.28 | 84.28 | 98.90 | 100.00 | 97.90 | 42.22 | 93.54 |

## VDBS

Developed by:

> J.L. Grolleau and L. Gelabert
> Reporting Systems Division
> Development and Co-operation Directorate
> O.E.C.D., 2 rue Andre-Pascal, 75775 Paris Cedex 16

Computer Makes:                               Operating Systems:

> Burroughs B6700                              B6800
>                                              B7800
>                                              Burroughs M.C.P. and Cande

Source Languages:

> Fortran for calculations
> Cobol for data management
> Algol for data retrieving and
>  data enciphering
> The different sub-routines are
> linked into a single program

Cost:

V.D.B.S. has been designed for the O.E.C.D. Development and Co-operation Directorate, and could be distributed under special agreement only.

Documentation:

Informal and limited because of the aids and the documentation provided by the system itself.  The system is self-teaching.  Descriptions appear in:

Working paper:  V.D.B.S., 1980, B.B. Stein Head Reporting Systems Division.
    Development Cooperation Directorate, O.E.C.D. 2 rue Andre-Pascal, 75775
    Paris. France.
Project Information System:  SQAD(79)106  Data Processing and Statisticals
    Services Directorate.  O.E.C.D.

## 4.06  <u>GTS</u>

### INTRODUCTION

GTS is a package for general-purpose tabulation, summarization and display of statistical tables. It has been under continuous development by the Bureau of the Census since 1976 for use on census and survey tabulation and publication. It is also being used by at least one other government agency.

### CAPABILITIES:  Data Processing and Display

GTS is designed to tabulate, summarize and display data from large surveys and censuses. Statements are coded in free-form format using an English-like syntax. The syntax includes a data dictionary facility. The user may construct a dictionary to describe his data records and to store frequently used GTS source statements. GTS has a limited editing facility. The values of data may be changed within a record, however records may not be added or deleted.

GTS builds a COBOL program based on the source statements to perform tabulations and produce table matrices. These table matrices may be summarized based on a hierarchical structure (usually geographic) defined by the user The user is allowed full control over the appearance of the displayed tables. During the display process, arithmetic calculations may be performed on rows and columns of the table. Alphanumeric text may be substituted for table cell contents.

### CAPABILITIES:  Statistical Analysis

GTS is not designed to provide statistical analysis capabilities. However by coding arithmetic calculations the user may generate descriptive statistics such as totals, means, medians, standard deviations, standard errors and percents.

### EXTENSIBILITY

The user may insert custom COBOL code or call subroutines written in other languages during the tabulation phase.

### INTERFACES WITH OTHER SYSTEMS

None.

### PROPOSED ADDITIONS IN NEXT YEAR

Output of photocomposed tables.
Increased machine independence.

### REFERENCES

Francis, I. and Sedransk, J.  "Comparing Software for Processing and Analyzing Survey Data". <u>Bulletin of International Statistical Institute</u>, 48, 1979.

SAMPLE JOB:  Input

```
@XQT GTS1*GTS.GTS
FILES INDATA     = (GTS1*GTSDATA-F)
      DICTIONARY = (GTS1*DDBENCH)
TABLE  AGE-BY-SEX  13 BY 5
SELECT IN-POPULATION-RECORD
     RECODE IN-AGE TO AGE-ROW
          (0 THRU 10 = 2)(THRU 20 = 3)(THRU 30 = 4)(THRU 40 = 5)
            (THRU 50 = 6)(THRU 60 = 7)(THRU 70 = 8)(THRU 80 = 9)
            (THRU 90 = 10)(THRU 100 = 11)(OTHERS = 1)
     RECODE IN-SEX TO SEX-COLUMN
          (IN-MALE-SEX   = 2)
          (IN-FEMALE-SEX = 3)
          (OTHERS        = 1)
     TALLY AGE-BY-SEX (AGE-ROW,12) BY SEX-COLUMN USING (1,IN-AGE)
ENDSELECT
FORMAT  AGE-BY-SEX  13 BY 5
     STUBWIDTH IS 20
     COLSPECS ARE (3,I,8)(2,D2,8)
     PAGENUMBERS ARE ON
     DATE IS 'April 1980'
LINE ADD 1 = 2 THRU 11  FOR 2 THRU 3
COLUMN ADD 1 = 2 + 3
LINE DIVIDE 12 = 12 * 10 / 1  FOR 1 THRU 3
LINE MEDIAN 13 = 2 THRU 11 * 10
        (0,11,21,31,41,51,61,71,81,91,100)  FOR 1 THRU 3
COLUMN PERCENT 4 = 2 * 10000 / CELL (1,1) FOR 1 THRU 11
COLUMN PERCENT 5 = 3 * 10000 / CELL (1,1) FOR 1 THRU 11
HEAD   1   1   T                                                        $$DAT
HEAD   1   2              E
HEAD   2   1   2              Table 2.  Age by Sex
HEAD   3   1   2                                      Number      Perce
HEAD   3   2              nt
HEAD   4   1   1                        Total   Male Female   Male  F
HEAD   4   2              emale
STUB   1   I   2   4      Total.....................
STUB   2   I   1   1       0 to 10 years.............
STUB   3                  11 to 20 years.............
STUB   4                  21 to 30 years.............
STUB   5                  31 to 40 years.............
STUB   6                  41 to 50 years.............
STUB   7                  51 to 60 years.............
STUB   8                  61 to 70 years.............
STUB   9                  71 to 80 years.............
STUB  10                  81 to 90 years.............
STUB  11                  over 90 years.............
STUB  12  D1  2           Average Age...............
STUB  13  D1              Median Age.................
FOOT   1   1   3              NOTE: This is test data
FOOT  50   1   0                                                          P
FOOT  50   2              age $$PAGE
```

SAMPLE JOB:  Output

                                                            April 1980

    TABLE 2.   Age by Sex

                                           Number            Percent
                             Total     Male   Female     Male    Female

          Total.............     666      329      337     49.40    50.60
      0 to 10 years......     278      124      154     18.62    23.12
     11 to 20 years......     134       67       67     10.06    10.06
     21 to 30 years......      99       50       49      7.51     7.36
     31 to 40 years......      65       35       30      5.26     4.50
     41 to 50 years......      49       29       20      4.35     3.00
     51 to 60 years......      18       14        4      2.10      .60
     61 to 70 years......       5        3        2       .45      .30
     71 to 80 years......       5        2        3       .30      .45
     81 to 90 years......       3        1        2       .15      .30
     over 90 years.......      10        4        6       .60      .90

    Average Age.........    20.4     21.6     19.3

    Median Age..........    15.1     17.0     13.2


        NOTE: This is test data

## GTS 1.8:  Generalized Tabulation System

Developed by:

    Systems Development Division
    Bureau of the Census
    Washington, D.C.  20233

Distributed by:

    Data User Services Division
    Bureau of the Census
    Washington, D.C.  20233

Computer Makes:

    UNIVAC 1100 series

Operating Systems:

    UNIVAC - Exec 8

Interfaced Systems:

    none

User Language

    English

Source Languages:

    COBOL primarily, FORTRAN - special I/O handlers (CENIO)
    ASSEMBLER - small operating system interfaces

Cost:

    Cost of materials.

Documentation:

    User's Manual, Internals Documentation, Release Documentation, Training-
Course and Materials.

Support:

    Ongoing support for the package is not provided by the Bureau.

## 4.07  CRESCAT

### INTRODUCTION

CRESCAT is a system for the analysis of real-time or sequential data.  It was developed by Kenneth Kaye and Stephen Muka at the University of Chicago and has so far been installed at two other universities.

### CAPABILITIES

CRESCAT offers a language in which investigators can search for any string patterns in large or small files, explore the contingencies among such patterns within each file, then tabulate, graph, and statistically reduce the extracted variables.  In batch jobs, the user edits, sorts, updates, creates and merges files, runs programs to answer any logically expressible question regarding the relations among events within files, and builds multi-dimensional arrays. These can be manipulated by algebraic formulas, or by functions calling the most common statistical operations (t-tests, frequencies, simple regression, etc.).  The system is used for complex data recoding, exploratory work with strings, computing conditional rates, response latencies, and contingent probabilities, Markov and semi-Markov analyses of either sequential or real-time data.  Transcripts of files and graphs of arrays are printed according to the user's own symbols and format.

### INTERFACES WITH OTHER SYSTEMS

Finished arrays can be output as sequential data sets in any desired format (for SPSS, SAS, Data-Text, etc.).

### REFERENCES

Published studies describing the use of CRESCAT include:

Kaye & Fogel, Temporal structure of mother-infant face-to-face communication.
    Developmental Psychology, 1980, 16, 454-464.
Kaye & Wells, Mothers' jiggling and the burst-pause pattern in neonatal sucking.
    Infant Behavior & Development, 1980, 3, 29-46.
Duncan & Fiske, Face-to-face Interaction.  Erlbaum, 1977.
Altmann,  Baboon Mothers & Infants.  Harvard, 1980.

SAMPLE JOB

```
FILE TEST29
A
B
A
B
A
A
A
A
ENDFILE


FILE TEST30
A
B
C
D
A
C
A
D
E
ENDFILE
```

THIS PROGRAM WILL FIND HOW FREQUENTLY EVENT A IS FOLLOWED
BY EVENT B BEFORE THE NEXT A
WE WONDER IF THIS RELATES TO NUMBER OF EVENTS IN FILE
PRETEND WE HAVE 30 FILES
WE WANT AN ARRAY CONTAINING 3 VARIABLES BY 30 SUBJECTS

```
        LET RESULTS = ARRAY(3,30)

        PROGRAM TESTPROG
        PATTERN TAB(4) REM=>SUBJNO IN ZFILE
        1--> PATTERN 'A' -->F(DONE)
        2--> LET ACOUNT = ACOUNT + 1
        PATTERN ('A' -->2) | 'B' -->F(DONE)
        LET BCOUNT = BCOUNT + 1 -->1
        DONE--> LET RESULTS<1,SUBJNO> = ACOUNT
        LET RESULTS<2,SUBJNO> = BCOUNT
        LET RESULTS<3,SUBJNO> = ZLAST
        LET ACOUNT=0 ; LET BCOUNT = 0
        ENDPROGRAM


        SUBJECT TEST1:TEST30
        RUN TESTPROG
        LET RESULTS<2,1:30> = RESULTS<2,1:30> / RESULTS<1,1:30>
        SAVE RESULTS
THIS IS THE PROBABILITY OF A "B" GIVEN AN "A"
        LET STAT(RESULTS<2,1:30>)
THIS IS ITS CORRELATION WITH TOTAL NUMBER OF EVENTS
        OUTPUT REGRESS(RESULTS<2,1:30>,RESULTS<3,1:30>)

        LET RESULTS<2,> = ROUND(RESULTS<2,>,.0001)
        OUTPUT RESULTS,'VARIABLES','SUBJECTS'
```

Results of Sample Job:

(This job cost $1.05 at the University of Chicago, including reading in 30 files of raw data, running the program "Testprog," computing basic statistics and regressing one variable (probability of a "B" given an "A") against another (number of events in the file, which happened in this case to be the number of the last event); and printing the output.

```
        SUBJECT TEST1:TEST30
        RUN TESTPROG
        LET RESULTS<2,1:30> = RESULTS<2,1:30> / RESULTS<1,1:30>
        SAVE RESULTS
THIS IS THE PROBABILITY OF A "B" GIVEN AN "A"
        LET STAT(RESULTS<2,1:30>)

NON-MISSING VALUES = 29 , MISSING VALUES = 1
MEAN = 0.361795 , VARIANCE = 0.1966755E-01 , S.D. = 0.1427237
MINIMUM = 0.1666666 , MEDIAN = 0.3333333 , MAXIMUM = 0.5555555

THIS IS ITS CORRELATION WITH TOTAL NUMBER OF EVENTS
        OUTPUT REGRESS(RESULTS<2,1:30>,RESULTS<3,1:30>)

REGRESS(SUBARRAY('RESULTS<2,1:30>'),SUBARRAY('RESULTS<3,1:30>')))


        R = 0.763           A = 4.0061    T = 6.131
      RSQ = 0.5819          B = 15.9953  DF = 28
    RESID = 3.8826                        P = 0.000


        LET RESULTS<2,> = ROUND(RESULTS<2,>,.0001)
        OUTPUT RESULTS,'VARIABLES','SUBJECTS'
```

VARIABLES ACROSS BY SUBJECTS DOWN.

|     | 1 | 2 | 3 |
|-----|---|---|---|
| 1   | 6 | 0.5000 | 9. |
| 2   | 6 | 0.1667 | 7. |
| 3   | 7 | 0.2857 | 9. |
| 4   | 3 | 0.3333 | 9. |
| 5   | 9 | 0.5556 | 16. |
| 6   | 6 | 0.5000 | 9. |
| 7   | 6 | 0.1667 | 7. |
| 8   | 7 | 0.2857 | 9. |
| 9   | 3 | 0.3333 | 9. |
| 10  | 9 | 0.5556 | 16. |
| 11  | 6 | 0.5000 | 9. |
| 12  | 6 | 0.1667 | 7. |
| 13  | 7 | 0.2857 | 9. |
| 14  | 3 | 0.3333 | 9. |
| 15  | 9 | 0.5556 | 16. |
| 16  | 6 | 0.5000 | 9. |
| 17  | 6 | 0.1667 | 7. |
| 18  | 7 | 0.2857 | 9. |
| 19  | 3 | 0.3333 | 9. |
| 20  | 9 | 0.5556 | 16. |
| 21  | 6 | 0.5000 | 9. |
| 22  | 6 | 0.1667 | 7. |
| 23  | 7 | 0.2857 | 9. |
| 24  | 3 | 0.3333 | 9. |
| 25  | 9 | 0.5556 | 16. |

CRESCAT:  Software system for the analysis of real-time or sequential data

Distributed by:

   Kenneth Kaye
   4810 S. Dorchester Ave.
   Chicago, Ill. 60615

Computers:                              Operating Systems:

   IBM 370                                  OS, VS, VM-CMS
   Amdahl 470
    (or any machine using 370
    Assembler language)

Source Languages:  SPITBOL, 370 Assembler.

Cost:

   There is a one-time charge of $500 for universities and $1500 for commercial use.

Documentation:

The manual, CRESCAT, can be ordered from the documentation manager, University
of Chicago Computation Center, 5737 University, Chicago, Ill. 60637.  Send
check for $10.

4.08  <u>PERSEE</u>

## INTRODUCTION

PERSEE is a package designed to manipulate data and files, do cross-tabulation and print data.  Its principal applications have been survey analysis, market research, file analysis (commercial, financial...).  It was developed originally by Jean Sousselier in 1970 (PERSEE 1).  PERSEE 2, which is the most widely used version, was developed in 1973-1974, and has been under continuous development since then.  PERSEE 3 has been developed from August 1978 to March 1980.  PERSEE must be used in batch, and is installed in 36 centers belonging to 24 companies.

## CAPABILITIES:  Data Processing

The <u>PREVAR program</u> is the primary way of creating a PERSEE file.  Input data can be any sequential file:  fixed or variable length, binary, EBCDIC, or packed; original data or old PERSEE file; simple or hierarchical data.

PREVAR handles variables of several types:  integer, double precision, literal, multivalued.

The user is given a large variety of statements, allowing arithmetic calculations, tests, grouping, deleting, reordering, logical statements, with automatic handling of missing data.

Output file is a PERSEE file, including a data dictionary, and used by all other PERSEE programs.

<u>MANIFIC</u> includes programs for sorting, merging and concatenating PERSEE files.

## CAPABILITIES:  Tabulation

The <u>TABUL and CUMUL</u> programs produce statistical tables with the following features:  crosstabulations with up to 4 dimensions, table breakdown, filters, weighting, means and standard deviation, several types of total, subtotals, and percents, CHI square calculation.  Unlimited number of tables.

<u>Editing</u> FLIPER is a report generator allowing user specified formats, and including an internal sort, up to 7 breakdown-levels.

## EXTENSIBILITY

The package has an EXIT facility.  User-written programs can be added to the system without difficulty.

## PROPOSED ADDITIONS IN NEXT YEAR

PERSEE 3 will be further developed in the next months: new free-format language, installation on new computers.

## REFERENCE

'Guide Europeen des produits logiciels', Centre d'experimentation de
Progiciels - 5, rue Monceau 75008, Paris, France.

## SAMPLE JOB

Define 2 variables (age and function) on a new file: define, label,
read them.   Group 'age', produce a crosstabulation.

## Input

```
**  PREVAR  **
  FUNC     2  1  5 1 FUNCTION
  AGE      2  1  5 2
LABEL
  1/TEACHERS      LECTURERS    /SCIENTISTS   TECHNOLOGIST/MEDICAL       /ACCOUNTANCY
  1/LAW           ROTHER       *
  2DUNDER 25      /25-34       FSUBTOTAL     UNDER 35      D35-49       /50-64
  2FSUBTOTAL      35-64        /65 OR OVER   RREJECT       *
PROGRAM
C    COMMENT----  READING  FUNC ------------------
        FUNC/10
C    -----    READING AND GROUPING  AGE  -------------
        AGE/12-13
        TRANS  AGE,AGE=1-24/25-34/35-49/65-99
        END
**  TABUL  **
T   PERSEE-STATIRO--SAMPLE PROBLEM
      FUNC  AGE                               22  1      1
```

Output

FERSEE-STATIRO—SAMPLE PROBLEM
FUNCTION

AGE

| | TEACHERS* LECTURERS* | SCIENTISTS* TECHNOLOGIST* | MEDICAL* | ACCOUNTANCY* | LAW* | OTHER* | TOTAL* |
|---|---|---|---|---|---|---|---|
| UNDER 25* | 85 * 9.5*<br>30.7 | 129 * 14.5*<br>26.3 | 116 * 13.0*<br>39.6 | 41 * 4.6*<br>46.6 | 445 * 49.9*<br>30.3 | 76 * 8.5*<br>21.7 | 892<br>30.0 100.0* |
| 25-34* | 24 * 8.8*<br>8.7 | 22 * 8.1*<br>4.5 | 37 * 13.6*<br>12.6 | 9 * 3.3*<br>10.2 | 157 * 57.5*<br>10.7 | 24 * 8.8*<br>6.8 | 273<br>9.2 100.0* |
| SUBTOTAL* UNDER 35* | 109 * 9.4*<br>39.4 | 151 * 13.0*<br>30.8 | 153 * 13.1*<br>52.2 | 50 * 4.3*<br>56.8 | 602 * 51.7*<br>40.9 | 100 * 8.6*<br>28.5 | 1165<br>39.2 100.0* |
| 35-49* | 14 * 5.5*<br>5.1 | 38 * 14.9*<br>7.8 | 30 * 11.8*<br>10.2 | 6 * 2.4*<br>6.8 | 144 * 56.5*<br>9.8 | 23 * 9.0*<br>6.6 | 255<br>8.6 100.0* |
| 50-64* | 14 * 8.4*<br>5.1 | 22 * 13.3*<br>4.5 | 17 * 10.2*<br>5.8 | 4 * 2.4*<br>6.2 | 91 * 54.8*<br>6.2 | 18 * 10.8*<br>5.1 | 166<br>5.6 100.0* |
| SUBTOTAL* 35-64* | 28 * 6.7*<br>10.1 | 60 * 14.3*<br>12.2 | 47 * 11.2*<br>16.0 | 10 * 2.4*<br>11.4 | 235 * 55.8*<br>16.0 | 41 * 9.7*<br>11.7 | 421<br>14.2 100.0* |
| 65 OR OVER* | 15 * 9.1*<br>5.4 | 34 * 20.7*<br>6.9 | 16 * 9.8*<br>5.5 | 3 * 1.8*<br>3.4 | 64 * 39.0*<br>4.4 | 32 * 19.5*<br>9.1 | 164<br>5.5 100.0* |
| TOTAL* | 277 * 9.3*<br>100.0 | 490 * 16.5*<br>100.0 | 293 * 9.9*<br>100.0 | 88 * 3.0*<br>100.0 | 1471 * 49.5*<br>100.0 | 351 * 11.8*<br>100.0 | 2970<br>100.0 100.0* |

PERSEE:  Programme d'Elaboration de Resultats Statistiques Extraits d'Enquetes

Developed by:                          Distributed by:

    STATIRO                                    Same
    145, avenue de Malakoff
    75116 Paris, France

Computer Makes:                        Operating Systems:

    PERSEE 1
    UNIVAC 1106/1108                           EXEC 8
    CYBER                                      SCOPE

    PERSEE 2
    IBM 360, 370                               OS - DOS
    IBM 3031-3033                              VS
    SIEMENS 4004                               BS1000
    UNIVAC 90                                  VS/9

    PERSEE 3
    HONEYWELL 6000                             GCOS
    PRIME 650                                  PRIMOS

Source Languages:

    PERSEE 1,2:  Assembler
    PERSEE 3  :  FORTRAN

Cost:

    Unlimited license fee:  $20000 to $40000 depending on version and modules
registered, including installation, 5 manuals, 2 years free of charge
maintenance.

Documentation:

PERSEE  Reference Manual
PERSEE  Technical Manual
MANIFIC Reference Manual
FLIPER  Reference Manual
CUMUL   Reference Manual

## 4.09  CENTS III

### INTRODUCTION

CENTS III is a special-purpose statistical software package which is used to manipulate data files, cross-tabulate individual observations, aggregate tabulations to higher levels, perform simple statistical measures, and format a publication-quality (camera-ready) tabular report.  The original Census Tabulation System (CENTS) was developed in 1970 by the International Statistical Programs Center (ISPC) of the U.S. Bureau of the Census to provide a method of rapid tabulation of population and housing data from a national census.  It was designed for use on small computers and was written in IBM Assembler Language Code (ALC).  Its companion package is COCENTS which is written in COBOL.  Since 1970 CENTS has been continuously developed and is now used for tabulating all types of statistical data and has been installed in over 50 computer installations overseas.

### CAPABILITIES:  Processing and Displaying Data

CENTS III can be used to extract or select certain sub-universes or samples from the data file or to recode individual data items prior to tabulation.

The approach to tabulation of data involves:  the preparation of tally blocks for the smallest observational unit or area desired, the consolidation (aggregation) of these tally blocks to the required publication levels or areas, and the display of the tally blocks in report form.

The system can operate on complex hierarchical files as well as simple flat files.  Observational units can be tabulated unweighted or weighted.  It can define new variables by grouping or re-ordering.  It can produce publishable tables from a data file in one run, allowing the user full control over the appearances of these tables.

The user instruction set is of a quasi-free-format design providing the ability to:  reference data stored in any form, create matrices and update their contents, use tag names for program referencing, and code instruction loops.

### CAPABILITIES:  Statistical Analysis

Basic statistics including totals, subtotals, percentage distributions (to one decimal place), ratios, means, and medians.

### INTERFACES WITH OTHER SYSTEMS

CENTS III can read files produced by many other packages including the edit package, CONCOR.

## PROPOSED ADDITIONS IN NEXT YEAR

CENTS III is undergoing extensive development in the next 12 months. It and its companion package, COCENTS, will be replaced by a single package, CENTS 4, which will provide significant improvement in the areas of: self-documenting, structured system source code; totally free-format user language; enhanced error protection and error message system; more flexible display capabilities; and additional members in the instruction set. CENTS 4 will be operational in both ALC and COBOL languages.

## SAMPLE JOB:

```
CENTRAL CENSUS TABULATION SYSTEM  VERSION ** 10/27/78 **

          DMN TAB2  11  18          MARITAL STATUS-RELATION TO HEAD
          DMN TAB3  13  12          CHILDREN EVER BORN
          DMN TAB4  12  26          SCHOOL ATTENDENCE
          DMN TAB5  10  24          HOURS WORKED
          DMN TAB6  12  24          ECONOMIC ACTIVITY
          DMN TAB7  13  28          MARITAL STATUS-AGE AND SEX
          DMN TAB8  11  22          NUMBER OF PERSONS BY ROOMS
          DMN TAB9  13  11          NUMBER OF ROOMS
          DMN TB10   3  15          RELATION TO HEAD/ECON BY TENURE

          ************
          *  TABLE 2
          ************
     29   EQ  IMAR   6P030          EXCLUDE UNDER 14 YEARS FROM TABLES 2 AND 3
          *
     30   NE  IPCQ  99P010          HOUSEHOLDS/COLLECTIVES
          *                         HOUSEHOLDS
     31   XRC IRELPOPL 3,4,5,6,7    RELATIONSHIP TO HEAD
     32   SKP P020
          *
   P010 TAG P010                    COLLECTIVES AND HOMELESS
     33   XRC IPCQPOPL 9,9,9,9,9,9,9,10,10,10,11
          *
   P020 TAG P020
     34   XRC ISEXSEX  0,9           SEX
          *
     35   XRC IMARMART 5,6,8,9,7,2  MARITAL STATUS
          *
     36   ADD SEX MARTSMAR          SEX AND MARITAL STATUS
          *
     37   TAL POPLSMARTAB2
          ************
          *  TABLE 9
          ************
   H010 TAG H010
          *
     13   XRC IND05URRU 6,10        URBAN/RURAL
     14   EQ  IVCT   7H020          HOUSING UNIT VACANT?
     15   XRC IVCTTENR 2,2,3
     16   SKP H030
          *
   H020 TAG H020
     17   XRC ITENTENR 0,0,1        NOT VACANT - OWN VS. RENTED
          *
   H030 TAG H030
     18   ADD TENRURRUTRUR          COMBINE TENURE AND URBAN/RURAL
     19   TAL TRURROOMTAB9
```

Sample Output

**TABLE 2. MARITAL STATUS BY SEX, RELATIONSHIP TO HOUSEHOLD HEAD, AND POPULATION IN COLLECTIVE QUARTERS, FOR PERSONS 14 YEARS OR OLDER: 1980**

| SEX AND MARITAL STATUS | PERSONS 14 YEARS OR OLDER | IN HOUSEHOLDS | | | | | | IN COLLECTIVE QUARTERS | | | HOMELESS POPULATION |
|---|---|---|---|---|---|---|---|---|---|---|---|
| | | TOTAL | HEAD | SPOUSE OF HEAD | SON OR DAUGHTER OF HEAD OR SPOUSE | OTHER RELATIVE | NON-RELATIVE | TOTAL | IN INSTITUTIONS | IN OTHER QUARTERS | |
| **TOTAL COUNTRY** | | | | | | | | | | | |
| TOTAL MALE | 70,975 | 67,554 | 50,024 | — | 14,065 | 2,179 | 1,286 | 2,362 | 1,109 | 1,253 | 1,059 |
| NEVER MARRIED | 20,054 | 17,598 | 2,451 | — | 13,304 | 1,064 | 779 | 1,468 | 544 | 954 | 888 |
| MARRIED | 45,764 | 45,275 | 44,367 | — | 267 | 531 | 110 | 448 | 230 | 218 | 41 |
| PERCENT | 100.0 | 99.9 | | — | .6 | 1.2 | .2 | 1.0 | .5 | .5 | .1 |
| SPOUSE PRESENT | 44,352 | 44,352 | 43,779 | — | 177 | 396 | 110 | 448 | 230 | 218 | 41 |
| SPOUSE ABSENT | 1,412 | 923 | 588 | — | 90 | 135 | 103 | 82 | 58 | 24 | 10 |
| SEPARATED | 1,013 | 921 | 526 | — | 181 | 111 | 88 | 185 | 171 | 14 | 9 |
| WIDOWED | 2,181 | 1,987 | 1,526 | — | 39 | 334 | 88 | 185 | 171 | 14 | 9 |
| DIVORCED | 1,963 | 1,773 | 1,154 | — | 274 | 139 | 206 | 179 | 136 | 43 | 11 |
| TOTAL FEMALE | 77,926 | 75,695 | 13,469 | 43,779 | 12,995 | 4,148 | 1,304 | 1,131 | 959 | 172 | 1,100 |
| NEVER MARRIED | 16,157 | | 2,378 | — | 11,616 | 1,191 | 762 | 395 | 274 | 121 | 1,049 |
| MARRIED | 45,789 | | | 43,779 | 555 | 660 | 174 | 123 | 107 | 16 | 21 |
| PERCENT | 100.0 | 99.7 | 1.0 | | | | .4 | | | | |
| SPOUSE PRESENT | 44,352 | 44,352 | — | 43,779 | 185 | 388 | 175 | 123 | 107 | 16 | 21 |
| SPOUSE ABSENT | 1,429 | 1,285 | 468 | — | 370 | 272 | 69 | 20 | 17 | 3 | 7 |
| SEPARATED | 1,751 | 1,724 | 1,275 | — | 203 | 177 | 169 | 540 | 512 | 28 | 17 |
| WIDOWED | 9,793 | 9,236 | 7,038 | — | 120 | 1,909 | 169 | 540 | 512 | 28 | 17 |
| DIVORCED | 3,012 | 2,953 | 2,312 | — | 301 | 211 | 129 | 53 | 49 | 4 | 6 |

**TABLE 9. NUMBER OF ROOMS IN UNIT BY TENURE AND VACANCY, FOR CONVENTIONAL HOUSING UNITS, URBAN AND RURAL: 1980**

| NUMBER OF ROOMS PER UNIT | CONVENTIONAL HOUSING UNITS | TOTAL | | | | URBAN | | | | RURAL | | | |
|---|---|---|---|---|---|---|---|---|---|---|---|---|---|
| | | OCCUPIED | | VACANT | | OCCUPIED | | VACANT | | OCCUPIED | | VACANT | |
| | | OWNER | RENTED AND OTHER | AVAILABLE | OTHER | OWNER | RENTED AND OTHER | AVAILABLE | OTHER | OWNER | RENTED AND OTHER | AVAILABLE | OTHER |
| **TOTAL COUNTRY** | | | | | | | | | | | | | |
| TOTAL | 68,725 | 39,907 | 23,586 | 2,159 | 3,073 | 27,197 | 19,343 | 1,667 | 896 | 12,710 | 4,238 | 492 | 2,177 |
| 1 ROOM | 1,317 | 102 | 957 | 99 | 159 | 54 | 862 | 85 | 39 | 48 | 95 | 14 | 120 |
| 2 ROOMS | 2,524 | 254 | 1,828 | 192 | 250 | 135 | 1,631 | 152 | 163 | 119 | 197 | 40 | 187 |
| 3 ROOMS | 7,599 | 1,236 | 4,759 | 469 | 523 | 730 | 4,779 | 400 | 167 | 522 | 576 | 69 | 356 |
| 4 ROOMS | 14,267 | 5,138 | 4,768 | 542 | 639 | 3,508 | 5,867 | 434 | 184 | 2,528 | 1,250 | 143 | 600 |
| 5 ROOMS | 17,267 | 10,683 | 2,308 | 227 | 370 | 7,609 | 1,695 | 155 | 130 | 3,074 | 1,601 | 108 | 245 |
| 6 ROOMS | 13,583 | 5,576 | | 64 | 176 | 7,653 | 465 | | 67 | 1,623 | 274 | 21 | 109 |
| 7 ROOMS | 6,575 | 5,576 | | 59 | 92 | 1,945 | 194 | 49 | 28 | 857 | 144 | 10 | 64 |
| 8 ROOMS | 3,291 | 2,802 | 338 | 40 | | 1,360 | 106 | 25 | 24 | 624 | 88 | 15 | 56 |
| 9 ROOMS OR MORE | 2,298 | 1,984 | 194 | | | | | | | | | | |
| MEDIAN | 5.0 | 5.6 | 4.0 | 4.1 | 4.3 | 5.7 | 3.5 | 4.0 | 4.5 | 5.4 | 4.5 | 4.4 | 4.2 |

## CENTS III, Version 3.5: Census Tabulation System III

Developed by:                              Distributed by:

    U.S. Bureau of the Census                 Same
    International Statistical
      Programs Center
    Washington, D.C.  20233

Computer Makes:                            Operating Systems:

    IBM System 360/370 series                 IBM - OS, VS, TSO, DOS-VS
    IBM 303X series
    WANG 2200/VS
                                           Interfaced Systems:

                                       CONCOR

Source Language:  IBM Assembler Language Code.

Cost:

    There is a one time charge of $150 for CENTS III.  This charge covers
expenses for providing an installation tape containing source modules, installation JCL, and computer readable documentation.  Hard copies of the system
reference manual, installation guide and procedures library are also included.

Documentation:

CENTS III CENTAL Reference Manual, Surveys and Evaluation Unit, International
    Statistical Programs Center, U.S. Bureau of the Census, Washington, D.C.,
    1978.
CENTS III  CENPREP Reference Manual, Surveys and Evaluation Unit, International
    Statistical Programs Center, U.S. Bureau of the Census, Washington, D.C.,
    1978.
CENTS III  Error Messages, ISPC, U.S. Bureau of the Census, Washington, D.C.,
    1979.

4.10  <u>COCENTS</u>

## INTRODUCTION

COCENTS is a special-purpose statistical software package which is used to manipulate data files, cross-tabulate individual observations, aggregate tabulations to higher levels, perform simple statistical measures, and format a publication-quality (camera-ready) tabular report.  The companion package to COCENTS, CENTS (<u>Cen</u>sus <u>Ta</u>bulation <u>Sy</u>stem) was originally developed by the International Statistical Programs Center (ISPC) of the U.S. Bureau of the Census to provide a method of rapid tabulation of population and housing data from a national census.  CENTS was designed for use on small computers and was written in IBM Assembler Language Code (ALC).  COCENTS (<u>CO</u>BOL <u>Cen</u>sus <u>Ta</u>bulation <u>Sy</u>stem) was written in COBOL which permits usage on a wider range of computer equipment.  Since 1972, COCENTS has been continuously developed and is now used for tabulating all types of statistical data and has been installed in over 70 computer installations overseas.

## CAPABILITIES:  Processing and Displaying Data

COCENTS can be used to extract or select certain sub-universes or samples from the data file or to recode individual data items prior to tabulation.

The approach to tabulation of data involves:  the preparation of tally blocks for the smallest observational unit or area desired, the consolidation (aggregation) of these tally blocks to the required publication levels or areas, and the display of the tally blocks in report form.

The system can operate on complex hierarchical files as well as simple flat files.  Observational units can be tabulated unweighted or weighted.  It can define new variables by grouping or re-ordering.  It can produce publishable tables from a data file in one run, allowing the user full control over the appearances of these tables.

The user instruction set is of a quasi-free-format design.

## CAPABILITIES:  Statistical Analysis

Basic statistics including totals, subtotals, percentage distributions (to one decimal place), ratios, means, and medians.

## INTERFACES WITH OTHER SYSTEMS

COCENTS can read files produced by many other packages including the edit package, CONCOR.

## PROPOSED ADDITIONS IN NEXT YEAR

COCENTS is undergoing extensive development in the next 12 months.  It and
its companion package, CENTS III, will be replaced by a single package, CENTS 4,
which will provide significant improvement in the areas of:  self-documenting,
structured system source code; totally free-format user language; enhanced
error protection and error message system; more flexible display capabilities;
and additional members in the instruction set.  CENTS 4 will be operational in
both ALC and COBOL languages.

## SAMPLE JOB

```
SPEC    DELT,TAPE,100,30
DMN     TAB2,18,11                   MARITAL STATUS-RELATION TO HEAD
DMN     TAB3,12,13                   CHILDREN EVER BORN
DMN     TAB4,26,12                   SCHOOL ATTENDENCE
DMN     TAB5,24,10                   HOURS WORKED
DMN     TAB6,24,12                   ECONOMIC ACTIVITY
DMN     TAB7,24,13                   MARITAL STATUS-AGE AND SEX
DMN     TAB8,22,11                   NUMBER OF PERSONS BY ROOMS
DMN     TAB9,11,13                   NUMBER OF ROOMS
DMN     TB10,15,3                    RELATION TO HEAD/ECON BY TENURE

*************
*  TABLE 2
*************
        EQ      IMAR,6,P030              EXCLUDE UNDER 14 YEARS FROM TABLES 2 and 3
*
        NEQ     IPCQ,99,P010             HOUSEHOLDS/COLLECTIVES
*                     HOUSEHOLDS
        XRC     IREL,POPL,3,4,5,6,7      RELATIONSHIP TO HEAD
        GOTO    P020
*
P010 NOP                    COLLECTIVES AND HOMELESS
        XRC     IPCQ,POPL,9,9,9,9,9,9,9,9,10,10,10,11
*
P020 NOP
        XRC     ISEX,SEX,0,9             SEX
*
        XRC     IMAR,MART,5,6,8,9,7,2    MARITAL STATUS
*
        ADD     SEX,MART,SMAR            SEX AND MARITAL STATUS
*
        TAL     TAB2,SMAR,POPL
*************
*  TABLE 9
*************
H010 NOP
*
        XRC     E1-5,URRU,6,10           URBAN/RURAL
        EQ      IVCT,7,H020              HOUSING UNIT VACANT
        XRC     IVCT,TENR,2,2,3
        GOTO    H030
*
H020 NOP
        XRC     ITEN,TENR,0,0,1          NOT VACANT - OWN VS. RENTED
*
H030 NOP    TENR,URRU,TRUR              COMBINE TENURE AND URBAN/RURAL
        TAL     TAB9,ROOM,TRUR
*
        GOTO    ENDC                     GET NEXT RECORD
```

Sample Output

## TABLE 2. MARITAL STATUS BY SEX, RELATIONSHIP TO HOUSEHOLD HEAD, AND POPULATION IN COLLECTIVE QUARTERS, FOR PERSONS 14 YEARS OR OLDER: 1980

| SEX AND MARITAL STATUS | PERSONS 14 YEARS OR OLDER | IN HOUSEHOLDS | | | | | | IN COLLECTIVE QUARTERS | | | HOMELESS POPULATION |
|---|---|---|---|---|---|---|---|---|---|---|---|
| | | TOTAL | HEAD | SPOUSE OF HEAD | SON OR DAUGHTER OF HEAD OR SPOUSE | OTHER RELATIVE | NON-RELATIVE | TOTAL | IN INSTITUTIONS | IN OTHER QUARTERS | |
| **TOTAL COUNTRY** | | | | | | | | | | | |
| TOTAL MALE | 70,975 | 67,554 | 50,024 | -- | 14,065 | 2,179 | 1,286 | 2,362 | 1,109 | 1,253 | 1,059 |
| NEVER MARRIED | 20,054 | 17,598 | 2,451 | -- | 13,304 | 1,064 | 779 | 1,468 | 514 | 954 | 988 |
| MARRIED | 45,764 | 45,275 | 44,367 | -- | 267 | 531 | 110 | 448 | 230 | 218 | 41 |
| PERCENT | 100.0 | 98.9 | 96.9 | -- | .6 | 1.2 | .2 | 1.0 | .5 | .5 | .1 |
| SPOUSE PRESENT | 44,352 | 43,929 | 43,729 | -- | 77 | 396 | 110 | 448 | 238 | 218 | 41 |
| SPOUSE ABSENT | 1,412 | 923 | 923 | -- | 90 | 135 | 103 | 82 | 58 | 24 | 10 |
| SEPARATED | 1,013 | 921 | 526 | -- | 181 | 111 | 88 | 185 | 171 | 14 | 9 |
| WIDOWED | 2,181 | 1,987 | 1,526 | -- | 39 | 334 | 206 | 179 | 136 | 43 | 11 |
| DIVORCED | 1,963 | 1,773 | 1,154 | -- | 274 | 139 | | | | | |
| TOTAL FEMALE | 77,926 | 75,695 | 13,469 | 43,779 | 12,995 | 4,148 | 1,304 | 1,131 | 959 | 172 | 1,100 |
| NEVER MARRIED | 17,589 | 16,145 | 2,376 | -- | 11,816 | 1,191 | 762 | 395 | 274 | 121 | 1,049 |
| MARRIED | 45,781 | 45,937 | 168 | 43,779 | 552 | 660 | 175 | 123 | 107 | 16 | 21 |
| PERCENT | 100.0 | 1.0 | | 43,779 | .4 | | .3 | .2 | .2 | | |
| SPOUSE PRESENT | 44,352 | 44,352 | 468 | 43,779 | 185 | 338 | 175 | 123 | 107 | 16 | 21 |
| SPOUSE ABSENT | 1,429 | 1,285 | | | 370 | 272 | 69 | 123 | 107 | | |
| SEPARATED | 1,751 | 1,724 | 1,275 | -- | 370 | 177 | 69 | 20 | 17 | 3 | 7 |
| WIDOWED | 9,793 | 9,236 | 7,038 | -- | 120 | 1,909 | 169 | 540 | 512 | 28 | 17 |
| DIVORCED | 3,012 | 2,953 | 2,312 | -- | 301 | 211 | 129 | 53 | 49 | 4 | 6 |

## TABLE 9. NUMBER OF ROOMS IN UNIT BY TENURE AND VACANCY, FOR CONVENTIONAL HOUSING UNITS, URBAN AND RURAL: 1980

| NUMBER OF ROOMS PER UNIT | CONVENTIONAL HOUSING UNITS | TOTAL | | | | URBAN | | | | RURAL | | | |
|---|---|---|---|---|---|---|---|---|---|---|---|---|---|
| | | OCCUPIED | | VACANT | | OCCUPIED | | VACANT | | OCCUPIED | | VACANT | |
| | | OWNER | RENTED AND OTHER | AVAILABLE | OTHER | OWNER | RENTED AND OTHER | AVAILABLE | OTHER | OWNER | RENTED AND OTHER | AVAILABLE | OTHER |
| **TOTAL COUNTRY** | | | | | | | | | | | | | |
| TOTAL | 68,725 | 39,907 | 23,586 | 2,159 | 3,073 | 27,197 | 19,348 | 1,667 | 896 | 12,710 | 4,238 | 492 | 2,177 |
| 1 ROOM | 2,324 | 102 | 1,928 | 199 | 54 | 862 | 85 | 39 | 48 | 195 | 14 | 120 |
| 2 ROOMS | 7,599 | 1,252 | 5,355 | 523 | 738 | 4,739 | 452 | 163 | 119 | 176 | 40 | 187 |
| 3 ROOMS | 14,266 | 5,836 | 5,079 | 567 | 784 | 3,508 | 4,829 | 424 | 167 | 522 | 1,250 | 69 | 386 |
| 4 ROOMS | 17,267 | 11,418 | 4,768 | 442 | 639 | 7,903 | 3,767 | 334 | 194 | 2,328 | 1,001 | 108 | 445 |
| 5 ROOMS | 13,583 | 10,683 | 2,308 | 227 | 370 | 7,609 | 1,695 | 155 | 130 | 3,515 | 613 | 72 | 240 |
| 6 ROOMS | 6,575 | 5,576 | 759 | 64 | 176 | 3,953 | 485 | 43 | 67 | 3,074 | 274 | 21 | 109 |
| 7 ROOMS | 3,291 | 2,882 | 338 | 59 | 92 | 1,945 | 194 | 49 | 28 | 1,623 | 144 | 10 | 64 |
| 8 ROOMS | 2,298 | 1,984 | 194 | 40 | 80 | 1,360 | 106 | 25 | 24 | 857 | 88 | 15 | 56 |
| 9 ROOMS OR MORE | 2,298 | 1,984 | 194 | 40 | 80 | 1,360 | 106 | 25 | 24 | 624 | | | |
| MEDIAN | 5.0 | 5.6 | 4.0 | 4.1 | | 5.7 | 3.9 | 4.0 | 4.5 | 5.4 | 4.5 | 4.4 | 4.2 |

COCENTS, Version 1.3:   COBOL Census Tabulation System

Developed by:                              Distributed by:

    U.S. Bureau of the Census              Same
    International Statistical
      Programs Center
    Washington, D.C.   20233

Computer Makes:                            Interfaced Systems:

    IBM 360/370 series                     CONCOR
    IBM 303X series
    IBM System/3
    IBM System/34
    UNIVAC 9400
    UNIVAC 1100 series
    CDC 3000 series
    ICL 1900 series
    ICL 2900 series
    NCR Century series
    NCR 8200
    FACOM 230 series
    DEC System/10
    Burroughs B1700
    Burroughs B3500
    Burroughs B6700
    Honeywell Bull 62
    Honeywell Bull 66
    Honeywell 200/1200
    HP 3000
    WANG 80/VS

Source Language:   COBOL

Cost:

    There is a one time charge of $150 for COCENTS.  This charge covers expen-
ses for providing an installation tape containing source modules, installation
JCL, and computer readable documentation.  Hard copies of the system reference
manual, installation guide and procedures library are also included.

Documentation:

COCENTS System Reference Manual, International Statistical Programs Center, U.S.
      Bureau of the Census, Washington, D.C., 1977.
COCENTS Error Messages, International Statistical Programs Center, U.S. Bureau
      of the Census, Washington, D.C., 1977.

## 4.11  CENTS-AID

INTRODUCTION

CENTS-AID is a generative COBOL system for tabulation, statistical analysis and data extraction.  Its principal applications have been in analysis of large-scale unit record files such as the 1970 U.S. Census Public Use Samples and the 1976 U.S. Survey of Income and Education.  CENTS-AID was originally developed in 1973 by Data Use and Access Laboratories (DUALabs) uder a contract with the National Institute of Child Health and Human Development, Center for Population Research.  It has been continually upgraded since then.  CENTS-AID 2.5 and 2.6 are used by about 100 installations.  CENTS-AID 3.0 is used by five test sites.  CENTS-AID 2.6 and 3.0 are the final versions of the system. CENTS-AID 3.0 was released for public use in August 1980.

CAPABILITIES:  Processing and Displaying Data

CENTS-AID can process simple sequential files and complex hierarchical files with as many as twenty-six different record formats.  Machine-readable file descriptions (Data Base Dictionaries) can be used but are not required. The user language is free-format with an English-like syntax.  The system can produce frequency distributions and cross-tabulations of up to eight dimensions. Tables can be labelled from information in the Data Base Dictionary or from user-supplied commands.  A table can include variables from several record types.

Data manipulation features allow redefinition of variables and construction of new variables through mathematical and logical expressions.  Weights can be applied.  Several retrieval commands allow selection of subsets of records which can vary throughout a particular run.

Tables are produced in a predetermined format that can be modified by the user.  Publishable tables can be produced in a single run, especially with CENTS-AID 3.0 which has commands to prepare detailed table headings.

The system design involves generating and compiling a COBOL program specific to the user's commands.  This permits processing of large command sets and large files with a single pass of the file.  The English-like command language, the compiler, and other enhancements were built around the Census Bureau's COCENTS system because of COCENTS' processing efficiencies.  One of CENTS-AID's initial design objectives was to reduce the cost of processing the 1970 U.S. Census Public Use Sample files which have over 2,000,000 records each.  Benchmark comparisons of cross-tabulations with early versions of CENTS-AID and other systems are provided in Hill et al. (1972) and Benenson (1979).

Additional features of CENTS-AID 3.0 include the processing of binary and packed decimal data and use of floating point decimals.  CENTS-AID 3.0 can also be used to create extract files with Data Base Dictionaries.  All the data manipulation and retrieval features can be applied in file creation runs.

CAPABILITIES: Statistical Analysis

Cross-tabulations can include descriptive statistics: percent, mean, median, variance, and chi square.

In addition, CENTS-AID 3.0 can produce Pearson correlation, covariance, sum of squares, sum of crossproducts, and means of variables with paired or listwise treatment of missing values. A correlation matrix can be written in the form required for input to SPSS subprograms REGRESSION and FACTOR. Statistical accuracy with large files is improved through use of double precision.

INTERFACES WITH OTHER SYSTEMS

Creates SPSS matrix files.

PROPOSED ADDITIONS IN NEXT YEAR

None anticipated.

REFERENCES

Benenson, H., Computer Analysis of Census and Government Research Data and Introduction to CENTS-AID New Jersey Educational Computer Network BAPLS-ACAD-018-00, (1979).
Francis, I., S.P. Sherman, and R.M. Heiberger, "A look at Languages for Tabulating Data from Surveys," Proc. Comp. Sci. & Statist: 9th Interface, (1976).
Hill, G., "Problems in Comparing Survey Packages (a Case Example)", Proc. Comp. Sci. & Statist: 9th Interface, (1976).
Hill, G., L.Brown, and K. Han, "Maximizing Access to the Public Use Samples," Review of Public Data Use, I, 1 (1972).

SAMPLE JOB

Recode the income variable (P37) to major catagories. Label the newly created variable. Select only records with a value greater than 13 on the age variable (P9). Establish a weight of 10,000 for records with a record type code (P120) of 3 (person records on a household/person file.) Produce a two-way table with row percents. The table uses Data Base Dictionary information for labelling the Relationship to Head of Household variable (P1) and for all format and technical specifications.

```
RECODE EARNINGS FROM P37 (THRU 49 = 0) (THRU 99 = 1) (THRU 149 = 2)
    (THRU 500 = 3) (THRU 999 = 4)
VAR LABEL EARNINGS PERSON's EARNINGS (0) $0 - 4,999
    (1) $5,000 - 9,999 (2) $10,000 - 14,999
    (3) $15,000 OR MORE (4) NO INCOME
SELECT IF P9 GT 13
IF P120 = 3 COMPUTE WEIGHT = 10000
OPTION TABLE PERCENT (ROW)
TABLE P1 BY EARNINGS
```

Output from Sample Job

TABLE T001: BASIC RELATIONSHIP BY PERSON'S EARNINGS

| BASIC RELATIONSHIP | PERSON'S EARNINGS | | | | | |
| | $0 - 4,999 | $5,000 - 9,999 | $10,000 - 14,999 | $15,000 OR MORE | NO INCOME | T O T A L |
|---|---|---|---|---|---|---|
| HEAD OF HOUSEHOLD | 12300,000 | 19820,000 | 9400,000 | 3560,000 | 18250,000 | 63330,000 |
| ROW % | 19.4 | 31.3 | 14.8 | 5.6 | 28.8 | 100.0 |
| WIFE OF HEAD | 14220,000 | 5140,000 | 390,000 | 30,000 | 23750,000 | 43530,000 |
| ROW % | 32.7 | 11.8 | .9 | .1 | 54.6 | 100.0 |
| SON,DAUGHTER OF HEAD | 11330,000 | 1660,000 | 150,000 | 10,000 | 13740,000 | 26890,000 |
| ROW % | 42.1 | 6.2 | .6 | .0 | 51.1 | 100.0 |
| RELATIVE OF HEAD | 1730,000 | 570,000 | 100,000 | 10,000 | 3590,000 | 6000,000 |
| ROW % | 28.8 | 9.5 | 1.7 | .2 | 59.8 | 100.0 |
| ROOMER,BOARDER,LODGR | 820,000 | 260,000 | 30,000 | 40,000 | 580,000 | 1730,000 |
| ROW % | 47.4 | 15.0 | 1.7 | 2.3 | 33.5 | 100.0 |
| PATIENT OR INMATE | 180,000 | 10,000 | | | 1750,000 | 1940,000 |
| ROW % | 9.3 | .5 | .0 | .0 | 90.2 | 100.0 |
| OTHER NOT RELATED | 3330,000 | 420,000 | 20,000 | 20,000 | 720,000 | 4510,000 |
| ROW % | 73.8 | 9.3 | .4 | .4 | 16.0 | 100.0 |
| T O T A L | 43910,000 | 27880,000 | 10090,000 | 3670,000 | 62380,000 | 147930,000 |
| ROW % | 29.7 | 18.8 | 6.8 | 2.5 | 42.2 | 100.0 |

CENTS-AID:   Census Tabulation System Aid

Developed by:                                    Distributed by:

    Data Use and Access Laboratories             Version 3.0
    (DUALabs)                                        International Data & Development,
    Suite 900                                          Inc. (IDD)
    1601 North Kent Street                           P.O. Box 2157
    Arlington, VA 22209                              Arlington, VA 22209

                                                 Version 2.6
                                               U.S. Department of Commerce
                                             National Technical Information
                                               Service (NTIS)
                                           5285 Port Royal Road
                                           Springfield, VA 22161
                                           (Order # PB-80-133002)

Computer Makes:                                  Operating Systems:

    IBM 360, 370                                 IBM-OS, DOS
    Honeywell 6000                               Honeywell - GCOS
    (Version 2.5)

                                               Interfaced Systems:

                                             SPSS

Source Languages:   COBOL.   Version 3.0 has a FORTRAN component for statistical
procedures.

Cost:

    Version 3.0:   $3,000 for government and non-profit organizations; $6,000
for all others.  Purchase includes load modules on tape, installation instructions, User Manual, and one year of maintenance.  Maintenance in subsequent
years cost $500 per year.

    Version 2.6:   $900.  Purchase includes source modules on tape, installa-
tion instructions, user Manual, and one year of maintenance.  Maintenance in
subsequent years cost $500 per year.

    Foreign sales available at higher cost.

    A separate system for creating Data Base Dictionaries (LEXICOGRAPHER) is
available for $500 from DUALabs or IDD. DBD's for major files also available.

Documentation:

Data Use and Access Laboratories:  CENTS-AID 2.6 User Manual, Arlington, VA,
    DUALabs: 1979.
CENTS-AID 2.6 Programmer's Notebook, Arlington, VA, DUALabs:  1979
CENTS-AID 3 Test User Manual, Arlington, VA, DUALabs:  1979.

4.12 <u>FILE</u>

## INTRODUCTION

FILE is a batch program, written by Jerry G. Gentry of the United States Center for Disease Control, for data management, tabulation, and statistical summary. It is not exported at present, but the developer would make it available at a nominal cost.

## CAPABILITIES:  Data Management and Display

Data Management options of the FILE program are designed to read data records from a sequential file and perform any or all of the following functions:  select or reject records from the input file on the basis of data in each record or its sequence in the file, sort the input file, reformat each record, edit and modify input records,  generate data including random and random normal numbers, print input records with optional titles, headings, spacing, date and page numbering, copy selected records to a sequential output file, merge selected records from the primary input file (Transaction File) with records from a secondary input file (Old Master) to produce a new output file (New Master), and use selected records from the Transaction file to correct or delete records from the Old Master file.

Two input files and two output files are optional.  Printed output files are created independent of other (primary and secondary) output files.  Printed listings of selected input records with formatting, titles, and headings are optional with the results of editing.  A printed update report is also optional and a summary report of input, output, and update counts is printed.

## CAPABILITIES:  Statistical Analysis

The Statistical Summary options of the FILE program provide for several commonly used statistical summary techniques.  These include:  frequency distributions, cross tabulations with counts, percent distributions, means, standard deviations, and percentiles.  These summaries are performed on data records from a sequential input file.

Features available at the option of the user are cross tabulations on up to nine variables with as many levels as desired on each variable limited only by available address space; means, and standard deviations, sums or percentiles of any numeric variable; titles, headings and labels on printed records defined by user; and choice in selection or omission of input records on the basis of data in the record.

## FILE 2.0

Developed by:

    Jerry G. Gentry
    Center for Disease Control
    1600 Clifton Rd.
    Atlanta, GA  30333

Computer Makes:

    IBM 370

Source Language:

    ALC/FORTRAN IV

Cost:

    Nominal

Distributed by:

    Same

Operating Systems:

    OS/MVS/VS2

## 4.13  OSIRIS 2.4

### INTRODUCTION

OSIRIS 2.4 is a generalised table generator including facilities such as data selection and derivation, data structuring into any number of tables, data presentation, and links to a data base for classifications and associated labels. It has been used in most of the applications of the Statistical Office of the European Communities and also for administrative work. It has been implemented by IREP, Grenoble University, France, for the Statistical Office of the European Communities, Luxemburg. It can be used in batch mode and is easily portable.

### CAPABILITIES:  Processing and Displaying Data

OSIRIS applications are written in a user-oriented, high level language accessible to statisticians. All descriptions can refer to an interactive Data Base for data types, classifications, labels and coefficients.

Input file description allows sequential files with fixed length records and classical data types and formats, extended by nomenclatures, i.e. stored lists of time qualified identifiers.

Data manipulation includes extensive selection facilities combining Boolean expressions, arithmetic expressions within a block-structured typed of description; use of arithmetic expressions to compute derived data; re-ordering and re-grouping of primary or derived values either through direct definition or through stored cross-classification time-qualified tables.

The table description allows cross-tabulation, concatenation and merging of breakdowns at any level. Several tables can be produced in one run. The lay-out facilities are: definition and positioning of titles, headings, labels with fixed or variable lengths, including use of stored labels associated to time-qualified identifiers. Tables can be printed by splitting in several pages and/or several files.

### CAPABILITIES:  Statistical Analysis

Basic statistics, including (weighted) means, (weighted) standard deviation, simple or crossed percentages, multi-level sub-totals and quantiles.

### EXTENSIBILITY

The user language allows the specification of new statistical procedures.

### INTERFACES WITH OTHER SYSTEMS

OSIRIS 2.4 can access files stored in the chronological Data Base of the Office.

## PROPOSED ADDITIONS IN NEXT YEAR

The system will undergo extensive development in the next 12 months:
extended capabilities for data editing and checking; provision of a catalogue
for application users including descriptions of data structures, data manipulations and tables; implementation of a general merge module to generate tables
from several ordered files referenced in the catalogue. The general aim of this
new release is to give users more flexibility to use the system in a more data
base oriented environment. Also system performances are under improvement.

## REFERENCES

Systeme OSIRIS (version 2.4) Manuel utilisateur, SOEC, Brussels, April 1980.

## SAMPLE JOB

Analysis of registration of new cars within EEC countries by country of
origin and capacity.  See output following page.

```
  SUR IMMATRIC
    PAYSDECL VENT OUVERT PAYSDECL //001,002,003,004,005,006,007,008,QCQ//;
    PAYSFABR VENT OUVERT PAYSFABR SUIVANT PAYS.NCP;
    ANNEE VENT ANNEE //1977,1976,1975,1974,1973,1972,1971,1970,1969,1968,
                   QCQ//;
    CYLINDRE VENT CYLINDRE //!<0,1500>!,!<1501,9999>!//;
    PRIX VENT PRIX //!<0,100000>!,!<100001,200000>!,!<200001,300000>!,
                   !<300001,9999999>!,QCQ//;
  NOMBRES ALLOUER NOMBRES;
  FIN
  IMMATRICULATION TABLEAU
1 PAYSDECL LIGNE HAUTEUR (ENTETE 6) CADRAGE (31 A 130)
  ENTETE ('OSEXEC2','ANALYSE DES IMMATRICULATIONS',
          'DES VOITURES NEUVES DANS LA CCE 1968 A 1977','PAGE',FOLIO,
          DATE,'PAYS DECLARANTE',//'ALLEMAGNE','FRANCE','ITALIA',
          'NEDERLAND','UEBL','UNITED KINGDOM','IRELAND','DANMARK',
          'CE9'//,'PAYS FABRIQUANTE')
  (2,30,59,105,110 'ZZ9',SAUT 1 2,7 'JJ/MM/AA',80,96,SAUT 3 2)
  INTERPOS PAGE
2 PAYSFABR LIGNE LIBEL (SUIVANT PAYS.NCP.(CODE,LIBNCPF))
  (2 A 4 REPETER,6 A 30 REPETER) INTERPOS 3 INTERCROIS 2
3 ANNEE LIGNE LIBEL (//'1977','1976','1975','1974','1973',
                        '1972','1971','1970','1969','1968','TOTALE'//)
  (10 A 15) INTERPOS 1 INTERCROIS 0
4 PRIX COLONNE LIBEL (//'MOINS 100000','100001 A 200000',
                        '200001 A 300000',
                        'PLUS 300000','TOTALE'//) (SAUT 4)
5 CYLINDRE COLONNE LIBEL (//'<1500 CC','>1500 CC'//) (SAUT 5)
6 SOMME NOMBRES P 8 'Z(6)9-';
```

OSEXEC2
DATE 06/11/PU

ANALYSE DES IMMATRICULATIONS DES VOITURES NEUVES DANS LA CEE 1968 A 1977     PAGE   1
PAYS DECLARANTE FRANCE

PAYS FABRIQUANTE

| | MOINS 10000 CC | | 10001 A 20000 CC | | 20001 A 30000 CC | | PLUS 30000 | | TOTALE | |
|---|---|---|---|---|---|---|---|---|---|---|
| | <1500 CC | >1500 CC | <1500 CC | >1500 CC | <1500 CC | >1500 CC | <1500 CC | >1500 CC | <1500 CC | >1500 CC |
| **001 FRANCE** | | | | | | | | | | |
| 1977 | c | 174 | 56 | 0 | 30 | 0 | 45 | 0 | 174 | 135 |
| 1976 | c | 174 | 56 | 0 | 30 | 0 | 45 | 0 | 174 | 135 |
| 1975 | c | 174 | 56 | 0 | 30 | 0 | 45 | 0 | 174 | 135 |
| 1974 | c | 0 | 90 | 0 | c | 0 | 45 | 0 | 174 | 135 |
| 1973 | c | 0 | 90 | 0 | 45 | 0 | c | 0 | 174 | 135 |
| 1972 | c | 0 | 90 | 0 | 45 | 0 | c | 0 | 174 | 135 |
| 1971 | c | 0 | 90 | 0 | 45 | 0 | c | 0 | 174 | 135 |
| 1970 | c | 0 | 90 | 0 | 45 | 0 | c | 0 | 174 | 135 |
| 1969 | c | 0 | 90 | 0 | 45 | 0 | c | 0 | 174 | 135 |
| 1968 | c | 0 | 90 | 0 | 45 | 0 | c | 0 | 174 | 135 |
| TOTALE | c | 522 | 798 | 0 | 727 | 0 | 225 | 0 | 1740 | 1350 |
| **002 BELG.-LUXBG.** | | | | | | | | | | |
| 1977 | c | 78 | 728 | 0 | 55 | 0 | 50 | 0 | 454 | 442 |
| 1976 | c | 78 | 728 | 0 | 55 | 0 | 50 | 0 | 454 | 442 |
| 1975 | c | 78 | 728 | 0 | 55 | 0 | 50 | 0 | 454 | 442 |
| 1974 | c | 0 | 787 | 0 | c | 0 | 50 | 0 | 454 | 442 |
| 1973 | c | 0 | 787 | 0 | 59 | 0 | c | 0 | 454 | 442 |
| 1972 | c | 0 | 787 | 0 | 59 | 0 | c | 0 | 454 | 442 |
| 1971 | c | 0 | 787 | 0 | 59 | 0 | c | 0 | 454 | 442 |
| 1970 | c | 0 | 787 | 0 | 59 | 0 | c | 0 | 454 | 442 |
| 1969 | c | 0 | 787 | 0 | 59 | 0 | c | 0 | 454 | 442 |
| 1968 | c | 0 | 787 | 0 | 59 | 0 | c | 0 | 454 | 442 |
| TOTALE | c | 234 | 3665 | 0 | 460 | 0 | 295 | 0 | 4540 | 4420 |
| **003 PAYS-BAS** | | | | | | | | | | |
| 1977 | c | 55 | 22 | 0 | 19 | 0 | 3 | 0 | 314 | 39 |
| 1976 | c | 55 | 22 | 0 | 19 | 0 | 3 | 0 | 314 | 39 |
| 1975 | c | 55 | 22 | 0 | 19 | 0 | 3 | 0 | 314 | 39 |
| 1974 | c | 0 | 36 | 0 | c | 0 | 3 | 0 | 314 | 39 |
| 1973 | c | 0 | 36 | 0 | c | 0 | c | 0 | 314 | 39 |
| 1972 | c | 0 | 36 | 0 | c | 0 | c | 0 | 314 | 39 |
| 1971 | c | 0 | 36 | 0 | c | 0 | c | 0 | 314 | 39 |
| 1970 | c | 0 | 36 | 0 | c | 0 | c | 0 | 314 | 39 |
| 1969 | c | 0 | 36 | 0 | c | 0 | c | 0 | 314 | 39 |
| 1968 | c | 0 | 36 | 0 | c | 0 | c | 0 | 314 | 39 |
| TOTALE | c | 165 | 318 | 0 | 57 | 0 | 15 | 0 | 3140 | 390 |
| **004 NF ALLEMAGNE** | | | | | | | | | | |
| 1977 | c | 0 | c | 0 | c | 0 | c | 0 | 67 | c |
| 1976 | c | 0 | c | 0 | c | 0 | c | 0 | 67 | c |
| 1975 | c | 0 | c | 0 | c | 0 | c | 0 | 67 | c |
| 1974 | c | 0 | c | 0 | c | 0 | c | 0 | 67 | c |
| 1973 | c | 0 | c | 0 | c | 0 | c | 0 | 67 | c |
| 1972 | c | 0 | c | 0 | c | 0 | c | 0 | | |

OSIRIS 2.4

Developed by:                                    Distributed by:

   Statistical Office of the                        Same
    European Communities
   UCL
   120 rue de la Loi
   1049 Bruxelles, Belgium

Computer Makes:                                  Operating Systems:

   IBM 370                                          OS/VS1
   ICL 2900                                         VME/B

                                      Interfaced Systems:

                                      SABINE, CRONOS

                                      User Language:

                                      French

Source Languages:

   PASCAL

Cost:

   Price of material and distribution.

Documentation:

Systeme OSIRIS (version 2.4) Manuel utilisateur,
SOEC, Brussels, April 1980.

## 4.14  GENERALIZED SUMMARY

### INTRODUCTION

The primary purpose of the Generalized Summary System is to summarize
detail data for non-probability surveys in the Statistical Reporting Service of
the United States Department of Agriculture.  Allowing the user maximum flexibility with matters such as data definitions and printed output were necessary.

Development of the Generalized Summary System began in December 1971 when
the general specifications were set down by members of the Methods Staff,
Livestock Branch and the Nebraska SSO.  System design was completed by February,
1972, and implementation on monthly Cattle on Feed, Poultry and Dairy was completed April 1, 1972 for Nebraska's surveys.

### CAPABILITIES:  Processing and Displaying Data

The Generalized Summary System will summarize data of any definition or
description as long as only two levels of summary totals are sufficient.  The
primary use of the system is for non-probability survey summaries where the
user has maximum control of input, computation, and output specifications.  It
cannot provide individual expansions necessary at the reporter (PSU) level, but
it can be used to expand data at the summary totals level.  Input media requirements are binary, fixed length records using floating point data fields.

The System will perform the following functions:

1.  Sum all data for all operators on input media into two levels of summary
    totals (e.g., Stratum and State, or District and State, or State and
    National, etc.)

2.  Expand all the summary totals above by an expansion factor which is governed
    by the user through control cards.

3.  Accept weights or historic parameters from cards.

4.  Allow user to control the processing of various computations to create new
    summary combinations, weights, indicators, etc.

5.  Compute the weighted indicator for the highest level of summary, (e.g.,
    District totals are weighted into State weighted totals).

6.  Allow user to extract necessary totals and direct them onto printed summary
    tables.

7.  Layout of printed summary tables are controlled through parameter and header
    cards.

8.  Capture final summary totals to be recycled into later summary processing.

The system can accept data directly from the Generalized Edit System (see
Chapter 3.01).  This program is not applicable to general public usage at
present.

## GENERALIZED SUMMARY

Developed by:

Systems Branch
ESCS, Statistics
U.S. Department of Agriculture
Washington, D.C.  20250

Computer Makes:

IBM 360/370
UNIVAC 1108

Language:

COBOL

Cost:

Not currently available.

Additions next 12 months:

None anticipated.

## 4.15 TPL

### INTRODUCTION

TPL is a computer system designed to select, restructure, crosstabulate, and display statistical data. The system was designed to solve problems connected with the collection of large volumes of data where crosstabulations must be produced in a wide variety of formats. The primary goal of TPL is to provide the individuals who need tables with the option of obtaining these tables directly. The syntax of the language allows the user to construct expressions representing the desired tables and auxiliary operations. The expressions are nonprocedural, compact, and written in English-like nomenclature. Users need not have experience with conventional computer programming languages.

### CAPABILITIES:  Processing and Displaying Data

There are two steps in table production when using TPL. First, the various types of data must be named and their lengths and acceptable values specified. These entries collectively are called a codebook. Codebook preparation is a one-time activity, after which many TPL users can reference the file for producing tables. Data to be tabulated may be stored in a wide variety of formats including hierarchical files. The second step uses TPL statements to reference codebook variables in producing tables. A TPL request indicates which data is to be tabulated and how it is to be formatted. The system can use arithmetic operations to calculate new variables and can delete, reorder, and regroup old variable values.

The Print Control Language (PCL) is an extension of TPL that allows the default table structures produced by TPL to be overridden. Through the use of PCL, extensive table formatting changes can be made without reprocessing the data. These formatting options include varying column widths, relabelling row and column titles, and adding footnotes. The phototype setting option, a major feature of PCL, prepares tables directly for publication without the need for manual typesetting.

### CAPABILITIES:  Statistical Analysis

Statements are available to perform a wide variety of computations, including averages, medians, minima, maxima, quantiles, relative time, and percentages.

### INTERFACES WITH OTHER SYSTEMS

The cell data can be reformatted using a special TPL to SAS procedure so that table cell values can be processed by SAS. TPL has also been interfaced directly with the TOTAL Data Base Management System so that multiple files can be processed. The linkages between TOTAL files are described in an Associative Codebook so that the user need not be concerned with the details of multiple-file processing.

### PROPOSED ADDITIONS IN NEXT YEAR

Work is underway to interface TPL directly to the RAPID data base management system from Statistics Canada. RAPID is based on the relational data

model and uses the transposed form of data storage.  The TPL/RAPID interface
will operate in a manner much like the TPL/TOTAL interface, i.e. user's requests
may implicitly combine data from several relations (data files) according to
linkage rules previously  established in an Associative Codebook.

REFERENCES

Francis, I., and S.P. Sherman,  "A Comparison of Software for Tabulating Survey
    and Census Data", <u>Proceedings of the American Statistical Association</u>,
    Statistical Computing Section, August, 1979.
Mendelssohn, R.C.,  "Major Enhancements to Table Producing Language,"  <u>Statis-
    tical Reporter</u>, No. 78-8, (May 1978).
Stitt, F.W., "Software for Data Management and Analysis of Large-Scale Biomedi-
    cal Studies", <u>Bulletin of the International Statistical Institute</u>, 48,
    1979.

SAMPLE JOB

```
USE ASA_DEMO CODEBOOK;
COMPUTE LABEL_DUMMY = 0;
DEFINE TOTAL_SEX ON SEX;
   /'Total Male' IF 0;
   /'Total Female' IF 1;
DEFINE AGE_GROUPS ON AGE;
   '14 to 19 Years' IF 14:19;   '20 to 24 Years' IF 20:24;
   '25 to 29 Years' IF 25:29;   '30 to 34 Years' IF 30:34;
   '35 to 39 Years' IF 35:39;   '40 to 44 Years' IF 40:44;
   '45 to 49 Years' IF 45:49;   '50 to 54 Years' IF 50:54;
   '55 to 59 Years' IF 55:59;   '60 to 64 Years' IF 60:64;
   '65 Years or Older' IF OTHER;
DEFINE EMPLOYED_CIVILIAN_LABEL ON LABEL_DUMMY;
   'Employed Civilians 14 Years or Older' IF ALL;
DEFINE WITH_JOB_NOT_WORKING ON EMPLOYMENT_STATUS;
     'With Job Not at Work' IF 2;
DEFINE AT_WORK_FILTER  ON  EMPLOYMENT_STATUS;
   'At Work by Usual Hours Worked per Week' IF 1;
DEFINE WORK_HOUR_GROUP ON HOURS_WORKED;
   '1 to 14 Hours' IF 0;   '15 to 29 Hours' IF 1;
   '30 to 34 Hours' IF 2;   '35 Hours' IF 3;
   '36 to 44 Hours' IF 4:5;  '45 to 59 Hours' IF 6;
   '60 Hours or More' IF 7;
TABLE T5 'Table 5. Hours Worked by Age and Sex, for Employed '
           'Civilians 14 Years or Older: 1980':
  TOTAL_SEX BY  AGE_GROUPS,
     EMPLOYED_CIVILIAN_LABEL THEN WITH_JOB_NOT_WORKING THEN
     AT_WORK_FILTER BY (TOTAL THEN WORK_HOUR_GROUP);
```

Output

Table 5. Hours Worked by Age and Sex, for Employed Civilians 14 Years or Older: 1980

| Age and Sex | Employed Civilians 14 Years or Older | With Job Not at Work | At Work by Usual Hours Worked per Week | | | | | | | |
|---|---|---|---|---|---|---|---|---|---|---|
| | | | Total | 1 to 14 Hours | 15 to 29 Hours | 30 to 34 Hours | 35 Hours | 36 to 44 Hours | 45 to 59 Hours | 60 Hours or More |
| Total Male | 47,970 | 1,257 | 46,713 | 2,100 | 2,576 | 2,440 | 2,269 | 28,419 | 4,549 | 4,360 |
| 14 to 19 Years | 3,461 | 82 | 3,379 | 944 | 893 | 305 | 125 | 970 | 92 | 50 |
| 20 to 24 Years | 4,877 | 105 | 4,772 | 316 | 432 | 293 | 237 | 2,878 | 373 | 243 |
| 25 to 29 Years | 5,660 | 119 | 5,541 | 94 | 152 | 254 | 276 | 3,639 | 636 | 490 |
| 30 to 34 Years | 4,886 | 104 | 4,782 | 64 | 113 | 181 | 236 | 3,052 | 581 | 555 |
| 35 to 39 Years | 4,869 | 109 | 4,760 | 44 | 105 | 192 | 242 | 3,000 | 592 | 587 |
| 40 to 44 Years | 5,157 | 124 | 5,033 | 61 | 105 | 219 | 199 | 3,206 | 600 | 643 |
| 45 to 49 Years | 5,357 | 116 | 5,241 | 65 | 105 | 224 | 223 | 3,484 | 555 | 585 |
| 50 to 54 Years | 4,722 | 160 | 4,562 | 66 | 94 | 215 | 218 | 3,012 | 476 | 481 |
| 55 to 59 Years | 3,978 | 144 | 3,834 | 60 | 116 | 202 | 205 | 2,521 | 356 | 374 |
| 60 to 64 Years | 2,909 | 111 | 2,798 | 94 | 144 | 173 | 174 | 1,797 | 189 | 227 |
| 65 Years or Older | 2,094 | 83 | 2,011 | 292 | 319 | 182 | 134 | 860 | 99 | 125 |
| Total Female | 29,258 | 1,071 | 28,187 | 2,574 | 3,845 | 2,485 | 3,195 | 14,851 | 646 | 591 |
| 14 to 19 Years | 2,517 | 69 | 2,448 | 767 | 566 | 162 | 161 | 756 | 26 | 10 |
| 20 to 24 Years | 4,325 | 149 | 4,176 | 286 | 432 | 313 | 529 | 2,493 | 87 | 36 |
| 25 to 29 Years | 2,894 | 111 | 2,783 | 188 | 309 | 290 | 328 | 1,572 | 53 | 43 |
| 30 to 34 Years | 2,477 | 94 | 2,383 | 173 | 324 | 205 | 248 | 1,318 | 64 | 51 |
| 35 to 39 Years | 2,636 | 90 | 2,546 | 220 | 388 | 245 | 286 | 1,293 | 61 | 53 |
| 40 to 44 Years | 3,111 | 117 | 2,994 | 189 | 418 | 260 | 339 | 1,628 | 76 | 84 |
| 45 to 49 Years | 3,238 | 121 | 3,117 | 161 | 420 | 299 | 357 | 1,715 | 78 | 87 |
| 50 to 54 Years | 2,901 | 105 | 2,796 | 164 | 329 | 254 | 341 | 1,571 | 58 | 79 |
| 55 to 59 Years | 2,454 | 84 | 2,370 | 150 | 257 | 207 | 315 | 1,306 | 73 | 62 |
| 60 to 64 Years | 1,587 | 71 | 1,516 | 101 | 187 | 143 | 193 | 794 | 46 | 52 |
| 65 Years or Older | 1,118 | 60 | 1,058 | 175 | 215 | 107 | 98 | 405 | 24 | 34 |

TPL: Table Producing Language, Version 5

Developed by:                                      Distributed by:

    U.S. Bureau of Labor Statistics      Same
    Division of General Systems
    441 G St. N.W.
    Washington, D.C. 20212

Computer Makes:                                    Operating Systems:

    IBM 360/370 or                IBM OS/MFT
      Compatible Hardware      MVT, VSI, MVS

                                   Interfaced Systems:

                                   SAS
                                   TOTAL Data Base Management System

Source Languages:   Primarily XPL, but also Assembler.

Cost:

    There is a charge of $300 for tape, installation instructions, and two sets of user manuals.

Documentation:

Mendelssohn, R.C.,  The Development and Uses of Table Producing Language.
TPL Version 5 Language Guide.
PCL Version 5 Language Guide.
TPL/PCL Version 5 Operations Guide.
(available from BLS)

## 4.16  RGSP

### INTRODUCTION

RGSP is a package for the tabulation of survey and other observational data of very varied types and has facilities for combining, condensing and otherwise manipulating tables from one or many batches of data.  It has been in use at agricultural research institutes in the U.K. since 1968, with various enhancements since then (F. Yates, J.D. Beasley, B.M. Church); over 20 agricultural institutes and the U.K. Overseas Development Administration are now linked to the Rothamsted computer.  It is also installed at 14 university and other computer centres.

### STRUCTURE

There are two main parts, which can be run independently or in sequence. Part 1 reads and validates the data and forms basic tables of totals and counts.  Part 2 provides facilities for arithmetic operations on tables, and for printing.  Communication between the two parts, and between different runs of Part 2, is by means of table files.

### PART 1:  CAPABILITIES

Part 1 comprises a compiler and a package of standard subroutines, called as required by a Fortran steering program written by the user.  This gives great flexibility for handling complex hierarchical structures, forming complicated derived variates, etc.  For straightforward jobs with up to 100 variates, including two-level hierarchies, no steering program is needed.

Data can be in any fixed format, including multiple punching, overpunching in the first column of a numerical field, and up to four missing-value symbols.  Validation facilities include sequence and associated reference checks, and checks on missing values, blank fields, invalid punching, and transgression of assigned limits.  Faulty records, together with any associated records, may be segregated in a separate file for correction.  Erroneous values, if not corrected, are automatically excluded from tables.

Variates are referred to by number in the instructions.  In addition to the usual provisions for grouping, weighting, and conditional entries to tables, values for several variates or multiply punched non-exclusive responses may be entered in different classes of the same table.  An updating facility enables late returns and corrections for errors in the data to be incorporated without program alteration or retabulation of the original data.

### PART 2:  CAPABILITIES

Part 2 operations are controlled by a simple problem-oriented language which contains powerful instructions for modifying, splitting and combining tables, with provision for loops, subroutines, etc.

The printing facilities are designed to give compact and well-arranged tables with minimum specification, while permitting more detailed control of contents and format when required; unnecessary decimal digits are not printed. The user can thus examine the whole of his tabulated material before deciding what is to be presented in a final report, and in what form. Tables classified by the same variates can be printed in parallel columns, with sparse classes automatically grouped under 'Others'.

The system of table files permits surveys covering several regions or repeated at intervals to be analysed batch by batch as the data become available, with later summarisation. Standard errors for virtually any type of design (including hierarchical data) can be calculated by Part 2 table functions from Part 1 tables of sums of squares and products.

## INTERFACES WITH OTHER PROGRAMS

Interfaces with GLIM and GENSTAT are provided for fitting linear and generalised linear models to tables. Subroutines are provided to enable other programs to read and write RGSP table files.

## EXTENSIONS PLANNED DURING NEXT YEAR

A new beginner's guide and a manual of examples; versions for minicomputers; graphical presentation of tabular matter.

## REFERENCES

Francis, I., Sherman, S.P. and Heiberger, R.M. (1976), Languages and programs for tabulating data from surveys. Proc. Comp. Sci. Stat.: 9th Ann. Sym. Interface.
Payne, R.W. and Nelder, J.A. (1976), Data structures in statistical computing. Proc. 9th Int. Biometric Conf., Vol II, 191-208.
Republic of Seychelles (1978). 1977 Census report.

## SAMPLE JOB

(extract) illustrating Part 1 tabulation instructions and Part 2 table operations (CML = combine levels).

| Part 1 | Part 2 |
|---|---|
| LIMITS | PROCESSING |
| 1  1,10 | READ T1...T5 |
| 2  0,3 | T11 = CML T1 X1 1...3,1+...3,= |
| 3  0,20 | 4...7,4+...7,8...11  (11,2) |
| 4  0,20 | T21 = PCT  T11  (01) |
|  | T2 = T2/T3 |
|  | T4 = T4/T5 |
| TABLES | *T21,T2,T4 |
| (1,2) | |
| C | |
| 3,4 A | |

Output:

TABLE 21  TYPES OF WATER SUPPLY (PER CENT)

| | NOT STATED | TAP | STAND PIPE | RIVER, WELL | HOUSE-HOLDS |
|---|---|---|---|---|---|
| DISTRICT 1 | | | | | |
| NORTH | 0.2 | 57.1 | 23.1 | 19.6 | 1903 |
| CENTRAL | 0.2 | 59.1 | 30.8 | 9.9 | 1349 |
| SOUTH | 0.4 | 54.3 | 20.2 | 25.1 | 1395 |
| ALL | 0.2 | 56.9 | 24.5 | 18.4 | 4647 |
| | | | | | |
| DISTRICT 2 | | | | | |
| EAST | 0.1 | 51.8 | 15.8 | 32.3 | 1665 |
| SOUTH | 0.4 | 31.5 | 17.4 | 50.7 | 1485 |
| WEST | 0.1 | 37.4 | 31.7 | 30.7 | 1380 |
| NORTH | 0.1 | 50.4 | 21.9 | 27.6 | 1982 |
| ALL | 0.2 | 43.7 | 21.4 | 34.7 | 6512 |
| | | | | | |
| DISTRICT 3 | 0.1 | 38.5 | 27.6 | 33.8 | 968 |
| DISTRICT 4 | 0.0 | 21.7 | 44.3 | 34.0 | 415 |
| DISTRICT 5 | 0.8 | 5.7 | 88.5 | 4.9 | 122 |
| | | | | | |
| ALL | 0.2 | 47.0 | 24.4 | 28.4 | 12664 |

RGSP:   Rothamsted General Survey Program

Developed by:                                      Distributed by:

    F. Yates, J.D. Beasley                          RGSP Secretariat
      and B.M. Church                              Rothamsted Experimental Station
    Rothamsted Experimental Station                 Harpenden
    Harpenden                                       Herts, AL5 2JQ
    Herts, AL5 2JQ                                  U.K.
    U.K.

Computer Makes:                                    Interfaced Systems:

    CDC 6000 series                                 GLIM
      CYBER series                                 GENSTAT
    IBM 360, 370                                    General interface subroutines
    ICL 2900 series                                  provided
    ICL System 4
    ICL 1900

Source Languages:   Primarily FORTRAN, a few assembler routines.

Cost:

    The yearly license fee for RGSP is £375 for universities (nil for U.K. universities); £275 for U.K. government organisations; and £650 for all others. The fee includes program source modules, load and/or object modules (if requested) and test data and results, all on magnetic tape.  It also includes maintenance and one copy of the installation and user's manuals.

Documentation:

Yates, F.,  The Rothamsted General Survey Program, Part 1, 1980.
Yates, F.,  The Rothamsted General Survey Program, Part 2, 1980.
Aitken, L.D.,  Introductory Guide to RGSP, 1976, (To be revised).

## 4.17  NCHS-XTAB

### INTRODUCTION

NCHS-XTAB is a crosstabulation program written by Gretchen K. Jones of the National Center for Health Statistics of the United States Public Health Service. The program is not exported at present.

### CAPABILITIES:  Displaying Statistical Data

The program was designed to produce small tables using weighted data from complex sample surveys. It allows up to 10 plates for the filter variable, up to 25 rows for the stub variable, and up to 20 columns for the spread variable (more than 10 columns creates a second plate). The program produces weighted and unweighted aggregates and weighted means, rates, or percents. Page (if the filter is used), row, and column totals are produced automatically. The stub may be crossed, and subtotals on the major and minor variables are included. The spread may also be crossed, but no subtotals are shown. The program allows for exclusions from the numerator, denominator, or both, and for recoding in defining the filter, stub or spread.

NCHS-XTAB:  National Center for Health Statistics
Crosstabulation Program

Developed by:                              Distributed by:

    Gretchen K. Jones                      Same
    National Center for Health Services
    Room 2-28 Center Building
    3700 East-West Highway
    Hyattsville, MD  20782

Machine:                                   Operating Systems:

    IBM 360/370                            OS

Language:

    PL/I

Cost:

    No charge.

4.18  <u>SURVEY</u>

INTRODUCTION

SURVEY, an interactive program available through Tymshare, Inc., was developed for the following purposes:

* In-house access to a report-producing program

* Quicker turnaround time for tabular reports, especially for additional tables

* High-volume, low-complexity questionnaire analysis.

<u>SURVEY</u>

<u>Distributed by</u>:

(through Tymshare, Inc.)
Frank James
Crossley Surveys
14th Floor
909 3rd Ave.
New York, New York  10022

<u>Computer Makes</u>:

PDP/9

<u>Source Language</u>:

Unknown

<u>Cost</u>:

Contract with Tymshare, Inc.

4.19  <u>SAP</u>


<u>INTRODUCTION</u>

SAP is a package for tabulation and statistical analysis of survey data.
It was developed in the Australian Bureau of Statistics in 1971-76.  It has
also been used in other Australian government departments.

<u>CAPABILITIES:  Processing and Displaying Data</u>

SAP is designed to read sequential files and simple hierarchical files.
It can process one file at a time.  Using SAP control cards, variables can be
recoded, range edits performed and selected records omitted from any or all of
the tables.  A user supplied subroutine can be used to perform more complicated
editing and data manipulation while the file is being read in.  When the tables
have been formed in core, they can be manipulated further by using a user-
supplied subroutine.

SAP can produce one-way, two-way, three-way or four-way frequency or
weighted tables.  The maximum number of variables is 100.  The tables are
printed in a fixed format with table headings and variable names only.
Totalling is automatic.

<u>CAPABILITIES:  Statistical Analysis</u>

Mean, standard error of the mean and standard deviation for rows of the
tables; chi-square (with or without Yates correction) and correlation coefficient; split halves variances.

<u>EXTENSIBILITY</u>

User-supplied fortran subroutines can be used.

<u>INTERFACES WITH OTHER SYSTEMS</u>

None.

<u>PROPOSED ADDITIONS IN NEXT YEAR</u>

None.

## SAMPLE JOB

Production of frequency tables with range edits on variables 3(PAY) and
4(TAX).  The master card (MA) describes the input file and says how many other
cards of each type are to be read.  The variable cards (1 in column 1) describe
the position of the variables and the type of variable.

```
USER    MATILDAKELLY    SOME SURVEY      CHECKTABLES

MA  2  6  6  0  0  2  000  60  80    0  80  1999

1 AGE   1    1U      0      5   4
1 YRS   2    IC      0      9   4
1 PAY   3    FUR     0     17   3
1 TAX   4    FUT     0     25   8
1 JOB   5    FU      0     33   8
1 LJOB  6    FC      0     41   8
3   4   0    4
3   3   0   11
7TAB1   1    2      0  0   12   15
7TAB2   3    4      0  0   12    5
```

## Partial Output from the Sample Job

```
                                                             11/04/73
         PAGE    4                 S U R V E Y   A N A L Y S I S   P R O G R A M

TABLE   TAB2   PAY   X   TAX

                TAX    0   1   2   3   4   TOT

        PAY

          0            1   0   0   0   0    1
          1            0   2   0   0   0    2
          2            0   1   0   0   0    1
          6            0   0   0   1   0    1
         11            1   0   0   0   1    2

        TOTAL          2   3   0   1   1    7
```

Note:  When statistics such as mean, etc. are requested, they are printed to
the right of the table under the appropriate headings.

### SAP, Version 2.0:   Survey Analysis Package

Developed by:                              Distributed by:

    Australian Bureau of Statistics        Same
    PO Box 10
    Belconnen ACT 2616
    Australia

Computer Makes:                            Operating Systems:

    CDC 3500                              MASTER
    CDC 3200/3300                         MSOS
    ICL 472

Source Language:   Fortran

Cost:

    Users are charged for the cost of providing them with a magnetic tape.

Documentation:

Computer Service Centre 32/33/3500 System Manual No. 32 Survey Analysis Program
    (SAP) Master Version 2.0, 1976.

### CHAPTER 5

SURVEY VARIANCE-ESTIMATION PROGRAMS


CONTENTS

INTRODUCTION

       As noted in the Introduction to Chapter 4, the tabulation programs used by
government agencies to process survey data are poorly equipped to calculate
standard errors of population estimates.  The programs in this chapter compute
estimates and standard errors from survey data already edited and tabulated.
All but two, CLUSTERS and SUPERCARP, were written for various government agencies with specific types of surveys in mind.  CLUSTERS was also written with
a specific type of survey in mind, namely the type used in approximately forty
surveys conducted by the World Fertility Survey.  It is portable to several
small computers.

       SUPERCARP is the only program that is being actively maintained, developed,
and exported as a general-purpose program for survey variance estimation and
analytical calculations.

       The performance of CLUSTERS, SUPERCARP and several other variance-
estimation programs was compared in an experiment conducted by Francis and
Sedransk (1979).  Since that time both programs have undergone improvements.

       The ratings by the respective developers on selected items are displayed in
Table 5.1.  Figure 5.1 compares developers' and users' ratings on all items,
while Figure 5.2 summarizes developers' ratings (D-scores) and users' ratings
(U-scores) on several relevant attributes.  These are explained in detail in
Section 1.4 of Chapter 1.  Figure 5.2 also points to other chapters which des-
cribe programs with strong variance-estimation capabilities.

       The remainder of this chapter contains descriptions of these variance-
estimation programs in the format described in Table 1.5.

## TABLE 5.1: RATINGS BY DEVELOPERS ON SELECTED ITEMS

### (i) Capabilities

Usefulness Rating Key:
3 - high
2 - moderate
1 - modest
. - low

| | Complex Structures | File Management | Consistency Checks | Probabilistic Checks | Compute Tables | Print Tables | Multiple Regression | Anova/Linear Model | Linear Multivariate | Multi-way Tables | Nonparametric | Exploratory | Robust | Non-linear | Time Series | Econometric |
|---|---|---|---|---|---|---|---|---|---|---|---|---|---|---|---|---|
| | 11 | 14 | 18 | 19 | 24 | 25 | 30 | 31 | 32 | 33 | 36 | 37 | 38 | 39 | 35 | 41 |
| 5.01 HES VAR X-TAB | . | 1 | . | . | 2 | 1 | . | . | . | . | . | . | . | . | . | . |
| 5.02 SPLITHALF | 1 | 2 | . | . | 1 | 1 | . | . | . | . | . | . | . | . | . | . |
| 5.03 MULTI-FRAME | 1 | 2 | . | . | 2 | 1 | . | . | . | . | . | . | . | . | . | . |
| 5.04 MWD VARIANCE | 1 | 2 | . | . | 2 | . | 1 | . | . | . | . | . | . | . | . | . |
| 5.05 GSS EST. | . | 1 | . | . | 1 | 1 | . | . | . | . | . | . | . | . | . | . |
| 5.06 CLFS VAR-COVAR | . | . | . | . | . | . | . | . | . | . | . | . | . | . | . | . |
| 5.07 CESVP | . | . | . | . | . | . | . | . | . | . | . | . | . | . | . | . |
| 5.08 KEYFITZ | . | . | . | . | . | . | . | . | . | . | . | . | . | . | . | . |
| 5.09 CLUSTERS | . | . | . | . | . | . | . | . | . | . | . | . | . | . | . | . |
| 5.10 SUPER CARP | . | . | . | . | . | . | 1 | . | . | . | . | . | . | . | . | . |
| 5.11 GENVAR | 1 | . | . | . | . | . | . | . | . | . | . | . | . | . | . | . |

## FOR SURVEY VARIANCE-ESTIMATION PROGRAMS

|  | Survey Estimates | Survey Variances | Select Sample | Math Functions | Operations Research | (ii) User Interface | | | | | | | | | | | | |
|---|---|---|---|---|---|---|---|---|---|---|---|---|---|---|---|---|---|---|
|  |  |  |  |  |  | Availability | Installations | Computer Makes | Mini Version | Core Requirements | Batch/Interactive | Stat. Training | Computer Training | Language Simplicity | Documentation | User Convenience | Maintenance | Tested for Accuracy |
|  | 43 | 44 | 45 | 47 | 48 | 49 | 50 | 51 | 52 | 53 | 54 | 55 | 56 | 57 | 58 | 59 | 60 | 61 |
| HES VAR X-TAB | 2 | 2 | . | . | . | 2 | 1 | 1 | 1 | . | . | 3 | 3 | . | 1 | 3 | 2 | 1 |
| SPLITHALF | 2 | 2 | . | . | . | . | . | 1 | 1 | 2 | . | 1 | 2 | . | 1 | 2 | 2 | . |
| MULTI-FRAME | 2 | 2 | 2 | . | . | . | . | . | . | . | 3 | . | . | . | 1 | 2 | 3 | 1 |
| MWD VARIANCE | 1 | 1 | . | . | . | . | . | 1 | . | 2 | . | . | 1 | . | 1 | 2 | 2 | 1 |
| GSS EST. | 3 | 2 | . | . | . | . | . | . | . | 1 | . | 1 | 2 | . | 1 | 1 | 1 | 1 |
| CLFS VAR-COVAR | 3 | 3 | . | . | . | . | . | . | . | . | . | . | . | . | 1 | 3 | 2 | 1 |
| CESVP | 3 | 3 | . | . | . | . | . | . | . | . | . | . | . | 3 | . | 1 | 1 | 2 |
| KEYFITZ | 2 | 3 | . | . | . | . | . | 1 | . | . | . | 1 | 2 | 1 | 1 | 1 | 2 | 2 |
| CLUSTERS | 2 | 2 | . | . | . | 2 | 2 | 2 | 1 | 2 | . | . | 3 | . | 3 | 1 | 1 | 1 |
| SUPER CARP | 2 | 1 | . | . | . | 2 | 2 | 2 | 1 | . | . | . | 3 | . | 2 | 2 | 3 | 2 |
| GENVAR | 2 | 3 | . | . | . | 2 | 1 | 2 | 1 | 3 | . | . | 2 | . | . | 2 | . | 1 |

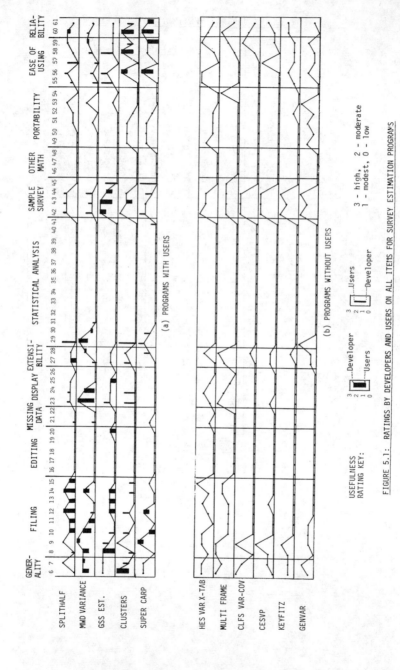

FIGURE 5.1:  RATINGS BY DEVELOPERS AND USERS ON ALL ITEMS FOR SURVEY ESTIMATION PROGRAMS

FIGURE 5.2 : SUMMARY D-SCORE AND U-SCORE FOR RELEVANT ATTRIBUTES

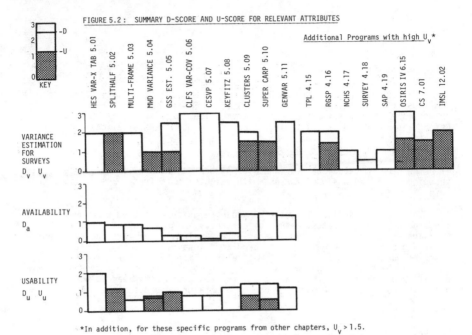

*In addition, for these specific programs from other chapters, $U_v > 1.5$.

## 5.01   HES VAR X-TAB

## INTRODUCTION

This program was written by Gretchen K. Jones of the National Center for Health Statistics (NCHS) of the United States Public Health Service. It is used to estimate variances for the Health Examination Survey (HES) of NCHS. It is being maintained and is available for export to other installations.

## CAPABILITIES:  Statistical Analysis

The program calculates variances using the balanced half-sample replication technique.  It was designed for use with weighted data from complex sample surveys.

Table output includes:  weighted and unweighted estimates of numerator and denominator, ratios (rates, means, or percents), standard errors, relative standard errors and relative variances of estimates, numerators and denominators, and standardized values.  The program allows for recoding, exclusions, labeling, and generates totals automatically.  It provides for a filter, stub (which may be nested), and spread (which also may be nested).  There is a maximum allowance for 10 filter plates, 25 stub rows, and 14 spread columns.  The programs are executed through the use of a catalogued procedure.

## REFERENCE :

Maurer, K., G. Jones, and E. Bryant, "Relative Computational Efficiency of the Linearized and Balanced Repeated Replication Procedures for Computing Sampling Variances," Proc. Section on Survey Research Methods, American Statistical Association, 1978.

### HES VAR-XTAB:   Health Examination Survey Variance and Crosstabulation Program

Developed by:                              Distributed by:

Gretchen K. Jones                              Same
NCHS, Room 2-28
3700 East-West Highway
Hyattsville, MD  20782

Machine:                                   Operating System:

IBM 360/370                                    OS

Language:

PL/I

Cost:

No charge.

## 5.02   SPLITHALF


## INTRODUCTION

SPLITHALF is a procedure to be used as a part of SPSS 7.0 for calculating variances using the Split Halves Variance technique.  It was developed by the Australian Bureau of Statistics in 1975-77 and is used in current production systems.

## CAPABILITIES:  Processing and Displaying Data

The capabilities of SPLITHALF are similar to those of the SPSS CROSSTABS procedure.

## CAPABILITIES:  Statistical Analysis

It produces tables of estimates, standard errors and standard error percents.  The types of estimates that can be produced are estimates of level and ratio estimates.  The methods of estimation available are unbiased estimation, stratum by stratum ratio estimation (with or without post-stratification), and across stratum estimation (with or without post-stratification).

## EXTENSIBILITY

None.

## INTERFACES WITH OTHER SYSTEMS

As for SPSS

## PROPOSED ADDITIONS IN NEXT YEAR

None.

## SAMPLE JOB

Calculation of standard errors using across stratum ratio estimation with post-stratification.

SPSS Control Cards to define variables, subfiles, etc.

```
RUN SUBFILES        EACH
SPLITHALF           CLASSVARS = AGE(1,4), SEX(1,2), TSLV(1,2),
                                STIL5Y(1,2)/
                    CLASSTABS = AGE BY SEX BY TSLV BY STIL5Y/
                                AGE BY SEX BY TSLV
                    TALLYVARS = PERSONS/
                    STRATVAR  = STRA (10000,99999)/
                    VARGROUP  = VG/
                    ESTIMATION = RATIOAS BENCHMARK = PERSONS
                    POSTSTRATIFIED BY POSTSTRA (1,34)/
OPTIONS             1
FINISH
```

## Partial Output From The Sample Job

```
SSS1 ADULT STANDARD ERRORS
FILE SSLFSYS (CREATION DATE = 80/03/13.)
SUBFILE  NSWMET
```

S P L I T H A L F   T A B U L A T I O N   O F
     AGE                                    BY SEX
CONTROLLING FOR,.
     TSLV                                   VALUE,.      1
USING TALLY VARIABLE.,   PERSONS

| | | SEX | | |
|---|---|---|---|---|
| | ESTIMATE | | | |
| | SE | | | ROW |
| | SE PCT | | | TOTAL |
| AGE | | 1 | 2 | |
| | 1 | 95287. | 101743. | 197030. |
| | | 4050.39 | 3604.92 | 5606.57 |
| | | 4.25 | 3.54 | 2.85 |
| | 2 | 78949. | 92926. | 171875. |
| | | 3979.94 | 3727.02 | 5486.57 |
| | | 5.04 | 4.01 | 3.19 |
| | 3 | 170375. | 179904. | 350278. |
| | | 6244.48 | 5809.16 | 8835.01 |
| | | 3.67 | 3.23 | 2.52 |
| | 4 | 435701. | 497736. | 933436. |
| | | 9185.95 | 11052.66 | 17685.37 |
| | | 2.11 | 2.22 | 1.89 |
| COLUMN | | 780311. | 872308. | 1652619. |
| TOTAL | | .12E+05 | .14E+05 | .22E+05 |
| | | 1.59623 | 1.61155 | 1.35121 |

SPLITHALF for SPSS 7.0

Developed by:                                    Distributed by:

  Australian Bureau of Statistics        Same
  PO Box 10
  Belconnen A.C.T.  2616
  Australia

Computer Makes:                             Operating Systems:

  CDC CYBER 72                                   KRONOS 2.1.2 LEVEL 411

Source Languages:

  ANSI FORTRAN
  CYBER 72 COMPASS for writing to scratch files

Cost:

  Users are charged for the cost of providing them with a magnetic tape.

Documentation:

Computer Services Division CYBER /2 Reference Manual, 1979.

## 5.03  MULTI-FRAME


### INTRODUCTION

The Multi-Frame Summary (Enumerative Survey Summary) System was developed by the Statistical Reporting Service of the United States Department of Agriculture.  It was designed as a generalized system for summarizing at various levels multi-frame probability surveys after editing had been completed.  It can compute estimates and standard errors at each summary level as well as combined estimates.

This system can interface with the Generalized Edit System (see Chapter 3.01).  It is undergoing enhancements and maintenance.  It is not applicable to general public use at present.


### MULTI-FRAME SUMMARY

Developed by:                                    Distributed by:

   Systems Branch                                   Same
   ESCS Statistics
   U.S. Department of Agriculture
   Washington, D.C.  20250

Machine:                                         Operating System:

   UNIVAC 1108                                      CSTS

Language:

   FORTRAN

Cost:

   Materials

## 5.04 MWD VAR


## INTRODUCTION

The MWD VARIANCE Program is used to design and analyse stratified, single stage, single phase surveys utilising either unbiased estimation or stratum by stratum ratio estimation.  It was designed primarily for internal use in the Central Office of the Australian Bureau of Statistics.  It was written by the Statistical Services Branch for a CDC 3500 under MASTER and would not be portable to other machines.  When the CDC 3500 is replaced, this VARIANCE program will be replaced.

## CAPABILITIES:  Processing Data

The program can process files of any size by splitting them into smaller files (strata?), processing them separately, then merging the results.

## CAPABILITIES:  Statistical Analysis

The program performs two broad tasks:

(1)  The program may be used to generate (for each stratum)
    (a)   an estimate of the population variance,
    (b)   an estimate of the sampling variance,
    (c)   an unbiased estimate of the total if population numbers are supplied, otherwise, the sample total,
    (d)   a stratum by stratum ratio estimate if benchmark totals are supplied, otherwise an estimate of ratio.

Note that this task will be useful in two different phases of a survey:

(i)   in the estimation of population variances (or exact population variance calculations if a population file is being used) for survey design purposes,
(ii)  the estimation phase of the survey.

(2)  The program may be used to perform simple linear regressions and correlations at the stratum level.  The results are presented in the format of an ANOVA table.  The purpose of this task is to calculate correlations or estimates of correlations to determine whether ratio estimation based on these two variables is feasible.

## MWD VARIANCE PROGRAM

Developed by:                          Distributed by:

   Statistical Services Branch          Same
   Statistical Users Services Div.
   Australian Bureau of Statistics
   Box 10, Belconnen, A.C.T.  2616
   Australia

Machine:                               Operating System:

   CDC 3500                             MASTER

Language:

   FORTRAN

Cost:

   A one-time cost of copying the tape.

5.05 <u>GSS</u>

## INTRODUCTION

The GSS ESTIMATION program is part of the GENERALIZED SURVEYS SYSTEM (GSS) designed to process and analyze generalized surveys at the Australian Bureau of Statistics (ABS). It has not been used outside the ABS and it is not portable to other operating systems. The reasons behind the design and development of a generalized survey system are described in the reference below.

## CAPABILITIES: Statistical Analysis

GSS was designed to cater surveys with the following characteristics:

1. Surveys must be single stage but may have 1 to 6 phases.
2. Estimates may be obtained using unbiased or ratio estimation or both. Across stratum estimation can be done.
3. Selection of units may be done by using techniques of equal probability, probability proportional to a function of size, both with or without replacement of units.
4. The survey is defined by type of strata which can be retained or collapsed (but not omitted) at phases other than the first.

The estimation phase of GSS performs the following functions:

1. Initialisation of internal tables to control the operation of the program, including processing of control cards and extraction of data from the GSS control files which contain data and flags set in previous GSS runs.
2. Processing the input SELF (selection file) to select the required data and create an intermediate tape file containing benchmark and survey data.
3. Calculation of cell primitives for original table cells and for crossed or collapsed cells (if applicable).
4. Calculation of smoothing primitives and variances. If detection of outliers is requested, calculation of modified estimates.
5. Creation of an output file of estimation results (ESTF) and update of the SIZF file with estimates of level.

GSS ESTIMATION can calculate estimates of

a. Level and variance of the estimate
b. Movement over time of estimand and its variance
c. Difference, sum or product of two estimands and the covariance between the estimands.

It allows for exclusion of sample units by the user and allows for the possibility that a unit may have become defunct since selection.

## REFERENCE

Foreman, E.K., <u>et.al</u>., "Why General Survey Systems - Australian GSS Progress and Prospects", <u>Aust. J. of Statis</u>. 18, 21-36, 1976.

## GSS ESTIMATION: General Survey System, Version 1, Estimation

Developed by:                                        Distributed by:

    Australian Bureau of Statistics          Same
    PO Box 10
    Belconnen, A.C.T. 2616
    Australia

Machine:                                             Operating System:

    CDC 3500                                  MASTER

Language:

    FORTRAN
    Compass (CDC 3500)

Cost:

    Not applicable.

## 5.06  CLFS VAR-COV

### INTRODUCTION

The Variance-Covariance System was developed to provide estimates of the sampling variance of estimates derived from the Canadian Labour Force Survey (LFS) and to obtain additional measures related to the quality and the performance of the survey.  The system is operational in Statistics Canada and is not currently distributed.

### CAPABILITIES:  Analysis of Statistical Data

The design of this system is sufficiently flexible to be adapted for surveys related to the LFS, i.e., surveys which are selected from the LFS frame such as Consumer Income, and Expenditure Surveys, and Labour Force Supplementary surveys.  In addition to providing various quality measures, the system may be a useful tool in analytic and evaluation studies aimed at improving the efficiency of the LFS procedures.

The system can estimate totals, linear combination of totals, ratios and differences between ratios.  Three types of weighting of estimates (subweighted, weighted, and unweighted) are available.  These estimates are obtained through the ratio estimation technique.  Second order estimates include variance estimates of ratios, variance estimates of differences of ratios, non-interview ratios and their variances.  These second order estimates are computed by means of the Keyfitz method.  The design effects (binomial factors) are also estimated.  All these can be computed for specified strata or PSU's.

### CLFS VAR-COVAR:  Canadian Labour Force Survey Variance-Covariance System

Developed by:

    Barry Bogart
    (formerly at Statistics
    Canada)

Distributed by:

    Census and Household Survey Methods
      Division
    Statistics Canada
    6-C-2 Jean Talon Building
    Ottawa, Ontario
    Canada K1A 0T6

Machine:

    AMDAHL 470

Operating System:

    OS/VS

Language:

    COBOL
    PL/I

Cost:

    Not currently distributed.

5.07  <u>CESVP</u>


## INTRODUCTION

The Consumer Expenditure Survey Variance Program (CESVP) was written to compute variances of estimates derived from the United States Bureau of Labor Statistics' Consumer Expenditure Survey. This batch program has been used internally only, has little documentation, and requires a user to have had considerable training in order to be effective. Although it is not exported at present and has limited maintenance, the developer would make it available free of charge.

## CAPABILITIES:  Statistical Analysis

This program computes estimates from stratified sample designs with two or three replicates per PSU or stratum, using the technique of balanced half or third samples. It will compute estimates of totals, means, ratios; variances of means, totals, and ratios; coefficients of variation, correlations. It will perform certain significance tests.

CESVP:  Consumer Expenditure Survey Variance Program

Developed by:                                    Distributed by:

  Ronald Lambrecht and Cathy Dippo                 Cathy Dippo
  Office of Survey Design                           Office of Survey Design
  Bureau of Labor Statistics                        Bureau of Labor Statistics
  GAO Building, Room 2122                            GAO Building, Room 2122
  441 G Street, NW                                   441 G Street, NW
  Washington, D.C.  20212                            Washington, D.C.  20212

Computer Makes:                                  Operating Systems:

  IBM 360                                           MVS

Source Language:

  PL/1

Cost:

  No charge.

5.08  <u>KEYFITZ</u>


INTRODUCTION

The Keyfitz Variance Estimation System was written by Robert Jewett for use by the United States Bureau of the Census to compute variances of estimates derived from the Current Population Surveys (CPS) and Annual Housing Surveys (AHS).

This batch program has only been used internally and, although it is fully maintained by the Bureau, it would not be practical for this program to be distributed elsewhere.

CAPABILITIES:  Statistical Analysis

The Keyfitz Variance System is used to calculate variances for data from the Basic CPS questions, from Supplemental CPS questions, or from AHS questions. It actually consists of three programs called the estimate, gamma factor and variance programs.  The estimate program uses as input a file specially prepared elsewhere in the Census Bureau and produces four files to be used by the variance program.  The gamma factor program reads two input files and produces a file of factors to be used in one of the second stage ratio estimation variance formulas in the variance program.

The variance program calculates variances for a maximum of one hundred population characteristics.  The estimate and variance programs are complex, while the gamma factor program is rather simple and inexpensive.  Total cost for running these three programs is about $400.  The variance program produces variances for the various components of variance and for the different levels of estimation.  Also, covariances, relative covariances, variance totals for self-representing and non self-representing areas and by geographic regions, are produced.  From the data files produced by the variance program additional statistics can be calculated, including quantities produced in the variance program and quantities only derivable from more than one data set.  Quantities which have been calculated between data sets are variances of averages over time, covariances and correlations, and variances of means.

The variance estimation method currently used is a revised form of a paired difference method of estimation.  The expression for a CPS estimator is linearized using a first-order Taylor Series approximation.

KEYFITZ:  Keyfitz Variance Estimation System

Developed by:                               Distributed by:

  Robert Jewett                             Same
  Bureau of the Census
  Statistical Methods Division
  Room 3742
  Suitland, Maryland 20233

Computer Makes:                             Operating Systems:

  UNIVAC 1100                               EXEC 8

Source Languages:

  FORTRAN V

## 5.09 CLUSTERS

### INTRODUCTION

CLUSTERS is a program to compute sampling errors taking into account the clustering and stratification of the sample design of a survey.  It was developed by Vijay Verma and Mick Pearce with contributions by several other staff members of the World Fertility Survey Headquarters staff in 1976.  Apart from 30 odd institutions in third world countries participating in the World Fertility Survey project, CLUSTERS has been distributed to over 10 other sites.

### CAPABILITIES:  Processing and Displaying Data

The program reads data according to a user supplied FORTRAN format statement.  The data must be in the form of a "rectangular" sequential file with no non-numeric characters in the field being referenced.  Comprehensive recode instructions are available for creating derived variables and for defining subclasses of the data.  Results may optionally be output to tape or disk to enable summaries or differently formatted reports to be produced.

### CAPABILITIES:  Statistical Analysis

Sampling errors for descriptive statistics are computed:  proportions, means, percentages, ratios and the differences in these.  CLUSTERS does not handle double ratio or linear combinations of ratios (except differences), nor more complex analytical statistics.  Sampling errors are computed for all variables over each subclass (and automatically over the whole sample) in the specified set.  The sample subclasses may be specified in pairs, in which case the difference and its standard error for each pair is calculated.  Subclasses may overlap.  The entire set of subclass-variable calculations may be repeated for separate geographical regions within a survey universe though these areas may not overlap.  Samples may be weighted.  The program will handle many different sample designs.

In addition to standard errors, CLUSTERS outputs two derived statistics namely the Design Effect (DEFT) and the Rate of Homogeneity (ROH).  They provide the basis for generalizing the computed results to other variables and subclasses of the sample and also possibly to other sample designs.

### PROPOSED ADDITIONS IN NEXT YEAR

Some changes in the control language are envisaged during the next year to make the program easier to use.  No changes are envisaged for the statistical capabilities, except for corrections where necessary.

## REFERENCES

Francis, I., and J. Sedransk (1979). "A comparison of software for processing
    and analysing surveys", Proceedings of the 42nd Session of the Interna-
    tional Statistical Institute, Manila.
Kalton, G., (1977). "Practical methods for estimating survey sampling errors",
    Proceedings of the 41st session of the International Statistical Institute,
    New Delhi.
National Fertility Surveys. A large number of First Reports of surveys con-
    ducted under the World Fertility Survey Programme have used and referred
    to the CLUSTERS program.
Verma, V., (1978). "CLUSTERS: A package program for computation of sampling
    errors for clustered samples", Statistical Commission and Economic Com-
    mission for Europe. Conference of European Statisticians: Meeting on
    Problems Relating to Household Surveys, Geneva.
Verma, V., Scott, C., and O'Muircheartaigh C., (1980). "Sample Designs and
    sampling errors for the World Fertility Survey", J. Royal Statistical
    Society, A, 143 P+4.

## SAMPLE JOB

Compute sampling errors from a sample of 6810 women for two estimates:
the percentage of women currently married, and the mean number of children ever
born. Separate estimates for rural and urban women are required.

|    | Variables | Location | Codes | Notes |
|----|-----------|----------|-------|-------|
| 1. | Domain | 17-18 | | Geographical divisions |
| 2. | Stratum | 19-22 | | |
| 3. | PSU | 23-26 | | Primarily sampling units |
| 4. | Area | 27-30 | | Area units by which data sorted |
| 5. | Weight | 31-34 | | |
| 6. | Currently married (CURM) | 145-146 | 0=No 1=Yes | Cases with values >2 to be excluded |
| 7. | Children ever born (NCEB) | 435-436 | 0=24 | Cases with values >24 to be excluded |
| 8. | Type of residence | 643-644 | 1=Urban(URB) 2,3=Rural(RUR) | |

## Input Control Cards

```
TITL      SAMPLING ERRORS FOR PROPORTION CUR. MARR. AND MEAN NO. CHILD
FORM   1  (T17,I2,4I4,T145,I2,T435,I2,T643,I2)
PROB      8    1    2    0    0    4    0    3    0    2    0    5    1
FACT         .00100    1   99    0
CLAS    URB        1RUR      1
RECO    1    0    0    0    0    0    8    0    1    0    0    0    0
RECO    1    0    0    0    0    0    8    2    3    0    0    0    0
VARI  CURM   0    1    1    1    0    0    0    0    0    0    0    0
RECO    1    0    0    1    1    6    6    2    0  100    1    0    0
VARI  NCEB   0    2 199999    0    0    0    0    0    0    0    0    0
RECO    1    0    0    0    0    0    7    0   24    0    0    0    0
RECO    3   -1    1    0    0    0    7    0    0    0    0    0    0
```

Partial output from example

## SAMPLE DETAILS

|       | READ AREA | AREA CARD FIELDS | | | | USED | | | | WGT | NO. OBS. |
|-------|-----------|-----|-----|-----|-----|------|-----|-----|-----|---------|----------|
|       |           | PSU | STR | DOM | WGT | AREA | PSU | STR | DOM |         |          |
| AREA  | 1 | 0 | 0 | 0 | 0 | 1 | 1 | 1 | 0 | .44300 | 11 |
| AREA  | 2 | 0 | 0 | 0 | 0 | 2 | 2 | 1 | 0 | .44300 | 11 |
| AREA  | 3 | 0 | 0 | 0 | 0 | 3 | 4 | 2 | 0 | .44300 | 7 |
| AREA  | 4 | 0 | 0 | 0 | 0 | 4 | 4 | 2 | 0 | .44300 | 7 |
| AREA  | 5 | 0 | 0 | 0 | 0 | 5 | 5 | 3 | 0 | .44300 | 11 |
| AREA  | 6 | 0 | 0 | 0 | 0 | 6 | 6 | 3 | 0 | .44300 | 7 |
| AREA  | 7 | 0 | 0 | 0 | 0 | 7 | 7 | 4 | 0 | .44300 | 12 |
| AREA  | 601 | 0 | 0 | 0 | 0 | 601 | 601 | 301 | 0 | 1.37200 | 7 |
| AREA  | 602 | 0 | 0 | 0 | 0 | 602 | 602 | 301 | 0 | 1.37200 | 22 |
| AREA  | 603 | 0 | 0 | 0 | 0 | 603 | 603 | 302 | 0 | 1.37200 | 12 |
| AREA  | 604 | 0 | 0 | 0 | 0 | 604 | 604 | 302 | 0 | 1.37200 | 16 |
| AREA  | 605 | 0 | 0 | 0 | 0 | 605 | 605 | 303 | 0 | 1.37200 | 20 |
| AREA  | 606 | 0 | 0 | 0 | 0 | 606 | 606 | 303 | 0 | 1.37200 | 21 |

NUMBER OF OBSERVATIONS =      6810   NUMBER OF THESE WITH ERRORS =    0

### DOMAIN  0  TOTAL

| | R | SE | N | WN | SER | SD | DEFT | ROH | SE/R | R-2SE | R+2SE | B |
|------|--------|------|--------|--------|------|--------|-------|------|------|--------|--------|------|
| CURM | 90.459 | .376 | 6810.0 | 6810.0 | .356 | 29.381 | 1.056 | .011 | .004 | 89.707 | 91.211 | 11.2 |
| NCEB | 3.942  | .042 | 6810.0 | 6810.0 | .035 | 2.857  | 1.208 | .045 | .011 | 3.859  | 4.026  | 11.2 |

## SUBCLASS RESULTS

### DOMAIN  0  TOTAL

| | R | SE | N | WN | SER | SD | DEFT | ROH | SE/R | R-2SE | R+2SE | B | CV |
|-----|------|------|--------|--------|------|------|-------|------|------|------|------|------|------|
| URB | .184 | .006 | 6810.0 | 6810.0 | .005 | .388 | 1.179 | .038 | .030 | .173 | .195 | 11.2 | .032 |
| RUR | .816 | .006 | 6810.0 | 6810.0 | .005 | .388 | 1.179 | .038 | .007 | .805 | .827 | 11.2 | .018 |

## RESULTS FOR SUBCLASS URB

| | R | SE | N | WN | SER | SD | DEFT | ROH | SE/R | R-2SE | R+2SE | B |
|------|--------|------|--------|--------|------|--------|-------|-------|------|--------|--------|------|
| CURM | 90.793 | .661 | 1800.0 | 1253.5 | .682 | 28.921 | .969  | -.006 | .007 | 89.471 | 92.114 | 10.6 |
| NCEB | 3.610  | .074 | 1800.0 | 1253.5 | .061 | 2.587  | 1.208 | .048  | .020 | 3.463  | 3.757  | 10.6 |

## RESULTS FOR SUBCLASS RUR

| | R | SE | N | WN | SER | SD | DEFT | ROH | SE/R | R-2SE | R+2SE | B |
|------|--------|------|--------|--------|------|--------|-------|------|------|--------|--------|------|
| CURM | 90.384 | .436 | 5010.0 | 5556.5 | .417 | 29.485 | 1.047 | .009 | .005 | 89.511 | 91.256 | 11.4 |
| NCEB | 4.017  | .048 | 5010.0 | 5556.5 | .041 | 2.909  | 1.175 | .056 | .012 | 3.921  | 4.114  | 11.4 |

## DIFFERENCES BETWEEN SUBCLASSES URB AND RUR

| | R | SE | N | WN | SER | SD | DEFT | ROH | SE/R | R-2SE | R+2SE | B |
|------|-------|------|--------|--------|------|--------|-------|-------|-------|--------|--------|------|
| CURM | .409  | .791 | 2648.5 | 2045.6 | .799 | 41.113 | .990  | -.002 | 1.934 | -1.173 | 1.992  | 11.0 |
| NCEB | -.407 | .088 | 2648.5 | 2045.6 | .074 | 3.784  | 1.197 | .043  | -.216 | -.583  | -.231  | 11.0 |

## CLUSTERS

**Developed by:**                          **Distributed by:**

  World Fertility Survey Headquarters    Same
  International Statistical Institute
  35-37 Grosvenor Gardens
  London SWIW OBS
  U.K.

**Computer Makes:**                        **Operating Systems:**

  IBM 360/370                           OS/DOS
  HP 3000                               MPE 11/111
  ICL 1900                              GEORGE 2/3
  ICL 2900
  CDC 6000                              SCOPE/KRONOS

**Core Requirement:**

  Approximately 50K bytes

**Source Languages:**

  FORTRAN

**Cost:**

    There is a one-time charge of $50 for universities, governments and non-profit organizations, and commercial use.

**Documentation:**

International Statistical Institute, Users' Manual for CLUSTERS, WFS/TECH 770, 1978.

## 5.10  SUPER CARP

### INTRODUCTION

SUPER CARP is a package for the analysis of survey data and (or) data observed subject to measurement error.  It is capable of estimating the sampling covariance matrices of totals, ratios, regression coefficients, and subpopulation means, totals and proportions for data collected in stratified multistage samples.

The package contains algorithms for estimating regression equations containing independent variables measured with error and an option for computing generalized least squares estimates for regression equations with nested error structure.

### CAPABILITIES:  Statistical Analysis

Compute the covariance matrices for totals, ratios, regression coefficients and subpopulation means, totals and proportions.  For subpopulations it is only necessary to enter the analysis variable and classification variable. The program automatically collapses a stratum with one PSU with the adjoining stratum.  The two-stage option automatically adjusts for PSUs with only one secondary unit.  The program has the capability to screen for missing observations, etc.

For regression equations, coefficients, t-statistics, $R^2$, and tests for groups of coefficients can be calculated.  Several options are available for equations with independent variables containing measurement error:  error variances known or estimated, reliabilities known or estimated, and error variances functionally related to the variables.  Tests for the singularity of the matrix of true values can be calculated.

The package is designed primarily for variance estimation and analytic calculations, is rather restrictive on input format and has little data management capability.  Input instructions are column coded.

### PROPOSED ADDITIONS IN NEXT YEAR

Options are being added to SUPER CARP to compute goodness of fit statistics of the chi-square type and tests of independence in two way tables for data collected in multistage stratified samples.  These will be available in the Fall of 1980.  Additional options for the errors-in-variables procedures will be added in the Fall of 1980.

### REFERENCES

Kaplan, B.A.,  "A Comparison of Methods and Programs for Computing Variances of Estimators from Computer Sample Surveys."  Unpublished M.S. Thesis, Cornell University, Ithaca, New York (1979).

Francis, I. and Sedransk, J.A.,  "Comparison of Software for Processing and Analyzing Surveys,"  Bulletin of the International Statistical Institute, 48, (1979).

## SAMPLE JOB

In this partial sample output the proportion of a variable Y1 in three
categories and the covariance matrix of the proportions are calculated.


EXAMPLE NO. 2        SUPER CARP   RELEASE 6-80     IOWA STATE UNIVERSITY



PROBLEM IDENTIFICATION--        EXAMPLE NO. 2


SAMPLE SIZE                        21

NUMBER OF VARIABLES                 3

INTERCEPT                          YES

DATA FORMAT(SURVEY TYPE)           FULL

NO. OF ANALYSES                     6

COLLAPSE WITH RATES

TWO STAGE WITH F.P.C.

DATA FORMAT USED:    (2I2,3F4.0,35X,F4.0)

Y1      =VAR( 1);    Y2      =VAR( 2);    CLASSIF=VAR( 3);    INTERCPT=VAR( 4);

DATA CONTAINS ONE-ELEMENT CLUSTERS

***STRATUM      5    FORMED FROM STRATA      4     5

***STRATUM      7    FORMED FROM STRATA      6     7


DATA LIST OPTION                FULL LST

EXAMPLE NO. 2          SUPER CARP   RELEASE 6-80     IOWA STATE UNIVERSITY


*** ANALYSIS  5

SUBPOPULATION PROPORTIONS

PROPORTIONS FOR VARIABLE Y1        ARE COMPUTED.


| CLASSIF | PROPORTION | STANDARD ERROR | T-STATISTIC |
|---------|------------|----------------|-------------|
| 1 | 6.0562879D-01 | 1.7399195D-01 | 3.4807862D 00 |
| 2 | 2.0983256D-01 | 1.0432103D-01 | 2.0114120D 00 |
| 3 | 1.8453865D-01 | 1.2095490D-01 | 1.5256815D 00 |


VARIANCE-COVARIANCE MATRIX OF PROPORTIONS

| 3.0273199D-02 | -1.3262994D-02 | -1.7010205D-02 |
|---------------|----------------|----------------|
| -1.3262994D-02 | 1.0882877D-02 | 2.3801171D-03 |
| -1.7010205D-02 | 2.3801171D-03 | 1.4630088D-02 |

Developed by:                          Distributed by:

    Department of Statistics              Same
    221 Snedecor
    Iowa State University
    Ames, Iowa 50010

Computer Makes:                        Operating Systems:

    IBM 360, 370                         IBM OS, TSO
    Others unknown

Source Languages:

    FORTRAN G

Cost:

    The cost for a tape containing examples, a tape listing and two manuals is $150.

Documentation:

Hidiroglou, M.A., Fuller, W.A., and Hickman, R.D., SUPER CARP, Iowa State
    University, Ames, Iowa, 1980.

## 5.11  GENVAR


### INTRODUCTION

GENVAR is a program for Generalized Variance Computation.  It was written
by Beverley Causey and Ralph Woodruff at the Statistical Research Division of
the United States Bureau of the Census.

### CAPABILITIES:  Statistical Analysis

As described in the reference below, GENVAR computes estimates of sampling
variances of non-linear functions of population sums from complex sample surveys, using a first-order Taylor series  expansion. The functions are supplied
by the user:  they must be twice-differentiable.  The within-stratum variance
can be calculated from an assumed simple random sampling design, or from a
user-supplied subroutine for any other design.

For each function the output includes the first-order estimate of the function, the estimated variance and standard deviation of this estimate, the
coefficient of variation, and estimates of the partial derivatives of the function.

This program is not actively maintained, and is not being further
developed.

### REFERENCE

Woodruff, R.S., and B.D. Causey, "Computerized Method for  Approximating the
    Variance of a Complicated Estimate,"  J. of Amer. Stat. Assoc., 71,
    315-321, 1976.

### GENVAR:  Generalized Variance Computation

Developed by:                                Distributed by:

    B. Causey                                    Same
    Statistical Research Division
    Bureau of Census
    Washington, D.C.   20233

Machine:

    IBM 360
    UNIVAC 1110

Language:

    FORTRAN

Cost:

    A one-time charge of $20.

CHAPTER 6


SURVEY ANALYSIS PROGRAMS


CONTENTS

INTRODUCTION

     The steps in the statistical "life cycle" which have been covered in pre-
vious chapters have been organizational or descriptive, not analytical.  Beginning with this chapter, and continuing through the rest of the book, we
encounter programs for building mathematical models to fit the observed data,
for making predictions about future data, or for making speculations or inferences about some underlying structure or process that produced the data.  We
call this simply "statistical analysis".

     All programs in this chapter have some capabilities for statistical analy-
sis, but in addition they virtually all have capabilities in data management,
editing, and tabulation.  These last three capabilities are all required in
processing data from large surveys, hence the title of this chapter.  In contrast, it can be seen from the taxonomy in Table 1.2 that while the programs in
Chapter 7 have uniformly strong capabilities for statistical analysis, many of
them lack strength in editing or data management.  Chapters 6 and 7 together
include all the well-known, statistical packages.

     The ratings by the respective developers on selected items are displayed
in Table 6.1.  Figure 6.1 compares developers' and users' ratings on all items,
while Figure 6.2 summarizes developers' ratings (D-scores) and users' ratings
(U-scores) on several relevant attributes.  These are explained in detail in
Section 1.4 of Chapter 1.  Figure 6.2 also points to other chapters which
describe programs with strong statistical analysis capabilities.

## TABLE 6.1: RATINGS BY DEVELOPERS ON SELECTED ITEMS

### (i) Capabilities

Usefulness Rating Key:
3 - high
2 - moderate
1 - modest
. - low

| | Complex Structures (11) | File Management (14) | Consistency Checks (18) | Probabilistic Checks (19) | Compute Tables (24) | Print Tables (25) | Multiple Regression (30) | Anova/Linear Model (31) | Linear Multivariate (32) | Multi-way Tables (33) | Other Multivariate (34) | Nonparametric (36) | Exploratory (37) | Robust (38) | Non-linear (39) | Bayesian (40) | Time Series (35) | Econometric (41) |
|---|---|---|---|---|---|---|---|---|---|---|---|---|---|---|---|---|---|---|
| 6.01 BTFSS | 3 | 3 | 1 | . | 3 | 1 | 1 | 1 | 1 | . | 1 | . | . | . | . | . | . | . |
| 6.02 EXPRESS | 3 | 3 | 3 | 1 | 3 | 3 | 2 | 1 | 1 | . | 3 | . | . | . | 1 | . | 2 | . |
| 6.03 EASYTRIEVE | 3 | 3 | 3 | 2 | 3 | 2 | 2 | 1 | 1 | 1 | 1 | 1 | 1 | 1 | 2 | 1 | 1 | 1 |
| 6.04 FOCUS | 3 | 3 | 2 | 2 | 3 | 3 | 3 | 2 | 2 | 1 | 1 | 1 | 1 | 1 | 1 | 1 | 1 | . |
| 6.05 SURVEYOR/SURVENT | 2 | 2 | 3 | 1 | 3 | 3 | 2 | 2 | 2 | 2 | 2 | 2 | 2 | 2 | . | . | . | . |
| 6.06 DATAPLOT | 2 | 2 | 2 | 3 | 3 | 1 | 2 | 1 | 1 | . | . | 1 | 3 | 2 | 3 | 2 | 3 | 1 |
| 6.07 PACKAGE X | 2 | 2 | 2 | 1 | 2 | . | 2 | 1 | . | 1 | . | 3 | . | . | . | . | . | 1 |
| 6.08 SCSS | 1 | 1 | 2 | 3 | 3 | 2 | 3 | 1 | 1 | 1 | 2 | 1 | 2 | 1 | . | . | . | . |
| 6.09 SPSS | 1 | 2 | . | . | 2 | 1 | 2 | 1 | 2 | . | 2 | 3 | . | . | . | . | . | . |
| 6.10 SOUPAC | 1 | 2 | 1 | . | 1 | . | 2 | 1 | 2 | 1 | 1 | 1 | . | . | . | . | 1 | 1 |
| 6.11 DATATEXT | 1 | 2 | 2 | . | 3 | 1 | 3 | 3 | 2 | 3 | 2 | 2 | . | 2 | . | . | 1 | . |
| 6.12 P-STAT 78 | 3 | 3 | 3 | 1 | 3 | 3 | 2 | 1 | 2 | 1 | 2 | 1 | . | . | . | . | . | . |
| 6.13 DAS | 3 | 3 | 2 | 2 | 3 | . | 3 | 1 | 3 | 2 | 3 | . | . | . | 1 | . | . | 1 |
| 6.14 SPMS | 3 | 3 | 2 | 2 | 1 | . | 2 | 1 | 2 | 1 | 2 | 3 | . | . | 2 | . | . | . |
| 6.15 OSIRIS IV | 3 | 3 | 3 | 1 | 3 | 2 | 2 | 2 | 2 | . | 3 | 3 | . | 1 | . | . | . | 1 |

## FOR SURVEY ANALYSIS PROGRAMS

### (ii) User Interface

| | Survey Estimates | Survey Variances | Select Sample | Math Functions | Operations Research | Availability | Installations | Computer Makes | Mini Version | Core Requirements | Batch/Interactive | Stat. Training | Computer Training | Language Simplicity | Documentation | User Convenience | Maintenance | Tested for Accuracy |
|---|---|---|---|---|---|---|---|---|---|---|---|---|---|---|---|---|---|---|
| | 43 | 44 | 45 | 47 | 48 | 49 | 50 | 51 | 52 | 53 | 54 | 55 | 56 | 57 | 58 | 59 | 60 | 61 |
| BTFSS | 1 | 1 | . | 2 | . | . | 1 | . | . | . | . | 3 | 3 | 2 | 1 | 2 | 1 | 1 |
| EXPRESS | 1 | 1 | . | 1 | 1 | 3 | 2 | 2 | 3 | . | 3 | 3 | 3 | 3 | 3 | 3 | 3 | 1 |
| EASYTRIEVE | 2 | 2 | 1 | 1 | 1 | 3 | 3 | 3 | . | 2 | . | 1 | 3 | 3 | 3 | 3 | 3 | 1 |
| FOCUS | 2 | 1 | 1 | . | . | 3 | 3 | . | . | . | 3 | 3 | 3 | 3 | 3 | 3 | 3 | 2 |
| SURVEYOR/SURVENT | 2 | . | 2 | . | . | 3 | 2 | 2 | 3 | 1 | 3 | 3 | 3 | 2 | 2 | 3 | 3 | 1 |
| DATAPLOT | 2 | 2 | 1 | 2 | 1 | . | . | 1 | 1 | . | 3 | 2 | 3 | 3 | 2 | 3 | 3 | 2 |
| PACKAGE X | . | . | . | . | . | 2 | 2 | 1 | 3 | 2 | 3 | 2 | 3 | 3 | 3 | 3 | 3 | 3 |
| SCSS | 1 | . | 1 | 1 | . | 3 | 3 | 3 | 0 | 1 | 2 | 2 | 3 | 3 | 3 | 3 | 3 | 2 |
| SPSS | 1 | . | . | . | . | 3 | 3 | 3 | 3 | 1 | . | 3 | 3 | 2 | 3 | 3 | 3 | 1 |
| SOUPAC | . | . | . | 3 | 1 | 2 | 3 | 2 | 1 | 1 | . | 1 | 2 | 1 | 2 | 2 | 2 | 1 |
| DATATEXT | . | . | . | . | . | 3 | 3 | 1 | 1 | . | . | 3 | 3 | 3 | 3 | 3 | 2 | 2 |
| P-STAT 78 | . | . | . | 2 | . | 3 | 3 | 3 | 3 | 1 | 3 | 2 | 3 | 2 | 2 | 3 | 3 | 2 |
| DAS | . | . | . | 3 | . | 2 | 1 | 1 | . | . | 3 | 2 | 2 | 3 | 3 | 3 | 3 | 2 |
| SPMS | 1 | 1 | 1 | 2 | . | . | . | 1 | 3 | 2 | . | 1 | 2 | 3 | 1 | 3 | 2 | 1 |
| OSIRIS IV | 3 | 3 | . | . | . | 3 | 1 | . | 1 | 1 | 3 | 2 | 3 | 2 | 2 | 3 | 3 | 1 |

The users of DAS gave it the second-equal highest rating on the $\Omega$-score in Table 13.1 which measures the degree of corroboration by users of the strengths claimed by developers. BTFSS, SPMS, EXPRESS, OSIRIS IV, and SPSS also received high $\Omega$-scores.

According to Figure 6.1 the users of SOUPAC think more highly of that pro-gram than does its developer, but the reverse is true for EASYTRIEVE, SURVEYOR/ SURVENT, DATAPLOT, and to a lesser extent DATATEXT.

The most widely-used of all statistical packages is SPSS, a Statistical Package for the Social Sciences. Yet despite its name its developer claims quite modest capabilities for statistical analysis, as do most developers in this chapter. The users of SPSS over-rate it on a number of items (see Figure 6.1), most noticeably on question 61: they were under the impression, rather trustingly, that the package had been extensively tested for accuracy, but the developer stated that this was not the case and claimed only that internal, unpublished tests had been performed.

The developer of SPSS over-estimated its ease of use as far as statistical and computing training are concerned (items 55 and 56): the developer felt that virtually no training in either area was needed, while the users agreed that a bachelor's degree in statistics was needed.

These differences may be due to inadequate documentation of options in the programs, lack of familiarity of the users with the program (although they were all suggested by the developers, and all expressed confidence in their ratings -- see Section 1.3.4), over-statements on the part of the developers, or some combination of these.

The remainder of this chapter contains descriptions of these survey analy-sis programs in the format described in Table 1.5.

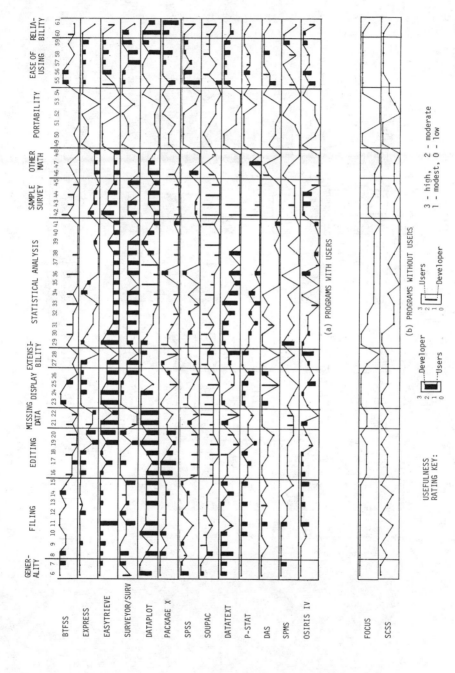

FIGURE 6.1: RATINGS BY DEVELOPERS AND USERS ON ALL ITEMS FOR SURVEY ANALYSIS PROGRAMS

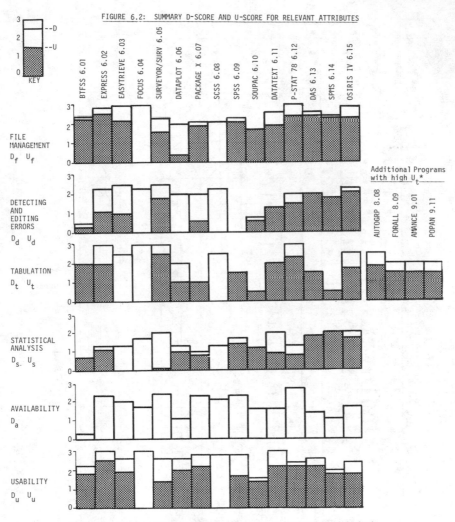

FIGURE 6.2: SUMMARY D-SCORE AND U-SCORE FOR RELEVANT ATTRIBUTES

Other chapters generally containing programs with strong Statistical Analysis
Capabilities are Chapters 7, 8, and 9 (see Figures 7.2, 8.2, and 9.2)

*In addition, for these specific programs, $U_t > 1.5$.

## 6.01   BTFSS -- PICKLE


### INTRODUCTION

PICKLE is a batch system for data management and general purpose statistical analysis.  It was originally developed by the Survey Research Center at the
University of California at Berkeley in the early 1970's and has been developed
and maintained by SRC since then.  PICKLE has been used primarily to analyze
and aggregate data.

### CAPABILITIES:  Data Processing and Display

The MS* routine reads in a standard case-ordered BCD data set and builds a
transposed file (ordered by variable) for later use.  This system file also
includes information on each stored variable such as type, width, missing data
values, and labels.  The advantages of the transposed system file are three-
fold:  new variables can be added at the end without rewriting the entire data
set, the whole data set need not be read by jobs which use only a subset of the
stored variables, and non-standard variables which do not have values for individual cases (such as correlation matrices) can be stored among the more usual
variables.

PICKLE's MFILE* command is used to link cases from different data sets.
Data can be aggregated or distributed from one data set to another.  The data
management procedures can be freely interspersed in a single job among the statistical procedures.  For instance, one might define a non-rectangular data
structure, perform arithmetic operations, read in new data, then go on to correlations and regressions.

PICKLE also contains a full set of variable construction and transformation
routines.  They include numerous ways of dealing with missing data on the input
variables.  PICKLE also includes a number of procedures that use multiple
response variables.

The TABLE* command is a 1-5 way cross tabulation program with optional
features including filters, weights, and bivariate statistics.

### CAPABILITIES:  Statistical Analysis

Common statistical procedures include univariate frequency distributions,
histograms, and statistics; bivariate crosstabulation and one way analysis of
variance; correlation analysis, multiple regression, and factor analysis.
Special capabilities include the computation of sampling variances and a matrix
algebra package.

### INTERFACES WITH OTHER SYSTEMS

PICKLE can write out SPSS system files.  OSIRIS dictionaries can be
converted to PICKLE file definitions.

## PROPOSED ADDITIONS IN NEXT YEAR

Due to changes in UCB's computer configuration, further development of PICKLE has been halted. A new and similar interactive system is being written for use on PDP 11s under UNIX (CSA-Conversational Survey Analysis).

## SAMPLE JOB

Generate a cross tabulation with vertical percentaging. The syntax is tree-field and only the first 2 characters of each keyword are required.

## Input Commands:

```
TA*
TV=(300,422)/ST=(VP)$
```

## Output:

```
TABLE    1              PAGE    1
VARIABLE DESCRIPTIONS:
                 NAME                                          USE

TV(      300)    CVL RIGTS PUSH TOO FAST         HORIZONTAL
TV(      422)    PTY R ALWAYS VOTED FOR          VERTICAL


TYPE OF PERCENTAGING IS    VERTICAL      (MINUS AR)
                  1       3       5       AR     TOTAL
                FAST    RIGHT   SLOW

1               149     246      72       25      492
        DEM     52.1    60.4    81.8     50.0     59.2

5               137     161      15       25      338
        REP     47.9    39.6    17.0     50.0     40.7

7                 0       0       1        0        1
        OTHER    0.      0.      1.1      0.       .1

AR              592     638      85      102     1417

TOTAL           878    1045     173      152     2248
                285     407      88       50      831
```

## BTFSS -- PICKLE, release 40:  Berkeley Transposed File Statistical System

Developed by:

  Survey Research Center
2538 Channing Way
University of California at Berkeley
Berkeley, California  94720

Computer Makes:

  CDC 6400,6600,7600

Operating Systems:

  CALIDOSCOPE, SCOPE

Interfaced Systems:

  SPSS

Source Languages:

  FORTRAN, COMPASS

Costs:

  The program has not been distributed to any other organization.

Documentation:

Baker, Margaret. User's Guide to the Berkeley Transposed File Statistical System, 2nd edition.  SRC Technical Report #1.  1976. $5.50.
Baker, Margaret and Lavender, George.  User's Guide to PICKLE Error Messages. SRC Technical Report #14.  1976.  $4.00.
Baker, Margaret.  User's Guide to PICKLE MFILE* and SFILE*.  SRC. 1978.Approx. $5.00 in xerox form.

## 6.02  EXPRESS

### INTRODUCTION

EXPRESS is an interactive data management and analysis system used to support marketing, financial, and general business decision making.  It combines graphics, report generation, data management, and statistical capabilities within one system.  EXPRESS has been developed and extended over the past ten years - it is now used by over 200 client companies.

### CAPABILITIES:  Data Management and Display

EXPRESS manages a data base in one or both of two formats:

* traditional records and fields
* a multidimensional array form in which data is organized by one or more dimensions such as time, corporate division, or geographic location.  These dimensions or organizing concepts are called SUBSCRIPTS.  The user can limit a subscript to a subset of values for reporting or analysis; the user can also establish relationships among subscripts (e.g. MONTHs aggregate into YEARs or CORPORATE DIVISIONs into PRODUCT GROUPs) to control detail in displays or analyses.  The aggregation facilities - averages, medians, etc. - can use these relationships and are, therefore, quite flexible.

EXPRESS can read data from tape, cards, disk, or the terminal.  The "Data Reader" subsystem includes facilities for range checks, consistency checks, etc.  EXPRESS contains security features at the individual data item level.

EXPRESS contains three report generators:  1) TABLE produces tables for ad hoc analysis, 2) DISPLAY is a single-command report generator, and 3) the REPORT GENERATOR produces completely customized, presentation-quality reports.

EXPRESS's plotting package can produce line charts, scatter plots, bar charts, and pie plots.  EXPRESS supports a full range of plotting devices - pen plotters and CRT's, in color and in black-and-white.

### CAPABILITIES:  User Extensibility

EXPRESS contains two facilities which allow a user to add commands or functions:

* A programming language that allows a user to combine and save EXPRESS commands.  Such a user-written program is then used like a built-in EXPRESS command.
* A dynamic loader that allows the user to load and invoke FORTRAN or PL/1 procedures from EXPRESS.  In many cases, FORTRAN programs can be used unchanged from within EXPRESS.

### CAPABILITIES:  Analysis

EXPRESS's analytical features are intended primarily for ad hoc analysis in a business setting.  The calculation and use of simple tools like averages, totals, medians, smoothed series, arithmetic and Boolean expressions are part of basic EXPRESS.  Specialized analytical features are provided in statistical packages which are fully integrated with EXPRESS's data management and display features.  The major such packages are:

* exploratory data analysis
* regression and best subset regression
* ARIMA analysis
* trend and S-curve fitting
* linear programming
* seasonal adjustment
* factor and cluster analysis
* solution of systems of equations, including goalseeking and very large (500+) sets of equations.

EXPRESS is completely integrated.  From a single data base, it is possible to use all the features and packages in EXPRESS.

REFERENCES

The following systems were developed using EXPRESS:

Brinkerhoff, Herbert N.  "Evaluating Effectiveness of Marketing Expenditures - A Point of Comparison,"  Presentation at 1978 A.N.A. Advertising Research Workshop.
McNurlin, Barbara C. "Sunmark Industries," EDP Analyzer vol. 18 No. 5, May 1980.

SAMPLE JOB

Produce a stem-and-leaf plot of average December oil use in northeastern cities.  Then produce a scatter plot of this use versus city population.  In the plot, include only those cities with small or moderate use rates.

        USG.CMF is usage for each CITY, MONTH, and FUEL
        POP.CY is population for each CITY and YEAR

→ limit fuel to oil
→ limit month to dec71, dec72, dec73, dec74, dec75
→ set avguse average (usg.cmf, city)
→ stemleaf avguse
→ limit city to avguse lt lm
→ plot scatter average (pop.cy, city) avguse vsize 30

Output:
```
    ->LIMIT FUEL TO OIL
    ->LIMIT MONTH TO DEC71, DEC72, DEC73, DEC74, DEC75
    ->SET AVGUSE AVERAGE(USG.CMF, CITY)
    ->STEMLEAF AVGUSE

            AVGUSE

    STEM-AND-LEAF DISPLAY,    OBSERVATION COUNT =  22

       (LEAF UNIT = 1000   SO 1 | 2 REPRESENTS 12000)

         4       2 | 0146
         8       3 | 0779
         8       4 |
         9       5 | 8
        (4)      6 | 0133
         9       7 | 38
         7       8 | 7
         6       9 |
         6      10 | 16
         4      11 | 7
         3      12 |
         3      13 |
         3      14 | 1

             HI | 1180, 4734

    ->LIMIT CITY TO AVGUSE LT 1M
    ->PLOT SCATTER AVERAGE(POP.CY, CITY) AVGUSE VSIZE 30
```

AVERAGE(POP.CY, CITY)
AVGUSE

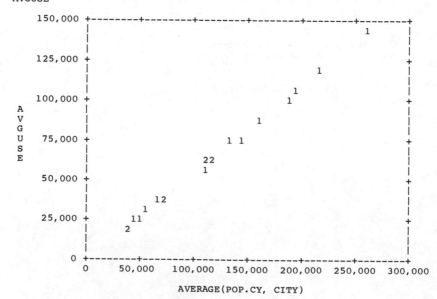

AVERAGE(POP.CY, CITY)

EXPRESS

Developed by:                              Distributed by:

  Management Decision Systems, Inc.          MDS as well as Tymshare, Inc.
  (MDS)                                      (Tymshare has offices throughout
  200 Fifth Avenue                           the US and Europe)
  Waltham, MA  02254
  Attn:  Walt Lankau

Computer Models:                           Operating Systems:

  IBM 360/370, 303x, 43xx                    IBM - VM/370, CMS
  Prime 400, 500, series 50                  Prime - PRIMOS

                                           Interfaced Systems:

                                             Fortran
                                             PL/1 (Prime only)

Source Languages:  AED, FORTRAN, limited assembler

Cost:

  EXPRESS is available on a time-shared bases from MDS or from TYMSHARE, Inc.
It can also be installed on a client's machine, either on a lease or perpetual
license basis.  A perpetual license costs $180,000 to $430,000 depending on the
features selected.

Documentation:

  Documentation includes both introductory and reference materials.

## 6.03 EASYTRIEVE

### INTRODUCTION

EASYTRIEVE is a systems software tool for file maintenance, information retrieval, and report writing. It can be used by non-systems people, and can retrieve any kind of record from any file structure, even complex data base management systems. It has over two thousand users in more than thirty countries, and is used by data processing staff, auditors, personnel managers, university financial aid officers, marketing managers, etc.

An EASYTRIEVE program, written in an English-like language, can use a few key words to call data and format a report. EASYTRIEVE programs are run in batch, and multiple jobs can be batched in a single program. EASYTRIEVE can also be run in a completely interactive mode using on-line systems such as TSO/SPF, CMS, ETSS II, ICCF, etc.

### CAPABILITIES: Processing and Displaying Data

Information retrieval: extracts all types of data from any files: sequential, ISAM, VSAM, or data base such as IMS, DLI, TOTAL, IDMS, etc; accesses data of fixed, variable, undefined, or spanned record formats; randomly accesses ISAM and VSAM files; randomly positions the start of sequential access in an indexed file; accesses variable location fields; defines nth dimensional arrays.

File Maintenance: allows an unlimited number of input and output files; creates and updates files; matches and merges files; adds, deletes, and reformats records; provides audit trail for file updates.

Information Analysis: selects data based on input, logic, and calculations; compares files, matches, and merges; provides conditional logic (IF, THEN, ELSE, GO TO), provides calculation capabilities; calls user programs and subroutines by use of standard linkage conventions; does table lookups; performs special tests (Alphabetic, Numeric, Blanks, Test under Mask, Sort Status; etc.); sorts on up to ten keys; control breaks on up to nine keys.

Reporting: produces unlimited reports from a single pass of the data; automatically formats reports (Titles, Column Headers, Centering, Spacing, Date and Page Numbers, Subtotal and Total Labels); provides customizing alternatives to all report format features; provides mailing labels of any size or format with one command; provides a user print command for outputting on any preprinted forms; provides a hexadecimal print command; produces summary reports and files; writes reports directly to microfiche tape.

### REFERENCES

Datapro and Auerbach reports.

EASYTRIEVE V8.1

Developed by:                               Distributed by:

    Ribek Corporation                       Pansophic Systems Inc.
    5600 Tamiami Trail North                709 Enterprise Drive
    Naples, Florida                         Oak Brook, IL  60521

Machine:                                    Operating Systems:

    IBM 360/370                             DOS and OS Systems
    UNIVAC                                  VS9
    SEIMENS                                 BS1000, BS2000
    FUJITSU
    HITACHI
    ITEL

Language:

    IBM ALC

Cost:

    One-time charge:  OS - $18,500; DOS - $14,500.  Other options available at
additional charge.

Documentation:

    EASYTRIEVE Reference Manual.

6.04  <u>FOCUS</u>

INTRODUCTION

    FOCUS/ANALYSE is a complete English language, non-procedural, Data Base Managment System, with its own report generator, graphics, and self-contained statistical analysis system.

<u>CAPABILITIES:  Data Management and Report Generation</u>

    The system is intended to be a high productivity tool for the manipulation of simple and complex files and their analysis with a moderately complex set of statistical analysis facilities.  This provides a superior environment to other statistical systems in that the data can be managed with full database convenience.  Little training is required to become effective in the use of the report generator which is employed for data selection and transgeneration.

<u>CAPABILITIES:  Statistical Analysis</u>

    Statistical analysis procedures include:  regression (multiple linear, step-wise, polynomial), descriptive statistics (deciles, quartiles, kurtosis, etc.), cross tabulation, analysis of variance, discriminant analysis, and factor analysis.  A new addition, TIMESER, provides a complete Time Series analysis capability including:  missing data, interpolation to create finer time periods, (e.g. months from quarters, etc.) all the standard lead, lag, and moving differences type functions, exponential smoothing, curve fitting and forecasting facilities (see terminal session on the following page.

<u>ADDITIONS IN NEXT 12 MONTHS</u>

    The following additions are planned during the next 12 months.

*    Enhanced modeling facilities
*    Analysis modules per user requests

## SAMPLE SESSION

```
ENTER COMMAND ( OR ? FOR HELP) ->fit hours
DO YOU WISH TO KEEP PREDICTED VALUES (TYPE "KEEP" OR "NOKEEP") ->keep
HOW MANY PERIODS DO YOU WISH TO EXTRAPOLATE ->6
DO YOU WISH TO KEEP RESIDUALS (TYPE "RESID" OR "NORESID") ->noresid
ENTER THE TYPE(S) OF EQUATION YOU WISH TO FIT ->exp power

NOTE:  TIME VARIABLE HAS DATE FORMAT; THUS FITTING IS AGAINST T.INDEX


EXPOTENTIAL:        Y =    67.31474  *  EXP(  .0684711 * T)

PREDICTED IN EF.HOURS

MEAN ABS ERR  = 31.17258  MEAN SQ ERR = .0823967  MEAN PCT ERR = .1313373
CORRELATION   = .8553630   R-SQUARED   = .7316458  T-STATISTIC  = 7.744742
F-STATISTIC   = 59.98116   VARIANCE    = .0898874  ST ERR OF EST= .2998123
DURBIN-WATSON = 2.087940


POWER:              Y =  43.55373 * (T ** .5656801)

PREDICTED IN PF.HOURS

MEAN ABS ERR  = 40.00346  MEAN SQ ERR = .0955443  MEAN PCT ERR = 1.119661
CORRELATION   = .8299553   R-SQUARED   = .6888258  T-STATISTIC  = 6.978537
F-STATISTIC   = 48.69995   VARIANCE    = .1042302  ST ERR OF EST= .3228471
DURBIN-WATSON = 1.214321

 >>graph file stathold
 >write hours pf ef across month
 >end

   NUMBER OF RECORDS IN GRAPH=    30  PLOT POINTS=  30

GIVE <CR> WHEN READY>
```

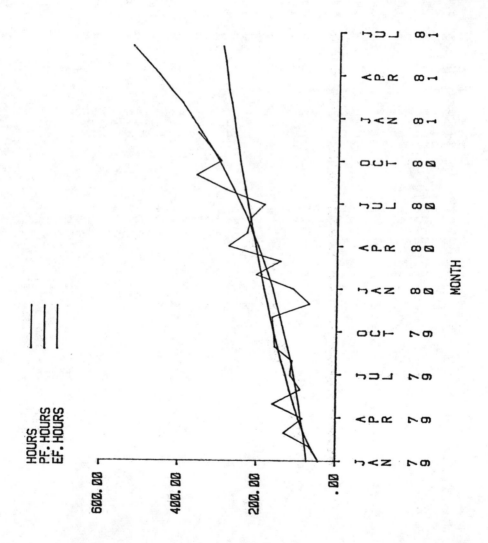

FOCUS/ANALYSE

Developed by:                                    Distributed by:

  Information Builders, Inc.                       same
  1250 Broadway
  New York, NY  10001

Machine:                                         Operating Systems

  IBM 370, 4300 series                             OS/TSO, CMS, CICS
                                                   DOS will be available in the future

Language:

  FORTRAN
  Assembler

Cost:

    For commercial use there is a License Purchase price of $43,000-$92,000
and a License Lease price of $1350-$2670/mo.  For non-profit and government use,
the price is the same as for commercial use.  For academic use,  there is a
20% discount.  Variation in the cost range depends on options requested.

                              6.05  SURVEYOR/SURVENT


INTRODUCTION

      SURVEYOR/SURVENT program was designed for use by survey researchers with
no computer experience in a business research environment.  It can be used
interchangeably in the batch or interactive mode, from a CRT/Teletype, RJE, or
on large processors.  Planned additions include extensions to the command
language, inclusion of free-form with fixed field commands, and ability to
accept defined as well as temporary variables.

CAPABILITIES:  Processing and Displaying Data

      The programs include:  MICROPUNCH, for punching data via CRT rather than
keypunching, SURVENT, for CRT telephone interviewing, SURVEYOR, for batch tabulation and editing, FASTAB, for interactive tabulation, and CLEANER, for interactive editing.  The system has a report generation ability.

                                SURVEYOR/SURVENT

Developed by:                            Distributed by:

    Computers for Marketing                  Same
    215 Market Street
    San Francisco, CA  94105

Computer Makes:                          Operating Systems:

    IBM 360/370                              OS
    Hewlett Packard 3000                     MPE
    UNIVAC 1100

Source Languages:

    FORTRAN
    PL/1
    SPL

Cost:

    $30,000 - purchase, $8,400 - annual lease.

6.06  DATAPLOT


INTRODUCTION

DATAPLOT is a high-level (free-format English-like syntax) language for: 1) graphics (continuous or discrete); 2) fitting (linear or non-linear); 3)general data analysis; 4) mathematics.

It was developed originally by James J. Filliben in 1977 in response to data analysis problems encountered at the National Bureau of Standards. It has subsequently been the most heavily-used interactive graphics and non-linear fitting language in use at NBS. It is a valuable tool not only for "raw" graphics, but also for manuscript preparation, modeling, data analysis, data summarization, and mathematical analysis. DATAPLOT may be run either in batch or interactively, although it was primarily designed for (and is most effectively used in) an interactive environment. DATAPLOT graphics may appear on many different types of output devices. Due to its modular design and underlying ANSI FORTRAN (PFORT) code, DATAPLOT is portable to a wide variety of computers.

CAPABILITIES: Processing and Displaying Data

All I/O is format-free. File names may be used directly in DATAPLOT READ/ WRITE command syntaxes. The many graphical, classical, and exploratory data analysis capabilities listed below serve as tools for the processing and editing of data files. Sorting and ranking capabilities exist.

Graphics capabilities include continuous display terminal plots (e.g., Tektronix), discrete (narrow-width or wide-carriage) terminal plots (e.g., TI 700), high-speed printer plots, high-quality secondary output plots (e.g., Calcomp); on-line interactive definition and plotting of functions; data plots; superimposed mixture of function and data plots; multi-trace plots; linear or log scale plots; plots with or without labels, titles, frames, tic marks, grid lines, legends, legend boxes, arrows, etc.; automatic hardcopying of plots; 3-d plots of functions and/or data; multicolored graphics; all of above for full data sets or subsets of data.

CAPABILITIES: Statistical Analysis

Fitting capabilities include interactive on-line model specification; fitting of linear, polynomial, multi-linear, and non-linear models; fitting may be linear/non-linear, weighted/unweighted, constrained/unconstrained; non-linear fitting without need of derivatives; pre-fit analyses for determination of non-linear fit starting values; exact rational function fitting; spline fitting; least squares smoothing; robust smoothing; automatic storage of predicted values/residuals from all fitting and smoothing operations; superimposed raw and predicted value plots; residual plots; fitting and smoothing over all data sets and subsets of data.

Graphical data analysis capabilities include box plots; complex demodulation plots; control charts; correlation plots; distributional frequency plots; histograms; lag plots; percent point plots; auto and cross periodograms; probability plots (24 distributions); probability plot corr. coef. dist. analysis plots (3 families); auto and cross spectral plots; scatter plots; pie charts; Youden plots; graphical ANOVA/ANOCOV; runs plots; 3-d dist. frequency plots; 3-d histograms; 4-plot per page univariate analysis; all of above for full data set or for subsets of data.

Classical statistical capabilities include elementary statistics (25 statistics); analysis of variance; tabulation of summary statistics; on-line definition and execution of functional transformations; cum. dist. functions (24 dist.); prob. density functions (24 dist.); percent point functions (24 dist.); random number generation (24 dist.); all operations may be over full data sets or subsets of data.

In addition, mathematical capabilities exist such as interactive on-line definition and concatenation of functions; functional analyses; exact analytic differentiation; root extraction; definite integration; convolution.

## EXTENSIBILITY AND INTERFACES WITH OTHER SYSTEMS

New statistics procedures are readily definable; macro capability exists.

DATAPLOT reads and writes files in a free-format fashion with the net result that they are typically directly linkable with other systems.

## PROPOSED ADDITIONS IN NEXT YEAR

Non-linear fitting for general norms; robust fitting; dynamic rotational 3-dimensional graphics.

## SAMPLE JOB

Carry out a non-linear fit on a Draper and Smith example. Read in data from a file DRAPER; carry out a non-linear fit; generate 3 plots: a superimposed plot of raw data and predicted values (with title and labels), a residual plot, and a normal probability plot of residuals.

```
COMMENT DRAPER AND SMITH NON-LINEAR EXAMPLE (PAGE 276)
.
READ DRAPER. Y X
.
LET ALPHA = .3
LET BETA = .02
FIT Y = ALPHA+(0.49-ALPHA)*EXP(-BETA*(X-8))
.
CHARACTERS X BLANK
LINES BLANK SOLID
TITLE DRAPER AND SMITH NON-LINEAR EXAMPLE (PAGE 276)
YLABEL PERCENT CHLORINE
XLABEL ELAPSED TIME IN WEEKS
PLOT Y PRED VERSUS X
.
YLABEL RESIDUALS
PLOT RES X

YLABEL
XLABEL NORMAL PROBABILITY PLOT
NORMAL PROBABILITY PLOT RES
```

## Output from the sample job:

```
LEAST SQUARES NON-LINEAR FIT
      SAMPLE SIZE N =        44
      MODEL-- Y = ALPHA+(0.49-ALPHA)*EXP(-BETA*(X-8))
      REPLICATION CASE
      REPLICATION STANDARD DEVIATION =        .9540735860-02
      REPLICATION DEGREES OF FREEDOM =         26
      NUMBER OF DISTINCT SUBSETS     =         18

ITERATION  CONVERGENCE  RESIDUAL  * PARAMETER
 NUMBER     MEASURE     STANDARD  * ESTIMATES
                        DEVIATION *
-----------------------------------*------------
   1--    .10000-01    .25031-01 *  .30000+00    .20000-01
   2--    .17086+00    .16816-01 *  .32435+00    .31817-01
   3--    .85430-01    .15643-01 *  .36357+00    .47300-01
   4--    .12814+00    .15293-01 *  .39489+00    .77453-01
   5--    .64072-01    .10973-01 *  .39133+00    .10049+00
   6--    .32036-01    .10913-01 *  .39015+00    .10166+0

      FINAL PARAMETER ESTIMATES      (APPROX. ST. DEV.)
      1 ALPHA           .390144     (  .5012-02)
      2 BETA            .101645     (  .1337-01)

      RESIDUAL    STANDARD DEVIATION =      .0109127270
      RESIDUAL    DEGREES OF FREEDOM =       42
      REPLICATION STANDARD DEVIATION =      .0095407359
      REPLICATION DEGREES OF FREEDOM =       26
      LACK OF FIT F RATIO =       1.8093 = THE  91.3253% POINT OF THE
      F DISTRIBUTION WITH     16 AND     26 DEGREES OF FREEDOM
```

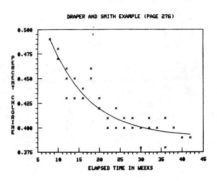

DRAPER AND SMITH EXAMPLE (PAGE 276)

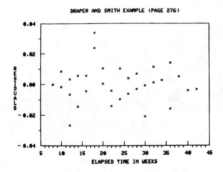

DRAPER AND SMITH EXAMPLE (PAGE 276)

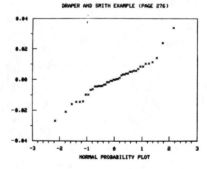

DRAPER AND SMITH EXAMPLE (PAGE 276)

DATAPLOT

Developed by:                                    Distributed by:

James J. Filliben                                Same
National Bureau of Standards
Statistical Engineering Laboratory
Administration Building A-337
Washington, D.C. 20234
Phone:  301-921-3651

Computer Makes:                                  Operating Systems:

Univac 1108                                      Exec 8

                                                 Interfaced Systems:

                                                 General

Source Languages:  ANSI FORTRAN/PFORT

Cost:

Although previously unavailable, DATAPLOT will be generally distributed as
of January 1, 1981.  Its underlying modular design and ANSI FORTRAN code are
design  features  which enhance portability to a wide variety of computers.
The one-time cost for DATAPLOT is approximately $800.  This includes source
code, load and segmentation modules, and test problems--all on magnetic tape.
It also includes implementation instructions and 1 copy of the various DATAPLOT
user manuals and guides.

Documentation:

Filliben, James J.  (1978).  "DATAPLOT--An Interactive System for Graphics,
    Fortran Function Evaluation, and Linear/Non-linear Fitting."  Proceedings
    of the Statistical Computing Section, American Statistical Association:
    1978.
Filliben, James J., (1979).  "Factors Affecting the Use of Statistical Graphical
    Software."  Proceedings of the Twelfth Interface Symposium on Computer
    Science and Statistics.  Toronto.
Filliben, James J.,  (1979),  "New Features in DATAPLOT--A Language for
    Graphics, Non-linear Fitting, Data Analysis, and Mathematics."  Pro-
    ceedings of the Statistical Computing Section,  American Statistical
    Association:  1979.
Filliben, James J.,  (1980).  " A Review of DATAPLOT--An Interactive High-Level
    Language for Graphics, Non-linear Fitting, Data Analysis, and Mathematics."
    Proceedings of the Statistical Computing Section,  American Statistical
    Association:  1980.

# 6.07  PACKAGE X

## INTRODUCTION

PACKAGE X is an integrated system for the management and analysis of data.
Developed originally by Dataskil Ltd. for the UK Government Statistical Services
it has been in regular use for many years.  Package X, now publically available,
is being used by a number of commercial organizations for applications varying
from the maintenance and analysis of personnel records or subscription lists to
the handling and analysis of industrial data.  Package X is fully supported by
an experienced maintenance and enhancement team.

Package X may be used conversationally, interactively, or in batch.  In
conversational mode the user is questioned in English by the system.  In interactive and batch modes the user uses an English-like control language.  The
package includes a programming language integrated with the system's data
handling and analysis facilities.  Programs written in this language may be
named and saved between runs.

## CAPABILITIES:  Processing and Displaying of Data

The Program language including arithmetic, output and IF statements, DO-
loops, and FOR-loops comparable in form and capability to those of Fortran and
BASIC can operate on data held in scalars, arrays and data matrices.  Missing
values are recognized.  The power of the system comes from the flexibility with
which the Program language facilities may iteract with the data management facilities for reorganizing and outputting data.  Data may be corrected and transformed and new sets of data may be formed by extraction from existing sets or by
amalgamation.  New data may be read as additions to existing data.

The system separates the operations of forming, testing and displaying
tables.  Tables with any number of dimensions may accumulate the value of a further variable as well as the usual frequencies.  Tests may be specified on two-
dimensional frequency tables.  Table display facilities include selecting,
combining or reordering rows and columns and the specification of fresh row and
column headings.  Transfer between tables and data matrices is a novel and
powerful facility.

Histogram display and plotting facilities using the terminal or printer
files are included.

## CAPABILITIES:  Statistical Analysis

Summaries include means, standard deviations, minimum and maximum values
and correlations whereas significance tests include the Normal-based tests, t,
F, chi-squared, Bartlett's, and some analysis of variance, as well as the
distribution-free tests, Wilcoxon/Mann-Whitney, David, Kruskal-Wallis and
Friedman.

Multiple regression, polynomial regression and stepwise regression are all
available.  The user at the terminal may elect to control the stepwise process.
Fitted values, residuals, coefficients of the fitted regression and values such
as the residual mean square may be transferred to variables, arrays and scalars
as appropriate.  Thus all these quantities can be made available for further
analysis.

The cases to be included in analyses and plots can be specified either by a list of case numbers or as a subset based on value selection. A subset defini- tion can be pre-declared and referenced by name. Normal rules for handling missing values apply.

## EXTENSIBILITY

Package X has full macro facilities, including editor, which together with the Program language allows many operations and analyses to be written by the user. For example, Almon regression has been provided by this means. The macro facility is also available to assist the user with his regular processing requirements.

## INTERFACES WITH OTHER SYSTEMS

Package X can read and write files in a variety of forms providing links with other applications software for survey analysis and statistics.

## PROPOSED ADDITIONS IN NEXT YEAR

Further data management facilities are planned including sorting but addi- tions will concentrate on extending text handling and labelling facilities and on fresh ways of presenting output of results of analyses and summaries.

## SAMPLE JOB

An interactive session is shown in which a table named SALLY is first defined and created. This is then displayed in its default form. The table is displayed again with both frequencies and accumulating variable and with two columns combined.

```
Command:  TABLE SALLY FORM AGE BY STATUS                        &
                  CATEGORIES AGE BOUNDARIES 25 40                &
                          STATUS MATCHES "MR" "MRS" "MISS"      &
          SUM SALARY


Response: TABLE SALLY IS NOW ACTIVE
```

Command:      TABLE DISPLAY

Response·   TABLE SALLY     FREQUENCIES

|              |        | --------- STATUS --------- | | |
|--------------|--------|-----|------|-------|
|              |        | !MR | !MRS | !MISS ! |
| ----- AGE ----- | | | | |
| ! FROM   UP TO <! | | | | |
|      | 25  | 2  | 1 | 5! |
| 25   | 40  | 11 | 7 | 4! |
| 40   |     | 9  | 1 | 0! |

Command:   TABLE DISPLAY FREQUENCIES AVERAGES              &
           INCLUDE STATUS (2 3) 1                          &
           LABELS STATUS "FEMALE" "MALE"                   &
           MARGINS AGE TOTALS PERCENTS                     &
                        STATUS TOTALS

Response:  TABLE SALLY    FREQUENCIES  AVERAGE  SALARY

|         |        | ------------ STATUS -------------- | | |
|---------|--------|--------|-------|--------|------|
|         |        | !FEMALE !MALE | !TOTALS | P.C. |
| ---- AGE ---- | | | | | |
| FROM    UP TO<! | | | | | |
|         | 25  | 6       | 2!      | 8  | 20.00 |
|         |     | !3433.33 | 3150.00! | | |
| 25      | 40  | 11      | 11!     | 22 | 55.00 |
|         |     | !4995.45 | 6640.91! | | |
| 40      |     | 1       | 9!      | 10 | 25.00 |
|         |     | !3000.00 | 8133.33! | | |
| TOTALS  |     | 18      | 22!     | 40 | |
| ACC TOTS |    | !78550.00 | 154750.0 | !233300.0 | |

## Package X   Version 5

Developed by:                                    Distributed by:

   Dataskil Ltd.                                    Dataskil Ltd.
   12-18 Crown Street                               Same address
   Reading
   Berkshire, RG1 2HP                      Owned by:
   England
                                     UK Government

Computer Makes:                                  Operating Systems:

   ICL 2900 Series                                  VME/B and DME
   ICL 1900 Series                                  G3,G2,G2+
   ICL 2903/2904                                    Any ETS or MTS Executive
   ICL ME29                                         TME

Source Language:

   Primarily Fortran with minimal assembler.  Some operating system dependencies for attaching files.

Cost:

   As at January 1981 first year licence fee is £6600.  Second and subsequent years licence fee is 20 per cent of the then current first year licence fee. licence fees are subject to review.  The fee includes 2 copies of the User Manual.  Installations receive enhanced versions as these become available.

Documentation:

   A comprehensive user manual is available.  A technical overview is also available, providing a concise description of the package.  Other documents describe the use of Package X for particular applications.  The HELP and INFormation pages are available on-line to assist the user at the terminal when required.

6.08  <u>SCSS</u>

INTRODUCTION

The SCSS Conversational System is an integrated, fully interactive statistical package for data analysis developed by SPSS Inc.  An elaborate series of
system prompts and pre-emptive commands enable the analyst to move around the
system at will.  SCSS is self documenting with a full range of help and tutorial
facilities to explain procedures.  In its fourth release, it is installed at
over 250 computer sites.  SCSS interfaces with SPSS and DEC System 1022.

CAPABILITIES:  File Facilities and Data Management

\* Data entry from the terminal, from raw data files, or from SPSS files;
transposed variable-oriented file structure for speed of access; transformed
variables may be passed from one workfile to another; no built-in limitation to
number of cases; no built-in limitation to number of variables; selection of
cases; weighting of cases.

\* Output of analysis to disk simultaneous with or instead of terminal output; output of raw data to disk; extensive variable and value labeling and missing value indicators; data transformations using arithmetic operators, logical
operators, and FORTRAN-like functions; recoding (revising) of variables permitted with simultaneous access to original values.

\* Facilities for collapsing or omitting categories during crosstabulation
and breakdown analysis; listing of values for all or selected cases; listing information for files, variables, matrices, and cases; listwise and pairwise deletion of missing data; instant notification of errors and opportunity for
corrections.

CAPABILITIES:  Statistical Analysis

\* Frequency tables, histograms, descriptive statistics, percentiles and
quantiles; crosstabulation, measures of association, and D-systems analysis;
tests of independent and paired-sample differences; description and testing of
subpopulation differences with automatic standardizing and rescaling.

\* Scatterplots, correlations, and casewise plots; partial correlation
analysis; Pearson product-moment correlation, with matrix storage facilities;
multiple linear regression analysis with analysis of residuals (forward selection, backward elimination, stepwise selection, forced entry, forced removal);
factor analysis.

SAMPLE JOB

The analyst enters regression through the CORRELATION procedure, requesting
default missing treatment (casewise deletion) and formatting.  When the system
asks how the analyst wants the correlation displayed, he requests the REGRESSION
procedure, choosing a stepwise regression of income on education, occupational
prestige, and sex.  After the system prints an initial display, the analyst asks
for summary statistics.  Finally, he requests an analysis of the residuals -
specifically a casewise plot of outliers, absolute value of the standardized
residual equal to or greater than 2.5, and a plot of the studentized residuals
against the studentized deleted residuals.

```
PROCEDURE?
corr rincome educ prestige sex
MISSING TREATMENT?
NUMBER OF CASES =  970

FORMAT?

DISPLAY?
/regression
...REGRESSION PROCESSING
SUMMARIES?
all
STATISTICS?
all
DEPENDENT VARIABLE?
rincome
READY TO BEGIN BUILDING EQUATION
DEPENDENT:   RINCOME - RESP INCOME FROM OWN JOB
EQUATION?
stepwise
```

| BK | MULTR | RSQ | ADJRSQ | F | SIGF | RSQCH | SIGCH | DEP: | RINCOME | BETAIN |
|----|-------|-----|--------|---|------|-------|-------|------|---------|--------|
| 1 | .3940 | .1552 | .1543 | 177.838 | .000 | .1552 | .000 | IN: | SEX | -.3940 |
| 2 | .5305 | .2815 | .2800 | 189.395 | .000 | .1263 | .000 | IN: | PRESTIGE | .3557 |

```
2 STEPS PERFORMED
PIN=.050 LIMIT REACHED.
DISPLAY?
summaries


DEPENDENT: RINCOME   2 VARIABLES IN.   LAST IN:  PRESTIGE

MULTIPLE R =     .53053     R SQUARE =        .28146     ADJ R SQUARE =  .27998
STD ERROR =     2.80607     R SQUARE CHG =    .12626     F CHG =       169.91977
SIGNIF F CHG = .00000       F =            189.39536     SIGNIF F =      .00000
```

| ANALYSIS OF VAR. | DF | SUM OF SQUARES | MEAN SQUARE |
|------------------|-----|---------------|-------------|
| REGRESSION | 2 | 2982.602 | 1491.30076 |
| RESIDUAL | 967 | 7614.167 | 7.87401 |

```
CONDITION NUMBER BOUNDS:  1.002, 4.008


VAR-COVAR MATRIX OF REGRESSION COEFFICIENTS (B)
BELOW DIAGONAL: COVARIANCE    ABOVE:  CORRELATION
```

| | SEX | PRESTIGE |
|---|-----|----------|
| SEX | .03281 | .04346 |
| PRESTIGE | 4.9364-05 | 3.9309-05 |

```
XTX MATRIX
```

| | SEX | PRESTIGE | RINCOME | EDUC |
|---|-----|----------|---------|------|
| SEX | 1.00189 | .04355 | .37850 | -.01425 |
| PRESTIGE | .04355 | 1.00189 | -.35567 | -.57257 |
| RINCOME | -.37850 | .35567 | .71854 | .02998 |
| EDUC | .01425 | .57257 | .02998 | .67267 |

```
DISPLAY?
/residuals
7 SCATTERPLOTS MAY BE REQUESTED
CASEWISE PLOT?
out=2.5 /identify sex
SCATTERPLOT?
*SDRESID *SRESID
SCATTERPLOT?
```

```
CASEWISE PLOT OF STANDARDIZED RESIDUAL
OUTLIERS = 2.5
                                      -5.5   -2.5  2.5    5.5
     #     SEX    RINCOME    *PRED    *RESID   0:.......:  :.......:0   *SRESID
    241   2.0000  13.0000    5.3562    7.6438   .          ..*       .    2.7272
    487   1.0000   2.0000   11.3017   -9.3017   .      *  ..         .   -3.3300
    796   2.0000  13.0000    5.6831    7.3169   .          ..*       .    2.6105
    862   1.0000   2.0000    9.0133   -7.0133   .        *..         .   -2.5023
   1069   2.0000  13.0000    3.8851    9.1149   .          ..  *     .    3.2555
   1192   2.0000  13.0000    5.3562    7.6438   .          ..*       .    2.7272
   1199   1.0000   2.0000    9.5854   -7.5854   .        *..         .   -2.7077
   1222   2.0000  13.0000    5.1928    7.8072   .          ..*       .    2.7856
   1495   1.0000   1.0000    8.2778   -7.2778   .        *..         .   -2.5960
9 OUTLIERS FOUND.
```

```
DISPLAY?
scattergram
```

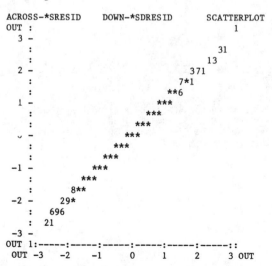

```
ACROSS-*SRESID    DOWN-*SDRESID        SCATTERPLOT
OUT :                                        1
  3 -
    :                                       31
    :                                      13
  2 -                                     371
    :                                    7*1
    :                                  **6
  1 -                               ***
    :                              ***
    :                             ***
  . -                           ***
    :                          ***
    :                         ***
 -1 -                       ***
    :                     ***
    :               8**
 -2 -            29*
    :          696
    :        21
 -3 -
OUT 1:-----:-----:-----:-----:-----:-----::
    OUT -3   -2   -1    0    1    2    3 OUT
```

The SCSS Conversational System, Version 4.2

Developed by:                                    Distributed by:

   SPSS Inc.                                         Same
   444 N. Michigan
   Chicago, Illinois 60611
   (312) 329-2400

Computer Makes and Operating Systems:   Interfaced Systems:

   IBM 360 (TSO,CMS)                             SPSS
   IBM 370 (TSO, CMS, MTS)                       DEC Sytem 1022
   Burroughs Large Systems
   DEC 10, 20 (except KAS)
   Univac 70/7; 90/60, 70 80
   Univac 1108
   Amdahl 470
   GEC 4000
   Honeywell 60/66, 600, 6000
   Honeywell 68/80 (Multics)
   Itel (TSO,CMS)
   Other Conversions Underway

Source Languages:

   Fortran 95%, Assembler 5%
   (No Fortran Compiler Needed)

Cost:

    Available on request from SPSS Inc.

Documentation and References:

Nie, Norman H. et al.  SCSS:  A User's Guide to the SCSS Conversational System.
   New York:  McGraw-Hill, 1980.
Norusis, M.J. and Chih-Ming Wang.  "The SCSS Conversational System", The
   American Statistician Vol. 34, No. 4, November, 1980.
Tagg, Stephen K.  "SCSS, A User's Guide to the SCSS Conversational System,"
   Quantitative Sociology Newsletter.  No. 25, Summer, 1980.
Tolbert, Marcia A.  (ed.)  Proceedings of the First Annual SPSS Users and
   Coordinators Conference.  Chicago:  SPSS (1977).

## 6.09  SPSS

### INTRODUCTION

SPSS is a full integrated computer package for data analysis and file management.  Currently in its 9th full release, it is installed in over 2500 sites in 60 countries.  The package runs on more than 30 different computers, models, and operating systems and is documented with a general manual, a primer, and an algorithm volume.

### CAPABILITIES:  Data Management, File Facilities, and Data Display

* Permanent or temporary data transformations
* Permanent or temporary case selection
* Permanent or temporary weighting
* Permanent or temporary random sampling of data
* Raw data input from cards, disk, or tape
* Creation, updating, and archiving of system files containing a complete dictionary of labels, print formats, and missing data indicators
* File capacity up to 5000 variables
* No built-in limitation to the number of cases
* Sorting of subfiles cases with automatic definition of subfiles
* Correlation, covariance, and factor matrix input and output
* Output of z-scores, residuals, factor scores, canonical variates, and aggregated files
* Report writing:  Automatic formatting, full range of summary statistics, composite functions across variables, multiple level breakdowns with summaries for all levels
* Color graphics:  pie charts, bar graphs, line and scatter plots, device independence.

### CAPABILITIES:  Statistical Analysis

* Frequency distributions, histograms, and descriptive statistics
* Multiway crosstabulations and measures of association for numeric or character data
* Tabulation of multiple response data
* Pearson, Spearman, and Kendall correlations
* Partial correlation
* Canonical correlation
* Analysis of variance with multiple classification analysis
* Stepwise discriminant analysis
* Multiple regression and output of residuals
* Manova
* Analysis of time series
* Bivariate plots
* Factor analysis
* Guttman scale analysis
* Nonparametric tests
* Survival analysis
* Paired and independent samples t-test
* Choice of treatment for missing values

## INTERFACE WITH OTHER SYSTEMS

*     DEC System 1022
*     OSIRIS
*     SCSS

## ADDITIONS TO SPSS IN NEXT YEAR

*     Generalized data handling of complex record structures
*     Greatly enhanced central system capabilities

## REFERENCES

Berk, Kenneth N. and Ivor S. Francis. "A Review of the Manuals for BMDP and SPSS." Journal of the American Statistical Association, 73, 361 (1978), pp. 65-98.

Holzer, Jean B. et al. Some Comparisons of SPSS and SAS. Gainsville, Fla: The Center for Instructional Research Computing Activities, University of Florida (1979).

Holzer, Jean B. and Michael Conlon. Some Comparisons of SPSS and SAS Revisited. Gainsville, Fla.: The Center for Instructional and Research Computing Activities. University of Florida (1980).

Kuiper, F. Kent and David Nelson. "Evaluation of Non Parametric Tests in SPSS and BMDP." Proc. Comp. Sci. Stat.: 10th Ann. Sym. Interface. Washington, D.C.: U.S. Govt. Printing Office (1979).

Tolber, Marcia A. (ed.). Proceedings of the First Annual SPSS Users and Coordinators Conference. Chicago: SPSS (1977).

## SAMPLE JOB

The user requests a stepwise regression of variable Y on X1 to X5. (Here only the second step of the regression is illustrated.) The user specifies statistics to be computed and the criteria of F-to-enter and F-to-remove. Also requested is a casewise plot of standardized residuals for all cases and a listing of: values on the dependent variable, predicted values, residuals, Studentized residuals, Studentized deleted residuals, Mahalanobis' distance, and Cook's distance.

```
GET FILE
NEW REGRESSION          VARIABLES=X1 to Y/
                        STATISTICS=DEFAULTS CHANGE ZPP HISTORY/
                        CRITERIA=PIN (.1) POUT (.15)/
                        DEPENDENT=Y/
                        STEPWISE/
                        CASEWISE ALL DEFAULTS SRESID
                        SDRESID MAHAL COOK/
```

```
'DEPENDENT VARIABLE..   Y       VERBAL MEAN TEST SCORE--ALL 6TH GRADERS

VARIABLE(S) ENTERED ON STEP NUMBER 2..   X4        MEAN TEACHER'S VERBAL TEST SCORE/

MULTIPLE R           0.94199                                        ANALYSIS OF VARIANCE        DF       SUM OF SQUARES      MEAN SQUARE
R SQUARE             0.88735              R SQUARE CHANGE   0.02772  REGRESSION                   2          570.49798        285.24899
ADJUSTED R SQUARE    0.87410              F CHANGE          4.18329  RESIDUAL                    17           72.42639          4.26038
STANDARD ERROR       2.06407              SIGNIF F CHANGE   0.0566

                                                                            F =     66.95395        SIGNIF F = 0.0000

------------------ VARIABLES IN THE EQUATION ------------------

VARIABLE        B           SE B        BETA      CORREL PART COR    PARTIAL         T     SIG T

X3            0.54156     0.05004     0.89611     0.92716  0.88092    0.93447     10.822  0.0000
X4            0.74989     0.36664     0.16937     0.33365  0.16650    0.44439      2.045   .0566
(CONSTANT)   14.58268     9.17541                                                 1.589   .1304

----------- VARIABLES NOT IN THE EQUATION -----------

VARIABLE    BETA IN   PARTIAL   MIN TOLER       T    SIG T

X1         -0.13550  -0.34441    0.72777    -1.467    .1616
X2          0.01148   0.01890    0.29581     0.076    .9407
X5         -0.06685  -0.11414    0.32232    -0.460    .6520

FOR BLOCK NUMBER  1   PIN =  .100 LIMITS REACHED.

* * * * * * * * * * * * * * * * * * * * * * * * * * * * * * * *

                                  SUMMARY TABLE
                                  -------------

STEP  MULTR   RSQ   ADJRSQ   F(EQU)   SIGF   RSQCH   FCH  SIGCH IN:  VARIABLE  BETAIN  CORREL   LABEL
 1   0.9272  0.8596  0.8518  110.230  0.000  0.8596  110.230 0.000 IN:   X3    0.9272  0.9272   SES COMPOSITE DEVIATE/
 2   0.9420  0.8873  0.8741   66.954  0.000  0.0277    4.183  .057 IN:   X4    0.1694  0.3336   MEAN TEACHER'S VERBAL TEST S
'DEPENDENT VARIABLE..   Y       VERBAL MEAN TEST SCORE--ALL 6TH GRADERS

CASEWISE= PLOT OF STANDARDIZED RESIDUAL

       -3.0       0.0        3.0
SEQNUM 0:.......::.........::0         Y        *PRED     *RESID    *SRESID   *SDRESID  *MAHAL    *COOK D
  1    .       *          .         37.0100   38.4291   -1.4191   -0.7345   -0.7241    1.4028   0.0254
  2    .          *       .         26.5100   26.5384   -0.0284   -0.0152   -0.0147    2.4334   0.0000
  3    .    *             .         36.5100   40.5269   -4.0169   -2.0548   -2.2993    1.0C60   0.1615
  4    .       *          .         40.7000   41.5884   -0.8884   -0.4600   -0.4490    1.4137   0.0100
  5    .         *        .         37.0472   37.0472   -0.0528   -0.0264   -0.0256    0.1464   0.0000
  6    .      *           .         33.9000   34.1164   -0.2164   -0.1416   -0.1377    7.6304   0.0055
  7    .        *         .         41.0400   40.7758    0.2642    0.0778    0.0747    1.0863   0.0292
  8    .          *       .         33.2454   33.2454    0.0774    0.0748    0.0849    1.5938   0.0184
  9    .           *      .         41.0100   39.8642    1.1458    0.5966    0.5849    5.5805   0.1679
 10    .            *     .         37.2000   35.5601   -1.6399   -0.9807   -0.9796    3.5631   0.1215
 11    .      *           .         23.3000   25.2482   -1.9482   -1.0811   -1.0869    1.2509   0.0680
 12    .           *      .         35.2000   32.7784   -2.4216   -1.2477   -1.2700    2.4443   0.0023
 13    .      *           .         34.9000   35.5458   -0.6458   -0.3232   -0.3145    0.5981   0.0007
 14    .        *         .         33.1000   33.4100   -0.3100   -0.1567   -0.1521    4.1926   0.1746
 15    .      *   *        .         22.7000   24.7933   -2.0933   -1.1876   -1.2032    0.64C1   0.0029
 16    .            *     .         39.7000   39.0889    0.6111    0.3009    0.3093    0.0038   0.0386
 17    .   *              .         31.8000   34.7780   -2.9780   -1.4804   -1.5389    2.1594   0.3661
 18    .            *     .         31.7000   27.2283    4.4717    2.3690    2.8080    1.5381   0.0211
 19    .           *      .         41.8521    1.2479    0.6485    0.6371                1.3845   0.0160
 20    .           *      .         41.0100   39.8783    1.1317    0.5854    0.5738
SEQNUM 0:.......::.........::0         Y        *PRED     *RESID    *SRESID   *SDRESID  *MAHAL    *COOK D
       -3.0       0.0        3.0
```

### SPSS, VERSION 9: Statistical Package for the Social Sciences

Developed by:                                      Distributed by:

    SPSS Inc.                                           Same
    444 N. Michigan Avenue
    Chicago, Il.  60611
    (312)  329-2400

Computer Makes:

    Nearly all computer environments, including
    IBM 360, 370, 4300, OS, DOS, CMS and all IBM compatibles
    Burroughs medium and large Systems
    CDC CYBER & 6000 Series
    Data General Eclipse and Nova
    DEC Systems 10, 20, VAX, PDP-11
    HARRIS 4, 7
    HEWLETT-PACKARD 3000
    Honeywell 60
    ICL 2900 Series
    Perkin-Elmer
    Prime 500-750
    Siemens BS 2000
    Univac 70, 90, 1100
    Other SPSS Conversions are available.  Contact SPSS Inc. for more
       information.

Source Languages:

    Fortran (95 percent) Assembler (5 percent) (Fortran compiler not needed to
run SPSS.)

Cost:

    Available on request.

Documentation:

Klecka, William R., Norman H. Nie, and C. Hadlai Hull.  SPSS Primer. New York:
    McGraw-Hill, 1975.
Nie, Norman H. et al.  SPSS: Statistical Package for the Social Sciences,
    2nd ed. New York:  McGraw-Hill, 1975.
Nie, Norman H. and C. Hadlai Hull.  SPSS Update 7-9.  New York:  McGraw-Hill
    1981.
Nie, Norman H. and C. Hadlai Hull:  SPSS Update:  New Procedure and Facilities
    for Releases 7 and 8.  New York:  McGraw-Hill, 1979.
Norusis, M.J.  SPSS Statistical Algorithms.  Chicago:  SPSS, 1979
SPSS Pocket Guide.  Chicago:  SPSS, 1979.

# 6.10  SOUPAC

## INTRODUCTION

SOUPAC is a general-purpose statistical package which is typically used to manipulate data, combine or separate data arrays, perform matrix algebra and do statistical analysis such as regression, factor analysis, analysis of variance, econometrics and spectral analysis.

It was developed at the University of Illinois in the early 1960's and has gone through several conversions, revisions and many enhancements. It has been in continuous use at the U. of I. and, without advertizing over 90 copies have been distributed since 1973. It is primarily batch oriented but may be run in an interactive environment.

## CAPABILITIES:  Processing and Displaying Data

The MATRIX and TRANSFORMATIONS programs in SOUPAC allow arithmetic, functional and logical manipulations of data. Data may be read from sources, external to SOUPAC (formatted or binary), combined with other sources of data, printed, punched, sorted, summarized, reordered, stored in binary, or output formatted. Variables or arrays may be generated and stored. SOUPAC makes continuous use of disk file storage for maximum retrievability of statistical and arithmetic results.

Some plotting and frequency counting capabilities exist. Also there is some very limited capability with character data. Output stored on SOUPAC disk may be printed with the user's own format if desired. Some variable labelling is optionally available.

## CAPABILITIES:  Statistical Analysis

Basic statistics, including means, standard deviation, skewness, kurtosis, and moving averages, with sub-grouping and missing data checks, are available. Also the one and two-way frequency counting program has percents, appropriate measures of association, with sub-grouping and missing data checks.

SOUPAC has:  correlation analysis with significance tests and sub-grouping; multiple regression with stepwise option, residual analysis, optional intercept and significance tests; t-test, one-way analysis of variance and co-variance and post-hoc tests; general analysis of variance for handling balanced designs: factorial, hierachical, repeated measures, and mixed designs, random and fixed factors, and unequal cell sizes; also discriminant analysis and classification. Some missing data correction is possible with these analyses. Several types of factor analysis and rotations, also factor scores, communality, and fixed rotation are available.

The following econometric analyses are in SOUPAC:  econometrics reduced form and residual analysis, k-class estimation, restricted least squares, stochastic restricted least squares, three-stage least squares, and seemingly unrelated regression.

The spectral analysis section has these techniques:  sample autocovariances and autocorrelations, cross-spectral analysis, fast Fourier transform, sample spectral density function, and general fast Fourier transform.

SOUPAC also has tests for goodness of fit, scale analysis, probit analysis, linear and quadratic programming, ranking and random number generation.

The matrix algebra program also has eigen-value extraction, three different inverters, Kronecker product and lag.

## EXTENSIBILITY

One may program one's own unique statistical procedure using SOUPAC's matrix algebra and other statistical results on SOUPAC's scratch disks, such as correlation matrices, factor matrices, predicted dependent variables, etc., all in the same job step. It is also possible to temporarily add to SOUPAC a new program written in FORTRAN, for example. This latter feature is used primarily for testing, but it has been used for production and classwork.

## INTERFACES WITH OTHER SYSTEMS

SOUPAC reads and writes card image and standard binary files, and therefore it can communicate with any other system that can read or write standard files.

## PROPOSED ADDITIONS NEXT YEAR

SOUPAC will receive a modification so that it will run smoothly under the IBM 370 MVS operating system. Also a CDC Cyber version will continue to be developed and may be released. The Manual of Program Descriptions is being rewritten. Minor changes will be made to the programs.

## REFERENCES

Anderson, Ronald E., and Edwin R. Coover. "Wrapping up the package: Critical thoughts on applications software for social data analysis". Computers and the Humanities, 7,2, November, 1972, p. 81 ff.

Kohm, Robert F., Thomas A. Ryan, Jr., and Paul F. Velleman, "Index of publicly available statistical software (microfiche)". Proceedings of the Statistical Computing Section, American Statistical Association, 1977.

Parsons, John. "Math software survey report". Proceedings of SHARE XL IV, SHARE, Inc., 1975, p. 1770ff.

Schucany, W.R., and Paul Minton, "A survey of Statistical Packages". Computing Surveys, 4, 2, June, 1972, p. 65ff.

Slys, William D., "An evaluation of statistical software in the social sciences". Communications of the ACM, 17, 4, June 1974, p. 326ff.

## SAMPLE JOB

```
1          MAT()(1).
    1          INP(S16)(S1)()(4)"(F1.0,2F2.0,F3.1)".
    2          LAB(S4)"DATA""CONSTANT""X1""X2""Y".
    3          MOV(S1)(S2/P(L)).  PRINT DATA
    4          PAR(S1)(S2)(4)(4).      Y
    5          COL(S1)(S3)(4).         X
    6          TRA(S3)(S4).            X'
    7          MUL(S4)(S3)(S5).        X'X
    8          INV(S5)(S3).            (X'X)INV
    9          MUL(S3)(S4)(S5).
   10          LAB(S4)"COEFFICIENTS".
   11          MUL(S5)(S2)(S3/P(F,L)).  (X'X)INV X'Y  IN F FORMAT
               END P
2          REG(S1)()(S2).  SAVE COR MAT ON S2
    1          VAR(2,4).
    2          MUL(2).          MULT REG AS ABOVE
               END P
           END S
```

## Partial output from MATRIX and REGRESSION programs:

```
PROGRAM NUMBER 1.
EXECUTING MATRIX PROGRAM


OUTPUT FROM MATRIX SUBPARAMETER CARD NUMBER 3.
CURRENT MATRIX OPERATION IS MOVE.
NUMBER OF ROWS =      12            NUMBER OF COLUMNS =      4          DOUBLE PRECISION              S2


DATA

              1              2              3              4
           CONSTANT        X1             X2             Y
     1   0.10000E 01   0.10000E 01   0.0           0.25000E 01
     2   0.10000E 01   0.10000E 01   0.0           0.35000E 01
     3   0.10000E 01   0.10000E 01   0.0           0.40000E 01
     4   0.10000E 01   0.10000E 01   0.0           0.20000E 01
     5   0.10000E 01   0.0           0.10000E 01   0.35000E 01
     6   0.10000E 01   0.0           0.10000E 01   0.50000E 01
     7   0.10000E 01   0.0           0.10000E 01   0.60000E 01
     8   0.10000E 01   0.0           0.10000E 01   0.55000E 01
     9   0.10000E 01  -0.10000E 01  -0.10000E 01   0.25000E 01
    10   0.10000E 01  -0.10000E 01  -0.10000E 01   0.35000E 01
    11   0.10000E 01  -0.10000E 01  -0.10000E 01   0.55000E 01
    12   0.10000E 01  -0.10000E 01  -0.10000E 01   0.45000E 01


OUTPUT FROM MATRIX SUBPARAMETER CARD NUMBER 11.
CURRENT MATRIX OPERATION IS MULTIPLY.
NUMBER OF ROWS =       3            NUMBER OF COLUMNS =      1          DOUBLE PRECISION              S3


COEFFICIENTS

              1
     1     4.00000
     2    -1.00000
     3     1.00000


EXECUTE TIME FOR MATRIX PROGRAM IS 1.97 SECONDS.
I/O REQUESTS USED (LOAD,EXECUTE) IS (14,227)
* * * * * * * * * * * * * * * * * * * * * * * * * * * * * * * * * * * * * * * * * * * * * * * *


PROGRAM NUMBER 2.
EXECUTING REGRESSION-CORRELATION PROGRAM


SAMPLE SIZE =        12
- - - - - - - - - - - - - - - - - - - - - - - - - - - - - - - - - - - - - - - - - - - - - - - -


EXECUTING SUBOPERATION  2.
MULTIPLE LINEAR REGRESSION


REGRESSION COEFFICIENTS (UNSTANDARDIZED)

                    4

     2      -0.10000D 01
     3       0.10000D 01


INTERCEPT

                    4

     A       0.40000D 01
```

SOUPAC:  Statistically Oriented Users' Package

Developed by:                          Distributed by:

Statistical Services/                      Same
Computing Services Office
University of Illinois
Urbana, Illinois  61801

Computer Makes:                        Operating Systems:

IBM 360/40+                                IBM OS/MVT, MFT, P CP
   370/145+                                OSVS, OSVS/SVS (includes
   3030 series                             VS1 and VS2)
   4300 series

Source Languages:  Primarily FORTRAN, assembler for input/output.

Cost:

The one time charge is $150 if the requestor provides a 9-track tape; and
$160 if we provide the tape.  The purchaser should specify the desired tape
density and whether the tape is to be standard labelled.  Orders should state
what machine and operating system is being used.  Later versions of the package
will be provided at the user's option for a charge of $75 or $85.

Documentation:

Computing Services Office of the University of Illinois.  SOUPAC Program
    Descriptions, volumes I and II, 1976.
    (Comes with package or order through:  Illinois Union Bookstore, 715 S.
    Wright, Urbana, Illinois 61801; about $4 per volume.)

## 6.11 DATATEXT

### INTRODUCTION

DATATEXT is a general-purpose data analysis system, consisting of an integrated set of data processing and statistical features, controlled by a natural language oriented toward social and behavioral research problems. The third generation IBM version was developed at Harvard University in 1968-72 by David J. Armor and Arthur S. Couch, and is currently available in a batch version that has been installed in over 100 research, university, and business computing centers.

### CAPABILITIES: Processing and Displaying Data

The DATATEXT language starts with a computer readable codebook which allows definitions of variables in terms of column location and card (record) numbers. A variable definition includes a variable label and specification of values and labels for categorical data, all contained in a single statement. A single instruction can also define an array of variables when values and value labels are identical. A special editing command will edit and validate all data, comparing actual data values with specifications in definition statements.

Variable transformations can also be specified within definition statements, including mathematical functions, logical IF expressions, recoding, across-item summation functions, and Z-score standardizing functions. Missing data (blank columns as the default) are handled automatically in all variable transformation expressions. Transformations can be applied to entire arrays, thereby executing the same operation for a list of variables; for example, $C(1-10) = A(1-10) + B(1-10)$.

Defined and transformed variables with associated labels can be saved as a standardized DATATEXT file; DATATEXT files can be used for more efficient processing within DATATEXT or for conversion to SPSS or SAS standardized files.

Special data processing features include input and transformation of alphabetic data, input and transformation of multiple-punch (column binary) data, unequal numbers of cards (logical records) per case, subfile processing, and merging/updating of two or more DATATEXT files. A utility is available for sorting DATATEXT files.

### CAPABILITIES: Statistical Analysis

Basic and advanced statistical procedures are available. Most routines have double precision and weighting options, and most standard analyses have the GROUP BY subgrouping feature that automatically partitions a data file. Few parameter restrictions exist; for example, unlimited number of cases; unlimited number of variables for correlations, regressions, and factor analysis; unlimited values for frequency distributions; unlimited cells (categories) for cross-tabs; and up to ten factors in ANOVA. All routines have a default procedure for missing data, and most routines offer several additional missing data options.

Statistical options include basic statistics, (means, sd's, higher moments); t-tests and F-tests for one-way ANOVA: frequency distributions; scatter plots; cross-tabs with many options including a multivalued variable feature; correlations with significance tests; factor analysis with rotations and factor scoring; regression analysis with residuals; and analysis of variance and covariance, with straightforward handling of repeated-measure (split-plot) and unbalanced (unequal cell size) designs. Factor scores and regression residuals can be output as DATATEXT files for merging with master files.

All statistical requests that fit within available memory are processed with one pass over the data. Therefore, efficient processing is available for large data files with multiple statistical requests.

## EXTENSIBILITY

User-written statistical routines can be added with moderate difficulty.

## INTERFACES

DATATEXT provides conversion routines for DATATEXT-to-SPSS and SPSS-to-DATATEXT standardized files. SAS provides DATATEXT-to-SAS conversion routines.

## PROPOSED ADDITIONS

A new users manual will be available in 1981, and a CDC version is being implemented at New York University.

## REFERENCES

Armor, D.J. "The DATATEXT System - An Application Language for the Social Sciences," AFIPS - Proceedings from the Spring Joint Computer Conference, v. 40, 1972.
Francis, I., S.P. Sherman, and R.M. Heiberger, "Languages and Programs for Tabulating Data from Surveys," Proceedings, Computer Science and Statistics: Ninth Annual Symposium on the Interface, Cambridge, 1976.
Heiberger, R.M. "The Specification of Experimental Designs to ANOVA programs," The American Statistician, Feb. 1981
Slysz, W.D. "An Evaluation of Statistical Software in the Social Sciences," Communications of the ACM, 17, June 1974.

## SAMPLE JOB

The program defines three variables, two by means of an array transformation (subtracting their values from 8). The third variable represents categorical data. A repeated measures (split plot) analysis of variance is requested.

```
*DECK        NIMH DRUG STUDY
*READ CARDS
*CARD FORMAT/UNIT = COL(1-4)
**
*RATING(1-2)=8-COL(52,56)=DOCTOR'S RATING/NURSE'S RATING(1-7=RANGE)
*DRUG        =COL(57)      =DRUG TREATMENT(1=PLACEBO/2=CHLOR./3=FLUPH./4=THIOR.)
**
*COMPUTE ANOVA (RATER BY DRUG)    REPEATED MEASURE(1)
*FACTOR(RATER) = (DOCTOR/NURSE)
*MEASURE(1)      = RATING(1-2) = IMPROVEMENT RATING
*START
```

## Partial Output from Sample Job:

TWO-WAY STATISTICS FOR MEASURE(1)        IMPROVEMENT RATING

| RATER | | DRUG<br>PLACEBO<br>1 | DRUG TREATMENT<br>CHLORPRO.<br>2 | FLUPHEN.<br>3 | THIORID.<br>4 | ROW<br>MARGINALS |
|---|---|---|---|---|---|---|
| DOCTOR | MEAN<br>EFFECT | 4.881<br>0.006 | 5.692<br>-0.035 | 6.000<br>-0.074 | 6.046 I<br>0.103 I | 5.655<br>0.104 |
| NURSE | MEAN<br>EFFECT | 4.661<br>-0.006 | 5.554<br>0.035 | 5.938<br>0.074 | 5.631 I<br>-0.103 I | 5.446<br>-0.104 |
| COLUMN MARGINALS | MEAN<br>EFFECT<br>SD<br>N | 4.771<br>-0.779<br>1.247<br>59.000 | 5.623<br>0.073<br>0.776<br>65.000 | 5.969<br>0.419<br>0.861<br>65.000 | 5.838 I<br>0.288 I<br>0.766 I<br>65.000 I | 5.550 |

ROW MARGINALS AND GRAND MEAN ARE UNWEIGHTED AVERAGES OF CELL MEANS.

UNWEIGHTED MEANS ANALYSIS OF VARIANCE TABLE FOR MEASURE(1)        IMPROVEMENT RATING

CLASSIFYING FACTORS
DRUG        DRUG TREATMENT
RATER
UNIT        SUBJECTS OR UNITS OF ANALYSIS

|   | SOURCE | SUM OF SQUARES | DF | MEAN SQUARE | F-TEST | SIGNIFICANCE |
|---|---|---|---|---|---|---|
|   | DRUG | 110.404 | 3 | 36.801 | 21.529*** | UNDER 0.001 |
| * | UNIT | 427.343 | 250 | 1.709 | NOT TESTED | |
|   | RATER | 5.534 | 1 | 5.534 | 11.005** | 0.002 |
|   | DRUG X RATER | 2.201 | 3 | 0.734 | 1.459 | 0.227 |
| * | RATER X UNIT | 125.714 | 250 | 0.503 | NOT TESTED | |
|   | TOTAL | 671.196 | 507 | 1.324 | | |

AN ASTERISK (*) MARKS THE EFFECT USED IN TESTING THE PRECEDING EFFECTS

DATATEXT

Developed by:                                      Distributed by:

    David J. Armor                                 Pro Systems
    The Rand Corp.                                 2940 Union Ave.
    1700 Main St.                                  Suite E
    Santa Monica, CA  90406                        San Jose, CA  95124
    (213)393-0411                                  (408)997-1776

Computer Makes:                                    Operating Systems:

    IBM Series                                     OS/VS or compatible

Source Languages:  Fortran IV, some IBM Assembler

Cost:

    There is an initial fee of $1,000 and an annual renewal fee of $500 for
Universities.  For government and non-profit organizations the initial fee is
$1,500 with an annual renewal fee of $600.  For commercial use, the initial fee
is $2,500 and the renewal fee is $1,000.

Documentation:

Armor, D.J. and A.S. Couch., The DATATEXT Primer, New York, Free Press, 1972.

## 6.12  P-STAT

### INTRODUCTION

P-STAT 78 is a large conversational system offering flexible file maintenance and data display features, crosstabulation, and numerous statistical procedures.  Its principle applications have been in areas such as demography,
survey analysis, research, and education.  P-STAT was originally developed by
Roald and Shirrell Buhler at Princeton University in 1962, and has been under
continuous development since then.  The system can be used interactively or in
batch mode, and is currently in use at over 100 installations around the world.

### CAPABILITIES:  Processing and Displaying Data

The P-STAT DATA program is the primary tool for creating a P-STAT system
file.  It is designed to detect and report any obvious errors in the input data.
A command is also provided which accepts free-format data from a terminal.
During an interactive session, an edit file is built which contains all commands
and data entered.  These can be checked for syntax, corrected, and executed
from within the P-STAT editor.  The file itself can be saved to be used again
in another session, or submitted to run as a batch job.

In P-STAT, many files can be active simultaneously, and many operations
performed in a single run.  Commands are provided to update files, to join
files in either a left/right or an up/down direction, sort files on row labels
or by up to 15 variables, and collate files which do not contain exactly the
same cases, or which have a hierarchical relationship.  There are also commands which permit aggregation across groups of related rows.  A typical P-STAT
run might include a number of these operations, perhaps combined with some
data modification, and the computing of correlations, regressions, and/or
crosstabulation.

In batch mode, tables, frequency distributions, listings with labels,
plots and histograms may be easily produced.  The TABLES command is particularly powerful in interactive mode.  Options include percentages, means,
side-by-side, nested, and n-way tables, and numerous formatting options to
provide for maximum clarity.  The double nested format (see example) can display, on a single surface, the inter-relationship of four variables.  Once a
table has been created, it can be modified conversationally without passing
through the data again.

### CAPABILITIES:  Statistical Analysis

Chi-square, F-tests, t-tests, means, and standard deviations are readily
available.  Commands are also provided to do correlations, regressions, principal components or iterative factor analysis, quartimax, varimax, or equimax
rotations, and backwards-stepping multiple discriminant analysis.  Other statistical procedures include MANOVA (multivariate analysis of variance), matrix
commands such as invert, and a command to produce oblique rotations.

EXTENSIBILITY

A user link facility is provided, as are routines which allow P-STAT files to be read from within any FORTRAN program.  A MACRO facility automates the execution of series of commands, and can be used to build instructional macros.

INTERFACES WITH OTHER SYSTEMS

P-STAT can read and write BMDP and SPSS save files.

PROPOSED ADDITIONS IN NEXT YEAR

* Character variables
* Report Generator
* Fully conversational regression program.
* Increased prompt and help messages.
* Additional plotting capabilities.
* P-RADE, a random access data enhancement for a data base capability within the P-STAT system.  P-RADE is now being tested.  It will be separately priced.

SAMPLE JOB

```
FIND = SURVEY2 / SD $

TABLES, IN = SURVEY2 (IF REGION .OUTRANGE. (1,4),
        DELETE), DES = SD, NO.HEAD $

LABELS = SEX (1) MALE (2) FEMALE/
        EDUCATION (1) GRADE SCHOOL (2) HIGH SCHOOL (3) COLLEGE/
        WORK.STATUS (1) FULL TIME (2) PART TIME (3) NOT WORKING/
        MARITAL.STATUS (1) MARRIED (2) NOT MARRIED/,

T = (P-STAT 78, A NESTED TABLE)
    (FEBRUARY 24, 1978)

SEX WITHIN EDUCATION BY WORK.STATUS WITHIN MARITAL.STATUS,
MEANS = SIBLINGS $

END $
```

SAMPLE JOB:


                    P-STAT 78.  A NESTED TABLE.
                      FEBRUARY 24, 1978


          CELL CONTENTS ARE....
             CELL COUNTS
          ---MEAN SCORE OF VARIABLE   ----SIBLINGS----

                 WORK.STATUS WITHIN MARITAL.STATUS

                 ....MARRIED.....  ..NOT MARRIED...

| EDUC ATION | SEX | FULL TIME | PART TIME | NOT WORK ING | FULL TIME | PART TIME | NOT WORK ING | ROW TOTALS |
|---|---|---|---|---|---|---|---|---|
| GRADE SCHOOL | MALE | 12 3.3 | 3 2.3 | 23 3.0 | 6 1.5 | | 7 1.4 | 51 2.6 |
| | FEMALE | 5 1.8 | 1 2.0 | 15 2.3 | 1 3.0 | 1 0.0 | 13 3.1 | 36 2.5 |
| HIGH SCHOOL | MALE | 41 2.4 | 4 2.3 | 15 2.5 | 11 0.9 | 6 0.0 | 16 1.3 | 93 1.9 |
| | FEMALE | 26 2.0 | 10 3.8 | 87 3.0 | 13 1.8 | 1 0.0 | 28 2.6 | 165 2.7 |
| COLLEGE | MALE | 39 2.3 | 6 0.8 | 10 1.2 | 13 0.5 | 1 0.0 | 7 0.3 | 76 1.5 |
| | FEMALE | 15 1.9 | 6 3.0 | 30 2.3 | 9 0.7 | 3 3.0 | 7 1.1 | 70 2.0 |
| TOTAL N MEAN | | 138 2.3 | 30 2.6 | 180 2.7 | 53 1.1 | 12 0.8 | 78 2.0 | 491 2.2 |

P-STAT 78

Developed by:                                        Distributed by:

P-STAT, Inc.                                             Same
P.O. Box 285
Princeton, New Jersey 08540

Computer Makes:                                      Operating Systems:

BURROUGHS 6700
CDC CYBER                                                NOS, SCOPE 3:4
DEC 10, 20.
HARRIS (in test)
HONEYWELL levels 66 & 68                                 GCOS, MULTICS, DTSS
HP 3000 (in test)
IBM 360/370/30xx/43xx                                    OS, DOS/VS, VM/CMS, TSO, VS1, MTS,MVS
ICL 2900 (in test)
SIGMA 7
UNIVAC 1106, 1108, 1110
UNIVAC 90
VAX

Interfaced Systems:

SPSS
BMDP

Source Language:

FORTRAN

Cost:

The yearly license fee for P-STAT is $1000 for universities; $5000 for the
first year, $2000 for renewal years for government and commercial installa-
tions.   The fee includes program source modules, load and/or object modules
(if possible), and installation instructions, all on magnetic tape.  It also
includes maintenance and one copy of the P-STAT 78 User's Manual.

Documentation:

P-STAT 78 User's Manual, Buhler, Shirrell and Roald, P-STAT, Inc., 1979.
A Pocket Guide to P-STAT 78, P-STAT, Inc., 1979.

## 6.13  DAS

### INTRODUCTION

The ICL 2900 Statistics System (DAS) consists of a statistical analysis subsystem, which provides a range of statistical operations, embedded in a high level language (ACL) which provides arithmetic and control statements and file assignment.  It was developed in 1976 for ICL 2900 series computers.  It can be used interactively or in batch mode.

### CAPABILITIES:  Processing and Displaying Data

Sets of data, flat or with a hierarchical structure, may be input from previously assembled files or directly at one time.  Extra records and extra variates may be added to or removed from previous sets of data; data records containing doubtful variate values may be deleted entirely or may just be omitted from analysis.  The value "missing" is an acceptable, distinct value for any variate, and statistical analyses take proper account of missing values.

All or some variates of the data set may be listed in a chosen format. Frequency and percentage tables of one, two or higher dimensions may be assembled and displayed.  Scatter diagrams, bar charts and probability plots of the original or transformed variates are available.  Several plots may be superimposed.

### CAPABILITIES:  Statistical Analysis

Basic statistics:  means, variances, correlations including auto- and partial correlations, histograms, distribution fitting.

Linear regression:  simple, multiple and polynomial regression; stepwise regression; two optimal regression methods; ridge regression; element analysis.

Analysis of variance for complete factorial designs, including crossed and nested factors; fixed and random effects, including mixed models; polynomial partitioning.

Principal components and factor analysis; group comparisons; discriminant analysis; cluster analysis; canonical analysis.

### EXTENSIBILITY

Any of the calculated statistics - mean, regression coefficients, cluster coordinates, and so on - may be captured in specified  scalars or arrays. Statements in the Applications Control Language (a high-level language including features reminiscent of Fortran and Algol 60) may be mixed in with statistical subsystem statements.  These two features allow the user to integrate his own ideas with the facilities of the statistics system - to rearrange the output, to do further calculations with the results of an analysis, or to introduce quite new methods to the system.

ACL-Statistics code developed in this way may be filed, as a whole program or as parameterised procedures, for repeated execution.

INTERFACES WITH OTHER SYSTEMS

ACL also links to the 2900 Matrix Handler, which provides a range of matrix operations, including matrix inversion, solutions of equations and eigenvalues.

PROPOSED ADDITIONS IN THE NEXT YEAR

None.  The system is fully maintained and bugs are corrected when discovered.

REFERENCES

Cooper, B.E.,  "Advances in statistical system design." J.R. Statist. Soc. A.
    140 (1977) 166-197.
Cooper, B.E.,  "Statistical and related systems." ICL Tech. J. 1 (1979)
    229-246.

SAMPLE JOB

Principal components analysis.  The data matrix is formed from an existing 2900 file.  Principal components analyses are done in turn on the correlation matrix and the covariance matrix, retaining the first three components each time.  The GRADUATE statement displays the component scores for each data point. One of the several pages of output is shown.

```
BEGIN PROGRAM
INVOKE STATISTICS
ACCEPT'SALESDATA'TYPE INPUT
ASSUME DEFAULT INPUT STREAM SALESDATA
READ DATA MATRIX IBA WITH VARIATES 'X'1:11
BASIC STATISTICS FOR ALL VARIATES
     COVARIANCE AND CORRELATION MATRICES
COMPONENTS ANALYSIS USING ALL VARIATES
     CORRELATION MATRIX
     RETAIN 3
     GRADUATE
REPEAT COVARIANCE MATRIX
END PROGRAM
FINISH
```

Partial Output:

```
INPUT VARIATES  X1   X2   X3   X4   X5   X6   X7   X8   X9
                X10  X11

              ****************************
              *                          *
              *  PRINCIPAL COMPONENTS ANALYSIS  *
              *                          *
              ****************************

ANALYSIS ON CORRELATION MATRIX


              CONTRIBUTION TO TOTAL VARIANCE
              ------------------------------

                EIGENVALUE    % OF VARIANCE    CUMULATIVE
                                EXPLAINED
                ----------    ------------     ----------

COMPONENT 1       3.7917        34.4700         34.4700
COMPONENT 2       2.7180        24.7092         59.1793
COMPONENT 3       1.6677        15.1612         74.3404

SUB-TOTAL         8.1774        74.3404

TOTAL            11.0000       100.0000


              COMPONENT COEFFICIENTS
              ----------------------

      COMPONENT 1    COMPONENT 2    COMPONENT 3
      -----------    -----------    -----------
X1       0.8869        -0.2476        -0.0756
X2      -0.6169        -0.6542         0.0173
X3      -0.5511        -0.1650        -0.3555
X4      -0.7981        -0.4574        -0.0682
X5      -0.4525         0.1587         0.7053
X6      -0.5783         0.5200        -0.4444
X7       0.4345         0.6314        -0.3692
X8      -0.5116         0.7555         0.0031
X9      -0.2801         0.8287         0.1738
X10      0.2931         0.2002         0.8170
X11     -0.7278        -0.0249         0.0636
```

## DAS:  ICL 2900 Data Analysis (Statistics) System

Developed by:                                    Distributed by:

    Dataskil Ltd.                                  ICL
    12-18 Crown Street
    Reading
    Berkshire RG1 2HP
    England

Computer Makes:                                  Operating Systems:

    ICL 2900 series,                               VME/B
      2960 and above

Source Languages:

    Primarily Fortran

Cost:

    Available on application to your local ICL sales office.

Documentation:

ICL Publication TP6733:  Analysing data using the 2900 Statistics System (1980).
ICL Publication TP 6873:  2900 Statistics Reference (1979).

## 6.14 <u>SPMS</u>

### INTRODUCTION

SPMS is an integrated statistical analysis system designed especially for clinical research. Once the user has entered the raw data into the SPMS data file, various related tasks, e.g., generation of new items, static and dynamic case selections, advanced statistical analyses, storage of computed results like factor scores, and so on, can be carried out in any sequence. It was developed by the Medical Informatics Group of the Tokyo Metropolitan Institute of Medical Science in 1976, and has been under continuous development since then. It can be used in batch, but is not yet implemented at any other site. However, it can be transferred to IBM-compatible machines by making some minor conversions of the system.

### CAPABILITIES: Processing and Displaying Data

The SPMS data file, based on the modified relational database model, can handle clinical time series records which have a kind of three dimensional data structure composed of case, item and time axis. Besides this, various data structures can be generated from original ones by appropriately combining "data structure transformation command," such as LINK, PROJECT, SORT and CONCATENATE commands.

SPMS can provide a variety of pre-analysis functions such as new item generations, case selection, item deletion and flexible definition of missing values. The CREATE, MAKE, UPDATE, DELETE and SELECTED BY commands do these functions.

It can produce publishable tables and figures. The SHOW and TABULATE command can print various tables. The PLOT command gives the user a great deal of flexibility for producing two-dimensional scatter plots, and also produce several graphs such as histograms, normal probability plots, moving average curves and individual variation plots.

### CAPABILITIES: Statistical Analysis

The COMPUTE and APPLY command do so-called "statistical analysis".

* Elementary Statistics --- Means, Standard deviation, skewness, etc.
* Parametric tests --- Homogeneity test of two means, two variances, two correlations or two populations, Normality test. Bartlett-M.
* Non-parametric tests --- Kolmogorov-Smirnov two sample, Kruskal-Wallis one-way ANOVA, Cochran-Q, Nann-Whitney-U, Friedman two-way ANOVA, etc.
* Contingency tests --- Yates correction, Fisher exact probability, Chi-squares and several measures of associations.
* Multivariate analyses --- Multiple regression, Discriminant analysis, Principal component analysis, Factor analysis with varimax rotation, Polynomial regression, Cluster analysis of cases, Hayashi quantification type 1-4, and non-linear least square curve fitting.
* Other methods --- Analysis of variance and covariance (fixed- and random-effects model), Clinical normal range setting (IRM) and random sampling.

## EXTENSIBILITY

The package has a statistical, printing and lexical anlysis subroutine library. Using these routines, user-made commands can be implemented. The EXECUTE command carries them out with connectivity to commands before and after in one job.

## INTERFACES WITH OTHER SYSTEMS

SPMS can produce the BMDP save file onto tapes.

## PROPOSED ADDITIONS IN NEXT YEAR

SPMS is not being extensively developed, however, a new test of survival curves is scheduled to be implemented, as well as log-linear model, multiple logistic analysis, and life-table methods. Improvement to the 'portability' of the package is also scheduled.

## REFERENCES

User Guide to Minvera, research report, General series No. 48, The Institute of Statistical Mathematics, Japan. 1980.

## SAMPLE JOB

```
     %   RELATION N+G-180 CONTAINS DATA OF S-UA AND SCORE.
     %   S-UA INDICATES SERUM URIC ACID LEVELS.
     %   SCORE IMPLIES CALCULATED SCORES SO THAT THE PLANE
     %   SPANNED BY S-UA AND SCORE DISCRIMINATES BEST BETWEEN
     %   HEALTHY SUBJECTS AND GOUTY PATIENTS.
     %   GROUP = 1 MEANS THE DATA STRATUM OF HEALTHY SUBJECTS.
     %   GROUP = 2 MEANS THE DATA STRATUM OF GOUTY PATIENTS.
REPEAT ( !N !C ) = ( ^T2 1) ( ^T3 2 ) ;
     PROJECT ^N+G-180 ON !N WITH ITEMS
          ^GROUP ^S-UA ^SCORE   SELECTED BY ^GROUP  ( !C ) ;
END REPEAT ;
     %   THE FOLLOWING PROCEDURES CREATE DATA  INDICATING
     %   95 PER CENT CONFIDENCE REGION OF HEALTHY SUBJECTS.
CREATE RELATION ^T1 (40) ;
IN RELATION ^T1 ;
     ASSIGN !V1 = 5.58 : !V2 = 1.12 : !V3 = 0.365 :
            !V4 = 12.24 : !V5 = 0.93 ;
     MAKE ^GROUP = 10 ;  MAKE ^ANGLE = FLOAT(INDEX) ;
     MAKE ^S-UA  = !V1 + !V2 / SQRT(2.0)
             *(SQRT(5.991*(1+!V3))*COS(3.14/20*^ANGLE )
             - SQRT(5.991*(1-!V3))*SIN(3.14/20*^ANGLE )) ;
     MAKE ^SCORE = !V4 + !V5 / SQRT(2.0)
             *(SQRT(5.991*(1+!V3))*COS(3.14/20*^ANGLE )
             + SQRT(5.991*(1-!V3))*SIN(3.14/20*^ANGLE )) ;
CONCATENATE ^T1 ^T2 ^T3 TO ^T4 ;
     %   THE NEXT PLOT COMMAND PLOTS THREE KINDS OF DATA,
     %   1. HEALTHY SUBJECTS,
     %   2. GOUTY PATIENTS, AND
     %   3. DATA INDICATING 95 PER CENT CONFIDENCE REGION
     %      OF HEALTHY SUBJECTS,
     %   ON THE SAME PLANE SPANNED BY S-UA AND SCORE.
IN RELATION ^T4 ;
     PLOT TYPE 1 GRAPH BETWEEN  ^S-UA  AND ^SCORE
          GROUPED BY ^GROUP  WITH PRINTED SYMBOL
               ( 1 -> 'N' , 2 -> 'G' , 10 -> '+' )
     WHERE X-RANGE IS (2,17)  Y-RANGE IS (8,18) ;
```

Partial Output from Sample Job:  The PLOT Command

```
*********   TIMS-STATISTICAL PACKAGE FOR MEDICAL SCIENCE     ( VERSION  50 )  JO
*********

                  RELATION : T4
                  GROUPED BY GROUP      ON   1->N  2->G 10->+


     SCORE                   ( X_UNIT =       .15 Y_UNIT =       .20 )
   18.00 +--------+---------+---------+---------+---------+---------+---------
         |
         |
   17.00 +
         |
         |
   16.00 +
         |
         |
   15.00 +
         |
         |                                     +  +  +  +
   14.00 +                          + N+              +
         |                       +                      +
         |                     +                          +
         |                  +                               N
   13.00 +            N                          G            +
         |        +          N                  N N N           +
         |      +          N                                   +
         |    +                          +                        G
   12.00 +  +                    +        N
         | +                   N                                +        G
         | +               NN           N N                   N  G  G     G   G
   11.00 + +                    N       N G +  N   G  G              G
         | +                              G  G   G                G
         |  +                 N               +      G
         |   +                        +                G  G  G     G
   10.00 +    + +            +  +    +         G  G  G     G
         |        + +                                       G
         |
    9.00 +                                     G
         |
         |
    8.00 +--------+---------+---------+---------+---------+---------+---------
           2.00     3.50      5.00     6.50      8.00      9.50      11.00    12
```

SPMS, VERSION 80:  Statistical Package for Medical Science

Developed by:                                    Distributed by:

  Medical Informatics Group                    Same
  The Tokyo Metropolitan Institute
   of Medical Science
  3-18-22, Hon-Komagome
  Bunkyo-ku, Tokyo
  Japan 113

Computer Makes:                                  Operating Systems:

  Hitac 8250                                     Hitac - NDOS, VOS1
      M-series

                                       Interfaced Systems:

                                       BMDP

Source Languages:  Primarily PL/I (95%), FORTRAN (1%) and Assembler (4%).

Cost:  Not yet determined.

Documentation:

SPMS, Statistical Package for Medical Science, 1st ed. Medical Informatics
    Group, The Tokyo Metropolitan Institute of Medical Science, 1980.

## 6.15  OSIRIS IV


### INTRODUCTION

OSIRIS IV is the current version of the OSIRIS software package developed at the Institute for Social Research, The University of Michigan.  It is designed as a general statistical and data manipulation package and is especially useful for processing and archiving data from large surveys.  Since its release in August, 1979, it has been installed at 30 computer sites both nationally and internationally.  OSIRIS IV is also accessible via TELENET.

### CAPABILITIES:  Processing and Displaying Data

An OSIRIS dataset is comprised of a dictionary describing the data and the actual data which can be of two forms, a standard fixed length rectangular dataset or a variable length hierarchical dataset.  The hierarchical dataset is composed of groups of variables, each group having its own record format and occupying identical or different levels of the hierarchy as other groups.  A hierarchical dataset may be retrieved in many different ways, with any level as the unit of analysis and with data from some or all of the groups.

OSIRIS IV data management facilities include:  matrix input and output and data correcting, updating, subsetting, weighting, sorting, and displaying.  OSIRIS IV can transform data values through arithmetic and logical operations both within and across records.  It can accept input and create output in any mode.

### CAPABILITIES:  Statistical Analysis

OSIRIS IV has the basic analysis features such as univariate and bivariate frequencies and associated statistics, multiple and dichtomous regression and correlation analysis, factor analysis, cluster analysis, and univariate and multivariate analysis of variance.  There are also special techniques for multivariate analysis of nominal and ordinal scale data, for searching among predictors for the greatest variance explanatory power, for calculating sampling errors for complex designs, and for replicated regressions.

### INTERFACES WITH OTHER SYSTEMS

OSIRIS IV is capable of interfacing with MIDAS and with SPSS.

### PROPOSED ADDITIONS IN NEXT YEAR

OSIRIS IV is undergoing constant development and expansion.  Plans for the next year include a direct access retrieval ability for better handling of network data structures, aggregation, configuration analysis, and multiple discriminant function analysis capabilities, an SAS interface, and an improved SPSS interface.  Additions to the documentation to be made within the next 12 months include an OSIRIS IV architecture and design technical document, a manual for novice users, a publication entitled Data Processing in the Social Sciences with OSIRIS IV, and a formula and statistical reference manual for OSIRIS IV.

REFERENCES

Kaplan, Bruce and Ivor Francis,  " A Comparison of Methods and Programs for
    Computing Variances of Estimates from Complex Sample Surveys,"  Proceed-
    ings, Section on Survey Research Methods, American Statistical Associa-
    tion, (1979).
Pelletier, Paula A., and Michael A. Nolte,  "Data Processing and Data Base
    Management Innovations--The Panel Study of Income Dynamics,"  Proceedings,
    Section on Survey Research Methods, American Statistical Association,
    (1978).
Pelletier Paula A., and Michael A. Nolte,  "Time and the Panel Study of Income
    Dynamics," in Data Bases in the Humanities and Social Sciences,  J. Raben
    and G. Marks, eds, (1980), pp. 53-60.
Roistacher, Richard C., "A Review of OSIRIS IV," in SIGSOC Bulletin of Associ-
    ation for Computer Machinery, January, 1980.

SAMPLE JOB

Produce nominal statistics on education by sex.

```
&TABLES DICTIN=BOOK.DI DATAIN=BOOK.DA
NOMINAL STATISTICS ON BOOK.DA
WT= V32
STATS=NOM P=(ROW%, UNWT) SUPPRESS=(ROWC,COLC)-
V=V35 S=V5 TITLE='EDUCATIONAL STATUS BY SEX'
&END
```

```
        *** TABLES -- CROSSTAB AND RANK ORDER STATISTICS ***

        JUL 31, 1980  NOMINAL STATISTICS ON BOOK.DA                      TABLES    1

        ILLEGAL CHARACTERS IN THE DATA WILL
        BE TREATED AS MISSING DATA 1

        THE DATA ARE WEIGHTED BY VARIABLE V32


            66 CASES PASSED FILTER



        ANALYSIS 1

        VARIABLE     V35     EDUCATION
        STRATA(ROW) V5      SEX

        EDUCATIONAL STATUS BY SEX

                    |ELEM SCH|HIGH SCH| COLLEGE|GRAD SCH|VOCA SCH|  TOTALS
                    |--------|--------|--------|--------|--------|
        MALES   |      150|     400|   4,500|   6,000|   4,500|  15,550
          Row  %|      1.0|     2.6|    28.9|    38.6|    28.9|   100.0
                    |--------|--------|--------|--------|--------|
        FEMALES |      300|  10,000|     400|   1,700|   3,000|  15,400
          Row  %|      1.9|    64.9|     2.6|    11.0|    19.5|   100.0
                    |--------|--------|--------|--------|--------|

        TOTALS        450    10,400   4 900    7,700   7,500   30,950
          Row  %      1.5     33.6    15.8     24.9    24.2    100.0

        TABLE OF UNWEIGHTED N'S

                    |ELEM SCH|HIGH SCH| COLLEGE|GRAD SCH|VOCA SCH|  TOTALS
                    |--------|--------|--------|--------|--------|
        MALES   |       3|      4|      8|      9|      8|    32
                    |--------|--------|--------|--------|--------|
        FEMALES |       6|     13|      4|      4|      7|    34
                    |--------|--------|--------|--------|--------|

        TOTALS         9      17      12      13      15      66

        GOODMAN AND KRUSKAL'S TAU A = 0.486   TAU B = 0.183

        MARGINALLY CORRECTED LEIK D(NOM-NOM)        0.524
        PROPORTION FREE OF MARGINAL RESTRAINTS =    0.759
                                        ROOT =      0.724

        LAMBDA    0.427    LAMBDA A    0.273   LAMBDA B    0.633
        ASE(L) =  0.0031    ASE(LA) =  0.0033   ASE(LB) =  0.0041
          Z(L) =137.1549     Z(LA) = 82.0698     Z(LB) =154.5348

        *****NORMAL TERMINATION OF TABLES    $  0.12    0.25 SECS
```

## OSIRIS IV

Developed by:                                    Distributed by:

The Institute for Social Research      Same
Survey Research Center
Computer Support Group
University of Michigan
P.O. Box 1248
Ann Arbor, MI  48106

Computer Makes:                                Operating Systems:

IBM 360/370 Series                         OS, VS1, VS2, MTS, MVS
AMDAHL 470                                 OS, MTS

Source Languages:

FORTRAN IV, IBM ASSEMBLER

Cost:

The first year license fee for OSIRIS IV is $2400 with an annual renewal fee of $1800 for commercial installations; $1600 with an annual renewal fee of $1200 for non-profit, government installations; $1200 with an annual renewal fee of $900 for academic institutions; and $900 with an annual renewal fee of $675 for Inter-University Consortium members.

The above fee includes the source and load modules for OSIRIS IV, one copy of the OSIRIS IV users manual, one copy of the OSIRIS IV subroutine manual, and one copy of OHDS:  An Introduction to the OSIRIS Hierarchical Data Structures Capabilities.  It also includes maintenance, new releases, and a reasonable amount of free consultation.

Documentation:

Andrews, F.M., Morgan, J.M., Sonquist, J.A., and Klem, L., Multiple Classification Analysis:  A Report on a Computer Program for Multiple Regression Using Categorical Predictors, 1974.
Andrews, F.M., and Messenger, R.C., Multivariate Nominal Scale Analysis:  A Report on a New Analysis Technique, 1973.
Andrews, F.M., Klem, L., Davidson, T.N., O'Malley, P.M., and Rogers, W.L., A Guide for Selecting Statistical Techniques for Analyzing Social Science Data, 1974.
Center for Political Studies, OHDS:  An Introduction to the OSIRIS Hierarchical Data Structures Capabilities, 1979.
Computer Support Group, OSIRIS IV:  Statistical Analysis and Data Management Software System, Survey Research Center, Institute for Social Research, Univ. of Mich., 1979.
Computer Support Group, OSIRIS IV Subroutine Manual, Survey Research Center, Institute for Social Research, Univ. of Mich., 1979.
Morgan, J.M., and Messenger, R.C., THAID:  A Sequential Analysis Program for the Analysis of Nominal Scale Dependent Variables, 1973.
Sonquist, J.A., Baker, Lauh E., and Morgan, J.N., Searching for Structure, late 1979, revised edition 1974.

CHAPTER 7

GENERAL STATISTICAL PROGRAMS

CONTENTS

INTRODUCTION

Chapters 6 and 7 contain all the well-known statistical packages.  The programs in Chapter 6 were not strong in statistical analysis (see the Taxonomy in Table 1.2), but had the capabilities to process and display data from large surveys.

Programs in this chapter are uniformly strong in most aspects of statistical analysis, hence the title of the chapter.  Many of them also have capabilities for data management and tabulation.  Indeed, based upon developers' claims, SAS is the overall champion program in its coverage of all operations in the "life cycle" of a statistical analysis.  This can be seen from the almost continuous line of solid bars opposite SAS in the taxonomy in Table 1.2, but more precisely it can be seen by looking at the overall D-scores in Table 13.2. These were defined in Section 1.5 of Chapter 1.  On this overall D-score SAS scored the highest value of 2.7 out of a maximum possible of 3.0, which is its developer's average rating on 27 items chosen to cover the whole range of operations in the "life cycle".  More importantly, SAS's users think almost as highly of this program as its developer does:  on the $\Omega$-score in Table 13.1 SAS places second equal.  This $\Omega$-score, also defined in Section 1.5, is the average of the users' scores on those items (from the 27 used for the D-score) on which the developer rated his program as 2 or better, in other words which the developer claimed as a significant purpose of his program.

These results can also be seen in Table 7.1, and in Figures 7.1 and 7.2, which are explained in detail in Section 1.4 of Chapter 1.  Briefly, the ratings by the respective developers on selected items are displayed in Table 7.1. Figure 7.1 compares developers' and users' ratings on all items, while Figure 7.2 summarizes developers' ratings (D-scores) and users' ratings (U-scores) on

## TABLE 7.1:  RATINGS BY DEVELOPERS ON SELECTED ITEMS

### (i) Capabilities

Usefulness Rating Key:
3 - high
2 - moderate
1 - modest
. - low

| | Complex Structures | File Management | Consistency Checks | Probabilistic Checks | Compute Tables | Print Tables | Multiple Regression | Anova/Linear Models | Linear Multivariate | Multi-way Tables | Other Multivariate | Non-Parametric | Exploratory | Robust | Non-linear | Bayesian | Time Series | Econometric |
|---|---|---|---|---|---|---|---|---|---|---|---|---|---|---|---|---|---|---|
| | 11 | 14 | 18 | 19 | 24 | 25 | 30 | 31 | 32 | 33 | 34 | 36 | 37 | 38 | 39 | 40 | 35 | 41 |
| 7.01 CS | 3 | 3 | 3 | 1 | 3 | 3 | 3 | 1 | 1 | 1 | 1 | 3 | 2 | 1 | 2 | 1 | 2 | 1 |
| 7.02 SAS | 3 | 3 | 3 | 3 | 3 | 2 | 3 | 3 | 3 | 3 | 2 | 2 | 2 | 2 | 3 | . | 2 | 2 |
| 7.03 OMNITAB 80 | 3 | 3 | 1 | 3 | 3 | 1 | 3 | 2 | 2 | 2 | 1 | 2 | 3 | 3 | 1 | . | . | 1 |
| 7.04 HP STAT PACKS | 1 | 2 | . | 1 | 3 | 1 | 3 | 2 | 2 | 2 | 1 | 3 | 2 | . | 3 | . | 2 | 1 |
| 7.05 MINITAB II | 3 | 2 | 1 | 2 | 3 | 3 | 3 | 2 | 1 | . | 1 | 2 | 3 | 2 | 1 | . | 3 | 1 |
| 7.06 BMDP | 1 | 1 | 2 | 3 | 2 | 1 | 3 | 2 | 3 | 3 | 3 | 2 | 1 | 2 | 3 | . | . | . |
| 7.07 NISAN | 2 | 2 | . | . | 2 | 2 | 3 | 3 | 3 | 2 | 3 | 3 | 2 | 2 | 2 | 3 | 1 | 2 |
| 7.08 GENSTAT 4.02 | 3 | 2 | 1 | 1 | 3 | 3 | 2 | 3 | 3 | 3 | 2 | 2 | . | 2 | 2 | . | . | 1 |
| 7.09 SPEAKEASY III | 3 | 2 | 2 | 1 | 3 | 3 | 3 | 2 | 3 | 1 | 1 | 1 | 1 | 2 | . | 3 | . | 3 |
| 7.10 TROLL | 2 | 3 | . | 3 | 2 | 2 | 3 | 3 | 2 | 2 | 2 | . | 3 | 2 | 3 | . | 2 | 3 |
| 7.11 IDA | 1 | 2 | . | 2 | 2 | 3 | 3 | 2 | 1 | 2 | 2 | 3 | . | 1 | 1 | 3 | 3 | 1 |

## FOR GENERAL STATISTICAL ANALYSIS PROGRAMS

### (ii) User Interface

| | Survey Estimates | Survey Variances | Math Functions | Operations Research | Availability | Installations | Computer Makes | Mini Version | Core Requirements | Batch/Interactive | Stat. Training | Computer Training | Language Simplicity | Documentation | User Convenience | Maintenance | Tested for Accuracy |
|---|---|---|---|---|---|---|---|---|---|---|---|---|---|---|---|---|---|
| | 43 | 44 | 47 | 48 | 49 | 50 | 51 | 52 | 53 | 54 | 55 | 56 | 57 | 58 | 59 | 60 | 61 |
| CS | 2 | 1 | 3 | 2 | 3 | 1 | 1 | 1 | . | 3 | 2 | 3 | 3 | 3 | 3 | 3 | 1 |
| SAS | 1 | 2 | 3 | . | 3 | 3 | . | 1 | 1 | 3 | 3 | 3 | 2 | 3 | 3 | 3 | 3 |
| OMNITAB 80 | 3 | 2 | 3 | 1 | 3 | 3 | 3 | 1 | . | 3 | 3 | 3 | 3 | 3 | 3 | 3 | 3 |
| HP STAT PACKS | 1 | . | 3 | 2 | 2 | 3 | . | 3 | 3 | 2 | 2 | 3 | . | 2 | 3 | 3 | 2 |
| MINITAB II | 1 | 1 | 3 | . | 3 | 3 | 3 | 3 | 2 | 3 | 3 | 3 | 3 | 3 | 3 | 3 | 3 |
| BMDP | 1 | 1 | 1 | . | 3 | 3 | 3 | 2 | 1 | . | 2 | 2 | 3 | 3 | 2 | 3 | 2 |
| NISAN | 2 | 2 | 2 | 1 | . | 2 | 1 | 1 | 1 | 2 | 2 | 2 | 3 | . | 1 | 1 | 1 |
| GENSTAT 4.02 | . | . | 2 | . | 3 | 3 | 3 | 3 | 1 | 3 | 2 | 2 | 2 | 2 | 3 | 3 | 3 |
| SPEAKEASY III | 1 | 3 | 3 | 3 | 3 | 3 | 2 | 3 | . | 3 | 3 | 3 | 2 | 2 | 3 | 3 | 1 |
| TROLL | . | . | 2 | 1 | 3 | 2 | . | . | . | 3 | 3 | 3 | 2 | 3 | 3 | 3 | 1 |
| IDA | . | . | 3 | . | 3 | 3 | 3 | 3 | . | 3 | 1 | 3 | 3 | 3 | 3 | 3 | 1 |

several relevant attributes.  Figure 7.2 also points to other chapters which
describe programs with statistical analysis capabilities.

Figure 7.2 emphasizes the extent of the strengths of these programs beyond
"statistical analysis" to include multi-way contingency table analysis, econometrics, time series, and mathematical functions.

A close competitor to SAS, both in terms of the overall developer's D-score
(2.4) and the users' $\Omega$-score (2.6) on Table 13.1, is CS, the Consistent System.
This is not from the same mold as the well-known statistical packages, but is
"an interactive computer system for the management, analysis, and modelling of
data."  Access to CS is more likely to be via a network to the MIT machine than
by installing the system on the user's machine.  Its portability is limited to a
Honeywell computer with the Multics operating system.

It should be noted that the portability of SAS is also "limited" to IBM
370 plug-compatible machines, but this limitation is of decreasing importance
judging by the number of machines already in this group.

Figures 7.1 and 7.2 provide an overall picture of the strengths of these
programs and in particular how well the users agreed with developers.  Agreement
was quite high for CS, SAS, GENSTAT, and BMDP, but we can see a major disagreement between the developer and users of OMNITAB.  Again we repeat our cautions
of the Introduction to Chapter 6 and of Section 1.3.2:  differences may arise
for several reasons.  Also in the case of OMNITAB only one of the three users
suggested by the developer responded to our repeated mailings.  But whatever
the reasons, it is clear that for this user, who was recommended by the
developer, the program is not as useful as the developer believes it to be.

The remainder of this chapter contains descriptions of these statistical
analysis programs in the format described in Table 1.5.

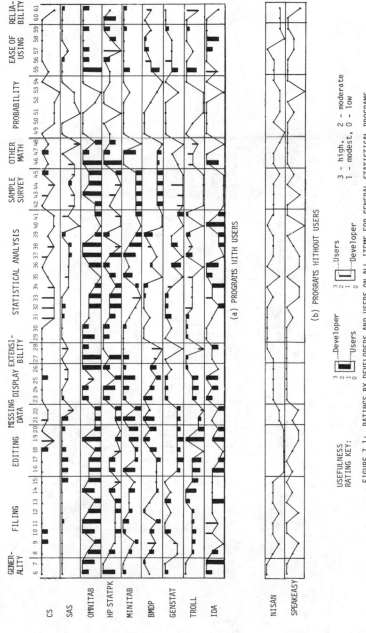

(a) PROGRAMS WITH USERS

(b) PROGRAMS WITHOUT USERS

USEFULNESS
RATING KEY:

3 — high,   2 — moderate
1 — modest, 0 — low

FIGURE 7.1:  RATINGS BY DEVELOPERS AND USERS ON ALL ITEMS FOR GENERAL STATISTICAL PROGRAMS

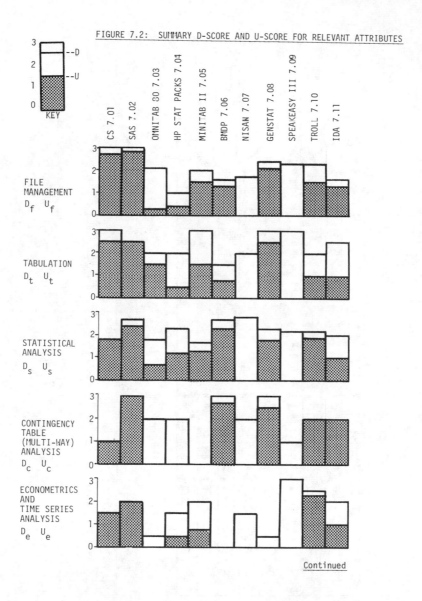

FIGURE 7.2:   SUMMARY D-SCORE AND U-SCORE FOR RELEVANT ATTRIBUTES

Continued

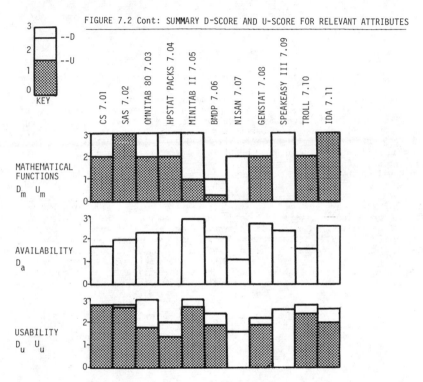

FIGURE 7.2 Cont: SUMMARY D-SCORE AND U-SCORE FOR RELEVANT ATTRIBUTES

Other chapters generally containing programs with strong Statistical Analysis Capabilities are Chapters 6, 8, and 9 (see Figures 6.2, 8.2, and 9.2).

7.01  <u>CS</u>

## INTRODUCTION

The Consistent System (CS) is an interactive computer system for the management, analysis, and modelling of data.  It has provisions for statistical
analysis, different types of data management, linear and network optimization,
indexing of natural language text, report writing, and graphical display of
information.  It was developed at the Massachusetts Institute of Technology
with the participation of researchers from Harvard University.  Development of
the core of the system took place between 1969 and 1976, development of library
components continues on an as-need basis.  It is installed at about six sites
at present.  The facilities at two of these sites are available through
regional or international networks to appropriate users.

The system is built around the notion that the non-programming user should
be able to explore and analyse data in a variety of ways, comparing the results
of different analyses when that is desirable, and without having to specify an
entire analysis in advance.  It is highly modular so that additions occur without disrupting established patterns of use and so that users can construct
macro-analyses as those are needed.  The user language is potentially different
for users of the tools of different disciplines.  All user languages are free
format and utilize English keywords.  While the system has been used in
teaching statistics, it was designed primarily for the experienced analyst with
potentially large datasets of potentially "dirty" data.

Because the non-programming user can combine existing facilities in a
variety of ways, the capabilities of the system for one user may differ widely
from its capabilities for another.  The comments that follow represent the
capabilities available in commands or command sequences that are explicitly
documented for those purposes, not the extended range of capabilities available
to the moderately experienced user.

## CAPABILITIES:  Processing and Displaying Data

The CS supports data input and management in two modes.  First, the system's self-describing files can be created directly by typing in, from cards or
tapes, or through a variety of data-generation procedures supported in the
system.  Facilities are available to edit, combine, sort, permute, and subset
multidimensional array files (with data matrix files as special cases) as well
as conventional files of text.  Second, the system contains a relational data
management subsystem with its own commands for database construction, query,
simultaneous reference to multiple datasets, aggregation operations, and the
like.  Naturally, the system contains commands to transform each of these two
forms into the other.  Many users of the system use the data management
environment to organize and prepare the data, and then pass orderly subsets of
them for statistical analysis.  The data management environment supports a
variety of data types, variable numbers of responses per subject, and complex
file organizations in addition to conventional data matrices and hierarchical
files.  Data are actually stored in the form represented, rather than being
"rectangularized" on input.

The system can cross-tabulate and form multidimensional arrays.  Those arrays may be displayed in several ways, or may be used as input to a variety of statistical procedures.  The cross-tabulation routines contain provisions for accumulating data sums and sums of squares as well as counts for input to statistical routines or reports that require them.

The system provides for scatterplots, histograms, time-series plots, auto-correlation plots, and schematic and stem-and-leaf plots on conventional terminals.  In addition, several types of plots are available for specialized graphical devices.  Contour and density plots are available, but not convenient at present.  Any collection of data values may be plotted, and the system's capabilities for transformation and generation of samples from a number of distributions make a variety of probability plots available to the user.

## CAPABILITIES:  Statistical Analysis

Among the CS's statistical facilities are provisions for basic descriptive statistics (including robust measures of locations and spread), exploratory data analysis (including most of the techniques proposed by Tukey), non-parametric statistics, and parametric significance tests and interval estimates.

Analysis of variance support nested designs, random and mixed models, repeated measures, and unequal cell sizes.  Multiple regression programs include an interactive stepwise procedure with which the user can control which variables are added and the selection criterion can differ from variable to variable if desired.  Residuals are available from regression and analysis of variance and can be passed to any statistical procedure as well as being displayed or plotted.

Multivariate facilities include the computation of several types of similarity and distance matrices.  Hierarchical clustering can then be performed using one of eight different clustering methods.  Principal component and factor analysis are also available, including a choice of estimation technique for the latter (e.g., maximum likelihood).

In addition to special time series display programs, there are programs to support time series computations, including autocorrelations and Box-Jenkins ARIMA models.

The system also supports other more specialized facilities, including linear programming, network optimization, automatic analysis of natural language text, and spatial (geographic) data management and modelling.

## EXTENSIBILITY

The system supports macro facilities that include parameterization, interception of terminal input and output, looping, and conditional flow of control. The organization of the mathematical and statistical components permit programming of new procedures in the macro-language.  The system also supports an interface to user-supplied FORTRAN (or equivalent) subroutines that permit most such routines to be used with other system facilities with no additional programming.  CS files can be passed into or out of APL workspaces, and CS programs can be invoked as APL functions.  Completely new user programs can be added without great difficulty.

INTERFACES WITH OTHER SYSTEMS

The CS contains a special version of TSP (TSP/Datatran, elsewhere in this volume). Information is automatically passed back and forth from that subsystem. An interface to BMDP save files is available; that interface is being expanded to accept BMDP files from other computer types.

PROPOSED ADDITIONS FOR NEXT YEAR

While the CS continues to grow and expand, it has been the long-standing policy to release new codes and capabilities when they are ready and fully tested. As a result, specific schedules for new developments are not released.

This example shows the use of the CS to perform parts of a regression analysis. The lines that begin with X are those typed by the user, other lines are printed by the computer. The excerpt shown here performs the regression, displays various output values, transforms one of the variables, and repeats the regression command. A more complete version of this example and an explanation of it appear in reference 2. "rgattr" is the name of the regression command that works on individual data attributes using the quadratic sweep technique. There are several other regression commands in the CS. To avoid misunderstandings, none of them are named "regression". Unless it is turned off by the user, the CS displays a "prompt" after each command completes execution. It is omitted here to save space.

```
Xeval rgattr(stack_loss on air_flow, water_temp, acid_conc
X    save(res:=residuals) incrsq) print coefs, rsq, anov

              coef  stand_err  dgfd   F_ratio    sig

constant    -39.920   11.896   17.000   11.261   0.004
air_flow      0.716    0.135   17.000   28.160   0.000
water_temp    1.295    0.368   17.000   12.387   0.003
acid_conc    -0.152    0.156   17.000    0.947   0.344


                 adj_rsquare
             rsquare

total_rsq    0.914   0.904
air_flow     0.846   0.846
water_temp   0.063   0.058
acid_conc    0.005   0.000


        sum_of_squares  dgf   mean_square  F_ratio  sig

total       2069.238   20.000   103.462      *       *
regression  1890.408    3.000   630.136   59.902   0.000
error        178.830   17.000    10.519      *       *

Xcm log_sl := log10(stack_loss)
Xeval rgattr(log_sl on air_flow, water_temp, acid_conc
X    save(res:=residuals)) print coefs,rsq

    (Output omitted)
Xplot_res res

    (Output omitted)
```

## CONSISTENT SYSTEM

Developed by:                                    Distributed by:

Laboratory of Architecture and          Renaissance Computing, Inc.
   Planning                              PO Box 699
Room 4-209                               Cambridge, MA   02139
Massachusetts Institute of Technology   (617) 491-0900
Cambridge, MA   02139

Computer Makes and Operating System:    Interfaced Systems:

Honeywell Multics (DPS-68, or            BMDP-79
   equivalent)                           IMSL

Source Lanugage:

ANSI/ISO PL/I, with some computational codes in FORTRAN

Cost:

License and support fees depend on the level of services required.  Contact
Renaissance Computing at the address above for further information.

Network access to the version running on the MIT machine is available at
standard rates to appropriate customers.

Documentation:

The CS is extensively documented in over 6000 pages of tutorial, reference,
and internal documentation.  A complete annotated list of available publications, and the publications themselves, are available from the Laboratory of
Architecture and Planning, at the address above.  The documents most likely to
be of interest to the readers of this book are:

Consistent System:  Program Summaries
Consistent System:  Annotated Examples
Consistent System:  Elementary Janus Manual
Consistent System:  Introduction for New Statistical Users
Consistent System:  Beginners Manual
Consistent System:  Handbook of Programs and Data

References:

Dawson, Ree and Christine Waternaux,  "Software Capabilities for the Analysis
     of Growth Data", Statistical Computing Section Proceedings of the American
     Statistical Association.  1978.
Dawson, Ree, John C. Klensin, and Douwe B. Yntema, "The Consistent System", The
     American Statistican, 35, 3 (August 1980), pp. 169-176.
Dawson, Ree and John C. Klensin, "User Extensions to Statistical Software",
     Paper given at the 1980 Joint Statistical Meetings, Houston, August 1980.
     To appear in Statistical Computing Section Proceedings of the American
     Statistical Association, 1980.
Hill, Claire and Paulann Balch, "On the particular applicability and usefulness
     of relational data base systems for management and analysis of medical
     data", Paper to appear in Fourth Annual Symposium on Computer Applications
     in Medical Care, 1980.

7.02  <u>SAS</u>

## INTRODUCTION

SAS is a software system that provides tools for data analysis: information storage and retrieval, data modification and programming, report writing, statistical analysis, and file handling. Since its beginnings in 1966, SAS has been installed at over 2000 installations worldwide, and is used by statisticians, marketing researchers, biologists, auditors, social scientists, business executives, medical researchers, computer performance analysts, and many others.

## CAPABILITIES: Data Processing and Display

The SAS language is free-format, with an English-like syntax. Language constructs like DO/END and IF-THEN/ELSE, array processing statements, and a macro facility are also included. Data can be introduced into the system in any form from any device. Data management features include creating, storing, and retrieving data sets. SAS can retrieve complex files containing variable-length and mixed record types, and hierarchical records. SAS reads BDAM, ISAM, and VSAM files too. SAS has utility procedures for printing, sorting, ranking, and plotting data; copying files from input to output tapes; listing label information from IBM tape volumes; listing, renaming, and deleting SAS data sets and partitioned data sets.

Report-writing capabilities include automatic or custom-tailored reports with built-in or user-specified formats, value labels, and titles. Printer graphics include pie, star, block, and vertical and horizontal bar charts, as well as scatter plots with overlay and overprint options.

SAS Institute also offers SAS/GRAPH, a product for device-intelligent interactive color computer graphics.

## CAPABILITIES: Statistical Analysis

SAS offers over 50 procedures for summary statistics; multiple linear or nonlinear regression; analysis of variance and covariance, multivariate analysis of variance; correlations; discriminant analysis; factor analysis; Guttman scaling; frequency and crosstabulation tables; categorical data analysis; spectral analysis; autoregression; two- and three-stage least squares; t-tests; variance component estimation; and matrix manipulation.

SAS Institue also offers SAS/ETS, a planning system for forecasting, modeling, and reporting.

## USER-WRITTEN PROCEDURES AND INTERFACES TO OTHER SYSTEMS

Users can write their own procedures. SAS also includes a BMDP interface procedure, as well as a procedure for converting BMDP, OSIRIS, and SPSS system files to SAS data sets.

PROPOSED ADDITIONS IN NEXT YEAR

SAS Institute's plans for the coming year include:
* IMS interface
* ability to reenter lines with syntax errors under TSO
* symbolic parameters in macros
* more explanation and examples for writing procedures
* procedure for transposing a SAS data set, changing observations into variables and vice versa
* procedure to compute ridge and incomplete principal component parameter estimates for linear regression models
* procedure to compute principal components of variables in a SAS data set.

REFERENCES

"A Datapro Report on SAS," Datapro 70, No. 70E-736-01a.  Delran, NJ:  Datapro
     Research Corporation, 1980.
"AUERBACH on SAS," AUERBACH Computer Technology Reports, No.660.5831.700.
     Pennsauken, NJ:  AUERBACH Publishers Inc., 1979.
Proceedings of the Third Annual SAS Users Group International Conference.
     Raleigh, NC:  SAS Institute Inc., 1978.
Proceedings of the Fourth Annual SAS Users Group International Conference.
     Raleigh, NC:  SAS Institute Inc., 1979.
Proceedings of the Fifth Annual SAS Users Group International Conference. Cary,
     NC:  SAS Institute Inc., 1980.

SAMPLE JOB

     For each level of CITY, the CHART procedure draws a three-dimensional building whose height represents the sum of the values of REVENUE.

```
PROC CHART;
    BLOCK DAY/DISCRETE TYPE=SUM  SUMVAR=REVENUE
         GROUP=WEEK  SYMBOL='XOA'  MISSING;
    TITLE  DAILY  REVENUE FROM BRANCH OFFICES FOR APRIL, 1979; BY CITY;
```

Output:

BLOCK CHART OF REVENUE

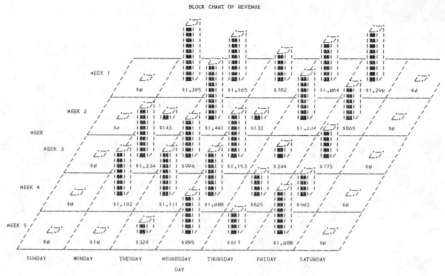

## SAS: Statistical Analysis System

Developed by:                              Distributed by:

   SAS Institute Inc.                      Same
   Box 8000, SAS Circle
   Cary, NC   27511

Computer Makes:                            Operating Systems:

   IBM 360/370/403x/43xxSeries             OS, OS/VS, VM/CMS
   Amdahl                                  MFT, MVT, VSI, VS2
   Itel                                    SVS, MVS
   National
   Two Pi
   Magnuson
   Hitachi
   Nanodata

Source Language

   PL/I
   IBM assembler

Cost (U.S. only)

   SAS is available for a one-year license fee of $4500 for corporate customers; $2000 each year thereafter.
     Government:  $4000/$2000
     Degree-granting institutions:  $1000/$500

   This price includes all maintenance, updates, one copy of all documentation, and limited consulting by telephone and mail.

Documentation

Helwig, Jane T., SAS Introductory Guide, Raleigh, NC:  SAS Institute Inc., 1978.
SAS Applications Guide, 1980 Edition.  SAS Institute Inc. Cary, NC:  SAS Institute Inc. 1980.
SAS Programmer's Guide, 1979 Edition.  SAS Institute Inc. Raleigh, NC:  SAS Institute Inc., 1979.
SAS Supplemental Library User's Guide, 1980 Edition.  SAS Institute Inc. Cary, NC:  SAS Institute Inc., 1980.
SAS User's Guide, 1979 Edition.  SAS Institute Inc. Raleigh, NC:  SAS Institute Inc., 1979.
SAS Views, 1980 Edition.  SAS Institute Inc. Cary, NC:  SAS Institute Inc., 1980.
SAS/ETS User's Guide, 1980 Edition.  SAS Institute Inc. Cary, NC:  SAS Institute Inc., 1980
SAS/GRAPH User's Guide, 1980 Edition.  SAS Institute Inc. Cary, NC:  SAS Institute Inc., 1980.

## 7.03  OMNITAB

### INTRODUCTION

OMNITAB 80 is a high quality integrated general-purpose programming language and statistical software computing system. The system enables the user to use a digital computer to perform data, statistical and numerical analysis without having any prior knowledge of computers or computer language. Simple instructions, in one's own native language, are written to obtain accurate results since reliable, varied and sophisticated algorithms for data analysis and manipulation are referenced. It may be used either interactively or in batch mode. OMNITAB II, a predecessor of OMNITAB 80, has been installed nationally and internationally. OMNITAB was conceived and developed by Joseph Hilsenrath and co-workers of the National Bureau of Standards in early 1960's. Since then it has gone through major modifications and expansions under the guidance of Dr. David Hogben and Mrs. Sally T. Peavy at the Statistical Engineering Division, Center for Applied Mathematics, National Bureau of Standards. OMNITAB 80 is transportable to any computer configuration sufficiently large to accommodate it.

### CAPABILITIES: Analysis of Data

OMNITAB 80 permits one to perform simple arithmetic, complex arithmetic, trigonometric calculations, data manipulation, special function calculations, statistical analysis and operations on matrices and arrays, alleviating the tedious task of formatting data for data input and output through free field input and readable and automatic printing of results. Because OMNITAB 80 is an integrated system, data and subsets of data can be analyzed in many ways within the same set of instructions. The system has extensive plotting, numerical analysis and matrix analysis capabilities. For interactive use, instructions exist for controlling size of page, amount of printing, and location of printing. When CRT's are used, printing can be stopped at the end of each page. Two instructions, CONTENTS and DESCRIBE, provide help for the interactive user.

### CAPABILITIES: Statistical Analysis

OMNITAB 80's statistical capability includes: elementary analysis (oneway and twoway analysis of variance), regression (least absolute residuals estimation and selection of best subsets), correlation analysis, cross tabulation of any 14 statistics, contingency table analysis, and over 100 instructions for probability densities, cumulatives, percentiles, probability plots and random samples. Almost all the statistical analysis instructions automatically (unless otherwise specified) provide comprehensive output.

The automatic output of the regression analysis instruction, among other information, will consist of the predicted values, the standard deviation of the predicted values, the residuals, the standardized residuals, diagnostics for identifying influential measurements, four plots of the standardized residuals, variance-covariance matrix, estimated correlations of the parameter estimates, estimates, accuracy and for the special equation $y_i = a_o + a_1 x_i$, a 95% confidence ellipse.

REFERENCES

ABRAMOWITZ, MILTON and STEGUN, IRENE (1964).  Handbook of Mathematical Func-
     tions, NBS Applied Mathematics Series 55, Superintendent of Documents,
     U.S. Government Printing Office, Washington, D.C.  20402.
BROWNLEE, K. A. (1965).  Statistical Theory and Methodology in Science and
     Technology, 2nd Edition, John Wiley and Sons, Inc.
HILSENRATH, J., ZIEGLER, G. G., MESSINA, C. G., WALSH, P. J. and HERBOLD, R. J.,
     (1966).  OMNITAB:  A Computer Program for Statistical and Numerical Analy-
     sis.  National Bureau of Standards Handbook 101, Superintendent of
     Documents, U.S. Government Printing Office, Washington, D.C.  20402.
     Reissued January 1968 with corrections.
NATRELLA, M. G. (1963).  Experimental Statistics.  National Bureau of Standards
     Handbook 91.  U.S. Government Printing Office.
WAMPLER, R. H. (1969).  An evaluation of linear least squares computer programs.
     Journal of Research, National Bureau of Standards, 73B, pages 59-90.
WAMPLER, R. H. (1970).  A report on the accuracy of some widely used least
     squares computer programs.  Journal of American Statistical Association,
     65, pages 549-565.

SAMPLE JOB

```
OMNITAB BROWNLEE 11.18 LACK OF FIT
LABEL STOP-DIST, MILES-PER-HR
SET STOPPING DISTANCE OF AUTO IN VARIABLE STOP-DIST
3.92  3.65  5.82  5.20  8.55  10.63  11.94
SET DATA FOR MILES-PER-HR
20.5  20.5  30.5  30.5  40.5  48.8  57.8
POLYFIT STOP-DIST WTS=1.0, DEGREE 1 FOR MILES-PER-HR
STATISTICAL ANALYSIS OF MILES-PER-HR DATA
STOP
```

Partial Sample Output

---

```
               LEAST SQUARES FIT OF RESPONSE, STOP-DIST,
        AS A POLYNOMIAL OF DEGREE  1. INDEPENDENT VARIABLE IS MILES-PER-HR
              USING    7 NON-ZERO WEIGHTS = 1.0000000
```

---

```
     DIAGNOSTIC INFORMATION FOR IDENTIFYING INFLUENTIAL MEASUREMENTS.
       I  = ROW, FOR  7 LARGEST VALUES, T(I) = STANDARDIZED RESIDUAL,
     H(I) = DIAGONAL OF HAT MATRIX,       D(I) = COOK STATISTIC,
   WSSD(I) = DANIEL-WOOD STATISTIC,       V(I) = VAR(YHAT) / VAR(RESIDUAL).
```

| I | T(I) | I | H(I) | I | D(I) | I | WSSD(I) | I | V(I) |
|---|------|---|------|---|------|---|---------|---|------|
| 4 | -1.74 | 7 | .554 | 7 | .61 | 7 | 126.68 | 7 | 1.24 |
| 6 | 1.22 | 1 | .333 | 6 | .30 | 1 | 58.42 | 1 | .50 |
| 7 | -.99 | 2 | .333 | 4 | .30 | 2 | 58.42 | 2 | .50 |
| 1 | .83 | 6 | .266 | 1 | .17 | 6 | 44.82 | 6 | .41 |
| 5 | .75 | 3 | .164 | 5 | .05 | 3 | 6.64 | 3 | .20 |
| 3 | -.25 | 4 | .164 | 3 | .01 | 4 | 6.64 | 4 | .20 |
| 2 | .11 | 5 | .163 | 2 | .00 | 5 | 6.20 | 5 | .19 |

---

```
       THE DURBIN-WATSON STATISTIC IS D =  2.0840896
```

Partial Sample Output, continued

ANALYSIS OF VARIANCE

-DEPENDENT ON ORDER INDEPENDENT VARIABLES ARE ENTERED, UNLESS VECTORS ARE ORTHOGONAL-

| TERM | SS=RED. DUE TO COEF. | CUM. RESIDUAL MS | D.F. | F(COEF=0) | P(F) | F(COEFS=0) | P(F) |
|---|---|---|---|---|---|---|---|
| 0 | 353.01201 | 10.875381 | 6 | 1692.346 | .000 | 1000.084 | .000 |
| 1 | 64.209316 | .20859323 | 5 | 307.821 | .000 | 307.821 | .000 |
| RESIDUAL | 1.0429662 | | 5 | | | | |
| TOTAL | 418.26430 | | 7 | | | | |

ESTIMATES FROM LEAST SQUARES FIT

| TERM | COEFFICIENT | S.D. OF COEFF. | RATIO | ACCURACY* |
|---|---|---|---|---|
| 0 | -1.1331589 | .50008449 | -2.27 | 8.00 |
| 1 | .23140149 | .013189163 | 17.54 | 8.00 |

RESIDUAL STANDARD DEVIATION = .45672008
BASED ON DEGREES OF FREEDOM    7 - 2 = 5

*THE NUMBER OF CORRECTLY COMPUTED DIGITS IN EACH COEFFICIENT USUALLY DIFFERS BY LESS THAN 1 FROM THE NUMBER GIVEN HERE.

OMNITAB 80

Developed by:                                      Distributed by:

   Dr. David Hogben                                   Mrs. Alice Dugan
   Mrs. Sally T. Peavy                                National Bureau of Standards
   National Bureau of Standards                       A323, Physic Bldg.
   Room A337, Bldg. 101                               Washington, D.C.  20234
   Washington, D.C.  20234

Computer Makes:                                    Operating Systems:

   UNIVAC-1108-1110                                   UNIVAC-EXE8
   IBM-4331                                           IBM-VM/CMS
   CDC-CYBER 173                                      CDC-NOS 1.4
   Hewlett Packard-HP 3000                            MPE III
   IBM-158                                            IBM-OS/VS1
   DEC-PDP 10                                         TOPS-10
   CDC-6600                                           KRONOS 2.1

Source Languages:  FORTRAN (PFORT)

Cost:

     There is a one-time charge of $1500 for domestic use and $1600 for foreign use.

Documentation:

HOGBEN, DAVID and PEAVY, SALLY T. (1977).  OMNITAB II User's Reference Manual
   1977 Supplement.  NBSIR 77-1276, Superintendent of Documents, U.S.
   Government Printing Office.  Washington, D.C.  20402
HOGBEN, DAVID, PEAVY, S.T., and VARNER, R.N. (1970).  OMNITAB II User's
   Reference Manual.  NBS Technical Note 552, Superintendent of Documents,
   U.S. Government Printing Office, Washington, D.C.  20402
KU, H.H. (1973).  A User's Guide to the OMNITAB Command "STATISTICAL ANALYSIS"
   NBS Technical Note 756, (out of print).
PEAVY, SALLY T., VARNER, RUTH N., BREMER, SHIRLEY G. (1970).  A Systems
   Programmer's Guide for Implementing OMNITAB II.  NBS Technical Note 550,
   (out of print).
PEAVY, S.T., VARNER, R.N., and HOGBEN, DAVID (1970a).  Source Listing of
   OMNITAB II Program.  NBS Special Publication 339, Superintendent of
   Documents, U.S. Government Printing Office, Washington, D.C.  20402.
VARNER, R.N., and PEAVY, S.T. (1970).  Test Problems and Results for OMNITAB
   II.  NBS Technical Note 551, Superintendent of Documents, U.S. Government
   Printing Office, Washington, D.C.  20402.

## 7.04  HP STAT PACKS

### INTRODUCTION

Hewlett-Packard's STAT PACKS is an integrated package developed specifically for the HP desktop computers.  The package uses a common front end, which
provides for considerable flexibility in data handling.  The basic statistics
and data manipulation front end have recently been enhanced to include additional features such as searching and selecting operators.  The programs are
interactive and use the CRT display to list a "menu" of options at appropriate
times.  A group of special function keys have been programmed to act as a
"command" language to pass the user directly to the specific operation or analysis that is desired.  The package's data analysis capabilities range from
elementary basic statistics to factor analysis procedures.  The analyses
included in the package are enhanced by extensive use of graphical output
either on the CRT graphics terminal or on hard copy plotters.  This package was
developed under the supervision of Thomas J. Boardman and has been used in the
Colorado State University Statistical Laboratory for several years.

### CAPABILITIES:  Processing and Displaying Data

Extensive data management capabilities are built into the HP Stat Library.
Several input modes are available including input directly from a graphics tab-
let device.  Friendly sorting, joining, subfile creation, recoding, and transforming are available to the user.  The user may select portions of the data
set by specifying those subfiles to be used and/or to select a portion of the
total data set.  Files may be stored on mass storage devices for future use.
These files may be easily catalogued by the user to check for specific
characteristics.

Extensive editing capabilities are available to the user.  In addition,
the data may be scanned for specified characteristics such as outliers or
missing values and those data sets may be edited or deleted.  A survey analyzer
package, which will include tabulation and cross tabulation routines, will be
developed during the 1980-81 for inclusion in the Stat Library.

One of the strong points of the library is the extensive use of graphical
tools throughout the package.  These are plotter quality graphics rather than
line printer plots because the graphics hardware is built into the CRT and
most users have flatbed plotters as well.  All images on the CRT can be dumped
to the internal thermal printer.  While all of the statistical analyses subprograms described below make use of graphics, the Statistical Graphics Pack
has most of the common graphics.  The Stat Graphics II subprogram under development will add an extensive array of EDA graphics as well as some of the most
useful multivariable plots.

### CAPABILITIES:  Statistical Analyses

A few of the statistical subprograms which are available are:

*       General Statistics:  most commonly used one, two and multisample
        parametric and nonparametric methods.

* Regression Analysis:  multiple linear regression, several selection
  procedures included stepwise regression, polynomial regression, and
  nonlinear regression--estimation procedures for one or more independent variables with user specified function.
* Analysis of Variance:  AOV for factorials balanced nested or mixed
  nested, two-way unbalanced factorials design and one-way classifications with or without covariate.
* Factor Analysis:  principal component analysis and factor analysis
  with choice of several oblique or orthogonal rotation schemes.
* Statistical Graphics Pack:  includes some of the common elementary
  forecasting and smoothing procedures as well as graphics routines to
  do bar charts and pie charts.

Four additional programs add complementary routines to those listed above.
These are:  Monte Carlo simulation; numerical analysis subroutines; linear programming package; a subprogram which provides either tabled values or right
tail probabilities for most of the common continuous and discrete distributions
including t, normal, lognormal, F, gamma, beta, chi-square, Weibull, Laplace,
logistic, and negative binomial, hypergeometric, Poisson, binomial.

## EXTENSIBILITY

New routines may be added by the users or the developers without difficulty.  The front end (Basic Statistics and Data Manipulation) provides an
effective starting point for any number of statistical or data manipulative
subprograms.

## INTERFACES WITH OTHER HP9845 PROGRAM PACKAGES

While this library is designed to operate only on the HP9845 desktop computer line, within that line some provisions have been to provide linkages
between several program packages such as the Data Base Management System called
QUERY 45.

## PROPOSED ADDITIONS FOR NEXT YEAR

The HP Stat Library is continually undergoing development to add new
enhancements to existing subprograms and, of course, to fix known anomalies.
Two new subprograms are in the development stages:  the first is a Stat Graphics
II collection of extensive new statistical graphical routines.  The second is a
survey analyses package designed to allow users to perform extensive tabulations on their data set.  The latter package will provide graphics output for
these tabulations.

## REFERENCES

Boardman, T.J., 1978.  "Developing an Integrated System for Data Exploration
    and Analysis on Desktop Computers." Proceedings of the Statistical
    Computing Section of ASA.
Boardman, T.J., 1980.  "Statistical Computing on Desktop Computers." The
    American Statistician, sent June 1980.

## SAMPLE JOB

Since this program operates interactively, it is not possible to show a
command language to create the Andrews Plot shown on the next page.  The data
is from Jolicoeur and Mosiman "Size and Shape Variation in the Painted Turtle--

A Principal Component Analysis," Growth, 1960, 24: 339-354. The three variables are length, width, and height of the carapace. There are 24 female and 24 male turtles in the data set. The Andrews function shows separation between the males and females from about zero PI to .6 PI, with the females on top of the males in the graph. Andrews plots and other multivariable plots can be useful in looking for patterns in a data set.

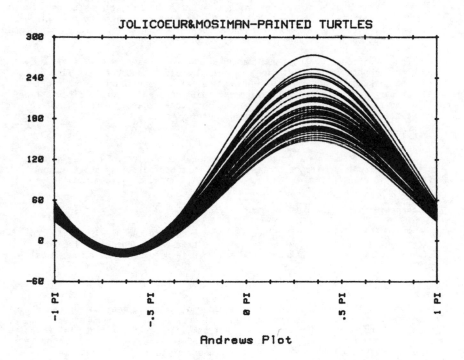

JOLICOEUR&MOSIMAN-PAINTED TURTLES

Andrews Plot

## HP STAT LIBRARY

Developed by:                                    Distributed by:

    T.J. Boardman                                David Deane, Marketing
    Statistical Laboratory                       Hewlett-Packard, D.C.D.
    Colorado State University                    3400 East Harmony Road
    Fort Collins, CO 80523                       Fort Collins, CO 80525

Computer Makes:

    Hewlett-Packard 9845 (System 45)
    Desktop Computer

Source Language:

    Hewlett-Packard extended BASIC

Cost:

The cost is approximately $1,500. The fee includes program source cartridges and the HP Stat Library Users Manual with annotated examples.

## 7.05  MINITAB II

### INTRODUCTION

Minitab is a general-purpose interactive statistical computing system.  It is designed for students and researchers with no computer experience, as well as for experienced data analysts.  A compatible batch version is available.  Minitab has been particularly useful for plotting, regression, general data manipulation and arithmetic.

### CAPABILITIES:  Processing and Displaying Data

Data sets are stored in a rectangular worksheet (rows by columns) in main memory.  The size of the worksheet varies with the installation, from just a few thousand numbers to several million.  Several data sets can reside in the worksheet at the same time.  Data sets can be subsetted (by case or variable), sorted, and concatenated; variables can be transformed, added, and deleted; values can be corrected, and recoded.  Missing data are handled automatically.

Minitab can display data in histograms, and in plots, including multiple plots on the same axes, plots with letters indicating group membership, time series plots, and probability plots.  In all cases, "nice" scales are automatically chosen unless the user specifies his own.

The TABLE command forms and prints multiway tables (from 1 to 11 factors).  Compact output can be produced by nesting several factors on the rows and/or columns.  The cells of the table may contain one or more of the following: counts and percents (i.e., crosstabulation), statistics such as means, medians, maxima, standard deviations, and the original data.  Printing of marginal statistics, information on missing values, and the use of case weights are all options.

### CAPABILITIES:  Statistical Analysis

Statistics such as means, medians, standard deviations, and correlations; t-tests, confidence intervals, chisquare tests, and non-parametric procedures.

Regression analysis, including weighted least squares, transformations, diagnostics for multicollinearity, outliers, and high leverage points.  Residuals, fitted values, and coefficients can be stored for further analysis.  One and two-way analysis of variance, general analysis of variance and covariance by regression.  Time series analysis, including autocorrelation, partial autocorrelation, and crosscorrelation plots, arima (Box-Jenkins) models, and regression on lagged variables.  Commands for EDA, including stem-and-leaf displays, boxplots, median polish, resistant line fits, and robust smoothing of time series.

### CAPABILITIES:  General

Data can be stored in a matrix, simple operations such as add, multiply, transpose, invert, find eigenvalues and vectors can be done.  Any set of Minitab commands can be stored for repeated execution (macros).  A simple looping capability is provided.  Output width can be varied.  Random data can be generated from a variety of distributions.  An on-line HELP facility provides general information on Minitab as well as details of a specific Minitab command. The Minitab source code is very modular, therefore it is easy to add local features.

PROPOSED ADDITIONS IN NEXT YEAR

Stepwise regression (including forward selection and backwards elimination as special cases); interface to BMDP files and to SPSS files; additional EDA commands; a new RECODE command; multiple columns on HISTOGRAMS, TTEST, and some other commands.

REFERENCES

Allen, I.F., P.F. Velleman, "The Handiness of Package Regression Routines," Proc. of Stat. Comp. Section, American Statistical Association, 1977, pp. 95-101.
Inglis, J., Book Review of Minitab Student Handbook by Ryan, Joiner, Ryan, JASA, 72, 1977, p. 932.
Jackson, J.E., Book Review of Minitab Student Handbook by Ryan, Joiner, Ryan, Technometrics, 20, 1978, p. 211.
Reid, A.J., and J.S. Lemon, "A Review of Interactive Statistics Packages in British Universities and Polytechnics," Proceedings of COMPSTAT, University of Edinburgh. August 1980.
Thisted, R.A., "User Documentation and Control Languages: Evaluation and Comparison of Statistical Computing Packages," Proc. Stat. Comp. Sec., American Statistical Association, 1976, pp. 24-30.
Velleman, P.F., J. Seaman, I.E. Allen, "Evaluating Package Regression Routines," Proc. Stat. Comp. Section, American Statistical Association, 1977, pp. 82-92.

SAMPLE JOB

Yield of a chemical process was measured at 10 temperatures (in degrees Fahrenheit) on each of 4 days. The program inputs the data from file TEST025, converts to degrees Celsius, does a plot, fits a line, and does some checks on the residuals.

```
NAMES C1='YIELD' C2='FTEMP' C3='DAY' C4='TEMP' C5='RESIDS'
READ 'TEST025'  'YIELD' 'FTEMP' 'DAY'
LET 'TEMP' = (5/9)*('FTEMP' - 32)
PLOT 'YIELD' VS 'TEMP'
REGRESS 'YIELD' ON 1 PREDICTOR 'TEMP', STORE 'RESIDS'
HISTOGRAM 'RESIDS'
PLOT 'RESIDS' VS 'TEMP'
BOXPLOTS 'RESIDS' BY 'DAY'
```

Part of the output from this job is given on the next page.

```
THE REGRESSION EQUATION IS
Y =    1148. + 0.586 X1

                                            ST. DEV.   T-RATIO =
          COLUMN       COEFFICIENT          OF COEF.   COEF/S.D.
          --             1147.55              11.11     103.27
X1        TEMP            0.586               0.206      2.85

THE ST. DEV. OF Y ABOUT REGRESSION LINE IS
S =       3.598, WITH ( 30- 2) = 28 D.F.

R-SQUARED = 22.5 PERCENT
R-SQUARED = 19.7 PERCENT, ADJUSTED FOR D.F.

ANALYSIS OF VARIANCE
  DUE TO      DF              SS        MS=SS/DF
REGRESSION     1           104.97        104.97
RESIDUAL      28           362.53         12.95
TOTAL         29           467.50

          X1          Y      PRED. Y    ST.DEV.
ROW     TEMP       YIELD      VALUE      PRED. Y    RESIDUAL    ST.RES.
 11     48.9    1168.601    1176.207      1.221      -7.607     -2.25R

R DENOTES AN OBS. WITH A LARGE ST. RES.


-- HISTOGRAM 'RESIDS'

   MIDDLE OF    NUMBER OF
   INTERVAL     OBSERVATIONS
     -2.0           1       *
     -1.5           3       ***
     -1.0           5       *****
     -0.5           2       **
      0.0           5       *****
      0.5           4       ****
      1.0           7       *******
      1.5           3       ***


-- BOXPLOTS 'RESIDS' BY 'DAY'
```

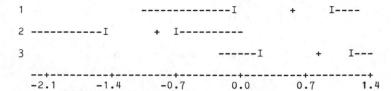

```
      1                      --------------I      +      I----

      2  -----------I        +   I----------

      3                            ------I       +      I---

      --+---------+---------+---------+---------+---------+
        -2.1      -1.4      -0.7      0.0       0.7       1.4
```

MINITAB, Version 80

Developed by:                                    Distributed by:

  Minitab Project                                  Same
  215 Pond Laboratory
  University Park, PA  16802
  Phone:  (814) 865-1595

Computer Makes:

  Amdahl (all)
  Burroughs 4000, 5000, 6000, 7000
  CDC (all)
  Data General Eclipse under AOS
  DEC 10, 20
  Harris (all)
  Hewlett Packard, HP 3000
  Honeywell 6000, under Multics, GCOS, DTSS
  IBM 360, models 40 and up          /
  IBM 370, models 115 and up         /
  IBM 4300                           /--under DOS/VS, OS, VS, VM,
  IBM 3030                           /   TSO, CMS
  Itel                               /
  PDP-11 11/03 and up under RSTS, RSX11, RT11, IAS
  PRIME 150, 250, 350 and up
  Tandem 16
  Texas Instrument TI 990
  Univac 70, 80, 90 and 1100
  VAX under VMS and UNIX
  Xerox Sigma

Source Language:  ANSI FORTRAN

Cost:

    Minitab is licensed for $1,000 per year, with a 50% discount for academic
installations.  Discounts are available for multiple machine sites.  This fee
includes (1) A magnetic tape with Fortran source (load modules are also
available for some brands), test problems, and the Reference Manual, (2) Updates
and limited telephone consulting, (3) One copy each of the Minitab Student
Handbook and the Reference Manual.

Documentation:

Ryan, T., B. Joiner, B. Ryan, Minitab Student Handbook, Duxbury Press, 1976.
Ryan, T., B. Joiner, B. Ryan, Minitab Reference Manual, Minitab Project, 1980.

7.06  BMDP

## INTRODUCTION

BMDP is a comprehensive library of general-purpose statistical computer programs that are integrated by a common English-based control language and self-documented save files for data and results.  Programs are available for beginners through advanced users.  Basic analyses can be made with a minimum of effort.  Many analysis features are not contained in any other widely distributed statistical system.  Emphasis is placed on integrating graphical displays with analysis and on assessing the plausibility that underlying statistical assumptions hold.  BMDP is available for large and small computers.  It is installed at over 1000 facilities throughout the world.  Programs are usually run in a batch mode, although many users set up and submit jobs from a computer terminal and then view output on the terminal.

## CAPABILITIES:  Processing and Displaying Data

Data can be entered into BMDP from formatted files, binary files, free-formatted files, BMDP files, and a user definable subroutine.  BMDP is interfaced with P-STAT, SAS, and SIR.  BMDP files can be sorted.  Several techniques are available for data screening.  Emphasis is given in BMDP on identifying cases (observations) with illegal and implausible values.  Values declared to be missing or out of range are automatically excluded from computations.  Missing values can be estimated by various regression and group mean techniques.  Data can be modified by a transformation processor that includes IF-THEN statements and a wide variety of mathematical functions.  Various methods are available for printing or analyzing all or part of a data file.  Two-way and multiway frequency tables (cross-tabulation) can be made.  Basic and detailed statistics can be obtained for subgroups of data.  Histograms, bivariate scatter plots, normal probability plots, cluster description plots, etc. are available.

## CAPABILITIES:  Statistical Analysis

Two-way and multiway frequency table analysis include a wide variety of statistics and loglinear models.  Linear regression features include simple, multiple, stepwise, all possible subsets, extensive residual analysis, detection of influential cases and multivariate outliers, principal component, stepwise polynomial, multivariate, and partial correlation.  Nonlinear regression includes derivative-based and derivative-free, stepwise logistic, and models defined by partial differential equations.  Analysis of variance features include t-tests and one- and two-way designs with histograms and detailed statistics for each cell, factorial analysis of variance and covariance including repeated measures, and balanced and unbalanced mixed models.  Multivariate techniques include factor analysis, multivariate outlier detection, hierarchical clustering of variables and cases, k-means clustering of cases, partial and canonical correlation, multivariate analysis of variance and discriminant analysis.  Other programs include nonparametric statistics, survival analysis, (including Cox-model covariates), and interactive Box-Jenkins analysis.

## EXTENSIBILITY

BMDP transformations can be supplemented by user defined FORTRAN statements.  These FORTRAN statements can include READ and WRITE statements, calls to subroutines, etc. It is also easy to replace or expand any BMDP subroutine by the same procedure. A sample BMDP program is provided to allow users to develop their own BMDP style programs.

## INTERFACES WITH OTHER SYSTEMS

The P-STAT, SAS, and SIR systems write BMDP files. P-STAT and SAS also read BMDP files. Documentation is also provided so that BMDP files can be read and written in user-developed programs. The FORTRAN interface described above allows interfaces with other systems also.

## PROPOSED ADDITIONS IN NEXT YEAR

Programs have been developed for release in early 1981 for spectral analysis, interactive Box-Jenkins analysis, model building for multiway frequency tables, Cox models (covariates) for survival analysis, multivariate analysis of variance and covariance including the multivariate approach to repeated measures, linear scores from preference pairs, and Boolean (binary) factor analysis. Several of these are distributed in preliminary form in 1980. A new manual will appear in early 1981. Extensive modifications are also being made to facilitate distribution of BMDP to different computer types including very small machines such as the LSI-11 version of PDP-11.

## REFERENCES:

Berk, K.N. and I. Francis. A review of the manuals for BMDP and SPSS. J. Am. Statist. Assoc. 73, p. 65 (1978).
Garcia-Pena, J. and S. Azen. The performance of BMDP3R in nonlinear estimation. Proceedings of the Computer Science and Statistics: 12th Annual Symposium on the Interface, pp. 349-353 (1979).
Muller, M.E. A review of the manuals for BMDP and SPSS. J. Am. Statist. Assoc 73, p. 71 (1978).
Wilkinson, L. and G.E. Dallal. Accuracy of sample moment calculations among widely used statistical programs. The American Statistician 31, p. 128 (1977).

## SAMPLE JOB

Perform two-sample t-test. Data are input from a self-documented BMDP file or free-formatted external file. Standard output and output from Robust option is made for either form of input. Output width is adjusted to a 72 character line length.

```
/ INPUT    VAR IS 9. FORMAT IS FREE.
/ VAR      NAMES ARE ID, BRTHPILL,  ..., ALBUMIN,... .
/ GROUP    CODES (2) ARE 1,2. NAMES(2) ARE NOPILL, PILL.
/ PRINT    LINE IS 72.
/ TEST     VAR IS ALBUMIN.
/ TEST     VAR IS ALBUMIN. ROBUST.
/ END
              ---- or ----
/ INPUT    UNIT IS 9.  CODE IS WERNER.
/ PRINT    LINE IS 72.
/ TEST     VAR IS ALBUMIN.
/ TEST     VAR IS ALBUMIN. ROBUST
/ END
```

```
                    ---- USUAL OUTPUT ----


************
* ALBUMIN *  VARIABLE NUMBER   7      GROUP      1 NOPILL   2 PILL
************                          MEAN         42.2799    40.3599
             STATISTICS    P VALUE   DF    STD DEV      5.5341     3.7625
                                          S.E.M.       1.1068     0.7525
T (SEPARATE)   1.43   0.159   42.3        SAMPLE SIZE     25         25
T (POOLED)     1.43   0.158   48          MAXIMUM      50.0000    47.0000
                                          MINIMUM      22.0000    34.0000
F(FOR VARIANCES)
    LEVENE     0.18   0.669    1,  48


                                                      X  X
                      HH  H                          XX  X
                      HHHH H                          XX XX
                      HHHH HHH                      X XXX XX X
          H           HHHHHH HHH                    XXXXX XXXX
        MIN--------------------MAX         MIN--------------------MAX


                  ---- OUTPUT FOR ROBUST OPTION ----


************
* ALBUMIN *  VARIABLE NUMBER   7      GROUP      1 NOPILL   2 PILL
************                          MEAN         42.2799    40.3599
             STATISTICS    P VALUE   DF    STD DEV      5.5341     3.7625
                                          S.E.M.       1.1068     0.7525
T (SEPARATE)      1.43   0.159   42.3     SAMPLE SIZE     25         25
T (POOLED)        1.43   0.158   48       MAXIMUM      50.0000    47.0000
T (TRIM SEP.)     2.16   0.037   44.0     MINIMUM      22.0000    34.0000
T (TRIM POOLED)   2.16   0.037   44       MAX(2ND)     49.0000    47.0000
F(FOR VARIANCES)                          MIN(2ND)     37.0000    34.0000
    LEVENE        0.18   0.669    1,  48  MX.ST.SC.     1.3950     1.7648
                                          MN.ST.SC.    -3.6645    -1.6903


                                                      X  X
                      HH  H                          XX  X
                      HHHH H                          XX XX
                      HHHH HHH                      X XXX XX X
          H           HHHHHH HHH                    XXXXX XXXX
        MIN--------------------MAX         MIN--------------------MAX
        CS.NO.MN.  37  CS.NO.MX.  43      CS.NO.MN.  10  CS.NO.MX.  12
```

## BMDP, November 1979 Release: Biomedical Computer Programs

Developed by:                                   Distributed by:

  BMDP Statistical Software                       Same
  Dept. of Biomathematics
  Univ. of California
  Los Angeles, CA 90024
  (213) 825-5940

Computer Makes:                                 Operating Systems:

  Amdahl                                          IBM -OS, VS, TSO, VM-CMS DOS-VS
  Burroughs                                       Many others.
  CDC 6000, Cyber
  Data General (in process)
  DEC 10/20
  Facom                                         Interfaced Systems:
  Hitac
  Honeywell                                       P-STAT
  HP 3000                                         SAS
  IBM                                             SIR
  ICL 2900/SYSTEM 4                               (Others through P-STAT and SAS)
  Interdata (Perkin-Elmer)
  Itel
  PDP-11/LSI-11
  Prime
  Siemens
  Telefunken TR 440
  Univac 70/90
  Univac 1108, 1110
  VAX 11/780
  Xerox Sigma 7

Source Languages:  Almost all FORTRAN except for some assembler for efficiency.

Cost:

    The yearly license fee (including the first year) for most systems is $500
for universities; $1000 for nonprofit organizations; and $1500 for all others.
The fee includes program source, load, object, test input, and test output
modules.  It also includes maintenance, user consulting, and user documentation.

Documentation:

Dixon, W.J. and M.B. Brown (eds.) BMDP Biomedical Computer Programs.
    Berkeley, Univ. of Calif. Press, 1979.
Forsythe, A.B. BMDP Pocket Guide.  Dept. of Biomathematics, Univ. of
    Calif. Los Angeles, 1980.
Forsythe, A.B.  BMDP Reference Card.  Dept. of Biomathematics, Univ. of
    Calif. Los Angeles, 1980.
Hill, M.A.  BMDP User's Guide.  Dept. of Biomathematics, Univ. of Calif.
    Los Angeles, 1979
BMDP Technical Reports (list available upon request from UCLA office)

## 7.07  NISAN

### INTRODUCTION

NISAN is a new interactive statistical package, being developed by a group of Japanese statisticians since 1978. The package is widely applicable in statistical analyses both confirmatory and exploratory, including newly-developed methods. The system is useful not only for trained statisticians in obtaining analytical insight in model-building and in choosing the optimal procedure of statistical analysis, but also for the beginner in statistical analysis through the use of HELP commands in NISAN.

### CAPABILITIES(1):  Processing and Displaying Data

NISAN has the following diverse features:  1) high portability, 2) interactive, 3) database, 4) inquiry, 5) documentation, and 6) extensive data investigation. The computer language used in NISAN system is 98% FORTRAN. Most STAT modules introduce several procedures of statistical analysis.

NISAN system files consist of a data file for analysis, and a document file for records. DATA modules control the data file, and DOC modules control the document file. Without any consideration of the file structures all computational modules can easily access any files through DATA and DOC modules. HELP modules have the complete inquiry functions for giving answers to the user's questions on interpretation and utilization of NISAN. It is also intended that even if a user is a beginner, he will easily be able to access the NISAN system with the help of HELP modules. The minimum equipment for accessing the NISAN system is a character-display or teletype terminal.

### CAPABILITIES(2):  Statistical Analysis

The major features of a statistical analysis are:  1) data investigations, 2) graphical representations, 3) generalized methods of optimal scaling, 4) various methods of cluster analysis and multidimensional scaling, and 5) the successive use of several alternative procedures in making a statistical inference.

In applying statistical methods in practice, it is very important to first investigate the features and properties of input data. Most statistical program packages have insufficient capabilities for data investigation, and therefore these features have been strengthened in NISAN, especially examination for multivariate normality, transformation of variables, checks for outliers, etc. This permits data investigation before statistical analysis. Visual examination of these graphical representations at every stage of the analysis occasionally offers important insights, valuable suggestions and interpretation not only of the input data but also regarding the analysis and results. These graphical features are available at the user's request. For categorical input data, the NISAN system includes generalized methods of optimum quantification, newly-developed for NISAN, as well as the ordinary nonparametric methods. Any of them can be chosen depending on external criteria and on the ordered relations among item-categories.

Procedures for cluster analysis and multidimensional scaling, also newly-developed, provide some exploratory methodology for data analysis. These can be applied depending on the quantity and properties of the input data and on the hierarchical or nonhierarchical structure of clustering.

Concerning statistical inference, methods of "testimating", "testi-predicting" and "testitesting" are included to handle the unspecified aspects of the mathematical models, e.g. pooling methodology, including the implied estimation, is available in ANOVA and MANOVA. Information on the accuracy and precision of estimates, the power of tests for various alternative hypotheses, and so on, can be obtained by applying a simulation function in the NISAN system. Thus, for the complete statistical analysis, different strategies can be considered and the performance of different procedures examined, in order to find the optimum procedure.

## PROPOSED ADDITIONS IN NEXT YEAR

The NISAN system is undergoing extensive development in the next year, including successive non-parametric procedures of various kinds, and extensive methods of cluster anlaysis. For data handling more HELP commands will be added.

## REFERENCES

Asano, Ch. et al.   "The Statistical Principle and Methodology in NISAN System,"
     Res. Rep, No. 88, Res. Instit. Fund, Infor. Sc., Kyushu University. (1978).

## SAMPLE JOB

Find the saved files and obtain a regression line of S22 on S15, S16, S17, and S19. Then make a scatter diagram between the estimates and observed data.

## Sample Output

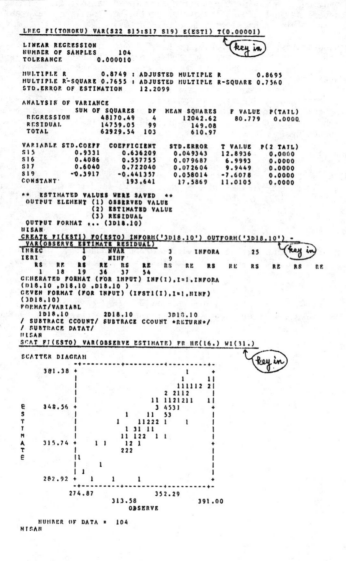

```
LREG FI(TOHOKU) VAR(S22 S15:S17 S19) E(EST1) T(0.00001)

LINEAR REGRESSION
NUMBER OF SAMPLES      104
TOLERANCE          0.000010

MULTIPLE R        0.8749 : ADJUSTED MULTIPLE R        0.8695
MULTIPLE R-SQUARE 0.7655 : ADJUSTED MULTIPLE R-SQUARE 0.7560
STD.ERROR OF ESTIMATION    12.2099

ANALYSIS OF VARIANCE
               SUM OF SQUARES   DF   MEAN SQUARES   F VALUE  P(TAIL)
  REGRESSION     48170.49        4    12042.62      80.779   0.0000
  RESIDUAL       14759.05       99      149.08
  TOTAL          62929.54      103      610.97

VARIABLE STD.COEFF  COEFFICIENT    STD.ERROR  T VALUE  P(2 TAIL)
  S15       0.9331     0.636209     0.049343  12.8936   0.0000
  S16       0.4086     0.557755     0.079687   6.9993   0.0000
  S17       0.6040     0.722040     0.072604   9.9449   0.0000
  S19      -0.3917    -0.441357     0.058014  -7.6078   0.0000
  CONSTANT           193.641      17.5869    11.0105   0.0000

** ESTIMATED VALUES WERE SAVED **
  OUTPUT ELEMENT (1) OBSERVED VALUE
                 (2) ESTIMATED VALUE
                 (3) RESIDUAL
  OUTPUT FORMAT ... (3D18.10)
NISAN
CREATE FI(EST1) FO(ESTO) INFORM('3D18.10') OUTFORM('3D18.10') -
VAR(OBSERVE ESTIMATE RESIDUAL)
INREC      1      NVAR       3      INFORA      25
IERI       0      NINF       9
   RS   RE   RS   RE   RS   RE   RS   RE   RS   RE   RS   RE   RS   RE
    1   18   19   36   37   54
GENERATED FORMAT (FOR INPUT) INF(I),I=1,INFORA
(D18.10 ,D18.10 ,D18.10 )
GIVEN FORMAT (FOR INPUT) (IFST1(I),I=1,NINF)
(3D18.10)
FORMAT/VARIABL
    1D18.10      2D18.10       3D18.10
/ SUBTRACE CCOUNT/ SUBTRACE CCOUNT *RETURN*/
/ SUBTRACE DATAT/
NISAN
SCAT FI(ESTO) VAR(OBSERVE ESTIMATE) FR HR(16.) WI(31.)

SCATTER DIAGRAM
        -+----------+----------+----------+-
  381.38 +                        1        +
         |                           1    1|
         |                     111112  2|
         |                 2 2112        |
         |             11 1121211    11  |
         |               3 4531           |
  348.56 +        1    11  53             +
E        |       1   11222 1      1       |
S        |        1 31 11                 |
T        |       11 122  1 1              |
T        |
M        |
A  315.74 +   1 1    12 1                 +
T        |          222                   |
E        |1                               |
         | 1                              |
         | 1                              |
  282.92 +  1   1    1    1               +
        -+----------+----------+----------+-
      274.87         352.29
           313.58         391.00
                OBSERVE

   NUMBER OF DATA =  104
NISAN
```

## NISAN: New Interactive Statistical ANalysis Package

Developed by:

Dr. Chooichrio Asano
Research Institute of
 Fundamental Information Science
Kyushu University 33
Fukuoka 812 JAPAN

Computer Makes:                                    Operating Systems:

   FACOM                                           FACOM M-190, OS-IV/F4, FACOM M-200
   IBM 370                                         OS, MVT/TSO

Source Languages:

   FORTRAN

Cost:

   Not yet determined.

7.08  GENSTAT

## INTRODUCTION

Genstat provides a high level language for data manipulation and statistical analysis.  It is mainly used for the analysis of experimental data, for fitting a wide range of linear and non-linear models and for finding pattern in complex data sets using multivariate and cluster analysis.  It was originated by John Nelder in 1965 and has been under continuous development since then. It can be used interactively or in batch and is installed at about 150 sites.

## CAPABILITIES:  Processing and Displaying Data

Data can be presented in many different formats and errors are detected and reported on.  Scalars, vectors, matrices and tables can be formed and manipulated.  Any structure with numeric values can be an operand in arithmetic expressions using a wide range of functions.  Up to 6-way tables of totals, means or counts can be formed and expanded to hold margins of means, totals, minima, maxima, variances or medians.

Tabular output in a wide range of layouts for both one-dimensional and multidimensional structures is provided.

Graphical output includes point plots, lines and histograms; also contour diagrams.

Temporary or permanent files can be transferred to and from disc or magnetic tape.

## CAPABILITIES:  Statistical Analysis

Classical least squares regression, with or without weights, can be carried out.  Independent variables can be fitted sequentially or selected automatically from a specified set.  For grouped data parallel lines can be fitted; also unbalanced experiments can be analysed.

The fitting of linear models is generalised by providing functions linking the mean to the value predicted from the model and four error distributions (Normal, Binomial, Poisson and Gamma).

Non-linear models can be fitted using an iterative procedure.

Designed experiments can be analysed including all balanced designs (e.g. randomised blocks, Latin Squares, Graeco-Latin Squares, and split-plots) and many partially balanced designs (e.g. lattices).

Procedures for cluster analysis, principal component analysis, canonical variate analysis, principal coordinate analysis and Procrustes rotation are available; also varimax and quartimax factor rotation.

Time series analysis and forecasting using seasonal ARIMA and transfer function models is provided.  Box-Cox transformation may be used and autocorrelation functions calculated.

## EXTENSIBILITY

Analyses not catered for by the standard procedures can be programmed using the full facilities of the Genstat language.  These procedures (macros) can be stored and recalled as required.

## INTERFACES WITH OTHER SYSTEMS

Genstat can transmit data to and from other programs.

## PROPOSED ADDITIONS IN NEXT YEAR

Graphical output on plotters and videos.

Extended operations on tables.

## REFERENCES

Bernard, G.  A comparison of three statistical packages:  Genstat, BMDP and
    SPSS.  COMPSTAT 1978, Proceedings in Computational Statistics, Physica-
    Verlag, pp. 445-451.
Federer, W.T. and Henderson, H.V.  Covariance Analysis of Designed Experiments
    using Statistical Packages.  Report of Biometrics Unit, Cornell University
    BU-346.
Francis, I., Heiberger, R.M. and Sherman, S.F.  Languages and Programs for
    Tabulating Data from Surveys.  Proceedings Ninth Interface Symposium.
    Prindle Weber and Schmidt (1976) pp. 129-139.
Heiberger, R.M.  Conceptualisations of Experimental Designs and their Specifi-
    cation and Computation with ANOVA programs.  Proceedings of the Statisti-
    cal Computing Section of the ASA (1976).
Hohwald, J. and Heiberger, R.M.  Two conceptualisations of discriminant analy-
    sis and their implementation in computer programs.  Proceedings Tenth
    Interface Symposium, U.S. Dept. of Commerce (1977).

## SAMPLE JOB

```
'REFERENCE"        AVCCOX(300)
''SPLIT PLOT DESIGN:   - EXPERIMENT TO COMPARE 6 BAKING TERMPERATURES AND 3
RECIPES FOR MAKING CHOCOLATE CAKES (COCHRAN AND COX:   EXPERIMENTAL DESIGNS,
2ND EDITION, 1957, p300)
       THERE WERE 15 REPLICATES (MADE AT DIFFERENT TIMES) CONSISTING OF ONE MIX
FROM EACH RECIPE.  EACH MIX WAS SUBDIVIDED TO FORM BATTER FOR 6 CAKES, WHICH
WERE RANDOMLY ALLOCATED TO THE 6 BAKING TEMPERATURES.  THE MEASUREMENT ANALYSED
BELOW IS THE ANGLE OF BEND, BETWEEN THE TWO HALVES OF A CAKE, AT WHICH BREAKAGE
OF THE CAKE OCCURED.''
'UNITS' $ 270
'VARIATE' Z=175,185...225
'FACTORS' RECIPE $ 3 : REP $ 15 : TEMP $ Z
'GENERATE' RECIPE,REP,TEMP
'READ' BREAKING
'HEADING' HB='' ANGLE OF CAKE''
'DESCRIBE' BREAKING $ ; HB
'BLOCKS' REP/RECIPE/TEMP
'TREATMENTS' RECIPE*POL(TEMP,1)
'ANOVA/PR=12' BREAKING
'RUN'
```

***** ANALYSIS OF VARIANCE *****

VARIATE: BREAKING ANGLE OF CAKE

| SOURCE OF VARIATION | DF | SS | SS% | MS | VR |
|---|---|---|---|---|---|
| REP STRATUM | 14 | 10204.24 | 56.24 | 728.87 | 35.605 |
| REP.RECIPE STRATUM | | | | | |
|   RECIPE | 2 | 135.09 | 0.74 | 67.54 | 1.578 |
|   RESIDUAL | 28 | 1198.47 | 6.61 | 42.80 | 2.091 |
| TOTAL | 30 | 1333.55 | 7.35 | 44.45 | 2.171 |
| REP.RECIPE.TEMP STRATUM | | | | | |
|   TEMP | 5 | 2100.30 | 11.58 | 420.06 | 20.520 |
|     LIN | 1 | 1966.71 | 10.84 | 1966.71 | 96.073 |
|     DEVIATIONS | 4 | 133.59 | 0.74 | 33.40 | 1.632 |
|   RECIPE.TEMP | 10 | 205.98 | 1.14 | 20.60 | 1.006 |
|     DEV.LIN | 2 | 1.74 | 0.01 | 0.87 | 0.043 |
|     DEVIATIONS | 8 | 204.24 | 1.13 | 25.53 | 1.247 |
|   RESIDUAL | 210 | 4298.89 | 23.69 | 20.47 | |
| TOTAL | 225 | 6605.16 | 36.41 | 29.36 | |
| GRAND TOTAL | 269 | 18142.96 | 100.00 | | |

GRAND MEAN                        32.12
TOTAL NUMBER OF OBSERVATIONS       270

***** TABLES OF MEANS *****

VARIATE: BREAKING ANGLE OF CAKE
GRAND MEAN    32.12

| RECIPE | 1 | 2 | 3 | | | |
|---|---|---|---|---|---|---|
| | 33.12 | 31.64 | 31.60 | | | |

| TEMP | 175.00 | 185.00 | 195.00 | 205.00 | 215.00 | 225.00 |
|---|---|---|---|---|---|---|
| | 27.98 | 29.96 | 31.42 | 32.18 | 35.84 | 35.36 |

| TEMP RECIPE | 175.00 | 185.00 | 195.00 | 205.00 | 215.00 | 225.00 |
|---|---|---|---|---|---|---|
| 1 | 29.13 | 31.53 | 30.80 | 33.53 | 38.67 | 35.07 |
| 2 | 26.87 | 29.40 | 31.73 | 32.13 | 34.47 | 35.27 |
| 3 | 27.93 | 28.93 | 31.73 | 30.87 | 34.40 | 35.73 |

***** STANDARD ERRORS OF DIFFERENCES OF MEANS *****

| TABLE | RECIPE | TEMP | RECIPE TEMP |
|---|---|---|---|
| REP | 90 | 45 | 15 |
| SED | 0.975 | 0.954 | 1.796 |

EXCEPT WHEN COMPARING MEANS WITH SAME LEVEL(S) OF:
  RECIPE                                        1.652

## GENSTAT: General Statistical Program

Developed by:

    Statistics Department,
    Rothamsted Experimental Station,
    Harpenden, Herts.,
    U.K.

Distributed by:

    Statistical Package Coordinator
    NAG Central Office,
    7 Banbury Road,
    Oxford.  OX2 6NN.  U.K.

Computer Makes:

| | |
|---|---|
| Burroughs 6700 | MCP |
| CDC 6000 series | |
|    Cyber series | NOS |
|    7600 | SCOPE 2.0 |
| DEC System 10,20 | TOPS 10,20 |
| Honeywell | GCOS |
| IBM 360, 370 | OS and some virtual memory systems |
| ICL 1900 series | GEORGE 2, 3 |
| ICL 2900 series | EMAS, VME/B, K |
| ICL 4-70/75 | Multijob |
| PRIME 400 or larger | PRIMOS |
| SIEMENS | BS2000 |
| Univac 1100 series | EXEC 8 |
| VAX | VMS |

Operating Systems:

Versions for MODCOMP, SEL, Telefunken, IRIS 80 and VAX under development.

Source Language:  FORTRAN (3 assembler routines)

Cost:

| | |
|---|---|
| Universities and government agencies in U.K. | £ 240 |
| Universities and government agencies outside U.K. | £ 340 |
| Other organizations | £ 540 |

Documentation:

Alvey, N.G., et. al.  GENSTAT, A General Statistical Program, Rothamsted
   Experimental Station, Harpenden, Herts., U.K., 1977.

## 7.09  SPEAKEASY

## INTRODUCTION

Speakeasy III is a general-purpose, high-level, user-oriented computing system designed to provide the non-computer specialist with effective access to the computing power available in large scale computer complexes. Although originally designed by Stanley Cohen in 1964 as a tool for physical scientists, the extensible structure has allowed it to evolve into a vehicle for interactive statistical and econometric studies. It is currently in use in over 100 different computer installations with known applications in education, engineering, management information sciences, financial planning and cartography. The single largest group of users is made up of economists including most of the major econometric forecasting services. Speakeasy can be used in batch or interactively. In the latter case, it functions as a super desk calculator with great power and with a very large vocabulary of well-known words.

## CAPABILITIES:  Processing and Displaying Data

The Speakeasy III System contains a wide variety of built-in data storage and data access techniques. These enable any user to retain or retrieve parts or all of the information constructed during an interactive session. In addition, the extensible nature of the system enables users to add specific accesses to existing data base systems available at a particular installation. General-purpose, FORTRAN-like read and write statements are also provided to ensure complete access to external data.

In a typical session, data is loaded into the user's workspace for analysis. Output in the form of tabular and graphical forms are produced by the Speakeasy commands TABULATE and GRAPH. The wide use of automatic controls enables a user to ignore any questions relating to output formats, and at the same time ensures that presentable results will be produced.

Specialized words for survey analysis enable a user to selectively and conditionally read information from large external files.

Speakeasy has a command language modeled on conventional mathematical notation. It provides users with a natural notation whose vocabulary closely parallels that used in their normal studies. With a vast number of built-in mathematical functions that operate on arrays as single entities, it is possible to reduce programming to a form closely related to the statement of the problem itself. The use of standard notation with familiar words leads to visual feedback that provides for error-free statements of problems.

Speakeasy contains a complete mathematical formulation for matrix algebra and all of the operations that are associated with that discipline. The procedures used for eigen analysis and other mathematical operations are state-of-the-art routines and provide users with assurances of correct numerical processing. Internal checks relating to structural consistencies provide independent automatic and extensive checks on consistency of formulation of problems.

CAPABILITIES:  Statistical Analysis

Although Speakeasy was not designed as a statistical analysis system, it contains most of the basic vocabulary for statistical operations.  In addition, the user community has contributed large numbers of additions to that vocabulary to tailor the sytem for specific needs.  In particular, many of the capabilities of F4STAT, TSP and some early BMD operations are available.  A package of operations specifically designed for economic modeling (FEDEASY) is also available.

The complete matrix algebra contained within the system enables a user to formulate specific operations in a direct and understandable way.  It is also possible for any user to add existing special-purpose FORTRAN routines to the system to answer specific needs.

EXTENSIBILITY

Speakeasy is designed for growth by the accumulation of operations placed in the libraries attached to the system.  The reference manual describes, in detail, how such additions are made.  A mechanism for distributing such additions to users at other sites is provided.

INTERFACES WITH OTHER SYSTEMS

Procedures exist in SAS to transfer information to and from Speakeasy.  An operation, GETSPSS, is available in Speakeasy to read information from an SPSS savefile.  Interfaces to many commercial data base systems exist.

PROPOSED ADDITIONS IN NEXT YEAR

Speakeasy III is undergoing major changes during 1981.  Time series as a new data type is being added and an extensive vocabulary for fundamental operations related to time series is being added.

REFERENCES

Ling, Robert F.  "General Considerations on the Design of an Intermediate System for Data Analysis,"  Comm. ACM, 23, 3 (March 1980), pp. 147-154.
Condie, James M.,  "A Speakeasy Language Extension for Economists,"  Proceedings 1977 Annual Conference ACM, (October 1977), pp. 425-427.
Cohen, Stanley.,  "Speakeasy - A Window Into a Computer,"  AFIPS - Conference Proceedings, Vol. 45, (June 1976), pp. 1039-1047.

SAMPLE JOB

An interactive session is shown below.  The lines with :_ are followed by
the user input.  All other lines are output from **Speakeasy**.

```
TSO SPEAKEASY III NU+   2:18 P.M. JANUARY 3, 1981

:_$ ------ DESK CALCULATOR MODE ----------

:_2*3/4+SIN(.34)*EXP(-.056)
2*3/4+SIN(.34)*EXP(-.056) =   1.8153
:_ANSWER+27
ANSWER+27 =  28.815

:_$ ------ SIMPLE TABULAR RESULTS ----------

:_X=23, 34.3 1e+2 23.44 56.3 , 323 89
:_Y=X**2-SQRT(X)
:_Z=ORDERED(X)
:_TABULATE X Y Z

       X            Y            Z
    *******    **********    *******
      23          524.2        23
      34.3       1170.6        23.44
     100         9790          34.3
      23.44       544.59       56.3
      56.3       3162.2        89
     323        104311        100
      89         7911.6       323

:_$ -------- MATRIX DEFS. AND ALGEBRA --------

:_X=MATRIX(3,3: 3.3 14.22 32.97 56.38 23.1 87.44  42.8)
:_VALS=EIGENVALS(X) ;VECS=EIGENVECS(X) ;TABULATE X VALS VECS

            X             VALS               VECS
  ********************    *******    ************************
    3.3   14.22  32.97    -25.598    .53273   .4628    .30571
   56.38  23.1   87.44    -12.471    .98256   3.1627   .84507
   42.8    0      0        64.489   -.89072  -1.5857   .20269

:_1/X
1/X  (A 3 BY 3 MATRIX)
    0          0        .023364
    .18149   -.068432   .076152
   -.047946   .029515  -.035183
:_X*ANSWER
X*ANSWER  (A 3 BY 3 MATRIX)
    1         2.7756E-17  -2.2204E-16
   2.2204E-16  1          -2.2204E-16
    0          0           1

:_$ ----- BASIC STATISTICAL OPS ---------

:_AVERAGE(VALS) ; STANDDEV(Z)   ; SUMSQ(Y)
AVERAGE(VALS)  =  8.8
STANDDEV(Z)    =  106.06
SUMSQ(Y) =  1.1055E10
:_CORRELATION(Y Z)
CORRELATION(Y Z) =  .076153

:_MULTIREG(Y Z)

MULTIREG(Y Z) (A 2 COMPONENT ARRAY)
   88.859      2.1176E-4
```

## SPEAKEASY III: Level OMNICRON

Developed by:

    Speakeasy Computing Corporation
    222 West Adams Street
    Chicago, Illinois   60606
    312.346-2745

Distributed by:

    Speakeasy Computing Corporation
    222 West Adams Street
    Chicago, Illinois   60606
    312.346-2745

Computer Makes:

    IBM 360, 370
    IBM compatible systems
    Fujitsu M series
    Burroughs 6700

Operating Systems:

    IBM OS, VS, TSO, VM/CMS, NCSS
    Burroughs 6700

Language

    Fortran

Cost:

    Yearly license maintenance fee $5600 for 1981.   Educational Institution
$2000 for 1981.

Documentation:

Speakeasy Reference Manual, S. Cohen and S.C. Pieper, Speakeasy Computing
    Corporation, 1980.
Lectures on Speakeasy by Dr. Gary D. Gordon, Speakeasy Computing Corporation,
    1980.

## 7.10  TROLL

### INTRODUCTION

TROLL is a software system that provides an interactive environment for modelling and data analysis. Free-format commands are entered directly to create or modify files, perform complex mathematical operations, and send output to a variety of devices. Command input may be interrupted at any point by a carriage return; TROLL will then prompt the user for the next command argument. When an error occurs during input of command arguments, processing is temporarily interrupted, an error message is displayed, and the user can continue from the point of error. On-line documentation is available to describe the format and usage of any command. Default conventions and global options are employed extensively to minimize necessary typing; commands are available to examine or reset any option. While originally designed for interactive use, TROLL can also be run in batch.

TROLL was originally developed at M.I.T. over the years 1966-71. It was subsequently extended at the National Bureau of Economic Research's Computer Research Center during 1972-1977. Thereafter, further development has occurred at the Massachusetts Institute of Technology at the Center for Computational Research in Economics and Management Science. TROLL is currently installed at some thirty computer centers in North America, Europe and Asia.

### CAPABILITIES:  Processing and Displaying Data

Each TROLL user has a personal filespace for the on-line storage of data, models, labels and macros. A user can also access other users' files or system libraries. Files may be constructed explicitly using TROLL's editing commands, created automatically from computational results, or loaded from external sources via a "database interface" facility. Multi-level archiving and search rules allow grouping of related files.

The TROLL data file is a time series or matrix, and includes information such as periodicity and start date. Facilities for calculation include all arithmetic operations, missing data identification and replacement, matrix decomposition, inversion and solution, time series smoothing, extrapolation and interpolation, and file-creation and reshaping functions.

Basic printing commands will display selected parts of specified files in a standard format under the control of user-modifiable output options. Special formatting functions are available to output data and labels according to a user-specified format. Some computational commands automatically display results. All the main display commands have off-line counterparts to route output to a high-speed printer.

Basic plotting commands produce time plots, scatter plots and histograms on either a standard terminal or a graphics device (usually Tektronix 4010 or 4662). Special plotting facilities include n-dimensional point cloud projections, box plots, probability plots, and hidden-line surfaces.

## CAPABILITIES:  Statistical Analysis

TROLL'S analysis facilities can be classified roughly into two area:  estimation and simulation.  Both use a common modelling language based on standard
algebraic notation.  Estimation methods include linear and non-linear ordinary
least squares, generalized least squares (e.g. autoregressive correction), distributed lags, two-stage least squares (with optional principal-components
analysis), ridge and robust regressions (e.g. iteratively reweighted least
squares), logit, probit, and tobit, and equation-system estimation (e.g. FIML,
three-stage least squares).  Simulation methods including full-system dynamic
simulation, periodically constrained simulation, elimination of selected equations, and single-equation forecasting.  Simulation models can include more
than a thousand simultaneous equations, and the equations need not be ordered
or normalized by the user.

## EXTENSIBILITY

A macro language allows users to create programs consisting of TROLL commands and macro control statements; some major TROLL subsystems are written as
macros.  Compiled code can be incorporated into TROLL by using the "database
interface" facility.  At the M.I.T. installation, special programs are available for creating TROLL functions from FORTRAN subroutines.

## PROPOSED ADDITIONS IN NEXT YEAR

New programs that will become available during the next year include:  a
complete ARIMA package; Box-Cox transformations; Bounded-Influence (robust)
regression; model linearization and state-space eigenanalysis; and a report
generator.

## REFERENCES

Belsley, D.A., E. Kuh and R.E. Welsch. Regression Diagnostics:  Identifying
    Influential Data and Sources of Collinearity, John Wiley and Sons, New
    York, 1980.
Choucri, N. and R. North. Nations in Conflict:  Population, Expansion and War,
    W.H. Freeman, San Francisco, 1975.
MacAvoy, P.W. and Pindyck, R.S.  The Economics of The Natural Gas Shortage,
    North-Holland, Amsterdam, 1975.

## SAMPLE JOB

The terminal session illustrates model-editing and estimation; user input
is in lowercase.  An archive in a system library is searched for the data.  An
equation is entered in the model and then estimated.  The Durbin-Watson statistic indicates autocorrelation in the error term, so the equation is
re-estimated using a first-order autoregression correction.  The equation is
then modified to include the dependent variable lagged one period as an explanatory variable, and the new version is estimated with ordinary least squares.

Sample Terminal Session

```
TROLL COMMAND: .search syslib_data_nber4; usemod consume;
NEW MODEL: CONSUME
TROLL COMMAND: .addsym coefficient c1 c2 c3; addeq top gc=c1+c2*gyd;
MODEDIT COMMAND: .prtdata search syslib specs comment gc gyd;

NBER4 GC   -   DATE REVISED: 7/07/80
   QUARTERLY DATA FROM 1946 1 TO 1980 1

   PERSONAL CONSUMPTION EXPENDITURES

NBER4 GYD   -   DATE REVISED:  7/07/80
   QUARTERLY DATA FROM 1946 1 TO 1980 1

   PERSN'L INCOME: DISPOSABLE PERSONAL INCOME

TROLL COMMAND: .period 4; reg 1950 1 to 1980 1; doeq top;

1:   GC = C1+C2*GYD

NOB = 121      NOVAR = 2
RANGE =  1950 1 TO 1980 1
RSQ =    0.99933      CRSQ =    0.99932      F(1/119) =  1.76E+05
SER =    9.6948       SSR = 11184.600       DW(0) = 0.38    COND(X) =       3.36

COEF            VALUE           ST ER          T-STAT

C1              -3.68142        1.61276        -2.28268
C2              0.92263         0.00220        419.77300

REG COMMAND: .regopt final rsq crsq ssr coef ster tstat rhol; gls auto; doeq
POSITION: .top;

RSQ =    0.99398      CRSQ =    0.99393      SSR =  3968.090

GLS PARAMETERS

RHO1    0.8344

COEF            VALUE           ST ER          T-STAT

C1              -5.11110        5.22309        -0.97856
C2              0.92675         0.00661        140.15800

REG COMMAND: .changeq /gyd/gyd+c3*gc(-1)/ 1; regopt final 2; doeq 1;

WARNING 2025
DATAMATRIX IS HIGHLY COLLINEAR

1:   GC = C1+C2*GYD+C3*GC(-1)

NOB = 121      NOVAR = 3
RANGE =  1950 1 TO 1980 1
RSQ =    0.99985      CRSQ =    0.99984      F(2/118) =  3.86E+05
SER =    4.6356       SSR = 2535.690        DW(0) = 1.96    COND(X) =     170.55

COEF            VALUE           ST ER          T-STAT

C1              -5.38519        0.77581        -6.94133
C2              0.09122         0.04146        2.20032
C3              0.92982         0.04635        20.06200
```

TROLL Release 9

Developed by:                              Distributed by:

   Center for Computational Research        Same
      Economics and Management Science
   Massachusetts Institute
      of Technology
   292 Main Street
   Room E38-200
   Cambridge, MA  02139

Computer Makes:                            Operating Systems:

   IBM 370                                     IBM-OS, VS, MVS/TSO, VM/CMS,
                                                 DOS, MTS

Source Languages:

   AED, FORTRAN, IBM 370 assembler

Cost:

   For universities and other non-profit organizations there is a one-time
fee of $500.00 and a yearly license fee of $1500.00.

DOCUMENTATION:

   Primary Manuals:

TROLL:  An Introduction and Demonstration, D0083, Fourth Revision, 1978, 58 pp.
TROLL Primer, D0046, Third Edition, 1979, 191 pp.
TROLL User's Guide, D0037, 1972, 468 pp.
TROLL Reference Manual (Standard System), D0062, Second Edition, 1979, 1045 pp.
   + 23 tab dividers.
User's Guide to the TROLL Macro Facility, D0090, Second Edition, 1975, 108 pp.

   All from:

                     Publications Office
                     Information Processing Services
                     M.I.T. Room 39-239
                     Cambridge, MA  02139

7.11  <u>IDA</u>

INTRODUCTION

IDA is a conversational package for general-purpose statistical data analysis.  Its statistical capabilities emphasize the tools associated with regression analysis and related model-building techniques.  IDA was designed with a high priority on user convenience to facilitate model identification and diagnostic checking.  IDA's commands are 4-letter mnemonics of standard statistical terms, e.g., HISTogram; the system prompts if more information is needed.  Complete on-line documentation and a bi-level prompt structure are designed to make the system useful for students and novices as well as experienced researchers.

IDA was developed originally by Robert F. Ling and Harry V. Roberts in 1972 at the Graduate School of Business, University of Chicago and has been under continuous development since then.  It is now distributed and undergoing further development by SPSS Inc.  IDA is designed for interactive use, and has been installed at more than 100 computer installations.

CAPABILITIES:  Processing and Displaying Data

IDA uses a rectangular data matrix:  users normally may specify up to 49 variables or 2310 observations.  (Installations may increase these dimensions at the systems level.)  Data manipulation features allow the user to splice data from one or more files; to delete temporarily or permanently; to change observations in place; to remove or add new variables; to save part or all of the data matrix on a user file at any point in the run; to locate and treat in a variety of ways observations with missing or wild values.

Facilities for data definition, data editing, and data transformation include:  terminal and/or file entry; sorting; random data generation; index variable generation; sampling of data matrix; case selection; data matrix redimensioning; arithmetic transformations; functional transformations with FORTRAN-like statements; categorical value creation; standardizing; lagging, leading, differencing, cumulating, and ranking.

In addition to providing multiple options for printing data, IDA's design emphasizes graphical data display and plotting on standard terminals.  Available plots include:  crosstabulation tables; tables of means; standardized sequence plots; multiple variable sequence plot; scatterplots; histograms; cumulative distribution function, and normal probability plot.

CAPABILITIES:  Statistical Analysis

Statistical capabilities emphasize regression and regression-related methods--including Box-Jenkins methods.  Model assumptions are automatically checked and users are warned if gross violations are detected.  IDA includes commands for data transformation and graphical and numeric diagnostic checks of model specifications that were designed to facilitate time-series analysis.  It can also perform tabular analysis and other standard statistical techniques.

Available operations include:  <u>Regression Analysis</u>:  automatic regression: all or selective subsets of variables; forward stepwise regression; backward stepwise regression; user-specified regression; regression-variable sweep

operation; analysis of variance; B-coefficient correlations and co-variances;
regression prediction and validation; automatic checking of residuals; user-
specified residual analysis; and summary statistics for regression results.
Time-Series Analysis:  Box-Jenkins estimation and forecasting; cumulative
periodogram; cross correlations; partial autocorrelation; autocorrelation; simulation of ARIMA processes.  Tabular Analysis:  crosstabulation; tables of means;
frequency distributions.  Summary and One Sample Statistics:  correlations;
covariances; partial correlations; autocorrelations; Box-Pierce statistic;
serial correlations; runs counts; Durbin-Watson statistic; homogeneous variance
test.  Probability Calculations:  normal; Student's-t; Chi-square; F; Poisson.

## EXTENSIBILITY

Many additional statistical techniques such as 2-stage least squares,
Cochrane-Orcutt transformation, Spearman's rank correlation, Mann-Whitney U test
can be carried out by combinations of basic IDA commands.  (The above techniques
and others are illustrated in the IDA User's Guide.)

The 1981 release includes a subsidiary "shell" program which interfaces
with the main IDA system and is designed to facilitate the addition of user-
written programs, e.g., new statistical procedures, specialized report generators, or data base interfaces, to the system.

## INTERFACES WITH OTHER SYSTEMS

IDA can read free-format character files, and can read and write both
unformatted binary files and character files formatted according to user-
specified FORTRAN format specifications.

## PROPOSED ADDITIONS IN NEXT YEAR

IDA will will undergo extensive development in the next 12 months.  Planned
additions include:  non-standardized scatterplot; MANOVA; installation-defined
expanded data matrix; development of a "shell" program to facilitate interfacing
user-written programs, e.g., specialized statistical routines, report-generator,
or data base interraces, with IDA; possible additional econometric commands;
possible EDA plots.

# SAMPLE JOB

The following interactive terminal session shows specification of a regression model, sweeping out a variable which does not contribute to the model, and some of the diagnostic tools available for residual analysis. User input is shown by the lower-case 4-letter commands which are preceded by ">". Annotations are preceded by "<====", user level.

```
> regr  <====REGRession

* DEP. VAR. : dlinc

HOW MANY INDEP. VAR. ? 2

INDEP. VAR.  1 : dlinv

INDEP. VAR.  2 : time

UPDATING CORR. MATRIX...
COMPUTING REGRESSION...
ANALYZING RESIDUALS...
CHECKING AUTO CORRELATIONS...

> coef  <====COEFficient displays some of the regression output

VARIABLE   B(STD.V)      B         STD.ERROR(B)    T

DLINV      0.8477     2.1484E-01    3.4129E-02    6.295
TIME       0.0138     2.1642E-04    2.1071E-03    0.103
CONSTANT   0          6.6248E-03    2.5487E-02    0.260

> sweep  <====SWEEp removes a variable from the model without
              respecification of the regression.

* VARIABLE TO BE SWEPT : time

ANALYZING RESIDUALS...
CHECKING AUTO CORRELATIONS...

> coef

VARIABLE   B(STD.V)      B         STD.ERROR(B)    T

DLINV      0.8507     2.1561E-01    3.2310E-02    6.673
CONSTANT   0          8.9672E-03    1.1043E-02    0.812

> summ  <====Additional regression output is displayed by SUMMary

            MULTIPLE R   R-SQUARE
UNADJUSTED    0.8507      0.7237
ADJUSTED      0.8411      0.7075

STD. DEV. OF RESIDUALS = 4.7625E-02
N = 19
```

```
> runs  <====RUNS count of residuals suggests actual number of runs above and
              below the mean conforms closely to random expectation

* VARIABLE ? residu

   OBSERVED NUMBER OF RUNS =   10
   EXPECTED NUMBER OF RUNS =   10.26
STANDARD-DEVIATION OF RUNS =    2.06
   (OBS.-EXP.)/(STD.-DEV.) =   -0.13

VARIABLE:  RESIDU

+----------+----------+----------+
+ = VALUE > MEAN
- = VALUE <= MEAN
MEAN =  0.0000E+00
FIRST ACTIVE ROW =   2
LAST ACTIVE ROW =   20

> auto  <====AUTOcorrelation is applied to residuals

* VARIABLE ? residu

* MAX ORDER ? 6

                     S.E.
            AUTO-  RANDOM
ORDER CORR. MODEL  -1  -.75 -.50 -.25  0  .25 .50 .75 +1     ADJ.B-P

1    C.049  0.212                     +   :*          +      0.5318E-01
2   -C.409  0.206               *     :          +          3.983
3   -0.007  0.200                     *:          +         3.984
4   -0.217  0.194              +      *'  :          +       5.238
5   -0.064  0.187                 +   :*        +           5.353
6    0.227  0.181                     + :     *      +       6.935

                   -1  -.75 -.50 -.25  0  .25 .50 .75 +1
                         * : AUTOCORRELATIONS
                         + : 2 STANDARD ERROR LIMITS (APPROX.)

VARIABLE : RESIDU
FIRST ACTIVE ROW =   2
LAST  ACTIVE ROW =   20
NUMBER OF OBSERVATIONS =   19

BOX-PIERCE STATISTIC FOR LAG  6:
UNADJUSTED =  5.179
ADJUSTED   =  6.935

RUNS OF AUTO ABOVE AND BELOW ZERO:
   OBSERVED NUMBER OF RUNS =   5
   EXPECTED NUMBER OF RUNS =   3.67
STANDARD-DEVIATION OF RUNS =   0.94
   (OBS.-EXP.)/(STD.DEV.) =   1.41

+++++
```

## IDA: Interactive Data Analysis and Forecasting System

Developed by:                                   Distributed by:

Robert F. Ling                                  SPSS Inc.
Dept. of Mathematical Sciences                  444 N. Michigan Avenue
0-104 Martin Hall                               Chicago, Illinois  60611
Clemson University
Clemson, South Carolina   29631                 Telephone: (312) 329-2400
                                                TWX#910-221-1396
Harry V. Roberts                                ANSWERBACK SPSS CGO
Graduate School of Business
University of Chicago
1101 East 58th Street
Chicago, Illinois  60637

Computer Makes:                                 Operating Systems:

IBM 360, 370                                    IBM - TSO, VM-CMS
DEC System 10, 20                               DEC - VMS, TOPS-10, TOPS-20
HP-3000                                         Data General - AOS, RDOS
Prime
Data General Eclipse,
  Nova, MV-8000
Honeywell (CP-6)
VAX 11/780, 11/750

Conversions Underway

PDP-11
UNIVAC-1100
and others

Source Languages:   Primarily FORTRAN, small amounts of assembler in some
                    versions.

Cost:

IDA is licensed on an annual basis. Fees provided on request. Substan-
tial discounts available to academic and not-for-profit organizations. The
annual fee covers complete maintenance services including updates, new releases,
all new and existing documentation, and consultation.

Documentation:

Ling, R.F., Roberts, H.V.  IDA: A User's Guide to the IDA Interactive Data
     Analysis and Forecasting System, New York:  McGraw Hill/
     Scientific Press, 1981.
Roberts, H.V., Ling, R.F.  Conversational Statistics with IDA - An Introduction
     to Data Analysis and Regression, New York:  McGraw Hill/
     Scientific Press, 1981.

SPECIFIC-PURPOSE INTERACTIVE PROGRAMS


CONTENTS

INTRODUCTION

While Chapters 6 and 7 contain general-purpose statistical packages, Chapters 8 to 12 contain programs written for more specific statistical purposes, as indicated by the taxonomy of Table 1.2.

Chapter 8 includes interactive programs with capabilities in several areas, but which lack broad statistical coverage or are not strong in data management and tabulation.

The ratings by the respective developers on selected items are displayed in Table 8.1.  Figure 8.1 compares developers' and users' ratings on all items, while Figure 8.2 summarizes developers' ratings (D-scores) and users' ratings (U-scores) on several relevant attributes.  These are explained in detail in Section 1.4 of Chapter 1.  Figure 8.2 also points to other chapters which describe programs with strong statistical analysis capabilities.

The most widely-distributed of these programs is GLIM which was written by a Working Party on Statistical Computing of the Royal Statistical Society.  It features a simple but powerful user language for fitting generalized linear models.

CADA is the only program which includes a major Bayesian component as well as classical components.  This Bayesian component assists the user in a conversational mode to specify his subjective probabilities and utilities, and provides the usual Bayesian and decision theoretic analyses.

## TABLE 8.1:   RATINGS BY DEVELOPERS ON SELECTED ITEMS

### (i) Capabilities

Usefulness
Rating Key:
3 - high
2 - moderate
1 - modest
. - low

| | Complex Structures | File Management | Consistency Checks | Probabilistic Checks | Compute Tables | Print Tables | Multiple Regression | Anova/Linear Model | Linear Multivariate | Multi-way Tables | Other Multivariate | Nonparametric | Exploratory | Robust | Non-linear | Bayesian | Time Series | Econometric |
|---|---|---|---|---|---|---|---|---|---|---|---|---|---|---|---|---|---|---|
| | 11 | 14 | 18 | 19 | 24 | 25 | 30 | 31 | 32 | 33 | 34 | 36 | 37 | 38 | 39 | 40 | 35 | 41 |
| 8.01 ISA | 2 | 2 | . | 1 | 2 | 2 | 2 | 3 | 3 | 2 | 2 | 1 | 3 | 2 | 1 | . | 2 | . |
| 8.02 GLIM | 1 | 1 | . | . | 2 | 1 | 3 | 2 | . | 3 | 1 | 1 | 2 | 3 | 2 | . | . | 1 |
| 8.03 ISP | 2 | 2 | . | 1 | 2 | . | 2 | 1 | 1 | 1 | 1 | . | 3 | 3 | 1 | . | 1 | . |
| 8.04 CMU-DAP | 1 | 2 | 1 | . | 1 | . | 3 | 1 | . | 2 | . | . | 3 | 2 | 2 | . | . | . |
| 8.05 RUMMAGE | 1 | . | . | 1 | 1 | 1 | 2 | 3 | 2 | 3 | . | . | 1 | 2 | . | . | . | . |
| 8.06 CADA | . | 1 | 1 | . | 1 | . | 2 | 2 | . | 1 | . | . | 2 | 1 | . | 3 | . | . |
| 8.07 SURVO | 2 | 2 | 1 | . | 1 | 3 | 2 | 1 | 2 | 1 | 2 | 1 | . | . | . | . | 3 | 1 |
| 8.08 AUTOGRP+ | 2 | 3 | 1 | . | 3 | 2 | . | 1 | . | 1 | 3 | 3 | 1 | 2 | . | . | . | . |
| 8.09 FORALL | 2 | 1 | 1 | 1 | 2 | 2 | 2 | 1 | 1 | 1 | 1 | 3 | . | 1 | 1 | . | 1 | . |
| 8.10 AQD | 1 | 2 | . | . | 2 | . | 2 | 1 | 2 | 1 | 2 | . | 1 | 2 | 2 | . | . | 2 |
| 8.11 STP | 1 | 1 | . | . | 2 | . | 2 | 1 | . | 1 | 2 | . | . | . | . | . | 1 | . |
| 8.12 STATPAK | 1 | 1 | . | . | 1 | . | 2 | 1 | 1 | 1 | 1 | 2 | . | . | . | . | 1 | . |
| 8.13 STATUTIL | 2 | 2 | . | . | 1 | 1 | 1 | 1 | . | . | . | 1 | 1 | 1 | . | 1 | . | . |
| 8.14 MICROSTAT | . | 2 | . | . | . | . | 2 | 1 | . | . | . | 2 | . | . | 1 | . | 1 | . |
| 8.15 A-STAT | . | 1 | . | . | 2 | 1 | 1 | 1 | | | | | | | | | | |

## FOR SPECIFIC INTERACTIVE STATISTICAL ANALYSIS PROGRAMS

(ii) User Interface

| | Survey Estimates | Survey Variances | Math Functions | Operations Research | Availability | Installations | Computer Makes | Mini Version | Core Requirements | Batch/Interactive | Stat. Training | Computer Training | Language Simplicity | Documentation | User Convenience | Maintenance | Tested for Accuracy |
|---|---|---|---|---|---|---|---|---|---|---|---|---|---|---|---|---|---|
| | 43 | 44 | 47 | 48 | 49 | 50 | 51 | 52 | 53 | 54 | 55 | 56 | 57 | 58 | 59 | 60 | 61 |
| ISA | 1 | 1 | 2 | 2 | . | 1 | 2 | 3 | 3 | 2 | 1 | 2 | 3 | 1 | 1 | 1 | 1 |
| GLIM | . | . | 1 | . | 3 | 3 | 3 | 3 | 2 | 3 | 1 | 3 | 2 | 2 | 3 | 3 | 3 |
| ISP | . | . | 2 | . | 2 | 2 | 1 | 3 | 2 | 2 | 2 | 2 | 2 | 1 | 3 | 3 | 1 |
| CMU-DAP | . | . | . | . | 3 | 2 | 3 | 1 | 2 | 2 | 2 | 3 | 2 | 2 | 2 | 1 | 2 |
| RUMMAGE | . | . | 1 | . | 3 | 2 | 3 | 2 | 1 | 3 | 1 | 3 | 2 | 1 | 2 | 3 | 2 |
| CADA | . | . | . | . | 2 | 2 | 3 | 3 | 3 | 2 | 1 | 3 | . | 2 | 2 | 3 | 1 |
| SURVO | . | . | 1 | . | 3 | 2 | 2 | 3 | 3 | 3 | 3 | 3 | 3 | 2 | 3 | 3 | 2 |
| AUTOGRP+ | . | . | 1 | 2 | 3 | 1 | . | 2 | . | 3 | 3 | 3 | 2 | 3 | 3 | 3 | . |
| FORALL | . | . | 2 | . | 1 | . | 1 | 1 | 1 | 3 | . | 2 | 2 | 2 | 2 | 1 | 1 |
| AQD | . | . | . | . | 3 | 2 | 2 | 2 | 2 | 3 | 3 | 3 | 3 | 2 | 3 | 3 | 1 |
| STP | . | . | . | . | 3 | 3 | . | 1 | 2 | 3 | 3 | 3 | 2 | 2 | 3 | 3 | 2 |
| STATPAK | . | . | . | 1 | 1 | 1 | 2 | 1 | 2 | 2 | 2 | 3 | 2 | 2 | 2 | 3 | 2 |
| STATUTIL | . | . | 2 | . | . | . | 2 | 3 | . | 3 | 2 | 3 | . | 1 | 2 | 3 | 2 |
| MICROSTAT | . | . | . | . | . | 2 | 2 | 3 | 3 | 2 | 2 | 3 | 3 | 2 | 2 | 3 | 2 |
| A-STAT | . | . | . | . | 2 | 3 | 1 | 3 | 3 | 3 | 3 | 3 | 3 | 2 | 2 | 3 | . |

AUTOGRP+ is a statistical data management package, which can handle large files interactively.  It is written mostly in CML (Conversational Modelling Language) which makes AUTOGRP+ extensible and potentially portable.  An unsupported version is available for $200 from the Center for Health Services, Yale University.

Several of these programs can be installed on microcomputers, in particular A-STAT, MICROSTAT, and probability STATUTIL and CADA.  We can expect many more statistical programs to be written for microcomputers in the coming years.

The remainder of this chapter contains descriptions of these specific-purpose interactive programs in the format described in Table 1.5.

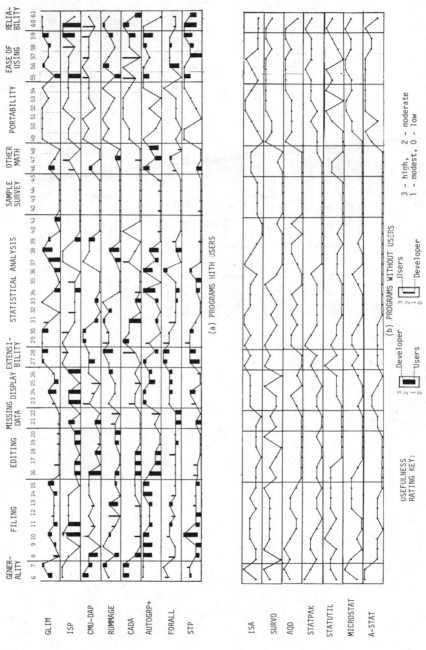

FIGURE 8.1: RATINGS BY DEVELOPERS AND USERS ON ALL ITEMS FOR SPECIFIC-PURPOSE INTERACTIVE PROGRAMS

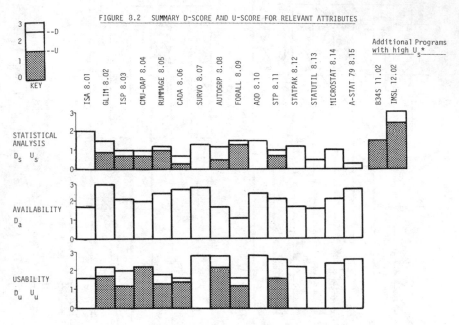

FIGURE 8.2   SUMMARY D-SCORE AND U-SCORE FOR RELEVANT ATTRIBUTES

Other chapters generally containing programs with strong Statistical Analysis Capabilities are Chapters 6, 7, and 9 (see Figures 6.2, 7.2, and 9.2).

*In addition, for these specific programs, $U_s > 1.5$.

8.01 <u>ISA</u>

## INTRODUCTION

ISA, a program for Interactive Statistical Analysis, was written in BASIC by J.N.R. Jeffers for small computers, with a broad range of statistical capabilities. It has only limited maintenance, additions or changes have not been considered, and it is not exported at present.

<u>ISA: Interactive Statistical Analysis</u>

<u>Developed by</u>:                                    <u>Distributed by</u>:

    J. N. R. Jeffers                              Same
    Director, Institute of Terrestrial
      Ecology
    Merlewood Research Station
    Grange-over-Sands
    Cumbria  LA11 GJU
    England

<u>Machine</u>:

    Interactive mini-computer. (Brand name not known.)

<u>Language</u>:

    BASIC

<u>Cost</u>:

    Not so far considered.

.8.02  GLIM

## INTRODUCTION

GLIM is an interactive program designed for the fitting of generalised linear models to data, although its uses are considerably wider than this.  It was developed by the Royal Statistical Society Working Party on Statistical Computing, beginning in 1971, and is installed at over 300 sites in 24 countries.  The program can also be used in batch mode.

## CAPABILITIES:  Processing and Displaying Data

GLIM can accept data from the terminal or from a file, in free or fixed format.  Data can be edited and displayed in tabular or graphical form or, by means of a macro, as a histogram.

## CAPABILITIES:  Statistical Analysis

The kernel of GLIM is the algorithm for the fitting of generalised linear models.  GLIM makes the fitting of classical linear models easy by using a specification whereby quantitative and qualitative x-variates are equally easy to describe and compound terms, such as interactions, can be simply built up. The system goes further, however, and provides a unified framework embracing not only such classical linear models but also loglinear models for contingency tables, logit and probit models for the analysis of proportions and models with gamma errors.  Specification of standard models is simple but powerful, allowing refinements such as prior weights and fixing of certain parameter values. A more general facility of GLIM, to do iteratively weighted least squares, can be used to program any generalised linear model and many robust regression procedures.  Such a procedure, once developed, can be stored as a macro for future use.

As well as its principal function of fitting generalised linear models, GLIM can be used

* as a calculator, operating on vectors and scalars with general arithmetic expressions, including many standard functions.  Macros may be defined and used as blocks in branching and looping; it is thus a programming language of considerable generality.
* for data exploration, to subdivide the data into subsets and to average values over those subsets, to check data visually for the presence of outliers and to calculate simple statistics.
* as a teaching aid.  Its use can remove the burden of getting the arithmetic right, thus allowing the student to think about the statistical aspects, and its flexibility allows scope for following through original ideas.  Standard examples can be prepared in advance and the output presented in text-book form if required.  Pseudo-random numbers can be generated to provide simulated data.

GLIM has extensive internal checks against faults in syntax, overrunning of space, incompatible data, etc.  Full diagnosis of a fault and suggested remedies are given.  The user can obtain information on the current configuration of GLIM, space allocation and availability, current program-control etc. Working space can be reclaimed by deleting unwanted structures.  Output from

GLIM may be written to a file for future input in another run or the program
state may be "dumped" so that it can be continued from the same stage at some
later date.

## EXTENSIBILITY

GLIM's macro facilities allow the user to develop and store his own procedures for future use.

## INTERFACES WITH OTHER SYSTEMS

GLIM can read data files produced by other systems provided that items are
either separated by spaces or new lines or are in a FORTRAN compatible format.

## PROPOSED ADDITIONS IN THE NEXT YEAR

Calculation of predicted values; choice of Givens or Gauss-Jordan method;
more efficient estimation in the presence of nuisance parameters; rewriting of
internal housekeeping.

## REFERENCES

Some reviews of the GLIM-3 manual are by
McCullagh, P. (1979) JASA, 74, 934-935.
Altman, D.G. (1980) Appl. Statist., No. 1 29, 98.
Armstrong, W. (1980) The Statistician, 29, 57-63.

## SAMPLE JOB

```
   GLIM 3.12 (C)1977 ROYAL STATISTICAL SOCIETY, LONDON

UNIT 5 OPENED
0019 £C
0020 £C
0021 £C
0022 £C           SET NO. OF UNITS AND GROUPING AS A 2-LEVEL FACTOR
0023 £UNITS 12 $FAC A 2
0024 £C           SET UP AND READ DATA MATRIX OF GROUP, COVARIATE
0025              AND BINOMIAL RESPONSE,  R OUT OF N
0026 £DAT A X R N  $READ
0027  1 2.6 34 42    1 3.6 51 52    2 2.3 31 115
0028  1 1.2  4 20    2 5.8 68 56    2 3.0 55  88
0029  1 0.1  2 64    2 7.3 71 75    1 1.8 13  32
0030  2 4.4 31 39    2 1.5 47 1 4   1 3.3 16  18
0031 £YVAR R $ERR B N
0032 £C           MODEL NOW SET UP  LOGIT LINK DECLARED BY DEFAULT
0033 £FIT : A 1
          SCALED
  CYCLE  DEVIANCE        DF
     3    318.5          11

          SCALED
  CYCLE  DEVIANCE        DF
     3    313.9          10
```

```
0034 £C            GROUP DIFFERENCES SMALL. TRY COVARIATE, IGNORING
0035               GROUPING, TO SEE IF SINGLE FITS
0036 £FIT X £
        SCALED
  CYCLE  DEVIANCE      DF
    4     65.62        10


0037 £C            KEEP COMMON SLOPE, BUT TRY DIFFERENT INTERCEPTS
0038 £FIT +A £
        SCALED
  CYCLE  DEVIANCE      DF
    4     65.18         9


0039 £C            FIT STILL BAD, SO TRY DIFFERENT SLOPES
0040 £FIT +A.X £
        SCALED
  CYCLE  DEVIANCE      DF
    4     1.958         8


0041 £C            FIT NOW GOOD, SO CHECK IF COMMON INTERCEPT
0042               IS ADEQUATE
0043 £FIT -A £
        SCALED
  CYCLE  DEVIANCE      DF
    4     43.06         9


0044 $C            IT IS NOT, SO RESTORE SEPARATE INTERCEPTS AND
0045               DISPLAY RESULTS
0046 £FIT +A $DIS E R £
        SCALED
  CYCLE  DEVIANCE      DF
    4     1.958         8


        ESTIMATE       S.E.       PARAMETER


    1   -3.949        0.5379      %GM
    2    2.005        0.2452      X
    3    2.896        0.5632      A(2)
    4   -1.490        0.2513      X.A(2)
  SCALE PARAMETER TAKEN AS       1.000

  UNIT  OBSERVED OUT OF  FITTED     RESIDUAL
    1      34      42    32.75      0.4658
    2      51      52    50.71      0.2600
    3      31     115    33.29     -0.4709
    4       4      28     4.931    -0.4621
    5      48      56    48.93     -0.3743
    6      55      88    54.62      0.8287E-01
    7       2      64     1.473     0.4393
    8      71      75    71.32     -0.1716
    9      13      32    13.31     -0.1101
   10      31      39    30.07      0.3558
   11      47     104    44.77      0.4417
   12      16      18    16.83     -0.7956


0047 £END
```

## GLIM: Generalised Linear Interactive Modelling

Developed by:

The Royal Statistical Society
Working Party on Statistical
 Computing
25 Enford Street
London W1H 2BH
U.K.

Distributed by:

The Numerical Algorithms Group
 Limited
7 Banbury Road
Oxford OX2 6NN
U.K.

Computing Ranges:

Burroughs 4800, 6700, 7700
CDC 6000, 7600, Cyber
CII IRIS
DEC System 10, 20, PDP-11, VAX 11
GEC 4080
Harris/4
HP 3000
Honeywell
IBM 360, 370 etc.
ICL 1900, 2900, System 4
NORD 10/100 (48 bit reals)
Perkin-Elmer (32 bit machines)
Prime 400 upwards
Siemens 4004
Telefunken TR 440
Univac 1100

Operating Systems:

Essentially all unless indicated
otherwise

GCOS, Multics

Source Language:

Primarily FORTRAN, some assembler on certain ranges.

Cost:

A single charge of £ 120 Sterling to academic sites, £ 240 to others. The
fee includes a source and compiled forms of the program on magnetic tape, a free
copy of the manual, installation instructions and a specimen newsletter.

Documentation:

User Manual, £4.00 + postage and packing.
Newsletter £ 2.50 per annum (2 issues).

8.03  <u>ISP</u>

## INTRODUCTION

ISP, the Interactive Statistical Processor, was developed to provide a flexible way of using an extensive collection of data analysis programs on a minicomputer. The requirements were that the data analyst should be able to enter, modify and save data on disk files, to re-express the data in various ways, to select subsets of the data in various ways, and to carry out various analyses of the results of these operations. The processor has been in daily use from the day it was first installed and has been used for a large number of analyses, often accounting for the largest single component of the daily use of the minicomputer on which it is run.

This program has been developed by Peter Bloomfield at Princeton University. The earliest version was developed by Ken Birman and later versions by David Donoho. Command programs have been contributed by many people.

ISP was developed for the Unix time-sharing system. A unified command syntax has been imposed by a command-interpreting "shell" program, wich communicates with user at his or her terminal and initiates execution of separate programs to carry out the required operations. Uniformity of these operational programs has been achieved by using a single structure for files and providing a library of subroutines for analyzing the standard syntax for specification of options.

Since the shell knows nothing about the programs that it executes, except for default places to find them, new commands may be added even during the course of a session. If a user wishes to give a single command that will cause a sequence of programs to be run (possibly a re-expression following by a plot of the residuals against the fit), a macro may be constructed to do this.

The program is portable to other PDP/11-UNIX systems, and is fully maintained and additions are being made. However due to manpower limitations it is not being actively distributed at present.

## CAPABILITIES: Processing Statistical Data

Simple commands exist to: create a text file; edit a text file, read a text file, converting to ISP variable; save an ISP variable or text file for future use, list contents of active, data, text, system areas, and print variable or string. A "let" command provides algebraic and manipulative capability.

## CAPABILITIES: Statistical Analysis

Data Analysis: boxplot schematic plots, stem and leaf displays, coded displays, five number summaries, robust estimates of location (biweight M-estimate), comparison (schematic) plots, scatterplots.

Robust Fitting: robust regression, oneway and twoway anova.

Least Squares: basic statistics of batches, correlation matrix, regressions, eigenvalues (real symmetric matrix), principal components, singular value decompositions, canonical correlations.

Time Series: Tukey's smoothers, Fast Fourier Transform, periodogram, cross periodogram, phase stuff.

REFERENCE

Bloomfield, P., "An Interactive Statistical Processor for the UNIX Time-sharing System," Proc. of Computer Science and Statistics:  Tenth Annual Symposium on the Inferface, 1977, pp. 2-8.


SAMPLE JOB

The following is a short example of a session with ISP, in which a set of data is typed in at the terminal, displayed and analyzed.  (Characters typed at the keyboard are underlined.  Lines typed at the keyboard are terminated by a 'carriage return'.  The symbol '^D' marks a control - D, which, when typed imme- diately after 'carriage return', indicates the end of input.)

The first command is to 'make' a file called 'junk', which contains the data.

```
                    *make junk
     1                  0
     2                  3
     3                  2
     4                  3
     5                  6
    ^D
```

The second command is to 'read' the data in 'junk' (i.e., convert to binary format), and place it in a variable called 'var'.  This line may be read as 'read junk onto var'.

*read junk > var

Since we want to use 'var' as an array of 5 lines each with 2 entries, the utility 'let' is invoked to reshape 'var'.

*let var = var (5,2)

Simple typewriter scatter-plots may be produced by 'scat'.  Since 'var' has only two columns, no options need be specified.  If 'var' had more columns, the command might be 'scat var {x=3;y=4}' (options are always enclosed in {}). The defaults are x=1 and y=2.

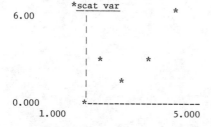

```
              *scat var
 6.00                          *
              |
              |
              |
              | *          *
              |
              |    *
              |
 0.000        *----------------------
      1.000                  5.000
```

The 'regress' command below fits a straight line (i.e., regresses column 2 on column 1 with a constant term). The symbol '>' is used as in the 'read' command above to indicate the disposition of output variables. Since 'regress' may produce more than one output, we specify 'res @ vresids' to indicate that the output known internally as 'res' is to be produced and placed in a variable called 'vresids'. Since 'res' is in fact the first output, 'regress var > vresids' would have the same effect (any other outputs will not be produced, since no variable name is given).

        *regress var > res @ vresids*

| variable | coeff. | corr. | t-stat |
|----------|--------|-------|--------|
| 1 | 1.20000 | 0.875190 | 3.13340 |

| | |
|-----------|-----------|
| intercept | -0.800000 |
| multiple r | 0.875190 |
| f-statistic | 9.81819 |

The output 'res', here in variable 'vresids', consists of two columns, the first containing the fitted values and the second containing the residuals (this is also true if 'regress' is used for multiple regressions).

ISP:   Interactive Statistical Processor

Developed by:                              Distributed by:

    Peter Bloomfield and others            Same
    Department of Statistics
    Princeton University
    Fine Hall, P.O. Box 37
    Princeton, New Jersey   98540

Computer Makes:                            Operation Systems:

    PDP/11                                 UNIX

Source Languages

    C, YACC, Ratfor, Fortran

Cost:

    A one-time charge of $200.

8.04  <u>CMU-DAP</u>

## INTRODUCTION

DAP is an interactive package for analysis and manipulation of statistical data.  It was designed to facilitate exploratory and graphical analysis of data. DAP was originally developed by Samuel Leinhardt in 1977 at Carnegie-Mellon University.  Earlier versions have been distributed to about a dozen installations.

## CAPABILITIES:  Processing and Displaying Data

DAP possesses a user workspace of currently active data and archives for permanent data storage.  Data can be entered into the DAP workspace from a terminal, a DAP archive, or an external file.  Workspace data can be written to a DAP archive or an external file.  The external file read-write capability allows DAP to be used in conjunction with other common statistical packages.

Data can be modified via a general algebraic function which allows a user to summarize a data vector, combine vectors algebraically, perform logical contrasts, evaluate inequalities, construct indicator variables, and perform exponential or logarithmic transformations.  Data can be manipulated by joining variables, excising values, and inserting or squeezing out n.a. values.  Data values can be located using various logical operators.

Data display capabilities include x-y plot, box and schematic plots, stem-and-leaf display, histogram, normal plots.

## CAPABILITIES:  Statistical Analysis

DAP contains procedures for performing exploratory as well as conventional statistical analysis.  Most of the procedures introduced by John Tukey in the book <u>Exploratory Data Analysis</u> are available, as are related newly developed procedures.  Many of DAP's diagnostic capabilities are based on Tukey's methods and allow the user to determine whether outliers, interactions, nonadditivity or nonlinearity are present in the data  and facilitate the choice of a data transformation to correct the problem.

Exploratory procedures include resistant linear fits, median polish for two way tables, extended number summaries, and non-linear smoothing for time series.

Other analytic features include random sampling from a variety of distributions, multivariate least squares regression, analysis of variance, log-linear fits for contingency tables, and various summary statistics, including mean, variance, median, scaled median of absolute deviations, trimmed means, and biweight.

## PORTABILITY

DAP is written in machine portable FORTRAN, with the exception of a few machine dependent sections (e.g. file handling and literal packing routines).

## INTERFACES WITH OTHER SYSTEMS

DAP can read and write FORTRAN formatted files, and so can interface with other packages that can read and write in FORTRAN format.

## PROPOSED ADDITIONS IN NEXT YEAR

Iteratively reweighted least squares, transformation tables, trimming.

## REFERENCES

Leinhardt, S., and Wasserman, S.S., "Exploratory Data Analysis: An Introduction to Selected Methods", Sociological Methodology 1979, edited by K. Schuessler, San Francisco: Jossey-Bass.

Leinhardt, S., and Wasserman, S.S., "Quantitative Methods for Public Management: An Introductory Course in Statistics and Data Analysis", Policy Analysis, 1978, Fall, pp. 549-575.

Leinhardt, S., and Wasserman, S.S., "Teaching Regression: An Exploratory Approach", The American Statistician, 1979, Vol. 33, pp. 196-203

Tukey, J.W., Exploratory Data Analysis, Reading, Massachusetts: Addison-Wesley Publishing Company, 1977.

## SAMPLE JOB

Determine a power transformation for symmetrizing data. A suggested power of approximately 0 corresponds to a log transformation. A boxplot of MORT.RATE reveals it to be asymmetric. The suggested transformation of approximately .1 indicates a log transformation. A boxplot of the logged variable reveals a high degree of symmetry.

```
BOXPLOT(MORT.RATE)
SYMMETRIZ(MORT.RATE)
LET LOG.MORT=LN(MORT.RATE)
BOXPLOT(LOG.MORT)
```

```
BOXPLOT(MORT.RATE)


   SCALE UNIT: 10.0

 0.0             160.0           320.0           480.0           640.0
 -+---------------+---------------+---------------+---------------+------

   VARIABLE:  MORT.RATE


      +--+------+
   <-I  +  O  I------------>    *          *                    *
      +--+------+
                OUTLIERS
        LOW                 HIGH
 INDEX  VALUE        INDEX  VALUE
      NONE             52:  300.0000
                        1:  400.0000
                       79:  650.0000


 DAP - TASK COMPLETE
 SYMMETRIZ(MORT.RATE)


   SYMMETRIZING POWER =     0.13764


 DAP - TASK COMPLETE
 LET LOG.MORT=LN(MORT.RATE)

   105 OBSERVATIONS IN RESULT.     4 MISSING.

 DAP - TASK COMPLETE
 BOXPLOT(LOG.MORT)


   SCALE UNIT: 0.075

 2.0             3.2             4.4             5.6             6.8
 -+---------------+---------------+---------------+---------------+------

   VARIABLE:  LOG.MORT


           +-----------+---------+
   <------------I           O+       I--------------------->
           +-----------+---------+


 DAP - TASK COMPLETE
```

CMU-DAP, Version 2.01:  Data Analysis Package

Developed by:                          Distributed by:

    Samuel Leinhardt                  Samuel Leinhardt
    School of Urban and Public Affairs    School of Urban and Public Affairs
    Carnegie-Mellon University        Carnegie-Mellon University
    Pittsburgh, PA  15213             Pittsburgh, PA  15213

Computer Makes:                        Operating Systems:

    DEC System 10, 20, VAX            Tops-20, VAX/VMS

Source Language:

    FORTRAN

Cost:

    Initial purchase fee of $500 for universities and non-profit government
installations, $1,000 for commercial installations.  Yearly update and main-
tenance fee of $200.

## 8.05  RUMMAGE

### INTRODUCTION

RUMMAGE is a program for solving linear model problems where either continuous or discrete, multivariate or univariate, weighted or unweighted data are considered. The system can handle balanced or unbalanced designs, missing data, and transformations. Expected mean square coefficients are provided for fixed, mixed, and random effects models.

RUMMAGE was developed in the Department of Statistics at Brigham Young University starting in 1970. It has been under continuous development since then. It can be used interactively or as a batch program, and it is presently installed in approximately 30 computer installations.

### CAPABILITIES:  Processing and Displaying Data

The input to RUMMAGE-II is in the form of a worksheet consisting of rows of variables. Rows of the worksheet correspond to observations or responses, and columns contain the dependent and independent variables and indices indicating levels of the factors under study. There is no restriction in RUMMAGE as to the number of data points which can be analyzed. The worksheet is processed row by row, as it is input, through use of transformation and editing commands. The commands consist of a command name followed by an argument list. The commands are free format, key-word oriented and in English words, or English-like sentences. RUMMAGE can be modified to interface with the user's data-base management system.

Plotting capabilities include the usual two-way plots, pseudo three-dimensional plots using either labelling or a third continuous axis, histograms, and time-ordered plots. The user may also overlay several plots on the same set of axes. In addition to the variables input by the user, the predicted values, residuals, or standardized residuals can be plotted. Several residual plots are produced automatically.

### CAPABILITIES:  Statistical Analysis

The common unifying point among analysis of variance, analysis of covariance, regression and contingency tables, etc. is the linear model. RUMMAGE uses the linear model in algebraic form to define the independent variables, effects, etc. The user defines each main effect in a model as being either fixed or random. The analysis of variance tables include expected mean square coefficients for each term in the model which may be used to find proper significance tests or estimate variance components.

For continuous data, both multivariate and univariate analyses are available. Correlation matrices can be calculated, as well as the observed cell means or marginal means for any margin of a multi-way table. Sequential and "adjusted" analysis of variance tables can be output. Estimated means and user specified linear combinations of the means are available for any model fit. For multivariate analysis, several test statistics are given as well as the basic eigenvalue structure of the hypotheses and error matrices.

For categorical data, both linear and log-linear models are available.
The IPF and generalized least squares procedures permit numerous linear or log-
linear models to be fit depending upon the assumptions desired by the user.
The logit command allows further flexibility in specifying models and submodels.
To exploit the similarities between continuous and discrete analysis and emphasize the fundamental linear model, much of the output for contingency table
analysis is in the same form as its counterpart in analysis of variance.

## FUTURE DEVELOPMENT

RUMMAGE is a major research tool for many consulting statisticians, and
development of new procedures and capabilities, as well as documentation in
the form of user's and programmer's manuals, will continue for the foreseeable
future.

## REFERENCES

Bryce, G.R., D.T. Scott, and M.W. Carter (1980). "Estimation and Hypothesis
    Testing in Linear Models -- A Reparameterization Approach to the Cell
    Means Model." Communications in Statistics -- Theor. Meth. A9(2): 131-
    150.
Searle, S.R., H.V. Henderson, and M.W. Carter (1979). "Annotated Computer Out-
    put for Analyses of Unbalanced Data: RUMMAGE." Biometrics Unit Bulletin
    BU-692-M, Biometrics Unit, Cornell University, Ithaca, New York.

## SAMPLE JOB

```
NOTE                    MIXED MODEL (REPEATED MEASURES)
NOTE                    UNIVARIATE ANALYSIS OF VARIANCE
NOTE
MODEL     Y(I,J,K) = T(I) + P(IJ) + V(K) + TV (IK) + E
FIXED     V    3   C3
RANDOM    P    4   C2
FIXED     T    3   C1
TEST      TV
TEST      V
TEST      T     P
ESTIMATE  T    ( )    P
TAPE   1
TABLE1.DAT
READ   C1 - C4
STOP
```

## Sample Output

THE DEPENDENT VARIABLE FOR THIS TABLE IS    C4

SEQUENTIAL ANALYSIS OF VARIANCE TABLE

| SOURCE | DF | SUM OF SQUARES | MEAN SQUARES | EXPECTED MEAN SQUARE COEFF. | | | |
|---|---|---|---|---|---|---|---|
| | | | | TMTS(T) | PATNTS(P) | VISITS(v) | TV |
| MEAN | 1 | 3182.662012 | | | | | |
| TMTS(T) | 2 | 0.822974 | 0.411487 | 10.909 | 3.000 | 0.000 | 0.000 |
| PATNTS | 8 | 5.065814 | 0.633227 | 0.000 | 3.000 | 0.000 | 0.000 |

THE  1 DEGREES OF FREEDOM LOST
ARE CONFOUNDED WITH THE TERMS ABOVE THIS MESSAGE

| SOURCE | DF | SUM OF SQUARES | MEAN SQUARES | EXPECTED MEAN SQUARE COEFF. | | | |
|---|---|---|---|---|---|---|---|
| VISITS | 2 | 17.716097 | 8.858048 | 0.000 | 0.000 | 11.000 | 0.000 |
| TV | 4 | 62.026875 | 15.506719 | 0.000 | 0.000 | 0.061 | 3.636 |
| REGR. | 16 | 85.631760 | 5.351985 | | | | |
| ERROR | 16 | 11.344828 | 0.709052 | | | | |
| TOTAL(C) | 32 | 96.976599 | | | | | |
| TOTAL | 33 | 3279.638611 | | | | | |

SIMPLE R-SQUARED = .883
R-SQUARED ADJUSTED FOR D.F. = .766

SEQUENTIAL HYPOTHESIS TESTS

| SOURCE | DF | | F-RATIO | | P LEVEL |
|---|---|---|---|---|---|
| TV | 4 | 15.506719 | = | 21.869658 | 0.000 |
| ERROR | 16 | 0.709052 | | | |
| VISITS | 2 | 8.858048 | = | 12.492810 | 0.001 |
| ERROR | 16 | 0.709052 | | | |
| TMTS(T) | 2 | 0.411487 | = | 0.649826 | 0.548 |
| PATNTS | 8 | 0.633227 | | | |

ESTIMATED MEANS
MEAN SQUARE USED IN STD. DEV. HAS    8 D.F.

| | SAMPLE | MEAN | STD. DEV. OF THE MEAN |
|---|---|---|---|
| T(1) | 12.00 | 9.65917 | 0.22971 |
| T(2) | 12.00 | 10.02167 | 0.22971 |
| T(3) | 9.00 | 9.76778 | 0.26525 |

. . .

*****************************************
* SEQUENTIAL ANOVA. The program expects to find the same *
* number of patients in each treatment level. Treatment *
* level three had only two patients and thus, the warning *
* message is produced. Note that the expected mean *
* square coefficients indicate the appropriate error term *
* for T is P. *
*****************************************

*****************************************
* SEQUENTIAL HYPOTHESIS TESTS. The test for TV and T are *
* exact F-ratios. The test for visits is exact for *
* hypothesis $H_3$ of section 3.5 but it might be desirable *
* to reorder the model to test $H_4$. *
*****************************************

*****************************************
* ESTIMATED AND OBSERVED MEANS. Since the full model is *
* being fit, the estimated and observed means are iden- *
* tical. *
*****************************************

RUMMAGE II:  Regression Using Many Models and General Estimation

Developed by:                                    Distributed by:

Department of Statistics                         Same
204 TMCB, Brigham Young University
Provo, UT  84602

Computer Makes:                                  Operating Systems:

Burroughs B3700, B3800                           IBM - OS, VS, TSO, VM-CMS DOS-VS
CDC 6000 series                                  CDC - NOS, SCOPE 3.4
  CYBER series                                   DEC - LINK  19 OVERLAY
DEC system 10, 20                                XDS Sigma 7
Honeywell 6000                                   UNIVAC 1108
IBM 360, 370                                     BURROUGHS 6700
Siemens 4004, 7000                               HONEYWELL 6000
Xerox Sigma                                      UNIVAC 8 and 10

                                                 Interfaced Systems:

                                                 SAS
                                                 ASPECT

Source Languages:  Primarily ANSI FORTRAN IV.  Assembly language for some
machines for specific machine dependent tasks.

Cost:

    The two year license fee for RUMMAGE is $500 for universities and $1000
for all others.  The fee includes program source modules, load and/or object
modules (if available), and installation instructions all on magnetic tape.  It
also includes maintenance and one copy of the RUMMAGE manual.  Other reports
and manuals are available.

Documentation:

Bryce, G.R. (1980).  "Data Analysis in RUMMAGE -- A User's Guide."  Statistics
    Department, Brigham Young University, August, 1980.
Carter, M.W. and A.S. Allen (1979).  "RUMMAGE Output with Annotation for
    Balanced, Unbalanced, and Missing Cell Data of Searle and Henderson."
    Brigham Young University Statistics Department Report Series SD-005-R.
Scott, D.T. (1980).  "Documentation of the RUMMAGE-II Numerical Algorithms."
    Brigham Young University Statistics Department Report Series SD-018-R.
Bryce, G.R., M.W. Carter and D.T. Scott (1980)  "Recovery of Estimability in
    Fixed Models with Missing Cells."  Brigham Young University Statistics
    Department Report Series SD-002-R.

## 8.06  CADA

INTRODUCTION

The Computer-Assisted Data Analysis (CADA) Monitor is a conversational, interactive computer software package of nearly 200 modules with a Bayesian, decision-theoretic viewpoint. It will lead a user step by step through a data analysis in much the same way that a computer-assisted instruction (CAI) program leads a student. The Monitor has as its primary function the teaching of statistical methods to students with minimal mathematical background. It also provides educational researchers, and others, with sophisticated methods of exploratory data analysis, linear models, data management, and decision-theoretic applications. The Monitor is hierarchically organized into nine component groups and is menu driven.

CAPABILITIES:  Processing and Displaying Data

Component Group 1 of the Monitor, Data Management Facility, provides data manipulation modules. Data may be entered from a terminal, stored in or retrieved from a permanent disk file, or retrieved from a catalog of data sets. The data entered may be displayed and edited interactively. There is a set of data transformation modules that permit most of the common unary and binary operations; all transformations may be executed conditionally.

Summary statistics, such as means, standard deviations, percentiles, etc., may be generated, as well as variance-covariance matrices and correlation matrices. The data may be grouped and reordered. There is also a matrix package that includes a sweep operator as well as the standard matrix operators and matrix management capabilities.

CAPABILITIES:  Statistical Analysis

Component Group 2 is Simple Parametric Models, including binary models, univariate normal models, multi-category models, and regression analysis. Component Group 3, Decision Theoretic Models, allows elicitation of utilities and selection/assignment applications.

Component Group 4, Bayesian Simultaneous Estimation, allows the estimation of proportions, means, or prediction in m groups. Component Group 5, Bayesian Full-Rank Analysis of Variance, provides a full-rank model I factorial analysis of variance and an analysis of repeated-measures designs. The latter is performed using the modules in Component Group 6, Bayesian Full-Rank Multivariate Analysis.

Component Group 7, Elementary Classical Statistics, provides frequency distributions, including contingency and expectancy tables; summary statistics; graphic displays, including histograms and normal probability plots; and classical regression. Component Group 8, Exploratory Data Analysis, provides the techniques of Tukey for viewing the data. Component Group 9, Probability Distributions, allows the calculation of percentiles, highest density regions, probability content of specified intervals, and graphs of eighteen distributions.

## EXTENSIBILITY

The Monitor is highly modularized in the Basic language using many standardized subroutines.  User-written modules may be added without difficulty.

## PROPOSED ADDITIONS IN NEXT YEAR

The Monitor is being significantly expanded, particularly in data management and multivariate methods.  A microcomputer translation is expected in approximately one year, but another major release is not expected until 1983.

## REFERENCES

Isaacs, G.L., and Novick, M.R.  The Bayesian Computer-Assisted Data Analysis (CADA) Monitor.  Proceedings of ACM SIGSCE-SIGCUE Joint Symposium.  New York: Association for Computing Machinery, 1976.

Novick, M.R.  High school attainment: An example of a computer-assisted Bayesian approach to data analysis.  International Statistical Review, 1973, 41, 264-271.

Novick, M.R.  A course in Bayesian statistics.  The American Statistician, 1975, 29, 94-97.

Novick, M.R.; Hamer, R.M.; and Chen, J.C.  The Computer-Assisted Data Analysis (CADA) Monitor, 1978.  The American Statistician, 1979, 33, 219-220.

Novick, M.R., and Jackson, P.H.  Statistical Methods for Educational and Psychological Research.  New York: McGraw-Hill, 1974.

## SAMPLE JOB

Since the Monitor is interactive and menu driven, there is no user language and a sample job cannot be conveniently displayed in the limited space. Following is a display from the Bivariate Exploratory Data Analysis Component.

SCHEMATIC PLOT    (PARALLEL SCHEMATIC PLOTS)

     FOR AN EXPLANATION TYPE '2', OTHERWISE TYPE '1' TO CONTINUE.?1


THERE WILL BE A SHORT PAUSE FOR CALCULATION.

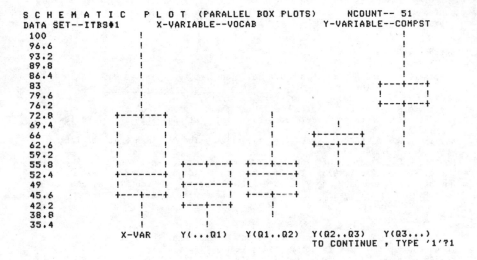

```
S C H E M A T I C    P L O T  (PARALLEL BOX PLOTS)       NCOUNT-- 51
DATA SET--ITBS#1       X-VARIABLE--VOCAB             Y-VARIABLE--COMPST
 100                !                                            !
 96.6               !                                            !
 93.2               !                                            !
 89.8               !                                            !
 86.4               !                                            !
 83                 !                                    +---+---+
 79.6               !                                    !   !   !
 76.2               !                                    +---+---+
 72.8         +---+---+                    !                     !
 69.4         !       !                    !             !       !
 66           !       !                    !       +--------+    !
 62.6         !       !                    !       +---+---+     !
 59.2         !       !                    !             !
 55.8         !       !  +---+---+    +---+---+           !
 52.4         +-------+  !       !    +-------+
 49           !       !  +-------+    !       !
 45.6         +---+---+  !       !    +---+---+
 42.2             !      +---+---+        !
 38.8             !          !            !
 35.4             !          !
              X-VAR      Y(...Q1)   Y(Q1..Q2)  Y(Q2..Q3)   Y(Q3...)
                                   TO CONTINUE , TYPE '1'?1
```

## CADA-1980:  The Computer-Assisted Data Analysis Monitor

Developed by:                                    Distributed by:

    CADA Research Group                              CADA Research Group
    348 Lindquist Center                             348 Lindquist Center
    The University of Iowa                           The University of Iowa
    Iowa City, Iowa  52242                           Iowa City, Iowa  52242

Computer Makes:                                  Operating Systems:

    HP2000                                           Basic
    HP3000                                           MPE III (BASIC/3000)
    Prime 750                                        PRIMOS (BASIC and BASICV)
    DEC PDP/11                                       RSTS/E (BASIC-PLUS)
    IBM 370                                          VS (VSBASIC)
    CDC CYBER                                        NOS (BASIC 3.4)

Source Languages:  Primarily Basic, with assembler subroutines in some systems for file manipulation.

Cost:

    A one-time charge of $600 (US) is assessed for CADA-1980.  There is a 50% discount for U.S. and Canadian educational and non-profit research institutions. Included are source modules, compiled modules (when appropriate), installation instructions, one copy each of the CADA Monitor (1980) Manual and the CADA Display Book, and maintenance until the next release.  Distribution is on magnetic tape for all systems.

Documentation

Novick, M.R.; Hamer, R.M.; Libby, D.L.; Chen, J.J.; and Woodworth, G.G.  The
    Computer-Assisted Data Analysis (CADA) Monitor (1980).  Iowa City, Iowa:
    The University of Iowa, 1980.
Libby, D.L.; Chen, J.J.; Woodworth, G.G.; Divgi, D.R.; and Mayekawa, S.I.,
    Display Book for the Computer-Assisted Data Analysis (CADA) Monitor (1980).
    Iowa City, Iowa:  The University of Iowa, 1980.

8.07   SURVO

INTRODUCTION

SURVO is a data processing system used to produce crosstabulations, correlations, regressions, factor analysis and time series analysis, etc. SURVO interfaces with SURDA, a data preparation system, and with MATRIX, a matrix calculation system. All matrices produced by one system can be transferred to another system. So these systems will assist one another. These three software products will support both education and research work in the social sciences.

SURVO was developed at the Computer Centre of University of Tampere in 1966, MATRIX in 1979 and SURDA in 1980. These systems are under continuous development according to the needs of the University. SURVO, SURDA and MATRIX can be used in interactive and batch modes and are installed in 5 computer installations in Finland.

CAPABILITIES:  Processing and Displaying Data

SURDA is used to create the primary data matrix for SURVO. SURDA can detect all values of variables falling above or below user-defined limits. A user can define an initial value for every variable which is automatically used to replace missing values of variables. SURDA has typical data-editor features to modify, sort, reorder or edit data matrices. SURDA will produce a summary of a data matrix, including names of variables, mean values, standard deviations, minima and maxima for every variable.

SURVO reads only numerical values of variables from a data matrix. New variables can be defined by arithmetic expressions, and different kinds of groups can be treated by logical expressions without any need to sort the data. The size of a data matrix is unlimited.

Statistical operations permit the transfer of data from operation to operation as requested. Tables, matrices and modified data matrices produced by SURVO can be processed by MARTIX, SURDA or user programs.

A SURVO program, an unlimited set of statistical operations, is defined by the user in a conversational mode. The results of a program can be referenced by another SURVO-program, so the user can proceed in a stepwise manner with statistical data processing.

CAPABILITIES: Statistical Analysis

In SURVO the user can define a set of statistical operations in a conversational mode. According to these definitions SURVO creates internal machine-independent code which will be processed immediately or put into the batch queue or sent to a remote computer.

$\chi^2$ SURVO's major statistical features include frequency tables, **t-tests**, $\chi^2$-test, fractiles, analysis of variance, analysis of covariance, means, standard deviations, moments, correlation for missing data, regression analysis, canonical analysis, partial correlations and covariances, multiple correlation, principal component analysis, factor analysis, regression coefficients for factors, transformation analysis, Bartlett-Box test, multiple discriminant analysis, cluster analysis, filtering time-series, auto- and cross-correlations, Box-Jenkins model for time-series, spectral analysis, estimation of lag structure, transfer function noise model, and forecasting function. All linear combinations can be treated as variables. This means that factor scores, regression scores, and different residuals can be handled.

SAMPLE JOB

```
NEXT OPERATION IS
! STDDEV IF SEX=0
  :
  :
NEXT OPERATION IS
! STDDEV IF SEX=1

USING VARIABLES
! WEIGHT LENGTH

OPTIONAL SPECIFICATIONS
! YES

THESE RESULTS ARE STORED UNDER THE NAME
! BOYS

THE TITLE OF THESE RESULTS IS
! MEANS AND STANDARD DEVIATIONS FOR BOYS

NEXT OPERATION IS
! TTEST

USING VARIABLES
! ALL

BETWEEN GROUPS
! GIRLS BOYS

THE TITLE OF THESE RESULTS IS
! TTEST OF WEIGHT AND LENGTH BETWEEN GIRLS AND BOYS
```

Partial Output

MEANS AND STANDARD DEVIATIONS FOR BOYS


                        STANDARD DEVIATIONS
                        *********************


NUMBER OF OBSERVATIONS:        55

STANDARD DEVIATIONS      BOYS

            VARIABLE        MEAN         STDDEV

            WEIGHT         3451.3        523.28
            LENGTH         50.236        2.0273


*********************************************************************


TTEST OF WEIGHT AND LENGTH BETWEEN GIRLS AND BOYS


                        T-TEST
                        ********


GROUPS ...............:    GIRLS AND     BOYS
NUMBERS OF OBSERVATIONS:     65 AND       55
DEGREE OF FREEDOM .....:    118.


VARIABLE      T          MEANS                  STDDEVS

   WEIGHT    2.16     3640.    3451.         438.2    523.3
   LENGTH    1.96     50.95    50.24         1.972    2.027


*********************************************************************

SURVO (SURDA, MATRIX)

Developed by:                              Distributed by:

  Computer Centre                            Same
  University of Tampere
  P.O. BOX 607
  SF 33101  TAMPERE 10
  FINLAND

Machine:                                   Operating Systems:

  Honeywell H1644                             TS, GECOS, EXEC8
  Honeywell H6000
  Honeywell H66/10
  UNIVAC 1108                                 OS1100
  UNIVAC 1100/11
  PDP 11/70
  (Implementation on CDC CYBER and NOS is under negotiation.)

User language

  SURVO uses English-like sentences.  The Users Guide is written in Finnish:
the English version will be prepared if the system is distributed abroad.

Source Languages:  FORTRAN, and a few assembler routines.

Cost:

  Case by case determination.

8.08  AUTOGRP+

## INTRODUCTION

AUTOGRP+ (pronounced auto-group-plus) is a general-purpose, interactive computer system for data management, statistical analysis, and report generation. Developed at Yale University (as AUTOGRP) for the analysis of health care data, AUTOGRP has been expanded and enhanced by Puter Associates, Inc. so that AUTOGRP+ is now a fully supported, highly extensible pattern analysis and decision support tool. AUTOGRP+'s comprehensive variable and group management facilities, flexible statistical models, and immediate response capability make this statistical software especially useful to managers and analysts who want to devise their own analysis strategies, test hypotheses, and get back immediate results. AUTOGRP+'s principal applications include survey analysis, trend analysis, risk analysis, control of output processes, health care and hospital management, crime pattern detection, insurance claims review, marketing research, and manpower and resource management.

## CAPABILITIES:  Processing and Displaying Data

AUTOGRP+ works with hierarchical as well as simple sequential data file structures stored on either tape or disk media. AUTOGRP+'s variable definition facilities permit users to:  Define each variable, Specify range of values, record location, and label re-coded variables; Specify dependent and weighted variables for statistical reporting; Use logical and arithmetic expressions to create new variables from combinations of existing variables.

AUTOGRP+'s group management facilities permit users to partition large populations of observations into meaningful, statistically well-behaved groups by letting users specify criteria for creating record subsets.

AUTOGRP+'s Variable Conversion System facilitates reformatting of data files, recoding of variables, calculation of new variables, and conversion from one computer representation scheme to another or from one recording medium to another. AUTOSEL+ lets AUTOGRP+ users specify complex criteria for selecting a subset of records residing on one storage medium (e.g., tape) and copying it to another, on-line medium for interactive analysis.

AUTOGRP+ produces user-defined reports ranging in complexity from a simple list of variable values to a multidimensional contingency table. AUTOGRP+ also generates multivariable profiles of several statistical measures such as average, ratio, variance, etc. In addition, AUTOGRP+ has the following tabular and graphic display capabilities:  Tabulates, for all observations in a specified group, values of any specified set of variables; Tabulates values and displays analysis of variance of a dependent variable for each value of an independent variable; Profiles interquartile ranges of a dependent variable for each value of an independent variable; Generates histograms including overlay histograms for two distributions; Generates and displays tree structures showing how groups are subdivided and the reductions in variance resulting from such subdivision; Produces scatter plots.

## CAPABILITIES:  Statistical Analysis

AUTOGRP+'s basic statistical analysis capabilities include count, mean, standard deviation, skewness, kurtosis, maximum and minimum values, and analysis of variance for any numeric variable in one or several groups of observations.  AUTOGRP+ also generates multivariable profiles of several statistical measures such as average, ratio, variance, etc.

In addition, AUTOGRP+ provides interactive command facilities for cluster analysis, decision tree analysis, parametric and nonparametric hypothesis testing, creating and testing distributions.  For instance, CLASSIFY recommends optimum ways of partitioning a set of observations into subsets based on the clustering of particular variable values as they affect some other variable. TREE analyzes reductions in variance and displays other statistics for successive subdivisions of an initial population of observations.  THEO constructs, from specified cumulative distribution functions, a theoretical distribution against which empirical distributions can be tested.  Facilities for creating and testing distributions include T-tests, F-tests, the Kolmogorov-Smirnov test, and the Wilcoxon, Mann-Whitney U-test.

## EXTENSIBILITY

AUTOGRP+ is a fully extensible user language capable of executing stored file commands.  AUTOGRP+ is supported by CML (Conversational Modelling Language) a high-level, algorithmic, hierarchical, block-structured, and fully extensible interactive programming language with special capabilities for programming discrete event simulation, linear optimization, and statistical analysis systems.

## INTERFACES WITH OTHER SYSTEMS

AUTOGRP+ can be interfaced with SAS, a complementary statistical reporting package.

## SAMPLE JOBS

Example 1:  Group Definition and Management.  The sample commands and output below illustrate how subsets of data may be created interactively  with AUTOGRP+.  Input commands are displayed in lower case type following the "==>" prompt symbol.  Resulting output appears below each command and is shown in upper case type.  In example (a), "all" refers to all records in the data file and "routine" refers to the set of hospital cost records for mothers with routine delivery problems.  In example (b), "complex" refers to the set of cost records for mothers with complex surgical delivery problems.  In example (c), "complex" is redefined to refer to the set of cost records for mothers less than 21 years of age with complex surgical delivery problems.

```
          a) ==>group routine=all which are "surgery 1-7"
             ROUTINE CONTAINS 562 OBSERVATIONS
          b) ==>combine complex=all without routine
             COMPLEX CONTAINS 76 OBSERVATIONS
          c) ==>remove age>21 from complex
             COMPLEX CONTAINS 12 OBSERVATIONS
```

Example 2:  PROFILE displays total cost for routine deliveries at each
hospital, showing values for the 25th, 50th, and 75th percentile.

==> profile routine using hosp order median

```
            COST (DOLLARS)
         --------------------                              HOSP
HOSP.  0  250  500  750  1000   COUNTS  MEAN    S.D.   MEDIAN  NAME
       |---------------------|
  1    |    *--M---*         |   1311   472.28  94.01  450.00  GENERAL
       |                     |
  2    |   *M-*              |   1497   576.11  52.90  500.00  UNIVERSITY
       |                     |
  3    |      *M---*         |   1578   698.87  82.44  610.00  MEMORIAL
       |                     |
  4    |       *M*           |   1444   720.17  49.35  640.00  LOCAL
       |                     |
  5    |        *M---*       |   1492   773.04  80.22  690.00  DOCTORS
       |                     |
  6    |        *M--*        |   1464   809.96  68.53  750.00  SAINTS
       |                     |
  7    |             *-M-*   |   1371   937.08  67.24  920.00  CITY
       |---------------------|

           (M = Median)
```

Example 3:  SEARCH, ORDER, REPORT command sequence tabulates patient
count, mean room cost, mean operating room cost, and mean laboratory cost
for each hospital.

==>search routine
==>order hospital
==>report count,mean roomcost,mean orcost,mean labcost

| HOSPITAL | COUNT | MEAN ROOMCOST | MEAN ORCOST | MEAN LABCOST |
|---|---|---|---|---|
| 1 | 1311 | 205.52 | 277.46 | 31.41 |
| 2 | 1497 | 396.48 | 379.42 | 34.13 |
| 3 | 1578 | 388.89 | 377.87 | 32.77 |
| 4 | 1444 | 396.80 | 381.133 | 35.57 |
| 5 | 1492 | 404.50 | 375.12 | 35.63 |
| 6 | 1464 | 404.38 | 378.10 | 35.47 |
| 7 | 1361 | 387.50 | 574.12 | 30.87 |
| TOTAL | 10147 | 395.04 | 377.64 | 33.75 |

AUTOGRP+   AUTOmated GRouPing System

Developed by:                          Supported and Distributed by:

   Ronald E. Mills, Ph.D.                 Puter Associates, Inc.
   Enes Elia                              345 Whitney Avenue
   Robert Leary                           New Haven, Connecticut  06511

Computer Makes:                        Operating Systems:

   IBM/370 or equivalent                  OS/MVS, MFT, MVT, VS1, VS2, CMS, TSO

Source Languages:                      Interfaced Systems:

   CML, Assembler                         SAS

Planned Additions in Next Year:

   1.  DOS multi-user system
   2.  Data editing/Data entry facilities
   3.  Report Writer
   4.  Additional mechanisms for dealing with secondary records
   5.  Additional statistical routines -- e.g., for correlation and regression
       analysis.

Cost:

   Puter Associates has established the following pricing structure for different categories of AUTOGRP+ users as follows:

|            | 1st Yr Licence Fee | Annual Fee |
|------------|--------------------|------------|
| Commercial | Negotiable | Negotiable |
| Non-Profit/govt. | " | " |
| Academic | " | " |

   Prices within each category are negotiable, depending on the level of support required.  AUTOGRP+ is also available from Puter Associates for a negotiable monthly fee or from General Electric (GEISCO:  MARK III Service) for
validated GE customers.

References:

R. Mills., R. Fetter, D. Riedel, and R. Averill,  "AUTOGRP:  An Interactive
   Computer System for the Analysis of Health Care Data," Medical Care  14
   (1976):  603-615.
A. Meyers, D. Brand, H. Dove, T. Dolan,  "A Technique for Analyzing Clinical
   Data to Provide Patient Management Guidelines," Am. J. Dis. Child. 132
   (January 1978):  25-29.

Documentation:

Theriault, K., Mills, R., Elia, E., Leary, R.,  The AUTOGRP+ Reference Manual.
   New Haven:  Puter Associates, 1979.

## 8.09  FORALL

### INTRODUCTION

FORALL is a package for general-purpose statistical analysis, such as non-orthogonal analysis of variance, estimation of botanical composition, and data manipulation and tabulation.  It was developed by John Kerr in 1975 and has been continuously developed since then.  It can be used interactively or in batch, and is installed on the CSIRO Cyber-76.

### CAPABILITIES:  Processing and Displaying Data

All data storage in core is in the form of named n-dimensional matrices, n=1,2,3,...and all results (and working arrays) are stored in such matrices. The names and sizes of these arrays are under user control.

The entire collection of such arrays may be output and saved as a file. Such files may be input in toto (and computation resumed from that point) or selectively.  As well, individual arrays or groups of arrays may be input or output.  Input may be in fixed field or free format.

Vector and matrix arithmetic is provided including mathematical, trigonometric, statistical and logical functions for manipulating data.

Arrays may be subdivided, concatenated, melded into matrices of higher dimension, or partitions.  The dimensional structure of an array may be redefined at any time.  Subsections of arrays may be referenced.

Data in error may be flagged, corrected or eliminated.

All arrays may be tabulated, and discrete or continuous data may be allocated to cells of another array and tabulated.

Line printer graphs are provided and, installation dependent, similar facilities are available on incremental plotters.

### CAPABILITIES:  Statistical Analysis

Basic statistics, including means, standard deviations, histograms, simple regression, unpaired t-tests, moments and derived statistics, functions for log-normal estimates, Bartlett homogeneity test, t, F and Chi-square probabilities, fractions of year from day, month and year data, simple sorting and ranking.

Non-orthogonal analysis of variance via multiple regression (and limited Manova) with procedures for automatic generation of variables for discrete factors.  Principal components, canonical correlation.  Simple one-way and orthogonal n-way ANOVA.  Correlation, segmented linear regression with unknown joint points, grouped regression, auto-covariance and spectral estimation via covariance function, filtering, estimation and manipulation of Markov chain transition matrices, estimation of botanical composition from field sampling, minimization.

## EXTENSIBILITY

The program is written in Fortran and designed to enable the user competent in Fortran to add algorithms with ease. It is thus the complement of those statistical packages which contain subroutines but no driving programs. Most extensions to FORALL in use or contemplated have been by incorporating algorithms in the published literature in a manner that enables their use in complete generality. An individual installation may add or subtract sections of the package to meet local requirements. FORALL has looping and macro facilities to give some of the capabilities of a high level language.

## PROPOSED CHANGES

Conversion to Fortran-V is proposed during 1981. Changes are made in a manner that preserves the usefulness of old jobs.

## REFERENCES

Hargraeves, J.N.G., and Kerr, J.D., "Botanal: A comprehensive sampling technique for estimating pasture yield and composition. II." Computational Package. CSIRO Division of Tropical Crops and Pastures, Technical Memorandum No.9 (1978).
Kerr, J.D., "FORALL: An Extensible Fortran System for Conversationally Accessing Subroutine Libraries." Software Practice and Experience (in press) ((accepted June 1980).

## SAMPLE JOB

Attach a short file with unspecified number of observations known not to exceed 100, transform the dependent value and run analysis of variance/ regression with one factor of three levels (X2) and two simple regressors. Recalculate the regression coefficients by matrix procedures.

```
COMMON ATTACH=2 DATAMY ID=NO1
IN 4 X1 X2 Y X3 NOBS=100 LUN=2
(F1,F2,F4.1,F2)
MAKE ASY=ASIN(SQRT(Y/20))
BCD DEP ASY F
BCD INDEP X2 X1 X3 F
FACREAD A X2 0 3 0 1 2
GENERATE
CANANOVA PRINT=RAF
TYPE VAL 0 1 2 F
TYPE TRANS 1 0 -1   0 1 -1 F
STRUCTURE TRANS ROWS=3 COLUMNS=2
TRANSFORM VEC=X2 TRAN=X2T
VECTOR N=SIZE X1
CREATE XO ROWS=N COLUMNS=1
VECTOR XO=1
MELD 4 XO X2T X1 X3 MATRIX=X
MAKE B=XX(INVERSE(XTX(X,X)),XTX(X,ASY))
LIST B
Q
```

```
NEXT COMMAND IS ---
GENERATE
VECTORS FOUND BY GENERATE ARE
          SFACTORNOA
NEXT COMMAND IS ---
CANANOVA PRINT=RAF
          8 OBS.TRANS
          8 OBS.CORRN
REGRESSION ESTIMATE COEFFICIENT  S.E. B   T-RATIO    EMS(ELIM) ASY
X2                  2-.05027631 .02523972-1.9919519   .00968682
X2                  3 .11945234 .02499427 4.7791894   .05576121
X1                  4-.02427353 .00785171-3.0914962   .02333250
X3                  5 .00900580 .01053417 .85491251   .00178430
INTERCEPT             .62275359           FISHER A  49.483250


PAGE   2 (CPU   .118)  30/07/80  14.23.49.
SUM SQS = REDUCTION ERROR SSQ AS FITTED.ANOVEC

          ANALYSIS OF (CO)VARIANCE      ASY          ASY
  SOURCE  QUAL  DF  SUM OF SQUARES         MEAN SQUARE      F-RATIO
MEAN       -2  1.         2.44261815   2.442618153632
TOTAL      -3  7.          .07893383    .011276260943
X2          0  2.          .04395483    .021977416115     9.00+
X1          0  1.          .02587075    .025870754021    10.60*
X3          0  1.          .00178430    .001784296745      .73
ERROR       1  3.          .00732394    .002441314534
ESTIMATES. SE.S DIFFERENCES 1,2.. BEFORE
ASY       X2          S.E. EST. S.E. DIFFERROR VEC
 .49364096           0 .02914468
 .66336961 1.0000000    .02898560  .04109979
 .47474122 2.0000000    .03689312  .04776009
ASY       X1          S.E. EST. S.E. DIFF
 .66179517           0 .03941527
 .65572678 .25000000    .03766571  .00196293
 .64965840 .50000000    .03593820  .00196293
          :          :        :         :
          :          :        :         :
          :          :        :         :
NEXT COMMAND IS ---
TYPE VAL 0 1 2 F
NEXT COMMAND IS ---
TYPE TRANS 1 0 -1 0 1 -1 F
NEXT COMMAND IS ---
STRUCTURE TRANS ROWS=3 COLUMNS=2
TRANSFORM VEC=X2 TRAN=X2T
VECTOR N=SIZE X1
CREATE X0 ROWS=N COLUMNS=1
NEXT COMMAND IS ---
VECTOR X0=1.
MELD  4 X0 X2T X1 X3 MATRIX=X
MAKE B=XX(INVERSE(XTX(X,X)),XTX(X,ASY))
LIST B
B          .62275359-.05027631 .11945234-.02427353 .00900580
NEXT COMMAND IS ---
Q
 .17400000 SECS. CPU
```

FORALL:  Package for Adding Statistical Algorithms

Developed by:

J.D. Kerr
CSIRO Division of Mathematics and Statistics
Private Bag No. 3
Indooroopilly Q. 4068, Australia

Computer Makes:                              Operating System:

CDC Cyber 76                                 Scope 2.1

Source Language:

FORTRAN-IV, ANSI standard with exceptions for tasks unavailable in the
standard.

Cost:

Copy of the Fortran souce code is available at the cost of copying plus
material, with no implied warranty.

Documentation:

FORALL manual maintained in machine readable form and freely available on
fiche.  Hargreaves & Kerr (1978) provides a detailed guide with annotated
examples for botanical composition use.

8.10  <u>AQD</u>

## INTRODUCTION

The AQD (Analysis of Quantitative Data) Statistical Collection is a package of more than fifty interactive programs for data analysis. It was developed for use in teaching and research at the Harvard Business School. The intended user is a person who wants to use existing statistical techniques to analyze data, not a computer specialist or a theoretical statistician. It is used in ten colleges and universities and is available on several time-sharing and in-house commercial systems.

## CAPABILITIES:  Processing and Displaying Data

AQD can process files consisting of a series of observations on a number of variables, whether arranged by variable or by observation. It can accept existing files in fixed or free format, or files can be typed by the user at the terminal. Large data files are stored with several values to a computer word and automatically unpacked when workfiles are created. Data can be manipulated by interactively defining transformations and adding them to a file, selecting observations from a file according to the value(s) of one or more variables, selecting variables, adding to each observation in a main file the values of "supplementary" variables contained in another file, sorting observations, and aggregating observations (e.g., reducing daily to monthly data).

Data can be tabulated, cross-tabulated, and graphed in the form of histograms, cumulative distributions, scatter diagrams, time-series plots, and boxplots.

Data can be rapidly selected from specially formatted versions of the Compustat Industrial file and the DRI Capsule Data Bank.

## CAPABILITIES:  Statistical Analysis

Basic statistics, including min and max, mean, standard deviation, skewness and kurtosis, fractiles, cumulative frequencies. Frequency distributions and counts of a variable conditional on values of up to two other variables. Averages, standard deviations, medians, median absolute deviations of a variable conditional on up to three other variables. Standard measures of association and significance tests, including chi-square, F, and t. Ratio estimation, one-two- and three-way analysis of variance with fixed effects, monotone ANOVA, and AID (Automatic Interaction Detection).

Correlation analysis, OLS and WLS regression, limited-influence regression, nonlinear regression, two- and three-stage least squares, regression with auto-correlated errors, distributed lags. The regression programs permit complete analysis of residuals, and are capable of providing a probabilistic forecast of a "natural" dependent variable when supplied with values of the natural independent variables even when the dependent as well as the independent variables in the regression model are transformations of natural variables. Fast algorithms used in the regression and related programs give full 7-digit accuracy on standard test data (Longley and Wampler).

Discriminant analysis and multivariate logit and probit analysis; principal-component, factor, and cluster analysis; multidimensional scaling.

Exploratory data analysis:  box-and-whisker plots, repeated running medians, robust and resistant estimation, median polishing, limited-influence regression.

EXTENSIBILITY

User-written Fortran code can be loaded with a skeleton main program which can present to the programmer the contents of an input workfile and permit him to write an output workfile or manipulate the contents of the input workfile in any way he wishes.

INTERFACES WITH OTHER SYSTEMS

Can output files in Hollerith which can then be read by other systems. AQD workfile format can be read directly by the Time-Series (Box-Jenkins) collection of Walter Vandaele.

PROPOSED ADDITIONS IN NEXT YEAR

Regression with ordinal or categorical dependent variable.

SAMPLE JOB

Given historical data on unit price and annual volume of silicon transitors sold, obtain an "experience-curve" relation by regressing log (price) on log (accumulated volume), and make a probabilistic forecast of price assuming next year's volume is 3000 (millions).  Responses by the user are underlined.

```
.R AQD
COPYRIGHT (C) 1972,1976 BY THE PRESIDENT AND FELLOWS OF HARVARD COLLEGE
PROGRAM? TRADEF
NAME OF DATA FILE? 1TRANS
LIST NAMES? Y
    1 YEAR(19--)
    2 VOLUME (MM)
    3 UNIT PRICE
CREATING NEW TRADEF FILE:
X4 = ? CUM X2
X5 = ? LOG X4
X6 = ? LOG X3
X7 = ? $
PROOFREAD? N
CORRECTIONS? N
ADDITIONS? N
NAME FOR VBL
    4? CUM VOLUME
    5? LOG CUM VOL
    6? LOG PRICE
PROOFREAD? N
CORRECTIONS? N
SPECS ARE ON YOUR DISK IN FILE 1TRANS.DTR.
PROGRAM? WFLQIK
NAME OF DATA FILE? 1TRANS
    18 OBSS ON   3 DATA VBLS:   3 TRANSF'NS.
KEEP ALL VBLS? 6 5
CRITERIA? N
VBLS NUMBERED AS FOLLOWS:
    1 LOG PRICE
    2 LOG CUM VOL
DATA ARE IN WFL 1:   18 OBSS ON   2 VBLS.
PROGRAM? REGRES
WORKFILE NO.? 1
INDICES? 1 2
OPTION? 1
DEP.VBL: LOG PRICE
NAT:                    EST        UNC       P(SGN)
    0 CONSTANT         2.844      0.102      1.000
    2 LOG CUM VOL      -.4630     .0128      1.000
EST.RES.SD   .1014E+00
SAM. R SQR   .988
(16 DEGREES OF FREEDOM)
OPTION? F
OUTPUT FILED.
PROGRAM? REGFOR
PREDICTED VALUE OF
VOLUME (MM)? 3000
DISTRIBUTION OF UNIT PRICE
PRINTOUT? 1
MEAN    =    0.167
STD DEV =    0.019
PRINTOUT? 2
FRACTILES
 .001    .01     .1    .25     .5    .75     .9    .99    .999
 1111   1252   1432   1536   1655   1783   1912   2188   2465
DEC.PT -4
```

AQD: Analysis of Quantitative Data

Developed by:                                    Distributed by:

    Robert Schlaifer                           Interactive Analysis
    Harvard Business School                    315 Morgan Hall
                                               Boston, MA  02163

Computer Makes:                                  Operating Systems:

    DECsystem 10                               TOPS-10
    DECsystem 20                               TOPS-20
    VAX 11/780                                 VAX/VMS

Source Languages:

    FORTRAN and assembly.

Cost:

    Yearly license fee is $1150 for degree-granting institutions, negotiable
for time-sharing vendors and other organizations.  The fee includes load modules
and object code, delivered on magnetic tape, and user and installation
documentation.

User Documentation*

Robert Schlaifer, User's Guide to the AQD Collection, Eighth Edition, 1981,
    (Case 9-173-095).
Program Cshflo:  Evaluation of Streams of Cash Flows (A nonstatistical program
    distributed with AQD) (Case 9-175-062, Rev. 8/80).
Data Banks Accessible via AQD (Case 9-175-265).
Protocols for the AQD Collection (Case 9-176-055).


* All documents can be ordered (by case number) from:

    Case Services
    23 Morgan Hall
    Harvard Business School
    Boston, MA  02163
    (617) 495-6249

8.11 <u>STP</u>

INTRODUCTION

STP is a large data manipulation and analysis package which is generally used for small to medium level analysis, including:  correlation, regression, factor analysis, analysis of variance and cross-tabulation.  Its primary use is in education and survey analysis.  It was developed originally by Richard Houchard in 1971 and underwent continuous development until 1976.  It can be used interactively or in batch and is installed in over 170 computer installations.

CAPABILITIES:  Processing and Displaying Data

STP allows data entry from the terminal, magnetic tape or disk files. Disk files may be either sequential ASCII or in a structured random access binary file called a BANK.  ASCII data may be either free format or a FORTRAN type format may be specified.  Names can be attached to the variables.  After data is entered, variables may be created, edited, transformed or deleted. Data is stored in the core image and the data area may be expanded to utilize all available core or virtual memory.  The data may be typed or displayed with scatterplot, bargraph or histogram to the terminal or lineprinter.

CAPABILITIES:  Statistical Analysis

Basic statistics include means, standard deviations, variance, median, mode, min, max, standard error of mean, skewness and coefficient of variation. Also available are:  frequency, cross-tabulation and percentiles.  Other statistics available include:  Z-scores, Kolomogorov-Smirnov, correlation, partial-correlation, Kendall tau, Spearman rank-order correlations, point biserial correlations, t-tests, correlated t, Mann-Whitney U, Wilcoxon rank, 1- and 2-way Anova, 1-way repeated measures ANOVA, analysis of covariance, regression, stepwise regression, t, F and Chi probabilities, Chi-squares and an exponential smoothing model.

EXTENSIBILITY

STP is written with a single level of overlay structure.  Each overlay is the routine for a single user command.  Each routine accesses the in-core data structure through the subroutine argument list and COMMON.  Adding additional routines to the program requires some simple programming.

INTERFACES WITH OTHER SYSTEMS

STP interfaces with BANK data files with conversion of integers to floating point and limited missing data capabilities.

PROPOSED ADDITIONS IN NEXT YEAR

No major changes are proposed in the next year.

## SAMPLE JOB

Enter a data set from the disk, get a simple description of it and run a regression on it.

```
WHICH COMMAND? FETCH
WHAT IS THE FILE NAME AND EXTENSION? STPILR.DAT
HOW MANY INPUT VARIABLES? 5
VARIABLES WILL BE READ ACCORDING TO FORMAT:
(20F)

WHICH COMMAND? DESC

THERE ARE    5 VARIABLES AND    6 OBSERVATIONS
VAR.         MEANS        STD.DEV.         VARIANCE
  1        11.16667      1.169045         1.366667
  2        22.00000      1.414214         2.000000
  3        31.16667      1.471960         2.166667
  4        42.33333      1.966384         3.866667
  5        53.33333      1.966384         3.866667

WHICH COMMAND? REGR
ENTER OPTIONS SEPARATED BY COMMAS

LIST THE INDEPENDENT VARIABLES?
1,2,3,4
WHICH IS THE DEPENDENT VARIABLE? 5

          ***** MULTIPLE LINEAR REGRESSION *****
SAMPLE SIZE     6
DEPENDENT VARIABLE:    5
INDEPENDENT VARIABLES:      1       2       3       4
COEFFICIENT OF DETERMINATION  0.98007
MULTIPLE CORR COEFF.   0.98998
ESTIMATED CONSTANT TERM    4.9830279

ANALYSIS OF VARIANCE
FOR THE REGRESSION
SOURCE OF VARIATION    DF    S. SQ.      M.S.          F       PROB
     REGRESSION        4    18.9479    4.73698      12.29    0.2104
     RESIDUALS         1     0.385398   .385398
     TOTAL             5    19.3333

        REGRESSION    S. E. OF   F-VALUE                 CORR.COEF.
VAR.    COEFFICIENT   REG. COEF. DF (1, 1)   PROB        WITH    5
  1    -0.2015290      .3139      .4122     0.6366       0.1450
  2     1.777829       .3683    23.30       0.1300       0.7192
  3     1.302905       .3255    16.02       0.1559       0.5298
  4    -0.6878439      .3142     4.792      0.2728       0.6379
```

STP, Version 4b:  STATPACK Statistical Package

Developed by:

Richard Houchard
Western Michigan University
Kalamazoo, Michigan 49001

Computer Makes:                          Operating Systems:

DECsystem-10,20                              DEC - LINK/OVERLAY

                                         Interfaced Systems:

                                             BANK

Source Languages:  Primarily FORTRAN, with some MACRO-10 assembler.

Cost:

    Only a handling fee of $10 is charged for STP.  If a magnetic tape is
supplied by WMU, an additional $18 is charged.  The fee includes source,
instructions and load control files for KA, KI, KL and DEC-20 processors, a
loaded version of a KL processor, all library files and 134 page machine-
readable document.

Documentation:

Houchard, R.A., STATPACK - Statistical Package, Library Program #1.1.4; W.M.U.,
    July 18, 1980.

## 8.12  STATPAK


### INTRODUCTION

STATPAK (U.S. Steel Statistics Package) is a set of 21 simple routines for performing interactive statistical and data analysis.  Most of its applications to date have been in engineering and the physical sciences.  STATPAK was first made available in 1973 and has been continuously expanded and modified since that time.  STATPAK is intended primarily for interactive terminal operation, but can be used in batch-type operation as well.

### CAPABILITIES:  Data Processing and Display

The STATPAK data matrix is entered into the program either from mass storage (free format or formatted) or by keyboard input, one record (or observation) at a time--a record consists of one value of each variable.  STATPAK will handle up to 40 variables and more than $10^5$ observations for most of the routines.  Any number of different  data sets and any number of appropriate routines may be used in a single execution of the program.

The user has the option of editing or transforming the data before each routine, and to save any new or modified data set on a local file before exiting from the program.  Observations may be added and/or deleted, and the available transformations include addition of or multiplication by a constant, natural logarithm, exponential, raising to a power, and squares and cross products.

The user command language is primarily free format and consists of answers to questions asked by STATPAK about the structure of the data and the routine that is to be used.  STATPAK will reject many inadmissable input values (such as a variable number outside the specified range) and allow the user to continue after displaying an appropriate error message.

Graphical capabilities of STATPAK include scatter diagrams with the option of one or several independent variables, and histograms, including a point plot of the cumulative frequency.  The user may optionally increase the screen width to take advantage of larger terminals.

### CAPABILITIES:  Statistical Analysis

Basic statistics, including means, standard deviations, maxima and minima, and simple correlations.  Regression analysis, including polynomial, multiple, and stepwise.  Significance tests, including Spearman rank correlation, chi-square for contingency table, t-tests (including pairwise), non-parametric analysis (including Kendall rank correlation and coefficient of concordance, Friedman ANOVA of ranks, Cochran Q test and Wilcoxon test).  Analysis of variance, including one-way (balanced or unbalanced) and two-way (factorial, nested, or split-plot).  Multivariate analysis, including canonical correlation, discriminant analysis, and factor analysis.  Time series, including auto- and cross-correlation, weighted moving average, and triple exponential smoothing.

Other:  Simultaneous linear equations; Linear programming.

EXTENSIBILITY

The program has sufficient flexibility to permit the addition of additional routines without exceeding the existing core limitations, but all such modifications require source-code changes.

PROPOSED IMPROVEMENTS FOR NEXT YEAR

Increase flexibility in transformations, develop interface with graphics hardware.

SAMPLE JOB

To start using STATPAK, the user must:

1.  Enter the timesharing system.

2.  Access the STATPAK program file and any needed system files.

3.  Access any data files needed.

4.  Execute the STATPAK program.

The accompanying sample job starts with step 4; the data file called STATSMP contains 22 rows (observations) and 6 columns (variables). Information typed in by the user is underlined; information output by the program is not underlined. If keyboard data input is desired, the response to the question "ON WHAT FILE ARE YOUR DATA" is INPUT.

```
COMMAND- STATPAK.

U. S. STEEL STATISTICS PACKAGE

DO YOU WANT A LIST OF AVAILABLE ROUTINES
?NO
WHICH ROUTINE
?MU
ENTER NUMBER OF OBSERVATIONS AND NUMBER OF VARIABLES
?22,6
ON WHAT FILE ARE YOUR DATA
?STATSMP
ENTER INPUT DATA FORMAT - TYPE 0 FOR FREE FORMAT
?0
DO YOU WANT TO TRANSFORM THIS DATA SET
?NO
DO YOU WANT A DATA LISTING
?NO
HOW MANY INDEPENDENT VARIABLES
?4
ENTER  4 VARIABLE NUMBERS
?1,2,3,4
DEPENDENT VARIABLE
?5

VAR.    MEAN      STANDARD    REGRESSION   STD.ERR.  CORRELATION COMPUTED
                  DEVIATION   COEFFICIENT  REG.COEF.   X VS Y    T-VALUE
  1    100.86     9.2649      -.40166E-02  .34169E-01  .0660     -.12
  2    38.545     3.4327      -.18280E-01  .12611      -.0577    -.14
  3    35.318     2.8011       .26849E-01  .15321      .0791      .18
  4    13.682     2.2967       .25413      .14181      .4303     1.79
DEPENDENT
  5    13.227     1.3778

INTERCEPT............................................ 9.9118
SQUARED MULTIPLE CORRELATION COEFFICIENT..........  .1874
STANDARD ERROR OF ESTIMATE....................... 1.3804

                ANALYSIS OF VARIANCE FOR THE REGRESSION

SOURCE OF VARIATION          D.F.   SUM OF SQ.   MEAN SQ.   F-VALUE
ATTRIBUTABLE TO REGRESSION     4    7.4715       1.8679       .98
DEVIATION FROM REGRESSION     17    32.392       1.9054
TOTAL                         21    39.864

TABLE OF RESIDUALS
?NO
ANOTHER MULTIPLE LINEAR REGRESSION
?NO

DO YOU WISH TO USE ANOTHER ROUTINE
?NO

    STOP
       .109 CP SECONDS EXECUTION TIME
COMMAND-
```

STATPAK: U. S. Steel Statistics Package

Developed by:                          Inquiries to:

  C.J. Davis, J.J. Geissler              F.A. Sorensen
  R.H. Heasley, F.A. Sorensen,           U.S. Steel Research Lab., MS 44
  U.S. Steel Research Lab                Monroeville, PA  15146

Computer Makes:                         Operating Systems:

  CDC Cyber series                      CDC - NOS/BE
  Honeywell 6000                        Honeywell - GCOS

Source Languages:

  FORTRAN Extended Version 4 (CYBER)
  FORTRAN IV (Honeywell)

Cost:

  Negotiable.

Documentation:

J.J. Geissler, F.A. Sorensen, R.H. Heasley, C.J. Davis, Users Guide for STATPAK-
  Time Sharing Statistical Programs, U. S. Steel Computer Documentation
  73-001 (1973).

## 8.13  STATUTIL


### INTRODUCTION

STATUTIL is made up of a set of linked workspaces (files) in APL, together with various utility functions for managing these workspaces. It has been developed by J.B. Douglas and various workers beginning about 1973, with continuing expansion based on user requests. It is installed on the University of New South Wales computer network; various partial versions of it have been run in Britain, Canada and the U.S.

Since STATUTIL is written in APL and uses its facilities explicitly, the user has access to standard APL facilities independent of the host system, including editing, debugging, library storage, error tracing, access to external files, etc.

It is designed for use by users not necessarily experienced with APL on moderate data sets of hundreds to thousands of items. Much is interactive, querying the user; but much can be bypassed by the experienced user and indeed can be run in batch mode.

### CAPABILITIES:  Processing and Displaying Data

Data can be entered directly, in response to a query or in anticipation, or can be read from existing files.

Data can be grouped in 1 or 2 dimensions and displayed; the original or grouped material can be used as the basis of the Statistical Analysis procedures. At any stage SAVE procedures are possible.

Graphing is flexible, including contour plots: any transformations can be provided for data or output from other procedures.

### CAPABILITIES:  Statistical Analysis

Basic statistics, including (weighted) means, medians, modes, (weighted) standard deviations, quartiles, geometric and harmonic means, arbitrary order moments, factorial moments, cumulants, factorial cumulants, etc. Frequency distributions, one and two way; histograms; stem and leaf displays; box plots; cumulative distribution functions. Correlation (simple and partial).

Confidence intervals for means, standard deviations, differences of means, ratio of standard deviations for normal parents. Non parametric intervals for medians.

Estimation for numerous discrete distributions (Poisson, binomial, negative binomial, Neyman Type A, Poisson-binomial, Poisson-Pascal, Thomas, double Poisson, logarithmic and some bivariate distributions) including estimated standard errors. Fitting of these distributions, with graphical display, goodness of fit, etc. Fitting the normal distribution with graphical display and goodness of fit.

Hypothesis testing on two way contingency tables including partitioning (chi-squared and likelihood ratio), Kolmogorov-Smirnov testing. Chi-squared goodness of fit testing.

Likelihood calculations, for 1 and 2 parameters, including graphical output.

Probability functions for continuous variates (beta, Cauchy, chi-squared, F, gamma, non-central-chi-squared, non-central F, negative exponential, normal (uni- and hi-variate) and t). Probability functions for discrete variates (those mentioned earlier under estimation).

Polynomial regression (with display of residuals and various options); multiple linear regression.

Random variable generators for various continuous and discrete random variables.

Functions for managing student records.

Utility (e.g. listing, and graphing (2 variable, and 3 variable contour)) functions. Mathematical functions (e.g. integration, minimization, solving equations).

## EXTENSIBILITY AND ADDITIONS

Users can extend or modify operating functions by direct editing in APL.

STATUTIL undergoes continuous extension according to user requests: its modular structure makes this simple and effective.

All documentation is on line, with hard copy listing generated by the user.

## SAMPLE JOB

Find the moment and maximum likelihood estimates for the parameters of a negative binomial distribution. The counts can be grouped via the frequency tabulating functions, and are here called *FREQ* (frequency in zero class, in the 1 class,...).

SIGN ON TO *APL* AND ENTER/COPY THE FREQUENCIES *FREQ*

```
)COPY *APLT ESTDISC NBML
NBML    FREQ
```

Output


          FREQ
187 185 200 164 107 68 49 39 21 12 11 2 5 2 3 1
          NBML FREQ

DATE: 1980 8 11 11 56 18

MOMENT FIT:  [KAO,RHO]   FOR  [KA,RH]
---------------------------------------------


N    = 1056      NxM1' = 2963
M1'  = 2.805871212
M2   = 6.404548862


KAO  =  2.1877, WITH S.E. =  0.2045
     CORR.[KAO,RHO] =  ⁻0.9586,    (GEN.VAR.)*1÷4  =   0.0853
RHO  =  1.2825, WITH S.E. =  0.1250



'2xLN RELATIVE LIKELIHOOD' =  16.98, PERHAPS WITH 13 DF

CPU TIME: 0 0 551



MAXIMUM LIKELIHOOD FIT:  [KAL,RHL]  FOR  [KA,RH]
------------------------------------------------------------


0
2.187723943 1.282552683
1
2.131010943 1.316685501
2
2.133713176 1.31501799

KAL  =  2.1337, WITH S.E. =  0.1772
     CORR.[KAL,RHL] =  ⁻0.9477,    (GEN.VAR.)*1÷4  =   0.0807
RHL  =  1.3150, WITH S.E. =  0.1152


'2xLN RELATIVE LIKELIHOOD' =  16.89, PERHAPS WITH 13 DF

CPU TIME: 0 1 637

## STATUTIL

Developed by:                                    Distributed by:

    J.B. Douglas                                    Same
    Department of Statistics
    School of Mathematics
    University of New South Wales
    P.O. Box 1
    Kensington, NSW 2033

Computer Makes:                                  Operating Systems

    Any system which supports
    APL - in its present form
    requiring dynamic copying
    for effective operation

Interfaced Language:

    Dependent on the host APL

Source Language:

    APL

Cost:

    Free

Documentation:

All on line; stapled listing available on request.

## 8.14 MICROSTAT

### INTRODUCTION

MICROSTAT is an interactive package for general-purpose statistical analysis on micro-computers. Its principal applications have been in education, research and small businesses. It was developed by Dr. J. B. Orris of Butler University in 1979. As of October 1980, over 60 packages have been distributed to various educational, corporate and research institutions.

MICROSTAT is designed for use on typical micro-computers with a CRT screen/keyboard, one or two disk drives, approximately 16K bytes of user memory and an optional printer. All of the programs are menu-oriented with the user making selections of options from lists displayed on the screen. The system was designed for ease of use and fail-safe operation. For example, many common errors have been 'flagged' to allow for corrective action rather than termination of the program.

### CAPABILITIES: File Building and Manipulation

All of the analysis programs that require data read files created by the DATA MANAGEMENT SUBSYSTEM. In this manner, the files are created and manipulated interactively, but are read in batch mode by the analysis programs.

The DATA MANAGEMENT SUBSYSTEM allows for the entry, listing, and editing of data. It can also destroy files, delete cases from a file, merge and vertically augment files. The transformations option allows for arithmetic, exponential, and logarithmic transformation of variables as well as linear transformation and summations. The sort option re-arranges a data file by sort-keys and the rank-order option replaces data with ranks. A lag-transformations option prepares time series data for subsequent analysis.

### CAPABILITIES: Statistical Analysis

Descriptive statistics (including sum, sum of squares and deviation sum of squares), frequency distributions with histogram, hypothesis tests for mean and proportions, analysis of variance, scatterplot, correlation matrix, simple and multiple regression, time series analysis (moving average, exponential smoothing), ten common non-parametric procedures, two-way cross tabulation, contingency table analysis, goodness of fit tests, factorials/permutations/combinations, probability distributions (binomial, Poisson, hypergeometric, exponential, normal and inverse normal, F, t, chi-square).

### PROPOSED ADDITIONS IN NEXT YEAR

MICROSTAT was originally written for North Star BASIC and North Star DOS. It has recently been converted to Microsoft BASIC-80 for the CP/M operating system. Conversions are in process for Cromemco and Apple system.

### REFERENCES

Orris, J.B., MICROSTAT: An Interactive Statistical Package for Micro-computers, The American Statistician, Vol. 34, No. 3, (1980). pp. 188-189.

SAMPLE JOB

   The Longley data as discussed in Beaton, et.al (see reference below) was
selected for sample output by the multiple regression option since the 'correct'
solution is available.  Six variables were selected as predictor variables.
Computation of residuals was requested as well as the Durbin-Watson statistic.
The total computation and display time was approximately 30 seconds.  As an end
of program option, the output was repeated on the printer without
re-computation.

Beaton, Albert E., Rubin, Donald B., and Barone, John L.,  The Acceptability of
   Regression Solutions:  Another Look at Computational Accuracy.  Journal of
   the American Statistical Association, Vol. 71, No. 353, pp. 158-163.

Output

```
----------------- REGRESSION ANALYSIS ------------------------

HEADER DATA FOR: LONGLEY,2   LABEL: JASA, V.62, P.819-841
NUMBER OF CASES: 16   NUMBER OF VARIABLES: 7   SIZE: 3 BLOCKS

-----------------------------------------------------------------
                  TEST OF MULTIPLE REGRESSION

INDEX         NAME        MEAN       STD.DEV.
  1          --X1-      101.681       10.792
  2          --X2-  387,698.440   99,394.939
  3          --X3-    3,193.313      934.464
  4          --X4-    2,606.688      695.920
  5          --X5-  117,424.000    6,956.102
  6          --X6-    1,954.500        4.761
DEP. VAR.:   --Y--   65,317.000    3,511.968


-----------------------------------------------------------------
DEPENDENT VARIABLE: --Y--

VAR.    REGRESSION COEFFICIENT   STD. ERROR   T(DF=  9)     BETA
--X1-              15.0479         84.9127        .177     .0462
--X2-               -.0358           .0335      -1.069   -1.0136
--X3-              -2.0202           .4884      -4.136    -.5375
--X4-              -1.0332           .2143      -4.822    -.2047
--X5-               -.0511           .2261       -.226    -.1013
--X6-            1829.1009        455.4679       4.016    2.4796
CONSTANT   -3482157.3000

STD. ERROR OF EST. =   304.855
        R SQUARED =      .995
        MULTIPLE R =     .998

                  ANALYSIS OF VARIANCE TABLE

   SOURCE     SUM OF SQUARES   D.F.   MEAN SQUARE   F RATIO
REGRESSION   184172400.000      6   30695400.000   330.283
RESIDUAL        836430.000      9      92936.667
TOTAL        185008830.000     15   '


         OBSERVED   CALCULATED   RESIDUAL
   1    60323.000   60055.666    267.334
   2    61122.000   61215.996    -93.996
   3    60171.000   60124.709     46.291
   4    61187.000   61597.131   -410.131
   5    63221.000   62911.305    309.695
   6    63639.000   63888.314   -249.314
   7    64989.000   65153.030   -164.030
   8    63761.000   63774.172    -13.172
   9    66019.000   66004.719     14.281
  10    67857.000   67401.602    455.398
  11    68169.000   68186.247    -17.247
  12    66513.000   66552.038    -39.038
  13    68655.000   68810.561   -155.561
  14    69564.000   69649.667    -85.667
  15    69331.000   68989.069    341.931
  16    70551.000   70757.774   -206.774

DURBIN-WATSON TEST =   2.5595
```

MICROSTAT

Developed by:                          Distributed by:

    J.B. Orris, Ph.D.                      ECOSOFT
    Butler University                      P.O. Box 68602
    4600 Sunset Ave.                       Indianapolis, IN  46268
    Indianapolis, IN  46208

Computer Makes:                        Operating Systems:

    Any micro-computer that will           North Star DOS
support North Star BASIC or            CP/M
Microsoft BASIC-80 and CP/M.
Systems must have at least 16K
bytes of memory space available
for the programs and data.

Source Languages:  North Star BASIC, Release 4 or later
                  Microsoft BASIC-80

Cost:

    The cost of the system is $250.  The manual is available for $15 plus
$3.00 shipping and handling.  The cost of the manual can be credited toward
purchase price.

Documentation:

    The MICROSTAT system comes with a 76 page manual that provides detailed
discussion of the operation of the system.  The manual includes test data and
sample output for all of the programs.

## 8.15  A-STAT

### INTRODUCTION

A-STAT 79 is a general-purpose statistical package for the Apple II micro-computer.  This system is a subset language of the popular main-frame package: P-STAT 78.  It is a package designed to optimize the power of the micro by employing many main-frame design concepts.  Data are processed from DOS files, increasing the capacity of the system to as many as 45 variables for each of 2,000 cases (one disk full of data).  This differs from the in-core processing approach of most current micro-based statistical programs.  A-STAT is ideal for market research, survey analysis, social and economic modeling, sumulations, teaching statistics, or any of countless research applications.

### CAPABILITIES:  Processing and Displaying Data

A-STAT has many of the same multi-file capabilities as P-STAT 78.  It uses DATA as the main way of creating system files.  A-STAT system files are stored in fixed format character records with accompanying dictionary files (like OSIRIS) for ease of access in the micro environment.  Dictionary files may be directly created to define large files.

A-STAT supports many of the variable transformation capabilities found in larger computer based systems.  Each procedure can access such operations and functions as ordinary arithmetic (+,-,*,/,**), case delete, sine, cosine, log, and random numbers.  Arithmetic can be based on the outcome of logical trans-formation language.

(IF INCOME .EQ. 1 SETX NEWINCOME TO OLDINCOME + OTHERVAR)

A-STAT can also create new permanently modified files with new variables and can merge these files either UP/DOWN or LEFT/RIGHT.  It can also produce formatted listings of data with its LIST procedure.

### CAPABILITIES:  Statistical Analysis

DATA:  for file definition and descriptive statistics (good N, sum, mean, standard deviation, minimum and maximum).

FREQ:  for frequency distributions (counts, percents, cummulative percents).

TABLES:  for bivariate frequency distributions (cell counts, row percent, column percent, total percent, expected values, cell chi-squares, overall chi-square, and exact probability of chi-square. )

CORRELATE:  for the creation of square correlation matrices (means, standard deviations, Pearson product-moment correlations, and good N).  Up to 15 by 15 matrices can be constructed without program modification.

REGRESSION: for multiple regression and path analysis of linear combinations of variables (simple correlations, standardized regression weights - beta's, non-standard regression weights and
intercept, standard error of regression weights, exact significance level of regression weights, multiple correlation
coefficient, R-square, R-square adjusted for shrinkage,
standard error of estimate, complete analysis of variance
table with sums of squares for regression and error, the
F-test for the equation, and the exact probability of F).

A-STAT automatically handles all missing data problems and adjusts statistics
accordingly.

## EXTENSIBILITY

A-STAT uses a very basic file structure and should be easy to adapt to
other existent routines.  Work is in progress on additional multivariate
procedures.

## INTERFACES WITH OTHER SYSTEMS

A-STAT currently interfaces with the Apple Computer Company's File Cabinet
program which is a popular data management system.  It is also soon to have an
interface to the Apple Plot programs from the same company.

## PROPOSED ADDITIONS IN NEXT YEAR

We are currently working on residual analysis enhancements and graphic
display routines.  SORT is already completed and being tested, and each of the
routines will be able to use a BY convention for subgroup summary and display
of information.  We will also be improving its multivariate capabilities and
matrix handling capacities.

## SAMPLE JOB

```
CORRELATE IN=SP80(C AGE - PAROCC) COR=SP80.COR$
REGRESSION COR=SP80.COR DES=SP80$
COLCOMP=AGE+SEX+SEMESTER+CUMGPA+HSCOMP+MSAT+VSAT+PAREDUC+PARECON+PAROCC
MSAT=PAREDUC+PARECON+PAROCC
*END
END$
```

Partial Output

JRUN ASTAT

A-STAT 79.4

??CORRELATE IN=SP80(C AGE SEX SEMESTER
MORE ?? CUMGPA COLCOMP HSCOMP MSAT VSAT
MORE ?? PAREDUC PARECON PAROCC)
MORE ?? COR=SP80.COR$

UPPER RIGHT -> CORRELATIONS
LOWER LEFT -> GOOD N'S

CORRELATIONS OF SP80.COR

|  | AGE | SEX | SEMESTER | CUMGPA | COLCOMP | HSCOMP | MSAT |
|---|---|---|---|---|---|---|---|
| AGE | 1 | -.0733 | .4034 | .0262 | -.0969 | -.2244 | -.3189 |
| SEX | 51 | 1 | -.3285 | .0631 | .0604 | .1134 | -.2919 |
| SEMESTER | 51 | 51 | 1 | -.0285 | -.0871 | -.0829 | .3454 |
| CUMGPA | 51 | 50 | 50 | 1 | .1663 | .1998 | .1155 |
| COLCOMP | 51 | 50 | 50 | 50 | 1 | .4979 | .154 |
| HSCOMP | 50 | 48 | 48 | 49 | 50 | 1 | .134 |
| MSAT | 49 | 49 | 49 | 49 | 48 | 48 | 1 |
| VSAT | 51 | 51 | 51 | 49 | 48 | 48 | 49 |
| PAREDUC | 51 | 51 | 51 | 50 | 50 | 50 | 49 |
| PARECON | 51 | 51 | 51 | 50 | 50 | 50 | 49 |
| PAROCC | 51 | 51 | 51 | 50 | 50 | 50 | 49 |

CORRELATIONS OF SP80.COR

|  | VSAT | PAREDUC | PARECON | PAROCC |
|---|---|---|---|---|
| AGE | -.3015 | .311 | .036 | .1219 |
| SEX | -.1442 | -.0158 | -.0421 | -.1172 |
| SEMESTER | .3753 | .2007 | -.0881 | .0599 |
| CUMGPA | .1171 | -.0641 | -.1449 | -.1432 |
| COLCOMP | .0143 | -.2151 | -.1691 | -.1959 |
| HSCOMP | .1533 | -.1052 | -.1187 | -.0868 |
| MSAT | .8694 | -.0987 | -.2319 | -.1915 |
| VSAT | 1 | -.1478 | -.3128 | -.3321 |
| PAREDUC | 49 | 1 | .2874 | .4819 |
| PARECON | 49 | 51 | 1 | .7018 |
| PAROCC | 49 | 51 | 51 | 1 |

??REGRESSION COR=SP80.COR DES=SP80$

??COLCOMP=AGE+SEX+SEMESTER+CUMGPA+HSCOMP+MSAT
  +VSAT+PAREDUC+PARECON+PAROCC

DETERMINANT=.0213193528

***NOTE: STATISTICS BASED ON N FOR EACH VAR

DEPENDENT VARIABLE=>COLCOMP
MINIMUM N=49

|  | BETA | SIMPLE R |
|---|---|---|
| AGE | .085 | -.097 |
| SEX | .101 | .06 |
| SEMESTER | .023 | -.087 |
| CUMGPA | .028 | .166 |
| HSCOMP | .489 | .498 |
| MSAT | .76 | .154 |
| VSAT | -.777 | .014 |
| PAREDUC | -.136 | -.215 |
| PARECON | .012 | -.169 |
| PAROCC | -.205 | -.196 |

DEPENDENT VARIABLE=>COLCOMP

|  | B-WT | STD-ERR | PROB |
|---|---|---|---|
| AGE | .022 | .045 | .627 |
| SEX | .176 | .245 | .518 |
| SEMESTER | .017 | .13 | .893 |
| CUMGPA | 5E-03 | .021 | .826 |
| HSCOMP | .804 | .216 | 1E-03 |
| MSAT | 4E-03 | 1E-03 | 1E-03 |
| VSAT | -4E-03 | 2E-03 | .043 |
| PAREDUC | -.131 | .143 | .633 |
| PARECON | -.016 | .228 | .944 |
| PAROCC | -.238 | .232 | .312 |
| INTERCEPT | .279 | | |

MULT, R= .64580333
R-SQUARED= .417061941
S.E. OF EST=.7521680I8
ADJUSTED R-SQR=.263657188

ANALYSIS OF VARIANCE TABLE

| SOURCE | DF | SS | MS | F-TEST | PROB |
|---|---|---|---|---|---|
| REG | 10 | 15.381 | 1.538 | 2.718 | .011 |
| RES | 38 | 21.499 | .566 | | |
| TOT | 48 | 36.88 | | | |

??END

## A-STAT 79

Developed by:

    Rosen Grandon Associates
    296 Peter Green Road
    Tolland, CT   06084

Distributed by:

    Rosen Grandon Associates
    296 Peter Green Road
    Tolland, CT   06084

    The Roper Center, Inc.
    User Services
    Box U-164R
    Storrs, CT   06268

Computer Makes:

    Apple II with 48K and Disk
    Soon:
    Commodore PET and CBM

    CP/M Micro-soft Basic

    Heath H89

Source Language:

    Applesoft Basic

    soon:
    Pure Micro-Soft Basic

Cost:

    A-STAT 79 complete with manuals (80 pages) and on disk is $125.00.

Documentation:

    A-STAT 79 has a 30 page user's manual and a 50 page complete language
reference manual.

CHAPTER 9


SPECIFIC-PURPOSE BATCH PROGRAMS


CONTENTS

INTRODUCTION

     Chapters 6 and 7 contain general-purpose statistical packages, and Chapter 8 contains some specific-purpose programs which operate  in the interactive mode.  This chapter also contains specific-purpose programs, but these operate in the batch mode.

     The ratings by the respective developers on selected items are displayed in Table 9.1.  Figure 9.1 compares developers' and users' ratings on all items, while Figure 9.2 summarizes developers' ratings (D-scores) and users' ratings (U-scores) on several relevant attributes.  These are explained in detail in Section 1.4 of Chapter 1.  Figure 9.2 also points to other chapters which describe programs with strong statistical analysis capabilities.

     The most well-known of these programs is LINWOOD/NONLINWOOD for fitting linear and non-linear equations to data, as described in the book by Daniel and Wood.  MULPRES provides a graphical procedure for interpreting the results of multiple regression analyses.

     Four of these programs, AMANCE, REG, TPD, and LSML have biometric origins and are used for various analyses of experimental data.  While POPAN and CAPTURE have zero ratings on all eight items used to represent "Statistical Analysis" in the taxonomy of Table 1.2, they do nevertheless have a specific statistical capability for calculating population estimates for mark-recapture data from (respectively) open and closed animal populations.  Both STATSPLINE and ALLOC use non-parametric methods to estimate densities.  ALLOC uses these densities for multigroup discriminant analyses.

     The remainder of this chapter contains descriptions of these specific-purpose batch programs in the format described in Table 1.5.

TABLE 9.1:   RATINGS BY DEVELOPERS ON SELECTED ITEMS

(i) Capabilities

Usefulness
Rating Key:
3 - high
2 - moderate
1 - modest
. - low

| | Complex Structures | File Management | Consistency Checks | Probability Checks | Compute Tables | Print Tables | Multiple Regression | Anova/Linear Model | Linear Multivariate | Multi-way Tables | Other Multivariate | Nonparametric | Exploratory | Robust | Non-linear | Bayesian | Time Series | Econometric |
|---|---|---|---|---|---|---|---|---|---|---|---|---|---|---|---|---|---|---|
| | 11 | 14 | 18 | 19 | 24 | 25 | 30 | 31 | 32 | 33 | 34 | 36 | 37 | 38 | 39 | 40 | 35 | 41 |
| 9.01 AMANCE | 1 | 2 | 1 | 2 | 2 | 2 | 3 | 3 | 3 | . | 2 | . | . | . | . | . | . | . |
| 9.02 MAC/STAT | 2 | 2 | . | . | 3 | 3 | 2 | 2 | 2 | 2 | 1 | . | . | . | . | . | . | . |
| 9.03 REG | 1 | 1 | 1 | 1 | 1 | 1 | 2 | 3 | 2 | 1 | . | . | . | 1 | 1 | . | . | . |
| 9.04 TPD-3 | 2 | 1 | . | 1 | 1 | . | 2 | 2 | 2 | 1 | 1 | 1 | . | . | 2 | . | . | 1 |
| 9.05 LSML76 | 1 | . | . | . | . | . | 2 | 3 | . | . | . | . | . | . | 1 | . | . | . |
| 9.06 LNWD/NLNWD | . | . | . | . | 2 | . | 3 | . | . | 1 | . | 2 | . | . | . | . | . | . |
| 9.07 ACPBCTET | . | . | . | . | . | . | 3 | 1 | . | . | . | . | . | . | 3 | . | . | . |
| 9.08 MULPRES | . | . | . | . | . | . | 1 | . | . | . | . | . | . | . | . | . | . | . |
| 9.09 STATSPLINE | . | 1 | 1 | . | 1 | . | . | . | . | . | . | . | . | . | 2 | . | . | . |
| 9.10 ALLOC | 1 | . | 2 | 1 | 2 | 2 | . | . | . | . | 3 | 3 | . | 2 | . | 2 | . | . |
| 9.11 POPAN | 1 | 3 | 1 | . | 2 | 2 | . | . | . | . | . | . | . | . | . | . | . | . |
| 9.12 CAPTURE | . | . | . | . | . | . | . | . | . | . | . | . | . | . | . | . | . | . |

## FOR SPECIFIC BATCH STATISTICAL ANALYSIS PROGRAMS

|  | Survey Estimates | Survey Variances | Math Functions | Operations Research | (ii) User Interface |  |  |  |  |  |  |  |  |  |  |  |  |
|---|---|---|---|---|---|---|---|---|---|---|---|---|---|---|---|---|---|
|  |  |  |  |  | Availability | Installations | Computer Makes | Mini Version | Core Requirements | Batch/Interactive | Stat. Training | Computer Training | Language Simplicity | Documentation | User Convenience | Maintenance | Tested for Accuracy |
|  | 43 | 44 | 47 | 48 | 49 | 50 | 51 | 52 | 53 | 54 | 55 | 56 | 57 | 58 | 59 | 60 | 61 |
| AMANCE | 1 | . | . | . | . | 1 | 2 | 3 | 3 | . | 1 | 2 | . | 3 | 1 | 1 | 1 |
| MAC/STAT | . | . | 2 | . | . | . | 1 | 1 | 2 | . | . | 2 | . | 2 | 3 | 1 | 1 |
| REG | 2 | . | 1 | . | 3 | 1 | 2 | . | . | . | 1 | 2 | 2 | 1 | 2 | 2 | 1 |
| TPD-3 | 1 | 1 | . | . | 2 | 1 | 2 | 3 | 1 | . | 1 | 2 | . | 1 | 2 | 2 | 1 |
| LSML76 | 2 | 1 | . | . | 2 | 3 | 3 | . | . | . | 1 | 3 | . | 1 | 3 | 3 | 2 |
| LNWD/NLNWD | . | . | 1 | . | 3 | 3 | 3 | 1 | 2 | 1 | 3 | 3 | . | 3 | 3 | 3 | 3 |
| ACPBCTET | . | . | . | . | 2 | . | . | . | . | . | 1 | 3 | . | 1 | . | 1 | 2 |
| MULPRES | . | . | . | . | . | . | 2 | . | 3 | . | 2 | 3 | . | 1 | 1 | 2 | 1 |
| STATSPLINE | . | . | . | . | 2 | 1 | 2 | 3 | 1 | . | 1 | 3 | 2 | 2 | . | 1 | 2 |
| ALLOC | . | . | . | . | 3 | 2 | 2 | 1 | 1 | . | 1 | . | . | 3 | 2 | 3 | 2 |
| POPAN | . | . | . | . | 2 | 1 | 1 | 1 | . | . | 2 | 2 | 3 | 2 | 3 | 1 | 1 |
| CAPTURE | . | . | . | . | 2 | 2 | 2 | 1 | 1 | 3 | 2 | 3 | 3 | 2 | 3 | 3 | 2 |

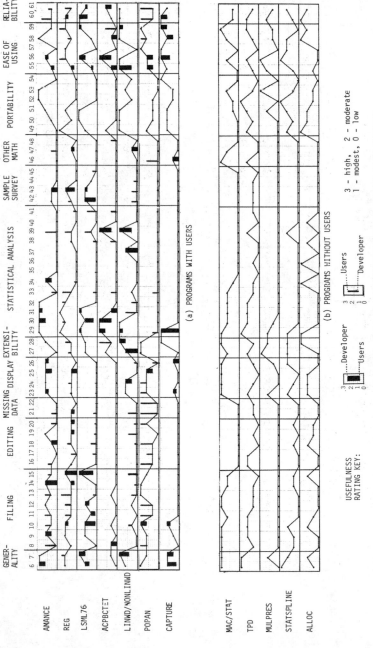

(a) PROGRAMS WITH USERS

(b) PROGRAMS WITHOUT USERS

USEFULNESS
RATING KEY:                3 - high,    2 - moderate
                           1 - modest,  0 - low

FIGURE 9.1:  RATINGS BY DEVELOPERS AND USERS ON ALL ITEMS FOR SPECIFIC-PURPOSE BATCH PROGRAMS

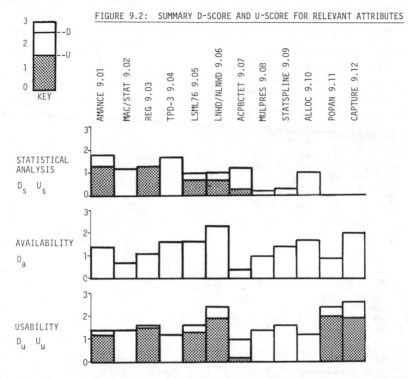

FIGURE 9.2:  SUMMARY D-SCORE AND U-SCORE FOR RELEVANT ATTRIBUTES

Other chapters generally containing programs with strong Statistical Analysis Capabilities are Chapters 6, 7, and 8 (see Figures 6.2, 7.2, and 8.2).

## 9.01  AMANCE

### INTRODUCTION

AMANCE is a statistical package which has been developed by biometricians of I.N.R.A. (National Institute of Agronomical Research) since 1975. It includes both descriptive and inferential methods of multivariate analysis, mainly based on linear model. Designed for minicomputers, it has been installed in over 16 installations.

### CAPABILITIES:  Data Processing and Display

The linkage of programs is based on a few standard files that permit: not to compute the same thing twice (i.e. correlation matrix is stored once for all); to separate data management from the computing process; to keep results that can be re-used as new data (ex. residuals, principal components, ...).

The first step is to create a source data file, in which transformations and data cleaning are supposed to be achieved by a specific routine. Besides the observed variable matrix, this sequential file contains descriptors such as:  names of variables and factors, levels of factors, weight and selection codes for each observation.

To use further programs it is only necessary to select a set of variables and factors and, if wanted, to declare the parameters of a standard filtering upon the observations.

The package offers elementary editions of two-dimensional cross-tables or scatter plots, adapted to structured data.

### CAPABILITIES:  Statistical Analysis

Basic uni- and bivariate statistics (histograms, correlation or covariance matrices) including the case of structured data (within and between covariances).

Principal components analysis (with option of orthogonalized regression), correspondence analysis, canonical correlation analysis. Designed in a descriptive way, these programs provide graphs of observations.

Regression analysis, with options for the step-wise procedure, ridge-regression, residuals examination.

Non-linear fitting.

Multivariate analysis of variance on factorial, non-orthogonal, crossed or nested designs, with fixed, random or mixed effects; variance components estimates; covariance of estimates; etc...

Multiple means comparisons by several methods.

Multivariate covariance analysis.

Discriminant factor analysis.

EXTENSIBILITY

New programs can be added by access to standard files.

PROPOSED ADDITIONS IN NEXT YEAR

Improved handling and coding of qualitative data, classification programs. New edition of the user's manual.

SAMPLE JOB

Perform a stepwise regression; produce residuals at the last step.

1) Declare 3 standard files A, B, C
   - source data are supposed to be stored in A
   - basic statistics are supposed to be stored in B (STAT 1 program has stored means, variances, correlation matrix for all variables of A)
   - residuals will be stored in C (which may contain previous results upon the same observations).

2) Execute REPRG program, with these parameter cards:
   - title
   - .11........ 1
     to compute and plot residuals, and provide correlations of estimators
   - 3 1 3 5 } to define the set of variables
   - 1 4

Partial Output

```
****************************
*  VARIABLE EXPLIQUEE :   4 *
****************************
```

VARIABLES ENTRANT DANS L'ANALYSE SUR   57 OBSERVATIONS

| NUMERO | NOM | MOYENNE | E.T. | MINIMUM | MAXIMUM |
|---|---|---|---|---|---|
| 1 | INTEMP | 72.772 | 1.069 | 70.000 | 75.000 |
| 3 | WINVEL | 8.316 | 3.357 | 3.500 | 17.400 |
| 5 | TEMPDIF | 37.158 | 10.358 | 11.000 | 62.000 |
| 4 | GASCONS | 15.165 | 8.205 | 0. | 35.200 |

```
               1     3     5     4

    1    1000
    3    -168  1000
    5     -27    83  1000
    4     -76   253   939  1000
```

PALIER NO. 2     VARIABLE INTRODUITE    3
```
**********************************
```

```
    S. TOTALE            3769.590
    S. MODELE            3437.965
    S. RESIDUELLE         331.624

    ECART-TYPE RESIDUEL                         2.478
    COEFFICIENT DE CORRELATION MULTIPLE          .955
    TEST GLOBAL D'AJUSTEMENT F( 2,  54)       279.910 **
```

EQUATION DE LA REGRESSION
..................

| COEFFICIENT | NO. VARIABLE | ECART-TYPE | T. STUDENT | BETA | VIF |
|---|---|---|---|---|---|
| -.1562E+02 * | CONST | .1429E+01 | -10.9255 ** * | * | |
| .4310E+00 * | X 3 | .9899E-01 | 4.3547 ** * | .1764 * | 1.0070 |
| .7319E+00 * | X 5 | .3208E-01 | 22.8123 ** * | .9240 * | 1.0070 |

```
                   NORME DU VECTEUR DE REGRESSION :      .9407
                   (DANS L'ESPACE DES PARAMETRES)
```

MATRICE DE CORRELATION DES ESTIMATEURS
.......................................

```
      3     5

3   1000   -83
5    -83  1000
```

VARIABLES NON ENTREES DANS L'EQUATION
...................................

NO VARIABLE,PUIS COEF. DE CORRELATION PARTIELLE (AVEC LA VARIABLE A EXPLIQUER,COMPTE

```
   1 -.07120     4 1.00000
```
                              TENU DES VARIABLES DEJA INTRODUITES)

AMANCE - Bibliotheque de programmes statistiques de l'I. N.R.A.

Developed by:                              Distributed by:

   Station de Biometrie                      Same
   I.N.R.A. - C.N.R.F.
   (Institut National de la
   Recherche Agronomique)
   Champenoux
   54280 Seichamps, France

Computer Makes:                            Operating Systems:

   PHILIPS P 880                          I. B. M. OS, DOS
   CII Mitra 15                           CII-HB SIRIS 8 (C 10)
   CII-HB Mini 6                          CII-HB GECOS 6 MOD 400
   SOLAR 16/40                            Overlay is required on minicomputers
   PRIME 650
   UNIVAC 90/30                       User Language
   I.B.M. 360/40, 360/50
   IRIS 80                                French
   (we maintain it on Mini 6 and Iris 80)

Source Languages:

   FORTRAN IV (including Define File)

Cost:

    For the present, the software is freely distributed.  The fee of $100
covers tape copying, manuals, and mailing.

Documentation:

Notices de description et d'utilisation de la programmatheque (documents
    Biometrie).
Manuel de la programmatheque AMANCE (to be published in early 1981).

9.02  MAC/STAT

INTRODUCTION

The MAC/STAT System is a package of programs intended to perform a wide variety of statistical computations and analysis on data arising in all fields of research, engineering and business.

The system has been organized to allow the user the greatest control and direction in the analysis of his data without the necessity of preparing numerous control cards or of repeated computer runs.

The MAC/STAT System of programs is totally integrated via a monitor routine. This allows the basic algorithms for computation and input parameters to be maintained throughout the range of analyses requested, thus minimizing the requirements for storage and computer processing time as well as input parameter preparation.

CAPABILITIES:  Data Processing

Data input may be on cards, magnetic tape written in either BCD or binary, or other storage media.  The system can repeatedly select a specified data set for multiple types of analyses.  Within the selection of a given data set, variable and observation selection can be specified.

CAPABILITIES:  Statistical Analysis

The MAC/STAT System consists of thirteen basic modules.  These are:

. MAC/STAT Monitor
. Random Variate Generation
. Data Preparation
. Plotting and Tabulation
. Correlation
. Linear Regression
. Univariate Variance Analysis Module
. Matrix Operations
. Multivariate Statistical Analysis
. Multivariate Variance Analysis Module
. Nonlinear Estimation
. Nonparametric
. Variable Reduction
. Add-on Modules

Output includes not only the usual computed values, but a number of computational checks to assure the user of the accuracy of the procedure involved.

## MAC/STAT: Statistical Data Analysis System

Distributed by:

University Software Systems
250 N. Nash St.
El Segundo, CA  90245

Machine:                          Operating Systems:

UNIVAC 1108                        Exec 2

Language:

FORTRAN IV

Cost:

Not available at present.

Additions in next 12 months:

None expected.

9.03  <u>REG</u>

## INTRODUCTION

REG is a statistical program typically used to analyse unbalanced data by the generalized linear model.  It is a least squares program written by a bio-metrician from 1971 to 1979.  It was designed to cope with the many and varied analyses required in agricultural research.  Use of REG has extended into health, education and social research.  It is a batch program installed in five installations (two are nationwide networks) in Australia.

## CAPABILITIES:  Processing and Displaying Data

REG obtains data either with the parameter cards or from a separate file (either a card image or sequential FORTRAN binary file).  When a record is read, it may be discarded or its contents transformed, edited and rearranged and then written to a scratch file.  It may be selectively printed or written to per-manent file.

The transformations include all arithmetic operations, common mathematical functions and several special functions needed to create design variables for various linear models.  They are specified in a simple free format manner identifying variables by position rather than name.

## CAPABILITIES:  Statistical Analysis

Basic statistics include the range, mean, variance, skewness and kurtosis of grouped data for selected variables.

Analysis of Variance, regression coefficients for univariate and multi-variate data fitted to linear models including unbalanced designs.  The models are specified by selecting the desired dependent and independent factors from those on the scratch file.  The Analysis of Variance may be either 'adjusting for preceding terms' or 'adjusting for all other terms' in the model.  A weighted analysis may be performed.  The analysis may be augmented with pre-dicted values, analysis of residuals or least square means with standard errors. With multivariate data, a repeated measures analysis may be performed or canonical variate coefficients (latent vectors from $HE^{-1}$) obtained.  Principal components of up to 30 variables may be obtained.  Variance Components for nested effects may be obtained (Henderson's method 3) utilizing absorption of the largest nested effect where needed.  The facility to obtain the sums of squares for a nominated set of all possible submodels is provided rather than automatic subset selection.

Binomial, multinomial, poisson and gamma data may be fitted to a linear model by iterative weighted least squares to obtain an analysis of deviance and the maximum likelihood solution.  Predicted values and standard errors on the transformed scale, predicted values and residuals on the retransformed scale may be obtained.

## EXTENSIBILITY

The user has complete control over the structure of the linear model; he is not restricted to any particular constraining method.

Sister programs exist for plotting, contingency tables, well-labelled tables of means and analysis of balanced data. These could but have not been incorporated because of the multitude of programs and packages which adequately cover these matters.  No specific interfaces with other systmes have been incorporated.

## PROPOSED ADDITIONS IN NEXT YEAR

No significant additions are planned for now as the principal author is undertaking postgraduate study.  A completely revised user manual (approx. 100 pages) is in preparation.  Implementation on one or two installations in New Zealand is likely.

## REFERENCES

Nicholls, Paul (1979) "Application of Iterative Weighted Least Squares to Binary Data" in Proceedings of the Regression Analysis Symposium of the New South Wales Statistical Society, Syndey, August, 1979.
Williams, Mrs. Jean. "A comparison of Multivariate Analysis of Variance output from REG and GENSTAT macro".  CSIRO GENSTAT WORKSHOP, National Measurement Laboratory, Lindfield, New South Wales.

## SAMPLE JOB

```
BEGIN A SIMPLE EXAMPLE OF REPEATED MEASURES MULTIVARIATE ANALYSIS
HEAD   DATA - BLOOD HISTAMINE LEVELS IN DOGS
COMMENT    REF:  COLE AND GRIZZLE, BIOMETRICS VOLUME 22   PAGES 810-828
READ 4
FORMAT(F1.Ø,3F5.2)
* 2(2) = ALOG1Ø(2(2))
C THE ABOVE TRANSFORMATION TAKES LOGS (BASE 10) OF VARIABLES 2, 3 AND 4
* 5 = SET(1;4)
C THIS SETS UP VARIABLES 5, 6 AND 7 AS CONSTRAINED TREATMENT VARIABLES
/TREATMENT CODE/LOG HISTAMINE LEVEL;3/TREATMENT;3
DATA
1   .20   .10   .08
1   .06   .02   .02
1 1.40   .48   .24
1   .57   .35   .24
2   .09   .13   .14
2   .11   .10   .09
2   .07   .07   .07
2   .07   .06   .07
3   .62   .31   .22
3 1.05   .73   .60
3   .83 1.07   .80
3 3.13 2.06 1.23
4   .09   .09   .08
4   .09   .09   .10
4   .10   .12   .12
4   .05   .05   .05
END OF DATA
AOV REM
FIT 2/3
```

Output

MON. 18 AUG 1980     A SIMPLE EXAMPLE OF REPEATED MEASURES MULTIVARIATE ANALYSIS
09:58:16             DATA - BLOOD HISTAMINE LEVELS IN DOGS

LOG HISTAMINE LEVEL
FACTOR  2  LEVELS  3

ANALYSIS OF VARIANCE TABLE      SOURCE  DROP  FILTER   WT   ABS   NR
ADJUSTING FOR ALL OTHER TERMS.  DATA      0     0       0    0    16

| SOURCE OF VARIATION | | DF HYP | GENERALIZED SUM SQUARE | LAMBDA | DF CHI | CHI-SQUARE | PERCENT PROB. |
|---|---|---|---|---|---|---|---|
| TIME * |  | 1 | 0.9546325E-02 |  |  | 12.542 | 0.189 |
| M * | TREATMENT | 3 | 0.1210146E 02 | 0.3197497 | 2 | 13.360 | 0.388 |
| TIME * | TREATMENT | 3 | 0.1569331E-01 | 0.3426615 | 3 | 19.648 | 0.320 |
| .M * | ERROR | 12 | 0.4149125E 01 | 0.1945055 | 6 |  |  |
| TIME * | ERROR | 12 | 0.3052435E-02 |  |  |  |  |
|  | TOTAL | 15 |  |  |  |  |  |

ERROR SUMS OF SQUARES AND CROSS PRODUCTS
              1          2          3
1  0.14096D 01
2  0.14561D 01  0.16788D 01
3  0.11881D 01  0.14199D 01  0.12308D 01
TEST OF UNIFORMITY.  CHI-SQUARE  20.685 WITH  4 DEGREES OF FREEDOM AND PROBABILITY  0.037

ERROR CORRELATION MATRIX
         1          2          3
1  1.00000
2  0.94653  1.00000
3  0.90206  0.98782  1.00000
TEST OF INDEPENDENCE.  CHI-SQUARE  66.552 WITH  3 DEGREES OF FREEDOM AND PROBABILITY  0.000

| | 1 | 2 | 3 |
|---|---|---|---|
| AVERAGE STND DEVN | 0.3463759 | 0.3740336 | 0.3202559 |
| ACTUAL STND DEVN | 0.3427300 | | 0.9245905 |
| PROPORTIONAL STND DEVN | 0.9894743 | 1.0798488 | |

FACTOR       LEVEL      REGRESSION COEFFICIENTS    FOLLOWED BY AVERAGE STANDARD ERROR

| CONSTANT TERM | | -0.6560830 | -0.7719936 | -0.8406301 | 0.0865940 |
|---|---|---|---|---|---|
| TREATMENT | 1 | 0.1513789 | 0.0964217 | -0.1682343 | 0.1499851 |
|  | 2 | -0.4224593 | -0.2937083 | -0.2117283 | 0.1499851 |
|  | 3 | 0.7131337 | 0.6964773 | 0.6190223 | 0.1499851 |

## REG, VERSION 80: Generalized Least Squares Program

Developed by:

 Biometrical Branch
 Department of Agriculture
 PO K220 Haymarket NSW 2001
 Australia

Distributed by:

 Chief Biometrician
 Same address.

Computer Makes:

 Burroughs B7700
 CDC Cyber 76
 Prime 500
 Univac 1106/1108

Operating Systems:

 Same

Source Languages:

 FORTRAN

Cost:

   A magnetic tape containing source modules and sample runs will be made available to universities, government departments and non-profit organizations for the cost of obtaining and writing the tape. Manuals may cost A$10. Commercial interests may negotiate. Notes on program structure will be made available with the tape to assist implementation.

Documentation:

Gilmour, A.R., Coote, B.G., Lill, W., "REG - A Program to fit the Generalized Linear Model by least squares." 2nd Ed. Biometrical Branch, Department of Agriculture, New South Wales. 1980.
Gilmour, A.R., "Department of Agriculture multivariate program", Statistical Computing Symposium N.S.W. Statistical Society, Sydney, August 1974.
Gilmour, A.R., and Cullis, B.R., "Generalized Linear Models" Regression Analysis Symposium of the NSW Statistical Society, August 1979.

9.04  <u>TPD</u>

INTRODUCTION

"TPD-3" ("Programmes Statistiques TPD, Version 3") is a package of Fortran main programs and subroutines for general-purpose statistical analysis. Its principal applications have been in the field of agricultural research (plant and animal productions, horticulture, forestry, rural engineering and economics, agricultural and food industries, etc.). It has been developed, since 1965, by the Faculty of Agricultural Sciences and the Agricultural Research Center, at Gembloux (Belgium), under the direction of Professor Pierre Dagnelie. It is oriented towards small- and medium-size computers, working in batch processing. Its documentation is written in French.

CAPABILITIES:  Processing and Displaying Data

"TPD-3" always uses completely standardized T̲itles, P̲arameters and D̲ata disks files, whatever the data format is. The "d̲ata" normally includes, for each variable, four 48-characters title lines, a maximum of 10 parameters, and the data themselves. "Version 3" normally considers a maximum of 500 variables and 100,000 data.

A set of general programs makes possible:
* entering data from cards, tapes, floppy disks, and terminals;
* writing or punching data on paper, cards, tapes, and floppy disks;
* doing data transformations of one variable at a time;
* computing functions of two or more variables;
* merging, ordering and sorting operations;
* computing and printing frequency tables and standard one- and two-dimensional parameters;    etc.

In every case, intermediate results are included in the standard T̲itles, P̲arameters, and D̲ata files.

CAPABILITIES:  Statistical Analysis

The two main sets of statistical programs are related to analysis of variance and analysis of correlation and regression, including some multivariate methods.

The analysis of variance programs cover one-, two-, three- and four-way standard analysis of variance (cross-classification and hierarchical models), plus several special situations, such as Latin square, split-plot designs, incomplete blocks designs, etc. Some of these programs include tests of equality of variances and/or multiple comparisons of means.

The correlation and regression programs cover standard correlation analysis, linear and non-linear regression, multiple regression, one- and two-way analysis of covariance, and multivariate methods such as simultaneous equations, principal component analysis, factor analysis, discriminant analysis, etc.

Other programs are:  tests of normality, Kruskal-Wallis and Friedman tests, etc.

EXTENSIBILITY

User-written programs can be added to the system without difficulty.

PROPOSED ADDITIONS IN NEXT YEAR

TPD is always undergoing new developments.  The work now in progress includes, among others, more multivariate methods.  Writing a new interactive version of some programs is also considered for next year.

SAMPLE JOB

One- and two-dimensional frequency distributions and parameters.

```
EXECUTION REGL

T100      P100      D100

    3    6
```

(REGL is the name of the linear regression program)
(T100, P100, and D100 are the names of a set of Titles,
 Parameters, and Data files)
(3 and 6 are the numbers of the variables "CIRCONFERENCE
 A1.30M" and "HAUTEUR TOTALE" within T100, P100, and D100).

```
*************************************************************************
F.S.A. ET C.R.A. GEMBLOUX (BELGIQUE) - PROGRAMMES STATISTIQUES T.P.D. - 30/07/80
REGL - REGRESSION LINEAIRE SIMPLE
*************************************************************************

305/8072 M. THILL                 X = LIM.INF.=  30   LIM.SUP.=  250   INTERVALLE=  10
CUBAGE CHENE                       Y = LIM.INF.= 900   LIM.SUP.= 3200   INTERVALLE= 100
X = CIRCONFERENCE A 1.30 M
Y = HAUTEUR TOTALE
```

(scatter-plot frequency matrix, X along top, Y down the side; TOT. column and TOT.* row)

| Y \ X | | TOT. |
|---|---|---|
| INF.* | | |
| 30* | | 22 |
| 40* | | 33 |
| 50* | | 22 |
| 60* | | 26 |
| 70* | | 37 |
| 80* | | 32 |
| 90* | | 47 |
| 100* | | 63 |
| 110* | | 57 |
| 120* | | 64 |
| 130* | | 62 |
| 140* | | 54 |
| 150* | | 48 |
| 160* | | 50 |
| 170* | | 51 |
| 180* | | 35 |
| 190* | | 32 |
| 200* | | 25 |
| 210* | | 33 |
| 220* | | 22 |
| 230* | | 22 |
| 240* | | 14 |
| 250* | | 14 |
| SUP.* | | 26 |
| TOT.* | | 891 |

Column totals (TOT.*): 3  9  17  13  16  24  25  21  39  62  76  83  93  89  68  66  56  26  23  18  24  17  12  9  2  *  891

| MOYX | MINX | MAXX | SCEX | VARX | ECTX |
|---|---|---|---|---|---|
| 131.86 | 22. | 389. | 0.32583567E 07 | 0.36610750E 04 | 60.51 |

| MOYY | MINY | MAXY | SCEY | VARY | ECTY |
|---|---|---|---|---|---|
| 2055.46 | 820. | 3240. | 0.19121634E 09 | 0.21484982E 06 | 463.52 |

| R | SPE | SCEY.X | VARY.X | ECTY.X |
|---|---|---|---|---|
| 0.6755 | 0.16860854E 08 | 0.10396733E 09 | 0.11694865E 06 | 341.98 |

| A | ECTA | B | ECTB | C | (1-R2)/(N-2) |
|---|---|---|---|---|---|
| 1375.13 | 27.48 | 5.1746 | 0.1895 | 7.6606 | 0.61160375E-03 |

FIN

TPD-3:  Programmes Statistiques TPD

Developed by:                              Distributed by:

Faculté des Sciences Agronomiques          Same
Statistique et Informatique
Avenue de la Faculté d'Agronomie 8
5800 Gembloux
Belgium

Computer Makes:                            User Language:

IBM 1130, 370                              French
ICL 2903-2904, ME29

Source Language:

Fortran

Cost:

One-time charge for universities, governments and nonprofit organizations: materials.

Documentation:

Carletti, G., Claustriaux, J.J., Dagnelie, P., Debouche, C., In, K., Oger, R., Rousseaux, G.,  Organisation d'une bibliothèque de programmes statistiques pour ordinateur.  Rev. Belge Stat. Inf. Rech. Opér. 12(4), 2-16, 1973.
Claustriaux, J.J., Oger, R., Programmes statistiques T P D (version 3): programmes d'analyse de la variance.  Gembloux, Fac. Sci. Agron., 10 p., 1977.
Dagnelie, P.,  Programmes statistiques T P D (version 3):  Presentation générale.  Gembloux, Fac. Sci. Agron., 12 p., 1976.
Debouche, C., In, K., Programmes statistiques T P D (version 3): programmes de corrélation et de régression.  Gembloux, Fac. Sci. Agron., 9 p., 1977.
Grayet, J.P., Rousseaux, G., Programmes statistiques T P D (version 3):  programmes généraux.  Gembloux, Fac. Sci. Agron., 12 p., 1977.

## 9.05  <u>LSML 76</u>

## INTRODUCTION

LSML76 is a general-purpose least squares program that is designed to analyze data under both the fixed linear model and under mixed models. The "usual" restrictions are imposed to obtain estimates of effects for discrete independent variables. Least squares means and appropriate standard errors are given for all sets of discrete independent variables. Variance and covariance components are estimated using Henderson's Method 3. An analysis of variance is given for each dependent variable and a linear contrast option may be used to obtain tests of significance among the estimated fixed effects. Differences among a set of random effects (either crossclassified or nested) that are confounded with fixed effects being considered may be recovered with a maximum likelihood option.

## CAPABILITIES

LSML76 is designed to handle both discrete and continuous independent variables directly without any special coding. The types of constants that may be fitted by least squares or maximum likelihood (when appropriate) are as follows: (i) crossclassified or main effects, (ii) nested effects, (iii) 2-factor interactions of crossclassified effects, (iv) 2-factor interactions of nested effects, (v) 2-factor interactions of crossclassified and nested effects, (vi) pooled partial regressions for continuous independent variables (linear, linear and quadratic or linear, quadratic and cubic) and (vii) interactions of crossclassified or nested effects with continuous independent variables (individual class regressions). By arbitrarily identifying cross-classified effects as nested effects one may obtain estimates of 2-factor interactions separately for each level of a third factor.

Although LSML76 is generally referred to as a mixed model program, analyses where all sources of variation (discrete and/or continuous independent variables) are fixed are readily computed (MTY = 1). The program is designed to complete six types of mixed model analyses directly using Method 3 of Henderson (MTY = 2, 3, 4, 5, 6, or 7). Models for these six types of analyses are given below (upper case letters refer to fixed effects and lower case to random effects):

<u>MTY</u>

$$2 \qquad y_{ijk} = \mu + a_i + F_j + e_{ijk}$$

$$3 \qquad y_{ijk\ell} = \mu + A_i + b_{ij} + F_k + e_{ijk\ell}$$

$$4 \qquad y_{ijk\ell} = \mu + a_i + b_{ij} + F_k + e_{ijk\ell}$$

$$5 \qquad y_{ijk\ell m} = \mu + A_i + b_{ij} + c_{ijk} + F_\ell + e_{ijk\ell m}$$

$$6 \qquad y_{ijk\ell} = \mu + a_i + B_j + (aB)_{ij} + F_k + e_{ijk\ell}$$

$$7 \qquad y_{ijk\ell m} = \mu + A_i + b_{ij} + C_k + (AC)_{ik} + (bC)_{ijk} + F_\ell + e_{ijk\ell m}$$

The F term in each model may include any of the types of fixed effects described above. Variance and covariance component estimates are given for each random set of effects in each analysis.  When appropriate, these may be used by LSML76 to compute estimates of genetic and environmental parameters; namely, heritabilities, genetic, phenotypic, and environmental correlations. The User's Guide for LSML76 explains how one may use this program to complete analyses under more complex mixed models.

Weighted least squares analyses may be completed with LSML76 under any of the model types when the variance-covariance matrix of the errors (e) is diagonal.

## PROCEDURE HARVEY

An early version of LSML76 that is designed primarily for the analysis of data under the general linear model is available in the Statistical Analysis System (SAS) as PROC HARVEY.  However, this version may be used to obtain estimates of variance and covariance components by a "direct" procedure using Method 3 of Henderson for sets of crossclassified or nested random effects.

## PROPOSED ADDITIONS IN NEXT YEAR

Instructions for a particular analysis are presently given to LSML76 on parameter cards which have fixed position alpha and numeric codes.  A revision is underway that will provide natural language commands for these instructions. Also, general surface fitting procedures are being incorporated into LSML76. The User's Guide will be revised.

## REFERENCES

Dillard, E.U., Oswaldo Rodriquez and O.W. Robison.  Estimation of additive and non-additive direct and maternal genetic effects from crossbreeding beef cattle.  J. Anim. Sci. 50:653-663.  1980.
Drewry, K.J.,  Sow productivity traits of crossbred sows.  J. Anim. Sci. 50:242-248, 1980.
Gaskins, Charles,  and D. Craig Anderson.  Comparison of lactation curves in Angus-Hereford, Jersey-Angus and Simmental-Angus cows.  J. Anim. Sci. 50:828-832.  1980.
Harvey, Walter R., Estimation of variance and covariance components in the mixed model.  Biometrics 26:485-504.  1970.
Polkinghorne, R.W., and W.R. Harvey,  Correlation between dam effects on body weight of dwarf intraline and of normal straincross broiler chickens. Poultry Science 59:1375-1384.  1980.
Searle, S.R., and H.V. Henderson.  Annotated computer output for analysis of unbalanced data:  SAS HARVEY.  Biometrics Unit, Cornell University. BU-659-M.  59 pp, Mimeo.  1980.

## SAMPLE ANALYSIS

Backfat thickness and carcass weight were obtained on 61 barrows and 40 gilts that were the progeny of 22 sires and 44 sows that belonged to two different lines of breeding.  The model underlying the analysis of variation in backfat thickness was as follows:

$$y_{ijk\ell} = \mu + L_i + s_{ij} + d_{ijk} + G_\ell + (LG)_{i\ell} + b_L(W_{ijk\ell m} - \bar{w})$$

$$+ b_{L\ell}(W_{ijk\ell m} - \bar{w}) + b_Q(W_{ijk\ell m} - \bar{w})^2 + b_{Q\ell}(W_{ijk\ell m} - \bar{w})^2 + e_{ijk\ell m}$$

## Partial Output

### LEAST-SQUARES ANALYSIS OF VARIANCE
### BK FAT

| SOURCE | D.F. | MEAN SQUARES | F | PROB | ERROR LINE |
|---|---|---|---|---|---|
| LINES | 1 | 0.4367 | 4.22 | 0.053 | S/L |
| S/L | 20 | 0.1035 | 0.46 | 0.958 | D/S/L |
| D/S/L | 22 | 0.2254 | 2.43 | 0.005 | REM |
| SEXES | 1 | 1.9468 | 21.01 | 0.000 | REM |
| LINES X SEXES | 1 | 0.1017 | 1.10 | 0.300 | REM |
| REGRESSIONS | | | | | |
| C WT    B   LINEAR | 1 | 2.1284 | 22.97 | 0.000 | REM |
| SEXES   RL C WT | 1 | 0.1331 | 1.44 | 0.236 | REM |
| C WT    B   QUAD | 1 | 1.2603 | 13.60 | 0.001 | REM |
| SEXES   RQ C WT | 1 | 0.5027 | 5.42 | 0.055 | REM |
| REMAINDER | 51 | 0.0927 | | | |

### VARIANCE AND COVARIANCE COMPONENT ESTIMATES
### FROM INDIRECT ANALYSES

K FOR RANDOM EFFECTS COMPONENT (D/S/L) = 2.0910
DEGREES OF FREEDOM = 22

SS, CP, MS, MCP, VARIANCE AND COVARIANCE COMPONENTS

| RHM | RHM | SS OR CP | MS OR COV | COMPONENTS |
|---|---|---|---|---|
| BK FAT | BK FAT | 4.95775 | 0.22535 | 0.06346 |

K VALUES (S/L) ARE:  K2 = 2.1197    K3 = 4.0204
DEGREES OF FREEDOM = 20

| | | | | |
|---|---|---|---|---|
| BK FAT | BK FAT | 2.06946 | 0.10347 | -0.03077 |

### LISTING OF CONSTANTS, LEAST-SQUARES MEANS AND
### STANDARD ERRORS[a] FOR PROBLEM NO. 1

| RHM NAME | INDEPENDENT VARIABLE | NO OBS. | EFFECTIVE NO. | CONSTANT ESTIMATE | LEAST-SQUARES MEAN |
|---|---|---|---|---|---|
| BK FAT | MU | 101 | 52.3 | 3.33841 | 3.33841 |
| BK FAT | LINES 2 | 62 | 31.6 | 0.07483 | 3.41324 |
| BK FAT | LINES 3 | 39 | 31.0 | -0.07483 | 3.26359 |
| ⁞ | | | | | |
| ⁞ | | | | | |
| BK FAT | LINES X SEXES  3  2 | 20 | 10.7 | 0.05028 | 3.57530 |
| BK FAT | RGRSN C WT  LINEAR | | | 0.03501 | 0.03501 |
| BK FAT | .1    SEXES  C WT RGN L | | | -0.00875 | 0.02626 |
| ⁞ | | | | | |
| ⁞ | | | | | |
| BK FAT | 2    SEXES  C WT RGN Q | | | -0.00158 | -0.00416 |

---

[a]Standard errors are not given here to save space.

## LSML76

Developed by:                                          Distributed by:

    Walter R. Harvey                                         Same
    Department of Dairy Science
    Ohio State University
    625 Stadium Drive
    Columbus, Ohio 43210

Computer Makes:                                        Source Languages:

    IBM 370                                              ALL FORTRAN
    UNIVAC 1108
    Burroughs 7700                                   Operating Systems:
    DEC PDP 10
    SIGMA 7                                              All which use FORTRAN programs
    CDC 6000 Series
    CEC CYBER Series                                 Interfaced Systems:
    ICL
    HP 2000-3000                                         None

Cost:

    The LSML76 package, which includes the FORTRAN IV Source Statements, test
examples, results of analyses of test examples and the User's Guide for LSML76,
may be obtained for a one-time charge of $75.00.

Documentation:

Harvey, Walter R.  User's Guide for LSML76.  Ohio State University,
    Columbus. Mimeo.  76 pp.  1977.

## 9.06  LINWOOD/NONLINWOOD

### INTRODUCTION

LINWOOD and NONLINWOOD are linear and nonlinear least-squares curve-
fitting programs.  Users manuals, glossary of terms, and interpretation of
results of examples are provided in <u>Fitting Equations to Data--Computer Analysis of Multifactor Data,</u> by Cuthbert Daniel and Fred Wood, Wiley Publisher,
1980, second edition.  The programs are designed for the analysis of both
global and interior characteristics of data--determining the influence of each
observation on the fit, assessing the plausibility of assumptions, searching
for influential subsets of variables, estimating measurement error to judge the
fit of candidate equations, providing statistics on the range and relative
influence of variables to recognize the strengths and limitations of the fit,
and for checking the validity of fitted equations as additional observations
become available.  The programs have been updated each year.  For the past six
years, they have been the most requested in both the SHARE library (double-
precision for IBM computers) and the VIM library (single-precision for CDC
computers).

### CAPABILITIES:  Statistical Analysis

LINWOOD has many options which allow the user to transform data into an
appropriate form, fits specified equations to the transformed data by linear
least-squares, and provides both statistics and plots to aid in evaluting the
fit.  A modified leaps and bounds Cp search technique determines rapidly if
smaller subsets of variables will represent the data equally well.  The program
allows the user to delete observations as well as to assign indicator variables
to given observations.  In addition to the usual statistics, the program calculates the maximum and minimum value of each variable as well as its range,
the relative influence of each variable, and the weighted squared standardized
distance of each observation from the centroid of all observations in influence
space.  Near neighbors are used to estimate the standard deviation of the
dependent variable, and hence, to judge the lack of fit of the equation.  Component and component-plus-residual plots are used:  (1) to choose the appropriate form of the equation, (2) to determine the distribution of the observations over the range of each independent variable, and (3) to ascertain the
influence of each observation on each term of the equation.  A table of functions related to the variance of the fitted value provides information on the
influence of the location of each observation in x-space.  Cross-verification
of coefficients can be made as additional observations become available.

The NONLINWOOD program allows the user to estimate the coefficients of a
nonlinear equation--nonlinear in the coefficients.  An iterative technique is
used; the estimates at each iteration are obtained by Marquardt's Maximum
Neighborhood Method which combines the Gauss (Taylor Series) Method and the
Method of Steepest Descent.

### EXTENSIBILITY

The read-data subroutine allows the user to program in FORTRAN special
data handling abilities such as calibration of recording instruments, table
step functions, etc.

## INTERFACE WITH OTHER SYSTEMS

The use of special formats allows data to be read from any system's files.

## PROPOSED ADDITIONS IN NEXT YEAR

Time share versions of the programs are being developed for easier access with such facilities.

## REFERENCES

Daniel, C., and F.S. Wood, with assistance of J.W. Gorman, Fitting Equations to Data, Computer Analysis of Multifactor Data, Second Edition, Wiley (1980).
Wood, F.S., "An Approach to Fulfilling Users' Needs in Fitting Equations to Data," Proc. Comp. Sci. Stat.: 8th Ann. Sym. Interface, J.W. Frane (ed.), UCLA, Los Angeles, California (1975).
Wood, F.S., "LINWOOD and NONLINWOOD Programs," Amer. Statistician, 34, No. 3, August (1980).

## SAMPLE JOB

Aid user in selecting form of equation as well as observing distribution of observations--example from Fitting Equations to Data.  Two independent variables, $x_1$ and $x_2$, have high covariance; the simple correlation coefficient between these variables is 0.75.  Equation fit is $Y = b_0 + b_1 x_1 + b_2 x_2$.  Component and component-plus-residual plots versus each independent variable requested (a 4 in Column 40 of the control card).  (We see in proper perspective a small curvature in $x_1$ (Figure 1), none in $x_2$ (Figure 2).  Distribution of observations is uniform, no far-out controlling observations or nested data.  A squared $x_1$ term would improve the fit of the equation.

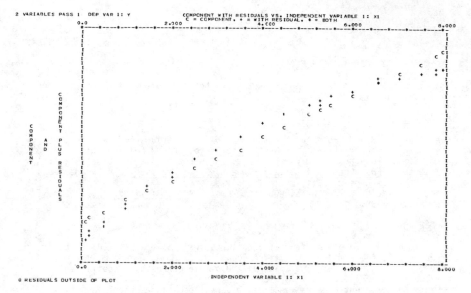

Figure 1

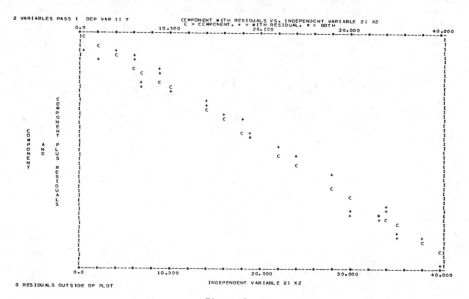

Figure 2

LINWOOD AND NONLINWOOD

Developed by:                                    Distributed by:

    Fred S. Wood                                 Program Libraries
    Standard Oil Company (Indianna)              (see below)
    200 E. Randolph
    Chicago, Illinois 60601

Computer Makes:                                  Libraries:

    Burroughs 4700-6700                          CUBE
    DECsystem 10, 20, PDP 11                     DECUS
    Honeywell 600/6000                           HLSUA
    IBM 360/370/3033, DOS-VS, VS2,               SHARE
      MVT, MVS
    AMDAHL 470
    UNIVAC 1108                                  UNIVAC
    Control Data 6400-6600                       VIM

Source Language:  FORTRAN IV

Cost:

    The one-time charge for each program on magnetic tape is \$35-\$65
depending on the library.  Included are installation instructions, source modules, sample test problems from book, printout of test problems, source module
of replacement program, replacement cards to reduce or increase dimensions of
LINWOOD or NONLINWOOD, and suggested job control language cards to run the programs.  New versions are available each year.

Program Libraries:

CUBE Library for Burroughs 4700 and 6700 computers, Department of Computer
    Science, U.S. Air Force Academy, Colorado 80840, Programs WIS/LINWOOD and
    WIS/NONLINWOOD.
DECUS Library for DECsystem 10, 20, and PDP 11 computers, One Iron Way, Marl-
    boro, Massachusetts 01752, Programs LINWOOD-10-257 and NONLIN-10-258,
    LINWOOD-11-419 and NONLIN-11-420.
HLSUA Library for Honeywell 600/6000 computers, Software Library Mail Station
    K16, Honeywell Information Systems, Post Office Box 6000, Phoenix,
    Arizona 85005, Programs GES-1206 and GES-1207.
SHARE Library for IBM 360/370 computers, Triangle University Computation Center,
    Post Office Box 12076, Research Triangle Park, North Carolina 27709,
    Programs 360D-13.6.008 and 360D-13.6.007.
UNIVAC Source for UNIVAC 1108 computers, Johnson Space Center Code TN74,
    Houston, Texas 77058, Programs LINWOOD and NONLINWOOD.
VIM Library for CDC 6400 computers, Vogelback Computer Center, Northwestern
    University, 2129 Sheridan Road, Evanston, Illinois 60201, Programs
    (Single Precision) LINWOOD and NLWOOD.

## 9.07  ACPBCTET

### INTRODUCTION

ACPBCTET is a small program designed to conduct the transformation of variables and to use the maximum likelihood method suggested by Box and Cox for the estimation of parameters in a single-equation model. Its principal application is in the determination of a correct functional form for a regression model. It was developed by Hui S. Chang in 1977 and can be used interactively or in batch. More than 150 statisticians, econometricians and institutions over the world have requested this program from the developer.

### CAPABILITIES:  Statistical Analysis

The principal capability of this program is the determination of a correct transformation parameter for the variables in a regression equation. By simply specifying an upper limit and a lower limit of the transformation parameter and the change of it each step, the user instructs the program to transform the data step by step by all the values of the transformation parameters desired and to conduct least-squares regression on each set of the transformed data. In addition to transforming both sides of an equation, the option of transforming only one side of an equation is also available.

The program accomodates up to 10 variables and 300 observations in a regression equation. No knowledge of programming is needed to use the program. The output of the program includes most of the statistics requested by statisticians and economists from regression, such as the estimated coefficients, standard errors of coefficients, t ratios, residuals, Durbin-Watson statistic, the analysis of variance table, correlation matrix of the variables, and the maximized log likelihood values for different transformation parameters.

### EXTENSIBILITY

The program can be revised to accomodate more variables and more observations or user-written programs can be added to the system without difficulty.

### REFERENCES

Box, G.E.P., and D.R. Cox.,  "An Analysis of Transformation", Journal of the Royal Statistical Society, Series B, 26 (1964), pp 211-243.
Hui. S. Chang, "Functional Forms and the Demand for Meat in the United States," The Review of Economics and Statistics, 59 (1977), pp. 355-359.
Huang, C.L., and L.C. Moon and H.S. Chang,  "A Computer Program Using the Box-Cox Technique for the Specification of Functional Forms", The American Statistician, 32 (1978), p. 144.

SAMPLE JOB

A regression equation has been estimated in which per capita consumption
of meat (X1) is the dependent variable, and per capita income (X2) and the
price of meat relative to the price of food (X3) are the independent variables.
In order to determine a correct functional form for the relationship, a transformation parameter with the values ranging from 1.6 to -1.6 and a change of
.02 each step was specified.  The data were then transformed automatically by
each of the 161 values of the transformation parameter specified.  A regression
equation was then estimated for each set of the transformed data and the
maximum likelihood value calculated.  A maximum likelihood estimate of the
transformation parameter can be determined and the null hypothesis that the
functional form is linear or log be tested.  For more details of the sample job,
see the article by the developer in The Review of Economics and Statistics.

Partial Output

*** BOTH SIDES OF THE EQUATION ARE TRANSFORMED ***

| LAMBDA | L-MAX | R_2/F | D-W | INTERCEPT | COEFF. OF X2 | COEFF. OF X3 |
|--------|-------|-------|-----|-----------|--------------|--------------|
| | | 0.9340 | 0.716 | 152.456088 | 0.039848 | -0.703860 |
| | | 234.37 | | ( 9.952 ) | ( 20.237 ) | ( -4.129 ) |
| 1.60 | -61.264 | 0.9068 | 0.633 | 1934.277560 | C.008537 | -0.845090 |
| | | 161.57 | | ( 8.267 ) | ( 16.902 ) | ( -3.231 ) |
| 1.40 | -59.432 | 0.9163 | 0.654 | 808.071043 | C.014300 | -0.793550 |
| | | 181.63 | | ( 8.757 ) | ( 17.887 ) | ( -3.489 ) |
| 1.20 | -57.494 | 0.9254 | 0.681 | 344.647122 | C.023899 | -0.746748 |
| | | 205.61 | | ( 9.317 ) | ( 18.993 ) | ( -3.787 ) |
| 1.00 | -55.451 | 0.9340 | 0.716 | 150.792077 | C.035848 | -0.703860 |
| | | 234.37 | | ( 9.951 ) | ( 20.237 ) | ( -4.129 ) |
| 0.80 | -53.312 | 0.9420 | 0.761 | 68.059748 | 0.066274 | -0.664181 |
| | | 268.83 | | ( 10.659 ) | ( 21.628 ) | ( -4.519 ) |
| 0.60 | -51.095 | 0.9493 | 0.817 | 31.852713 | C.109930 | -0.627116 |
| | | 309.95 | | ( 11.414 ) | ( 23.172 ) | ( -4.960 ) |
| 0.40 | -48.835 | 0.9559 | 0.888 | 15.475791 | 0.181832 | -0.592170 |
| | | 358.45 | | ( 12.124 ) | ( 24.863 ) | ( -5.451 ) |
| 0.20 | -46.589 | 0.9616 | 0.974 | 7.718154 | 0.299879 | -0.558941 |
| | | 414.41 | | ( 12.536 ) | ( 26.670 ) | ( -5.986 ) |
| -0.00 | -44.439 | 0.9665 | 1.075 | 3.776161 | C.493055 | -0.527116 |
| | | 476.57 | | ( 11.997 ) | ( 28.530 ) | ( -6.548 ) |
| -0.20 | -42.504 | 0.9704 | 1.189 | 1.533467 | C.808119 | -0.496458 |
| | | 541.44 | | ( 8.903 ) | ( 30.332 ) | ( -7.110 ) |
| -0.40 | -40.924 | 0.9733 | 1.307 | 0.014216 | 1.320243 | -0.466798 |
| | | 602.62 | | ( 0.133 ) | ( 31.915 ) | ( -7.628 ) |
| -0.60 | -39.850 | 0.9753 | 1.414 | -1.273027 | 2.149859 | -0.438032 |
| | | 651.33 | | ( -14.341 ) | ( 33.087 ) | ( -8.048 ) |
| -0.80 | -39.404 | 0.9762 | 1.495 | -2.622920 | 3.489258 | -0.410103 |
| | | 678.77 | | ( -25.052 ) | ( 33.679 ) | ( -8.319 ) |
| -1.00 | -39.638 | 0.9763 | 1.538 | -4.268573 | 5.644499 | -0.382999 |
| | | 679.90 | | ( -29.456 ) | ( 33.606 ) | ( -8.410 ) |
| -1.20 | -40.521 | 0.9754 | 1.538 | -6.455927 | 5.101242 | -0.356736 |
| | | 655.90 | | ( -30.677 ) | ( 32.904 ) | ( -8.323 ) |
| -1.40 | -41.955 | 0.9737 | 1.501 | -9.458127 | 14.627763 | -0.331344 |
| | | 612.91 | | ( -30.422 ) | ( 31.704 ) | ( -8.085 ) |
| -1.60 | -45.559 | 0.9682 | 1.298 | -13.829643 | 23.433867 | -0.306796 |
| | | 502.90 | | ( -27.939 ) | ( 28.665 ) | ( -7.349 ) |

## ACPBCTET: A Computer Program for Box-Cox Transformation and Estimation Technique

Developed by:                                   Distributed by:

  Hui S. Chang                                      Same
  Department of Economics
  University of Tennessee
  Knoxville, Tennessee 37916

Computer Makes:

  IBM 360,370
  DEC Systen 10, 20
  Etc.

Source Languages: FORTRAN

Cost:

  A one time charge of $15.  The fee includes the source program in punched card form, input cards for a sample run, the computer output for the sample run, and instruction for supplying input cards.

Documentation:

Hui S. Chang.,  "A Computer Program for Box-Cox Transformation and Estimation Technique", Econometrica, 45 (1977), p. 1741.

9.08  MULPRES


INTRODUCTION

     MULPRES presents the results of multiple regression analysis in graphical
form.  Scientists and engineers are accustomed to graphical presentation of the
relationship between two variables, and can frequently learn something from the
graph which would not have been apparent from the original data.  Data about a
dependent variable and two independent variables can, in principle, be presented
as a 3-dimensional model.  When there are more than two independent variables,
no complete presentation is possible.

     MULPRES presents a series of graphs, one for each independent variable,
which may be considered either as views through a model in n+1 dimensions
viewed parallel to the hyperplane which represents the regression equation or
as a series of graphs of the dependent variables against each of the independent variables, but showing the effect of each of these variables as if they
were acting alone, each of the others being "adjusted" to have a fixed value.
The facility is helpful when fitting empirical regression models to data with
multiple regression coefficients in the 0.5-0.9 range, in order to recognise
non-linear effects, and also when presenting results of regression analysis
(not necessarily in linear functions) to readers with limited statistical
experience.

     The principle was first used by John Aston about 1960.  Improvements and
modifications have been made from time to time.  The current version of the
program has been used by researchers in a number of departments in the ori-
ginating University, to present various types of laboratory and industrial data.

CAPABILITIES:  Processing and Displaying Data

     The current form of the FORTRAN version is designed to read data from
cards (or card images) prepared for ICL 1900 Statistical Package XDS3, but
would require only minor modfication to read data in other formats.  Additional
cards specify:  how the variables are to be manipulated; the coefficients from
the previously-computed regression equation; and the graphical lay-out required.

     The main output is a series of graphs, each showing a regression line
together with all the data points.  There is one graph for each independent
variable and an additional one for observed-against-predicted values for each
data point.  The co-ordinates of the points are listed, as well as being
plotted.

CAPABILITIES:  Statistical Analysis

     The program does not carry out statistical analysis - that has already
been done by a regression program.  It can, however, manipulate the variables
to provide the commonly-used transformations.  The presentation procedure is
able to handle variables which appear in the regression equation in two forms,
(usually x and $x^2$), or as the product of two variables (e.g. $x_2 * x_4$).

REFERENCES

(The first includes a brief description.  The other two are typical of many
which present graphs produced by the program, but without specific reference to
it):

Aston, J.L., and Muir, A.R.,:  "Factors affecting the Life and Drop Forging
     Dies", J. Iron and Steel Inst. 1969, Vol. 207, pp. 167-176.
"Dimensional Tolerances in Castings", Second Report of Technical SubCommittee
     TS71, Brit. Foundryman, 1971, Vol. 64, pp. 364-379.
Balogun, S.A.,  "Relevant Factors in Cavity Press Forging Die and Process
     Design", Metals Technology, 1974, Vol. 1, pp. 501-505.

SAMPLE JOB

     Data from an industrial electro-deposition process has been used.  The
regression equation included 5 significant independent variables (x's).  The
graphs (2 out of the 5 are reproduced) show the effect of each x on y, with
line and points adjusted to the values expected if the other 4 x's had been
held at fixed values.

```
7 8 6 0 0 0
5 7 2 3 4 5 9
1 1 1 1 1 1 1 1            Set graph scales and regression equation.
56 80
55 3.0 36 35 12
85 6.6 60 65 30
70 5 50 50 20
-0.0865 0.987016 -0.111074 -0.20235 0.208016 80.777
0
```

Insert data here

```
              -9.9 0 0 0 0 0 0 0 0 0
3
22
TEMPERATURE   CELSIUS
TEMPERATURE/COPPER IN SOLUTION
20
COPPER IN SOLUTION
TEMPERATURE/ZINC IN SOLUTION
16
ZINC IN SOLUTION                        Set graph titles etc.
TEMPERATURE/CYANIDE IN SOLUTION
7
CYANIDE
TEMPERATURE/HYDROXYL IN SOLUTION
12
HYDROXYL ION
TEMPERATURE/CURRENT DENSITY
15
CURRENT DENSITY
```

Part of Graphical Output

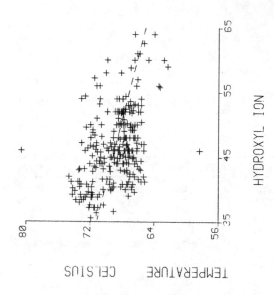

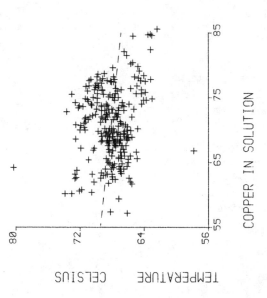

## MULPRES:  Multiple Regression Presentation Program

Developed by:                                    Distributed by:

    Dr. J. L. Aston                              Same
    Dept. of Metallurgy and
     Materials Engineering
    University of Aston in
     Birmingham
    Gosta Green
    Birmingham B4 7ET

Computer Makes:                                  Operating Systems:

    ICL 1904S with CALCOMP Plotter               GEORGE 3
    HP2000 (No plotting)                         ACCESS

Source Languages:                                User Language:

    FORTRAN                                      Unrestricted
    BASIC (No plotting)

Cost:

    Listing of FORTRAN source program - $250.  Listing of BASIC program - no charge.

Documentation:

    Description of principle and outline user instructions available without charge from above address.

## 9.09  STATSPLINE

### INTRODUCTION

The STATSPLINE programs compute and plot "density" curves derived from observed data; the theory (see below) uses derivatives of cubic splines, supplemented with "reflection/absorption" boundary conditions. The Histospline program transforms a histogram to a density curve and also, optionally, provides a subdivision of the histogram based on this curve. The Dataspline program similarly transforms a set of (perhaps weighted) individual observations to a density curve. Here each observation is replaced by a "unit" histogram with just one filled cell:  this is centred at the observation, has content 1, and is of width = SPAR (the spline parameter); thus the data-points are replaced by a histogram with possible overlapping cells.

### DETAILS

A unit histogram on $(-\infty, \infty)$ (which in Dataspline will represent an observed data-point) is replaced by a unit "histospline" having zero integrals over the empty cells and unit integral over the one filled cell, and maximum first-order smoothness. The final histospline is to depend additively on the data, so it will be formed by adding together shifted unit histosplines (called "delta-splines"). The reflection-type boundary conditions impose the condition that the data lives naturally on (i) the whole line, or (ii) a half-line, or (iii) an interval, or (iv) a circle. The absorption-type boundary conditions allow one to force zero density at one or more boundary points.

The purpose of the programs is to throw light on the form of the density generating the data in those regions where that density is not negligibly small. In the low-density regions the ordinate of the density curve will be small, and may fluctuate in sign. The programs are <u>not</u> intended to be used for precision estimation of individual ordinates.

A device is included to assist judgement in the choice of the value for the spline parameter SPAR. Each program can use a printer-plot subroutine to give a rough plot of the density curve on a line printer. They are also designed to use a plotter to give more accurate plots of the histogram, data values, and density curves. Parameters to guide the transformations and the form of the output are given by means of keywords with associated values.

### PROPOSED ADDITIONS IN NEXT YEAR

Two-dimensional extensions exist in prototype form and it is planned to include these when development is complete.

STATSPLINE

Distributed by:

    John A. Lambert
    Computing Centre
    University of Newcastle
    NSW  2808  Australia

Computer Makes:                          Operating Systems:

    IBM 370                              OS, MVT
    ICL 1900                             George 3
    PDP 11                               RSTS/E
    VAX 11/780                           VMS

Source Language:

    FORTRAN

Cost:

    The one-time charge for universities, governments and nonprofit organizations, and commercial use is \$A50.00 (materials).

Documentation:

Boneva, L.I., Kendall, D.G., Lambert, J.A., & Silverman, B.W.,  The STATSPLINE
    program (Version 3)  (Obtainable from the distributor named above.)

References:

Boneva, L.I., Kendall, D.G., & Stefanov, I.,  Spline transformations:  three
    new diagnostic aids for the statistical data analyst (with discussion).
    J. Roy. Statist. Soc.  (B)33(1971) 1-70.
Kendall, D.G.,  Pole-seeking brownian motion and bird navigation.  J. Roy.
    Statist. Soc.  (B)36(1974) 365-417.
Silverman, B.W.,  Density ratios, empirical likelihood, and cot death.  Appl.
    Statist. 27(1978) 26-33.
Silverman, B.W.,  Choosing the window width when estimating a density.
    Biometrika, 65(1978) 1-11.

## 9.10  ALLOC

### INTRODUCTION

ALLOC is a package for non-parametric multigroup discriminant analysis. Its most characteristic feature is that the densities within each group are estimated non-parametrically using kernel functions (Parzen density estimators). The aim of discriminant analysis is to study for several groups their differences (or overlap) in variability of a p-dimensional measurement vector X. The distributional assumptions, made to describe this variability, are of fundamental importance. ALLOC offers an alternative to parametric analyses in situations where the distributional assumptions of a parametric analysis are either violated or difficult to check (e.g. equality of covariance matrices in the usual linear discriminant analysis).

### CAPABILITIES:  Statistical Analysis

The present package is designed for continuous variables only (see, however, additions in the next release). Some descriptive statistics like means and standard deviations are displayed. The main output consists, however, of the posterior probabilities which for each sample element X are calculated, and of classification matrices. Prior probabilities and losses (utilities) can be taken into account. There is an option for forced classification of a sample element (to the group with the maximal posterior probability) as well as for classification of doubt (no posterior probability above a threshold). The analysis might be performed with a fixed dimension of the variables as well as with a forward stepwise selection of variables using the error rate as selection criterion. However, this selection of variables branch has to be considered as prohibitive for large scale problems, due to the combination of kernel density estimation and the selection criterion.

### PROPOSED ADDITIONS IN THE NEXT RELEASE (spring 1981)

A second, enlarged, release of ALLOC is expected to be available in spring 1981. Additional features of the 1981 release are the extension from continuous data to mixed data, see (4); the introduction of the variable kernel density estimate, which is especially appropriate for irregular and heavily tailed distribution, see (2), (5); and the calculation of a measure of atypicality in order to detect outlying observations. Moreover, the input will be changed into keyword-style, in order to make interactive use of the package easier. The input-control will be extended.

### REFERENCES

Aitchison, J., and C.G.G. Aitken.  "Multivariate binary discrimination by the kernel method,"  Biometrika, 63 (1976), 413-420. (1)

Breiman, L.W., Meisel and E. Purcell.  "Variable kernel estimates of multivariate densities,"  Technometrics, 19 (1977), 135-144. (2)

Habbema, J.D.F., and J. Hermans.  "Selection of variables in discriminant analysis by F-statistic and error rate,"  Technometrics, 19 (1977), 487-493.  (3)

Habbema, J.D.F., J. Hermans, and J. Remme.  "Variable kernel density estimation in discriminant analysis,"  Proc. Comp. Stat. COMPSTAT 1978. L.C.A. Corsten and J. Hermans (eds.), Physica Verlag, Wien (1978).(4)

Raatgever, J.W., J.D.F. Habbema and J. Hermans.  "Estimation of the kernel
     width in the variable kernel model," Technical Report. Dept. of medical
     statistics, University of Leiden (1978). (5)
Remme, J, J.D.F. Habbema and J. Hermans.  "A simulative comparison of linear,
     quadratic and kernel discrimination," J. Statist. Comput. Simul. 11 (1980)
     87-106. (6)
Renaker, A.C., and S.F. Larson.  "Bias in Wilks'Λ in stepwise discriminant ana-
     lysis," Technometrics, 22 (1980), 349-356.  (7)

SAMPLE JOB

   - arbitrary text
   - input parameters (fixed columns)
   - sample sizes
   - priors (optional)
   - losses (optional)
   - format cards (F-format)
   - data

Input for ALLOC-3, version October 1975

| Control cards | Format | Columns | Variables | Code | Comments |
|---|---|---|---|---|---|
| 1 | I5 | 1-5 | N-JOBS | | Total number of runs |
| 2 | 20A4 | | TEKST | | Arbitrary text |
| 3 | 10I5 | 1-5 | NG | | Total number of groups |
| | | 6-10 | NTG | | Number of training groups |
| | | 11-15 | NV | | Number of variables |
| | | 20 | NSIG | 0 | Vector (length NTG of by the program |
| | | | | 1 | Vector of smooth-ness pa |
| | | 25 | MLOSS | 0 | Loss matrix element (i, |

## Partial Output

```
     TEST   ALLOC-3
   INPUT PARAMETERS :
        3     3     3     0     0     1     0     1     0     4     0.800

   DATA OF TRAINING GROUPS


   1  1           24.000        1.000        90.000
   1  2            5.000        0.0         285.000
   1  3            2.000        4.000         7.000
   1  4            9.000        2.000       434.000
   1  5            5.000        5.000        48.000
```

NUMBER OF GROUPS      3      TRAINING GROUPS      3      TEST GROUPS      0

NUMBER OF VARIABLES      3

THRESHOLD PROBABILITY FOR ALLOCATION OF DOUBT :  0.800

NUMBER OF CASES/GROUP

| GROUP | 1 | 2 | 3 |
|---|---|---|---|
| NUMBER | 20 | 17 | 20 |

PRIOR PROBABILITY FOR EACH GROUP
     1:  0.3333      2:  0.3333      3:  0.3333

POSTERIOR PROBABILITIES

| IDENTI-FICATION | OBJ. NR. | TRUE GROUP | ALLO-CATED GROUP | | | | | |
|---|---|---|---|---|---|---|---|---|
| 1  1 | 1 | 1 | D | 1: 0.5379 | 2: 0.4541 | 3: 0.0080 |
| 1  2 | 2 | 1 | D | 1: 0.3927 | 2: 0.6062 | 3: 0.0011 |
| 1  3 | 3 | 1 | D | 1: 0.0929 | 2: 0.1480 | 3: 0.7591 |
| 1  4 | 4 | 1 | 1 | 1: 0.9011 | 2: 0.0989 | 3: 0.0000 |

ALLOCATION MATRIX

| ALLOCATED GROUP | TRUE GROUP | | |
|---|---|---|---|
| | 1 | 2 | 3 |
| 1 | 4 | 0 | 1 |
| 2 | 0 | 3 | 0 |
| 3 | 1 | 1 | 2 |
| D | 15 | 13 | 17 |
| TOTAL | 20 | 17 | 20 |

ALLOCATION MATRIX EXPRESSED IN PERCENTAGES

| ALLOCATED GROUP | TRUE GROUP | | |
|---|---|---|---|
| | 1 | 2 | 3 |
| 1 | 20.0 | 0.0 | 5.0 |
| 2 | 0.0 | 17.6 | 0.0 |
| 3 | 5.0 | 5.9 | 10.0 |
| D | 75.0 | 76.5 | 85.0 |

ALLOC, Version 1975

Developed by:                                      Distributed by:

    Department of Medical Statistics              Centraal Reken Instituut
    University of Leiden                          P.O. Box 9512
    Wassenaarseweg 80                             2300 RA LEIDEN
    P.O. Box 9512                                 The Netherlands
    2300 RA  LEIDEN
    The Netherlands

Computer Makes:                                    Operating Systems:

    IBM 370/158                                   MVS
    Amdahl V 7 B                                  MVS
    CDC Cyber                                     NOS

Language:

    Fortran

Cost:

    $100 for universities; $500 for others.  The fee includes the program
source text and a user's manual.

Documentation:

J. Hermans and J.D.F. Habbema, Manual for the ALLOC discriminant analysis pro-
    grams, (1976), Department of Medical Statistics, University of Leiden.

Contents:

    *      An outline of discriminant analysis (19 p.)
    *      Theoretical background of the ALLOC programs (5 p.)
    *      A comparison with other programs (3 p.)
    *      Input description for the computer programs (13 p.)
    *      Input and output of test examples (12 p.)
    *      References (2 p.)

## 9.11  POPAN

### INTRODUCTION

POPAN is designed to be an analysis and data maintenance system for mark-recapture data from open animal populations (subject to dilution or losses or both; analyses for closed populations are better handled by the programs of Otis et al. 1978).  The system is intended for uniquely marked animals whose complete capture history and (optional) attributes (age, sex, species, etc.) are known.  Full allowance is made for removals (loss-on-capture) and experimentally controlled additions (injections) of animals.  Program capabilities include:  data maintenance and selection of data subsets; calculation of statistics and estimates; and simulation of sampling experiments.  Tasks are specified by user-written paragraphs and keywords, similar in structure to that used by BMDP.  Capabilities are listed below by paragraph name.

### CAPABILITIES:  Data Maintenance and Selection

CREATE allows user to specify attribute and sample time codes and descriptions.  Raw data is read in and edited (range and error checks) and used to create a POPAN file (like an SPSS or SAS file, with a descriptive header and case (animal) data in compressed form).  ADD allows addition of new animal histories or additional attribute and capture data to existing histories in a POPAN file.  MODIFY allows alteration or deletion of history or header information on an existing file.  LIST causes display of a file in raw data form (string of capture times of each animal) or binary form (string of bits, $b(i)$, $i = 1...K$) where $b(i) = 1$ if caught at time $i$, $b(i) = 0$ if not caught).  SELECT specifies the existing POPAN file to be used for subsequent LIST, STATISTICS, or ANALYSIS.  The user can select the entire file or specify subgroups based on attribute classes or sample times or both (see example below).

### CAPABILITIES:  Statistical Analysis

STATISTICS permits accumulation of count data for user-specified classes of capture histories (see example below), using a very general syntax (verbs include SEEN, FIRSTSEEN, LASTSEEN, JUSTSEEN, NEXTSEEN, LOST, INJECTED).

ANALYSIS provides 16 pre-programmed Jolly-Seber type analyses to estimate abundance, survival rate, and/or births.  Options allow for death or birth or both or neither and include 2 pooling options to increase precision by combining information among sample times.  Automatic deletion of null sample times and full error-trapping is carried out.  All estimates include estimated standard errors.

SIMULATE allows simulation of sampling experiments with user specified number of animals, capture, birth and survival rates.  Simulated histories are passed to ANALYSIS and may be saved.  Replication is allowed.  Simulation with the usual Jolly-Seber assumptions satisfied are useful for planning experiments and predicting precision.  Violation of assumptions (tag loss, non-permanent emigration, heterogeneity in capture or survival rate) may be simulated to assess bias in the estimates.

## EXTENSIBILITY

User-written ANALYSIS programs can be invoked and may access a current or saved STATISTICS table, and/or ANALYSIS results.

## INTERFACES WITH OTHER SYSTEMS

None, although STATISTICS can produce frequency counts required for the Otis et al. closed population analyses.

## PROPOSED ADDITIONS IN NEXT YEAR

A TEST paragraph for goodness-of-fit tests and model selection.

## REFERENCES

Jolly, G.M., Book Review of POPAN2 - A data maintenance and analysis system for mark-recapture data by A.N. Arnason and L. Baniuk, Biometrics 35(2) (1979) p. 527.

Otis, D.L., K.P. Burnham, G.C. White, and D.R. Anderson. Statistical inference from capture data on closed animal populations. Wildl. Monographs 62 (1978).

## SAMPLE JOB

```
C SELECT A PREVIOUSLY CREATED FILE (ON UNIT 12) OF VOLE CAPTURE
C HISTORIES...24 CAPTURE TIMES AND 3 ATTRIBUTES (FOR GENUS (A1=M/C),
C AGE (A2=1/2), AND SEX (A3=M/F/BLANK). SELECT A SUBSET OF THESE.

    SELECT:
       TITLE = '1974 ADULT MICROTUS WITH SAMPLE TIME SUBSET' ;
       INPUT = 12 ;  WRITE = BOTH ;

C OMIT FIRST WEEK AND MIDWEEK AM AND PM SAMPLES.  GROUP AM AND PM
C SAMPLES FOR REMAINING (ONCE A WEEK) SAMPLES.

       BEGIN = 5 ;  END = 22;  OMIT = (7,8,11,12,15,16,19,20) ;
       GROUP = (5,6),(9,10),(13,14),(17,18),(21,22) ;

C SELECT ONLY ADULT MICROTUS AND SAVE (UNSORTED)

       ATTRIBUTE = (A1.EQ.'M' .AND. A2.EQ.'2') ;
       SAVE = ASIS;  DATASET = 13 ;/

    ANALYSIS:
       TITLE = 'FIRST ANALYSIS ALLOWING BIRTH AND DEATH' ;
       DILUTION = PRESENT ;  LOSSES = PR ;  POOLING = NONE ;/

C COLLECT STATISTICS FOR TEST OF NO DILUTION (BIRTHS) IN THE PRESENCE
C OF LOSSES ... SEE K.H. POLLOCK ET AL.  BIOMETRICS 30:77-88 (1974).

    STATISTICS:
       TITLE = 'TABLE OF COUNTS FOR TEST OF NO DILUTION GIVEN LOSSES' ;
       NUMBER = 4 ;  SCAN = FULL ;
       SYMBOL = ' S(I) ' ;  DESCRIPTION = 'MARKED CAPTURES AT TIME I' ;
       CONDITION = 'SEEN AT (I) AND SEEN BEFORE (I)' ;
       SY=' T(I) ' ; DE='MARKS KNOWN ALIVE AT I+1: JOLLY"S Z(I)+R(I)' ;
       CO = '(SEEN BEFORE (I) AND NOT SEEN AT (I) AND SEEN AFTER (I)) OR
             (SEEN AT (I) AND SEEN AFTER (I) )' ;
       SY = ' N(I) ' ;  DE = 'SAMPLE SIZE AT TIME I' ;
       CO = 'SEEN AT (I)' ;
       SY=' C(I) ' ; DE='DISTINCT SIGHTINGS AFTER I: JOLLY"S Z"(I)+R(I)' ;
       CO = '(SEEN BEFORE (I) AND NOT SEEN AT (I) AND SEEN AFTER (I)) OR
             (FIRSTSEEN AFTER (I)) OR (SEEN AT (I) AND SEEN AFTER (I))' ;/
```

Partial Output from SAMPLE JOB:   The SELECT and STATISTICS Paragraphs.

```
***********************
*                     *
*  SELECT             *     SUMMER 1974 VOLE DATA FROM BURN AREA
*  PARAGRAPH #    1   *     1974 ADULT MICROTUS WITH SAMPLE TIME SUBSET
*                     *
***********************
```

| IDENTIFIER TYPE | # ATTRIBUTES | BEGIN | END | LSEL | # HISTORIES | FILE ORDERED |
|---|---|---|---|---|---|---|
| NUMERIC | 3 | 1 | 24 | 24 | 449 | YES |

MAP OF NEW NUMBERING FOR SAMPLE TIMES

```
   OLD NUMBER                                NEW NUMBER

   OX(I)        SDES(I)      ABS. TIMES      NX(I)     NEW SDES(I)     ABS. TIMES
     1        JULY  1/AM      1.0000            0        OMITTED          0.0
     2        JULY  1/PM      1.5000            0        OMITTED          0.0
     3          NULL          5.0000            0        OMITTED          0.0
     4        JULY  5/PM      5.5000            0        OMITTED          0.0
     5        JULY  8/AM      8.0000            1        JULY  8/AM       8.2500
     6        JULY  8/PM      8.5000            0        GROUP AS   1     0.0
     7        JULY 12/AM     12.0000            0        OMITTED          0.0
     8        JULY 12/PM     12.5000            0        OMITTED          0.0
     9        JULY 15/AM     15.0000            2        JULY 15/AM      15.2500
    10        JULY 15/PM     15.5000            0        GROUP AS   2     0.0
    11        JULY 19/AM     19.0000            0        OMITTED          0.0
    12        JULY 19/PM     19.5000            0        OMITTED          0.0
    13        JULY 22/AM     22.0000            3        JULY 22/AM      22.2500
    14        JULY 22/PM     22.5000            0        GROUP AS   3     0.0
    15        JULY 26/AM     26.0000            0        OMITTED          0.0
    16        JULY 26/PM     26.5000            0        OMITTED          0.0
    17        JULY 29/AM     29.0000            4        JULY 29/AM      29.2500
    18        JULY 29/PM     29.5000            0        GROUP AS   4     0.0
    19        AUG.  2/AM     33.0000            0        OMITTED          0.0
    20        AUG.  2/PM     33.5000            0        OMITTED          0.0
    21        AUG.  5/AM     36.0000            5        AUG.  5/AM      36.2500
    22        AUG.  5/PM     36.5000            0        GROUP AS   5     0.0
    23        AUG.  9/AM     40.0000            0        OMITTED          0.0
    24        AUG.  9/PM     40.5000            0        OMITTED          0.0
```

```
        OLD LSEL (ORIGINAL DATA) =    24

        NEW LSEL (REDUCED DATA)  =     5
```

ATTRIBUTE SELECTION CONDITION WAS:
   AT =(A1.EQ.'M' .AND. A2.EQ.'2')

```
***********************
*                     *
*  STATISTICS         *     SUMMER 1974 VOLE DATA FROM BURN AREA
*  PARAGRAPH #    1   *     1974 ADULT MICROTUS WITH SAMPLE TIME SUBSET
*                     *     TABLE OF COUNTS FOR TEST OF NO DILUTION GIVEN LOSSES
***********************
```

STATISTICS TABLE
================

| I | S(I) | T(I) | N(I) | C(I) |
|---|---|---|---|---|
| 1 | 0 | 17 | 32 | 117 |
| 2 | 9 | 22 | 41 | 90 |
| 3 | 13 | 27 | 35 | 73 |
| 4 | 16 | 26 | 42 | 46 |
| 5 | 26 | 0 | 46 | 0 |

```
# HISTORIES SCANNED USING FULL        SCAN          =        132

# HISTORIES REJECTED ON ATTRIBUTES                  =        247

# HISTORIES REJECTED FOR NO CAPTURES IN (BEGIN,END) =         70

STATISTICS TABLE NOT SAVED...EXECUTION CONTINUING
```

POPAN -2:  POPulation ANalysis. latest revision released:  December 31, 1979.

Developed by:                             Distributed by:

    A.N. Arnason & L. Baniuk           Charles Babbage Research Centre
    Department of Computer Sciences    Box 370
    University of Manitoba             St. Pierre, Manitoba  ROA 1V0
    Winnipeg, Manitoba R3T 2N2
    Canada

Computer Makes:                           Operating Systems:

    IBM 360, 370 and                  IBM - OS/VS, MVS, MT
      compatible machines               - DOS/VS
    (Amdahl V7, V8)

Source Languages:  FORTRAN (IBM)

Cost:

    A one-time cost of $US 125 is charged for the installation package (for
all classes of users).  Package includes tape containing source, load modules,
JCL procs, test programs, data, and output plus installation instructions and
user manual.  User manual is available separately at $US 25.00.

Documentation:

Arnason, A.N. and L. Baniuk.  POPAN-2:  A data maintenance and analysis system
    for mark-recapture data.  Charles Babbage Research Centre, Box 370,
    St. Pierre, Manitoba.  1978  viii + 269 pp.

## 9.12  CAPTURE

### INTRODUCTION

CAPTURE is a large applications program which is used to calculate capture-recapture population estimates for closed populations. Its principal applications are to wildlife and fisheries work. It was developed originally by personnel of the Utah Cooperative Wildlife Research Unit, Utah State University, in 1976-77, and has been updated slightly in 1980. The program is mostly for batch use, and is currently installed on 4 machine brands.

### CAPABILITIES:  Capture-Recapture Estimates

Program CAPTURE is designed to calculate many of the population estimates and hypothesis tests for closed population capture-recapture data useful to the biologist. TASK READ CAPTURES is used to read a set of data, and perform inital summaries of animal movement. TASK MODEL SELECTION is then used to select the model which best fits the data based on the observed variation of capture probabilities. TASK POPULATION ESTIMATE then calculates an estimate based on the model selected.

Capabilities to simulate capture-recapture data and to estimate animal density are also available.

### EXTENSIBILITY

The program is written in a fairly modular form, but adding new procedures would require a good knowledge of FORTRAN and some understanding of the program structure.

### PROPOSED ADDITIONS IN NEXT YEAR

The current FORTRAN IV version will be translated to FORTRAN 77, and additional flexibility of data input will be added.

### SAMPLE JOB

This program titles the output, reads a set of capture data with identifying information, calculates a test for closure, selects the appropriate model for the data, and calculates a population estimate and density for the grid.

```
TITLE='EXAMPLE INPUT FOR TABLE II'
TASK READ CAPTURES   OCCASIONS=10 SUMMARY FILE=17
DATA='DATA FROM ERIC LARSEN, PEROMYSCUS MANICULATUS, &
     PARACHUTE CREEK, COLORADO.'
TASK CLOSURE TEST   OCCASIONS=3-10
TASK MODEL SELECTION    OCCASIONS=3-10
TASK POPULATION ESTIMATE APPROPRIATE REMOVAL OCCASIONS=3-10
TASK DENSITY ESTIMATE APPROPRIATE INTERVAL=15  &
     CONVERSION=10000 INNER GRID X=4-7 Y=4-7 OCCASIONS=3-10
MIDDLE INNER GRID X=3-8 Y=3-8 OCCASIONS=3-10
MIDDLE OUTER GRID X=4-9 Y=4-9 OCCASIONS=3-10
ENTIRE GRID X=1-10 Y=1-10 OCCASIONS=3-10
END OF GRID DEFINITIONS
```

## Sample Output

The following is the output generated for one of the population estimates.

FREQUENCIES OF CAPTURE, F(I)

| I= | 1 | 2 | 3 | 4 | 5 | 6 | 7 | 8 |
|----|---|---|---|---|---|---|---|---|
| F(I)= | 85 | 33 | 21 | 12 | 6 | 1 | 0 | 0 |

COMPUTED JACKKNIFE COEFFICIENTS

|   | N(1) | N(2) | N(3) | N(4) | N(5) |
|---|------|------|------|------|------|
| 1 | 1.875 | 2.625 | 3.250 | 3.750 | 4.125 |
| 2 | 1.000 | .357 | -.625 | -1.696 | -2.661 |
| 3 | 1.000 | 1.000 | 1.372 | 2.098 | 2.964 |
| 4 | 1.000 | 1.000 | 1.000 | .848 | .535 |
| 5 | 1.000 | 1.000 | 1.000 | 1.000 | 1.036 |

THE RESULTS OF THE JACKKNIFE COMPUTATIONS

| I | N(I) | SE(I) | .95 CONF. | LIMITS | TEST OF N(I+1) VS. N(I) CHI-SQUARE(1 D.F.) |
|---|------|-------|-----------|--------|-------------------------------------------|
| 0 | 158 | | | | |
| 1 | 232.4 | 11.81 | 209.2 | 255.5 | 35.957 |
| 2 | 274.9 | 18.84 | 238.0 | 311.8 | 12.878 |
| 3 | 303.4 | 25.80 | 252.9 | 354.0 | 6.197 |
| 4 | 324.0 | 32.78 | 259.8 | 388.2 | 3.693 |
| 5 | 338.7 | 39.20 | 261.9 | 415.5 | 0.000 |

AVERAGE P-HAT = .1153
INTERPOLATED POPULATION ESTIMATE IS  323 WITH STANDARD ERROR
   32.3504
APPROXIMATE 95 PERCENT CONFICENCE INTERVAL 259 TO 387

HISTOGRAM OF F(I)

FREQUENCY   85   33   31   12   6   1   0   0
-------------------------------------------------
EACH EQUALS        9 POINTS

```
       81      *
       72      *
       63      *
       54      *
       45      *
       36      *   *
       27      *   *
       18      *   *   *
        9      *   *   *   *   *
-------------------------------------------------
```

Developed by:                           Distributed by:

    Gary C. White                           SHARE Program Library Agency
    Los Alamos Scientific Laboratory        P.O. Box 12076
    MS-495                                  Research Triangle Park
    Los Alamos, NM USA 87545                North Carolina, USA 27709

Computer Makes:                         Operating Systems:

    Burroughs 6700                          IBM - OS, VS, TSO
    CDC 6000 series                         CDC - NOS, LTSS
        7000 series                         Burroughs 6700
        CYBER series
    IBM 360, 370

Source Languages:  FORTRAN IV, 77

Cost:

    A one time charge of $40.  The fee includes program source modules and 13 example data sets taken from Otis et al. (1978).

Documentation and References:

White, G.C., K.P. Burnham, O.L. Otis, and D.R. Anderson.  User's Manual for
    Program CAPTURE.  Utah State University Press, Logan, UT. 40 p. (1978)
Otis, D.L., K.P. Burnham, G.C. White, and D.R. Anderson.  Statistical Inference
    from Capture Data on Closed Animal Populations.  Wildlife Monograph
    No. 62:1-135 (1978).

CHAPTER 10


MULTI-WAY CONTINGENCY TABLE ANALYSIS PROGRAMS


## CONTENTS

## INTRODUCTION

As noted in the Introduction to Chapter 8, Chapters 6 and 7 contain general-purpose statistical packages, while Chapters 8 to 12 describe specific-purpose programs.  The seven programs in this chapter were written with just one purpose in mind, the analysis of multi-dimensional contingency tables.

The ratings by the respective developers on selected items are displayed in Table 10.1.  Figure 10.1 compares developers' and users' ratings on all items, while Figure 10.2 summarizes developers' ratings (D-scores) and users' ratings (U-scores) on several relevant attributes.  These are explained in detail in Section 1.4 of Chapter 1.  Figure 10.2 also points to other chapters which describe programs with multi-way table analysis capabilities.  In particular, it can be seen that a number of the additional programs on Figure 10.2 and the more general-purpose programs in Figure 7.2, most notably GLIM and BMDP, can analyze multi-way tables with as much flexibility and with considerably more convenience, than these special-purpose programs.

Two programs in this chapter offer unusual features.  GUHA searches for the sources of dependencies in the tables and thereby generates hypotheses, and performs hierarchicial decompositions.  MLLSA fits latent structure models to these multi-way contingency tables.

The remainder of this chapter contains descriptions of these multi-way table analysis programs in the format described in Table 1.5.

## TABLE 10.1: RATINGS BY DEVELOPERS ON SELECTED ITEMS

### (i) Capabilities

Usefulness Rating Key:
3 - high
2 - moderate
1 - modest
. - low

| | Complex Structures | File Management | Consistency Checks | Probabilistic Checks | Compute Tables | Print Tables | Multiple Regression | Anova/Linear Model | Linear Multivariate | Multi-way Tables | Other Multivariate | Nonparametric | Exploratory | Robust | Non-linear | Bayesian | Time Series | Econometric |
|---|---|---|---|---|---|---|---|---|---|---|---|---|---|---|---|---|---|---|
| | 11 | 14 | 18 | 19 | 24 | 25 | 30 | 31 | 32 | 33 | 34 | 36 | 37 | 38 | 39 | 40 | 35 | 41 |
| 10.01 GUHA | . | . | 1 | . | . | . | | . | . | 2 | 2 | . | . | . | . | . | . | . |
| 10.02 CATFIT | . | 1 | . | . | 2 | 1 | | . | . | 3 | . | . | . | . | . | . | . | . |
| 10.03 TAB-APL | 1 | . | . | . | 1 | 1 | | . | . | 2 | . | . | . | . | . | . | . | . |
| 10.04 MULTIQUAL | 2 | . | . | . | . | . | | . | . | 3 | . | . | . | . | . | . | . | . |
| 10.05 C-TAB | . | . | 1 | . | 1 | 1 | | . | . | 2 | . | . | . | . | . | . | . | . |
| 10.06 ECTA | . | . | . | . | . | 1 | | . | . | 3 | . | . | . | . | . | . | . | . |
| 10.07 MLLSA | . | . | . | . | . | . | | . | . | 2 | 1 | . | . | . | . | . | . | 1 |

## FOR ANALYSIS OF MULTI-WAY TABLES PROGRAMS

(ii) User Interface

| | Survey Estimates | Survey Variances | Math Functions | Operations Research | Availability | Installations | Computer Makes | Mini Version | Core Requirements | Batch/Interactive | Stat. Training | Computer Training | Language Simplicity | Documentation | User Convenience | Maintenance | Tested for Accuracy |
|---|---|---|---|---|---|---|---|---|---|---|---|---|---|---|---|---|---|
| | 43 | 44 | 47 | 48 | 49 | 50 | 51 | 52 | 53 | 54 | 55 | 56 | 57 | 58 | 59 | 60 | 61 |
| GUHA | 1 | . | . | . | 2 | 2 | 2 | 2 | . | . | 1 | 2 | . | 2 | 2 | 3 | 1 |
| CATFIT | . | . | . | . | 3 | 1 | . | 3 | 3 | 2 | 3 | 3 | 2 | 1 | 3 | 1 | 1 |
| TAB-APL | . | . | . | . | 2 | 3 | 2 | 3 | 3 | 2 | 1 | 3 | . | 1 | 1 | 1 | 2 |
| MULTIQUAL | . | . | . | . | 2 | 3 | 3 | 1 | 1 | . | 1 | 3 | 2 | 1 | 3 | 3 | 2 |
| C-TAB | . | . | . | . | 2 | 1 | 2 | . | 2 | . | 1 | 3 | . | 1 | 1 | 1 | 2 |
| ECTA | . | . | . | . | 2 | 3 | 3 | 1 | 2 | . | 1 | 3 | . | 1 | 2 | 1 | 2 |
| MLLSA | . | . | . | . | 2 | 2 | 2 | . | . | . | 1 | 2 | . | 2 | 1 | 3 | 2 |

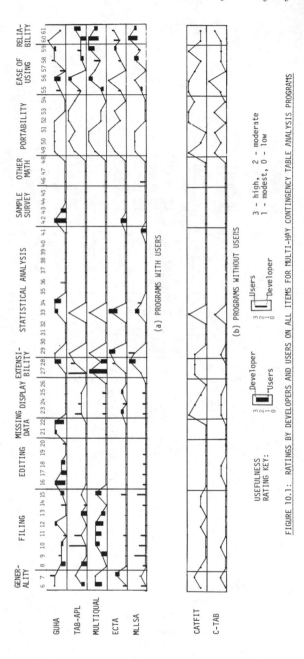

FIGURE 10.1: RATINGS BY DEVELOPERS AND USERS ON ALL ITEMS FOR MULTI-WAY CONTINGENCY TABLE ANALYSIS PROGRAMS

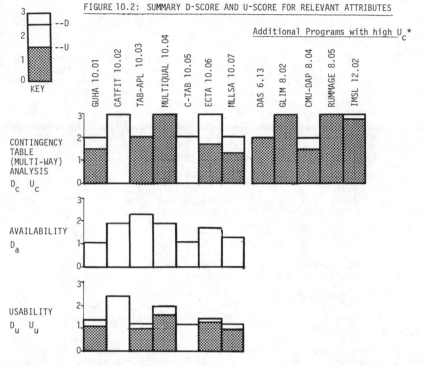

FIGURE 10.2:  SUMMARY D-SCORE AND U-SCORE FOR RELEVANT ATTRIBUTES

Another chapter generally containing programs with strong
Contingency Table Analysis Capabilities is Chapter 7 (see
Figure 7.2).

*In addition, for these specific programs, $U_c > 1.5$.

10.01  <u>GUHA</u>

## INTRODUCTION

GUHA is intended as a program package complementary to some general statistical packages such as BMDP or SPSS.  Its primary function is the analysis
of multidimensional data of categorical type.  All procedures are intended for
exploratory data analysis in complex situations (i.e. for generating hypotheses
to be tested in further studies.)  It is currently used in social, biomedical
and industrial research, mainly for analysing complex questionnaires.  GUHA was
developed in the late sixties on MINSK 22 computers.  In the last three years
it was completely rewritten and innovated and it is still under continuous
development.  The present version can be used in batch and is installed at over
twenty computer installations in Czechoslovakia and USSR.

## CAPABILITIES:  Processing and Displaying Data

The GUHA INPUT program is used to create GUHA system files.  It is designed
to detect and report errors in the input data and variables description.

The DICHOT procedure transforms categorical or continuous variables into
two-valued variables needed for ASSOC and IMPL programs.

The REPORT program rewrites under user's control results of ASSOC and IMPL
programs.  (Results are extensive and complex).

## CAPABILITIES:  Statistical Analysis

No basic statistics; only specific procedures for handling categorical and
multiway data.

The ASSOC program searches for positive association between variables
derived from multiway two-valued tables.  (The version for general categorical
data is under development).  The scope of the search and comprehension of
results is under the user's control.

The IMPL program searches for high conditional probabilities of some
derived variables conditioned by others (the same data types as ASSOC).

The CORREL program searches for high conditional rank correlations between
continuous variables conditioned by derived two-valued variables.

The COLLAPS program searches for the source of dependency in general two-
way contingency tables.

Automatic handling of missing data items.  All search procedures are
optimized using logical and AI means.

## EXTENSIBILITY

User-written programs can be added to the system without difficulty.

## INTERFACES WITH OTHER SYSTEMS

GUHA can read BMDP files.

## PROPOSED ADDITIONS IN NEXT YEAR

English documentation: English-based, free field, alpha-numeric user command language for all programs; enhancements of procedures ASSOC and IMPL. Rewriting of COLLAPS program for multiway tables: inclusion of Wermuth's procedure TASEL. New procedure for user-defined transformations of input data and updating of GUHA files. New procedure DRANK for searching groups defined by derived two-valued variables having maximal distance w.r.t. some continuous variables.

## REFERENCES

Abrham, J., S. Tuma, P. Stepanek, "Waist narrowing of the bladder in children," Radiol. Clin. Biol. (Basel), 39 (1970), 348-356.
Geizerova, H., D. Hejl, Z.Grafnetter, M. Santrucek, "Prevalence of ischaemic heart disease in a man aged 50-54 years; processing by the GUHA method," Review of Czechoslovak Medicine, 20 (1974), 85-95.
Pauckova, V. "A mathematical approach to the results of experimental studies," Scripta fac. sci. nat. UJEP Brunensis, Biologia, 3 (1973), 1-6.
Renc, Z., K. Kubat, J. Kourim, "An application of the GUHA method in medicine," Int. J. Man-Machine Studies, 10 (1978), 29-35.

## SAMPLE JOB

Find dependencies in an RXC table. Perform hierarchical decomposition, permute rows and columns and print the resulting table (with adjusted residuals). Input data frequencies.

```
// EXEC COLLAPS
//SYSIN DD *
    TITLE='SNEE' 'S DATA : EYE/HAIR COLOR.'
    PRINT='BATH,ALL'     HIER='YES'
    NAME1='EYE COLOR'
    NAME2='HAIR COLOR'
    NAMES1='BROWN,HAZEL,GREEN,BLUE'
    NAMES2='BLACK,BRUNETTE,RED,BLONDE'
    SIZE1=4     SIZE2=4    N=3  ;
```

Partial Output:

SOLUTION:
    3 HYPOTHESES WERE FOUND.
------------------------------------------------------------------------------

  (X)F ... LEVELS OF F- (Y)G ... LEVELS OF G-  2 BY 2 TABLE:
  EYE COLOR                 HAIR COLOR
------------------------------------------------------------------------------

HYPOTHESIS NO. 1:

BASIC FORM:

|  |  |  | (Y)G | -(Y)G |  |
|---|---|---|---|---|---|
| 3-GREEN | 4-BLONDE | (X)F | 110 | 169 | 279 |
| 4-BLUE |  | -(X)F | 17 | 296 | 313 |
|  |  |  | 127 | 465 | 592 |

COMPLEMENTARY FORM:

| 1-BROWN | 1-BLACK | CHI-SQUARE | = | 101.1694 |
| 2-HAZEL | 2-BRUNETTE | CHI | = | 10.0583 |
|  | 3-RED | PSI | = | 0.4134 |
|  |  | YULE Q | = | 0.8378 |

------------------------------------------------------------------------------

ADJUSTED STANDARDIZED RESIDUALS (SIGN SCHEMA):
==================================================

  F - EYE COLOR / G - HAIR COLOR

|  | 4 BLONDE | 1 BLACK | 2 BRUNETTE | 3 RED |
|---|---|---|---|---|
| 3 GREEN | 0.73 | -2.29 | -0.51 | 2.58 |
| 4 BLUE | 9.97 | -4.25 | -3.40 | -2.31 |
| 1 BROWN | -8.33 | 6.14 | 2.16 | -0.10 |
| 2 HAZEL | -2.74 | -0.58 | 2.05 | 0.99 |

### GUHA, Version 79.1:   General Unary Hypotheses Automaton

Developed by:                                          Distributed by:

    GUHA group                                          Same
    Czechoslovak Cybernetics
    Society, CSAV -
    Center of Biomathematics
    CS 142 20 Prague
    Videnska 1083

Computer Makes:                                        Operating Systems:

    IBM 360, 370                                        OS,VS
    EC 1040
    M 4030
    ICL System 4 (restricted
    version)

Source Languages:   FORTRAN (with assembler for bit string operations), PL/1.

Cost:

    Distributed on exchange basis.

Documentation:

Hajek, P.,  T. Havranek,  Mechanizing Hypothesis Formation, Springer-Verlag,
    Berlin-Heidelberg-New York, 1978.
Hajek, P.(ed.)  Special issue devoted to the GUHA method, Int. J. Man-Machine
    Studies, 10 (1978), no.1.
Hajek, P., T. Havranek, M. Chytil,  Metoda GUHA automatickeho generovani
    hypotez, Academia Prague (in print).

10.02 <u>CATFIT</u>

## INTRODUCTION

This interactive program was written by James A. Davis for the analysis of multidimensional contingency tables, including the estimation of D-system parameters, and odds-ratios.  It operates on the Dartmouth Time Sharing System, and on a PDP 11/70 under UNIX.  In the Dartmouth version it interfaces with IMPRESS.

<u>CATFIT</u>

<u>Developed by</u>:

    James A. Davis
    Department of Sociology
    William James Hall
    Harvard University
    Cambridge, MA   02138

<u>Computer Makes</u>:

    PDP 11/70

<u>Source Languages</u>:

    BASIC

<u>Cost</u>:

    A one-time charge to cover costs, between $50 and $100.

<u>Distributed by</u>:

    D. Garth Taylor
    National Opinion Research Corp.
    Chicago, Illinois
    312-753-1300

<u>Operating Systems</u>:

    UNIX

## 10.03 <u>TAB-APL</u>

### INTRODUCTION

TAB is a workspace of APL programs for the interactive fitting of log-linear models to tabular data. It was developed by John Fox in 1978, and has been distributed to over 100 users.

### CAPABILITIES: Statistical Analysis

The TAB workspace consists of three main functions, ENTERTABLE, TABLE, and LOGLIN, together with a number of subfunctions, documentation functions, and global variables. ENTERTABLE facilitates keyboard entry of frequency tables and variable labels. TABLE constructs frequency tables from case-by-variable data sets. LOGLIN is a conversational program for fitting and testing loglinear models; it prints out tables of observed and expected frequencies, model parameters, and other statistics. Program activity is directed by keyboard input and by global APL variables that may be set at the user's option. The conversational nature of the program makes it suitable for users who are unfamiliar with APL. The TAB workspace includes functions that print out user documentation.

### TAB-APL

Developed by:

    John Fox
    Department of Sociology
    York University
    4700 Keele Street
    Toronto, Ontario
    M3J 1P3

Distributed by:

    Same

Computer Makes:

    Developed for DEC-10
    other computers with
    similar APL interpreters

Operating Systems:

    Any supporting an APL interpreter.

Source Language: APL

Cost: Free

Documentation: supplied and on line.

Sample Terminal Session for TAB


      *LOGLIN FIENBERG3Δ2*
*LOGLINEAR MODELS FOR CONTINGENCY TABLES    J.FOX [10/78]*
*MAXIMUM CYCLES = 10    CONVERGENCE CRITERION = 0.1    STANDARD START-VALUES*
*NUMBER OF VARIABLES = 3  SHAPE OF INPUT TABLE: 2 2 2*
*PRINT INPUT TABLE? [Y/N]*
*Y*

```
 VAR A=  1
ROW |   COLUMN
VAR B | VAR C
                1            2
     1         15           18
     2         48           84
--------------------------
 VAR A=  2
ROW |   COLUMN
VAR B | VAR C
                1            2
     1         21            1
     2          3            2
```

*SPECIFY MODEL: MARGINALS TO BE FIT*
*ENTER MARGINALS , ONE PER LINE ('OUT' WHEN DONE)*
□:
      1 2
□:
      1 3
□:
      2 3
□:
      OUT
*CONVERGENCE AFTER 4 CYCLES (12 ITERATIONS)*
*MAXIMUM ABSOLUTE CELL DIFFERENCE ON LAST CYCLE = 0.07488324748*

*L.R.CHISQ =      2.71    G.F.CHISQ =      3.02    DF = 1*

*PRINT: [1]EXPECTED FREQS [2]CELL RESIDS [3]EFFECTS? [E.G.,YNY]*
*NNN*

*ANOTHER MODEL? [Y/N]*
*N*

## 10.04  MULTIQUAL


INTRODUCTION

The MULTIQUAL program implements the highly general form of logit and log-linear analysis described in Chapter 8 of Bock, R.D., Multivariate Statistical Methods in Behavioral Research, New York: McGraw-Hill, 1975, and in Bock, R.D., Multivariate Analysis of Qualitative Data, Chicago, International Educational Services, (in press).  The program fits models to frequency data, evaluates the goodness of fit, estimates effects and their standard errors, and computes expected proportion under the model both in the cells and in the margins of cross classifications.  In version II of the program, core storage and computing time required have been greatly reduced, a provision for restricting effects has been added, and the output formats improved.  The program operates in batch using an improved version of the MULTIQUAIL command language.

CAPABILITIES:  Statistical Analysis

The MULTIQUAL program  fits logit and log-linear models by means of pre-liminary minimum logit chi-square estimation, followed by Newton-Raphson itera-tions to attain the maximum likelihood solution.  Models are constructed from abstract design and model formulas and fitted in a hierarchical manner.  Both goodness of fit chi-squares and components of chi-square for testing hypotheses concerning terms added to the hierarchical model are calculated.  The user has control over the number of iterations, including stopping at the stage of the minimum logit chi-square estimates.  Estimates and standard errors are given for the solution in all cases.

The program specializes to five (5) types of analysis of qualitative data: 1) tests a specified proportion, 2) complex contingency table analysis, 3) bino-mial response relations, 4) multinomial response relations, and 5) models for ordered response categories.  For any fitted model, cell proportions and mar-ginal proportions can be estimated as aids to interpretation.  Where there is lack of fit, standardized and adjusted residuals of cell proportions may be calculated.

The program provides for the case of structural zeros in the frequency table and also for restricting specific effects in the model to be zero.  The program accepts original data in the form of attributes, which may be coded arbitarily, or summary data in the form of frequencies.  Data may be in either fixed or column-free format, or in a binary file.  Results and many intermediate calculations may be saved in formatted or unformatted form.

REFERENCES

Bock, R.D., (1975) Multivariate Statistical Method in Behavioral Research, New
    York:  McGraw-Hill, 1975.
Bock, R.D., (1980) Multivariate Analysis of Qualitative Data, Chicago:  Inter-
    national Educational Services (in press).

SAMPLE JOB

The following problem is a test of homogeneity of association in a 2 X 2 cross-classification of responses from four populations arranged in a 9-lence sampling design.  This is a multinominal response relations problem.

## Job Listing

```
>TITLE TEST OF HOMOGENEITY OF ASSOCIATION
>COMMENT DATA FROM ASHFORD AND SOWDEN (1970).  "MULTIVARIATE
PROBIT ANALYSIS,' BIOMETRICS,26,535-546.
       NUMBERS OF SUBJECTS IN NINE AGE GROUPS REPORTING
SYMPTOMS, BREATHLESSNESS AND/OR WHEEZE.
       THERE ARE 9 SAMPLE GROUPS AND 4 RESPONSE CATEGORIES.
       THE AGE*BREATHLESSNESS*WHEEZE TERM WILL BE DELETED FROM THE S-R
       MODEL.
>PROCEDURE MULTINOMIAL;
>S-WAY;
   9,AGE,'20-4','25-9','30-4','35-9','40-4','45-9',
'50-4','55-9','60-4';
>R-WAYS;
     2,BREATH,YES,NO;
     2,WHEEZE,YES,NO;
>S-MODEL   I+AGE;
>R-MODEL BREATH+WHEEZE+BREATH*WHEEZE;
>DELETE AGE*BREATH*WHEEZE;
>S-ESTIMATE DF=2;
>R-ESTIMATE   ALL=1;
>S-CONTRAST;
     AGE,P,2;
>S-TEST   TERMS=(I,AGE)
>R-TEST TERMS=(WHEEZE,'BREATH*WHEEZE')
>MARGIN EXPECTED PROPORTIONS LOGITS;
     AGE*BREATH,AGE*WHEEZE;
>LIST FULL WIDTH=80,NSIG=4;
>SAVE OBSERVED EXPECTED PROPORTIONS ESTIMATES SE COVMAT;
>INPUT GROUPS IGNORE=1;
20-24 1 9 7 95 1841
25-29 2 23 9 105 1654
30-34 3 54 19 177 1863
35-39 4 121 48 257 2357
40-44 5 169 54 273 1778
45-49 6 269 88 324 1712
50-54 7 404 117 245 1324
55-59 8 406 152 225 967
60-64 9 372 106 132 526
>STOP
```

MULTIQUAL II: Log-Linear Analysis of Nominal or Ordinal Qualitative Data
by the Method of Maximum Likelihood

Developed by:                               Distributed by:

R. Darrell Bock                             International Educational Services
Department of Behavioral Science            1525 E. 53rd Street - Suite 824/829
University of Chicago                        Chicago, Illinios  60615
Chicago, Illinois  60637

George R. Yates
Texet
Chicago, Illinois  60615

Computer Makes:                             Operating Systems:

IBM 360/370                                 IBM - VS2, SVS, MVS
CDC 6000 Series                             CDC - NOS, SCOPE
BURROUGHS                                   BURROUGHS - MCP III
DEC 10                                      DEC 10 - TOPS 10
UNIVAC                                      UNIVAC - EXEC 8

Source Language: FORTRAN IV

Cost:

The one-time license fee for MULTIQUAL II source code is $518 for univer-
sities; $673 for government and non-profit organizations; and $673 for all
others.

Documentation:

Bock, R.D., MULTIQUAL II, Log-Linear Analysis of Nominal or Ordinal Qualitative
Data by the Method of Maximum Likelihood.

Bock, R.D., Multivariate Analysis of Qualitative Data, Chicago, International
Educational Services, (in press).

10.05  <u>C-TAB</u>

## INTRODUCTION

C-TAB is a program  written by Shelby J. Haberman for analyzing multi-dimensional contingency tables.  The program computes maximum likelihood estimates, standardized residuals, asymptotic variances and chi-square statistics for hierarchical log-linear models through iterative proportional fitting and iterative adjustment.

### C-TAB:  Contingency Table Analysis

<u>Developed by</u>:                          Distributed by:

    Shelby J. Haberman                    International Educational Services
    5734 S. University                    1525 E. 53rd St.
    Chicago, IL  60637                    Chicago, IL  60615

<u>Machine</u>:

    IBM 370
    CDC
    UNIVAC

<u>Language</u>:

    FORTRAN

<u>Cost</u>:

    One-time charge of $350.

10.06 <u>ECTA</u>

### INTRODUCTION

ECTA is a program designed to assist in the analysis of cross-tabulated data. Specifically, the program is able to fit log-linear models to cross-tabulated data using iterative proportional fitting. There is a large amount of statistical literature on the log-linear model; Goodman (1978) and Bishop, Fienberg, and Holland (1975) are just two of several excellent references in this area. The documentation for ECTA assumes that the user has become familiar with this statistical technique. Robert Fay wrote the initial version of the program in 1973 while a graduate student in Statistics at the University of Chicago.

### CAPABILITIES:  Statistical Analysis

The hierarchical log-linear models fitted by ECTA are stated in terms of the main effects and interactions among the variables that are cross-classified. The user supplies the tabulated data as part of the card input to the program. The program fits models requested by the user and calculates the appropriate Pearson and likelihood ratio chi-square test statistics. Optionally, the fitted values for the cells will be displayed with the original data.

Upon request, ECTA will compute and display the estimated log-linear parameters. Approximate standard errors are also given. (These standard errors represent an estimate for the asymptotic variance under the "saturated" model only, but generally they will approximate the standard errors under the fitted model as well.)

The program will also allow the user to fit tables with "structural" zeros, cells that are deliberately zero because of the design of the analysis.

There are no specific limits on table size or the number of dimensions, except the overall restriction imposed by available computer storage. The documentation details the few changes required to increase available storage to enable ECTA to handle fairly large tables, in the range of several thousand cells.

### PROPOSED ADDITIONS

The original developer is preparing a version of ECTA for the analysis of data from complex sample surveys, using replication methods. The new program is being written in FORTRAN-77.

### REFERENCES  (Citing ECTA)

Goodman, Leo A. (1978)  <u>Analyzing Qualitative/Categorical Data</u>.  Cambridge, Mass.:  Abt Associates Inc.
Upton, Graham J.G. (1978) <u>The Analysis of Cross-Tabulated Data</u>.  New York: John Wiley & Sons.

### ECTA:  Everyman's Contingency Table Analysis

Developed by:                                Distributed by:

    Robert Fay                                   Department of Statistics
    Statistical Methods Division                 University of Chicago
    U.S. Bureau of the Census                    1118 E. 58th Street
    Washington, D.C.  20233                      Chicago, Illinois  60637

Computer Makes:                              Operating Systems:

    IBM                                          Requires only standard FORTRAN
    CDC                                          support.
    Univac
    others
    (program in generally portable FORTRAN)

Source Language:

    FORTRAN

Cost:

    Commercial                            $35
    Non-profit, government, academic   $35
    (Foreign higher)

Documentation:

    From Distributor.

## 10.07 MLLSA

### INTRODUCTION

MLLSA is designed to estimate a wide range of unrestricted and restricted latent structure models for contingency tables involving polytomous variables. It employs an iterative-proportional-scaling algorithm to derive maximum likelihood estimates. It can check for the identifiability of parameter estimates and assign respondents into latent classes. The program is useful for analysis involving discrete indicators of underlying traits, for scaling models, and for path models involving both observable and unobservable variables.

### CAPABILITIES: Processing and Displaying Data

MLLSA requires a contingency table as input; it is not capable of reading raw data files. It is not a program for displaying data in the form of plots.

### CAPABILITIES: Statistical Analysis

The parameters of the latent class model, including the latent class proportions and the conditional probabilities ("recruitment probabilities"). Chi-square statistics (goodness-of-fit and likelihood-ratio), index of dissimilarity between observed and estimated expected frequencies. Assignment of respondents into latent classes, as well as indexes which measure the quality of the assignment. Degrees of freedom for both identifiable and unidentifiable models, the rank of a Jacobian matrix used to determine identifiability. Statistics for the independence model (useful in many contexts as a baseline).

### EXTENSIBILITY

Latent class models for tables of any dimension can be considered by altering the DIMENSION statements.

### INTERFACES WITH OTHER SYSTEMS

Not presently available.

### PROPOSED ADDITIONS IN THE NEXT YEAR

*     Improve efficiency by adding subroutines for special cases.
*     Add a subroutine that will automatically calculate start values for the parameter estimates.

### REFERENCES

Clogg, C.C., "New Developments in Latent Structure Analysis," in D. Jackson and E. Borgatta (eds.), Factor Analysis and Measurement in Sociological Research. New York: Sage (1980).
Clogg, C.C., "Latent Structure Models of Mobility," American Journal of Sociology 86, forthcoming (1981).

Clogg, C.C., "Some Latent Structure Models for the Analysis of Likert-Type
    Data," Social Science Research 8:243-272 (1979).
Clogg, C.C. and D.O. Sawyer. "A Comparison of Some Alternative Models for
    Analyzing the Scalability of Response Patterns," in S. Leinhardt (ed.),
    Sociological Methodology 1981. San Francisco: Jossey-Bass (1981).
Goodman, L.A., "Exploratory Latent Structure Analysis Using Both Identifiable
    and Unidentifiable Models." Biometrika 61:2150-231 (1974).

SAMPLE JOB

    The following input cards can be used to estimate the unrestricted 2-class
model.  See the Goodman reference above.

   Model H1 (Unrestricted 2-Class), Stouffer-Toby Data, 1974 Biometrika.

```
    4 2    216
    2 2 2 2
   (8F3.0)
    42 23  6 25   6  24   7  38
     1  4  1  6   2   9   2  20
   .25      .75
   .75      .25       .10       .90
   .90      .10       .30       .70
   .90      .10       .35       .65
   .95      .05       .70       .30
```

Partial Output from Sample Job

```
FINAL LR      2.720 PRSN     2.721   INDEX OF DISSIMILARITY      0.0387
FINAL LATENT CLASS PROBABILITIES
0.2796724  0.7203276
FINAL CONDITINAL PROBABILITIES
LATENT CLASS=           1        2        3        4
1.         1.         0.7687   0.1322
1.         2.         0.2313   0.8678
2.         1.         0.9262   0.3538
2.         2.         0.0738   0.6462
3.         1.         0.9394   0.3294
3.         2.         0.0606   0.6706
4.         1.         0.9931   0.7135
4.         2.         0.0069   0.2865
ASSIGNMENT OF RESPONDENTS TO LATENT CLASS J GIVEN MANIFEST RESPONSE I
CELL=   OBSERVED= EXPECTED=  ASSIGN TO CLASS=   MODAL P=
  1        42.       41.83          1           0.9591
  2        23.       23.30          1           0.5183
  3         6.        6.32          1           0.5060
  4        25.       21.47          2           0.9551
  5         6.        6.07          2           0.5733
  6        24.       23.63          2           0.9670
  7         7.        6.56          2           0.9685
  8        38.       41.81          2           0.9985
  9         1.        0.97          2           0.7101
 10         4.        4.59          2           0.9816
 11         1.        1.28          2           0.9825
 12         6.        8.24          2           0.9992
 13         2.        1.42          2           0.9872
 14         9.        9.18          2           0.9994
 15         2.        2.55          2           0.9994
 16        20.       16.77          2           1.0000
PERCENT CORRECTLY ALLOCATED =   90.31 NUMBER CORRECTLY ALLOCATED=
195.07    LAMBDA=  0.6536

NUMBER OF ESTIMATED PARAMETERS=  9  DEGREES OF FREEDOM IF IDENTIFIED = 6

COLUMN RANK=   9 DEGREES OF FREEDOM=   6
```

### MLLSA:  Maximum Likelihood Latent Structure Analysis

Developed by:                              Distributed by:

   Leo A. Goodman                          Clifford C. Clogg
   Department of Sociology
   University of Chicago
   1126 E. 59th Street
   Chicago, Illinois 60637

   Clifford C. Clogg
   Department of Sociology
   Penn State University
   University Park, Pa. 16802

Computer Makes:                            Operating Systems:

   IBM 370                                 IBM-OS, VS, MVS
   CDC
   ICL

Source Language:  FORTRAN

Cost:

   One-time charge of $20.  Fee includes manual and example outputs.

Documentation:

Clogg, C.C.,  "Unrestricted and Restricted Maximum Likelihood Latent Structure
    Analysis:  A Manual for Users,"  Population Issues Research Center, Working
    Paper 1977-04, University Park, Penna.  16802.

CHAPTER 11


ECONOMETRIC AND TIME SERIES PROGRAMS


CONTENTS

INTRODUCTION

    This chapter describes another seven specific-purpose programs as distinct
from the general-purpose packages of Chapters 6 and 7.  The specific purpose is
the estimation of econometric and time series models.

    The ratings by the respective developers on selected items are displayed in
Table 11.1.  Figure 11.1 compares developers' and users' ratings on all items,
while Figure 11.2 summarizes developers' ratings (D-scores) and users' ratings
(U-scores) on several relevant attributes.  These are explained in detail in
Section 1.4 of Chapter 1.  Figure 11.2 also points to other chapters which
describe programs with strong econometric capabilities.  Chapter 7 particularly
contains several programs with strengths in this area.

    It can be seen from the taxonomy of Table 1.2 that some of these programs
also include other statistical procedures, most notably linear and non-linear
modelling.  QUAIL estimates a multinomial logit model.  KEIS/ORACLE combines an
economic data base system with its capabilities for estimating large econometric
models.  Two versions of TSP are included, one having been adapted for the
DATATRAN language.

    A casual reader may wonder whether the names of some of these programs,
particularly ORACLE and SHAZAM, reflect in any way the Delphic nature of econometric predictions.  Indeed throughout this book there appear several other
programs with names suggestive of supernatural origins and implications:  ISIS,
OSIRIS (two programs), LEDA, and TROLL.  This reader might also wonder whether
such names as SPEAKEASY, CAPTURE, and QUAIL are reminders of the inherent dan-
ger associated with incautious use of packaged statistical programs.

    The remainder of this chapter contains descriptions of these econometric
and time series programs in the format described in Table 1.5.

TABLE 11.1:   RATINGS BY DEVELOPERS ON SELECTED ITEMS

(i) Capabilities

Usefulness Rating Key:
3 - high
2 - moderate
1 - modest
. - low

| | Complex Structures | File Management | Consistency Checks | Probabilistic Checks | Compute Tables | Print Tables | Multiple Regression | Anova/Linear Model | Linear Multivariate | Multi-way Tables | Other Multivariate | Nonparametric | Exploratory | Robust | Non-linear | Bayesian | Time Series | Econometric |
|---|---|---|---|---|---|---|---|---|---|---|---|---|---|---|---|---|---|---|
| | 11 | 14 | 18 | 19 | 24 | 25 | 30 | 31 | 32 | 33 | 34 | 36 | 37 | 38 | 39 | 40 | 35 | 41 |
| 11.01 TSP/DATATRAN | 1 | 2 | 1 | . | . | . | 3 | 2 | 2 | . | 2 | 2 | . | . | 2 | 2 | 3 | 3 |
| 11.02 B34S | . | 3 | 1 | . | . | . | 3 | 3 | . | . | . | . | . | . | 3 | 2 | 3 | 3 |
| 11.03 PACK | . | . | 1 | . | . | . | 3 | . | . | . | . | . | . | . | 1 | . | 3 | 2 |
| 11.04 SHAZAM | 1 | 1 | . | . | 1 | . | 3 | 2 | 2 | . | 1 | . | . | 1 | 2 | . | 1 | 3 |
| 11.05 TSP | 1 | 1 | . | . | 1 | . | 2 | . | 1 | . | . | . | . | . | 2 | . | 1 | 3 |
| 11.06 QUAIL | 3 | 1 | 1 | 1 | 1 | 1 | 1 | . | . | 1 | . | . | . | . | . | . | 1 | 3 |
| 11.07 KEIS/ORACLE | . | 2 | 1 | . | 1 | 2 | 2 | . | . | . | . | . | . | . | . | . | 1 | 2 |

## FOR ECONOMETRIC ANALYSIS PROGRAMS

### (ii) User Interface

| | Survey Estimates | Survey Variances | Math Functions | Operations Research | Availability | Installations | Computer Makes | Mini Version | Core Requirements | Batch/Interactive | Stat. Training | Computer Training | Language Simplicity | Documentation | User Convenience | Maintenance | Tested for Accuracy |
|---|---|---|---|---|---|---|---|---|---|---|---|---|---|---|---|---|---|
| | 43 | 44 | 47 | 48 | 49 | 50 | 51 | 52 | 53 | 54 | 55 | 56 | 57 | 58 | 59 | 60 | 61 |
| TSP/DATATRAN | . | . | 2 | . | 1 | . | 2 | . | . | 3 | 2 | 3 | 3 | 1 | 3 | 1 | 1 |
| B34S | . | . | . | . | 2 | 2 | 2 | . | . | . | . | 3 | . | 2 | 2 | 3 | 2 |
| PACK | . | . | . | 1 | 3 | 3 | 3 | 1 | . | . | . | 3 | . | 2 | 2 | 3 | 2 |
| SHAZAM | . | . | 1 | . | 3 | 3 | 2 | 1 | . | 3 | 2 | 3 | 2 | 2 | 3 | 3 | 3 |
| TSP | . | . | 1 | . | 3 | 3 | 3 | 2 | . | . | 2 | 3 | 2 | 2 | 2 | 3 | 2 |
| QUAIL | . | . | 1 | . | 3 | 3 | 2 | 1 | . | 1 | 1 | 2 | 2 | 2 | 3 | 1 | 1 |
| KEIS/ORACLE | . | . | . | . | 1 | 1 | 1 | 1 | 1 | 1 | . | 3 | 2 | 3 | 3 | 3 | . |

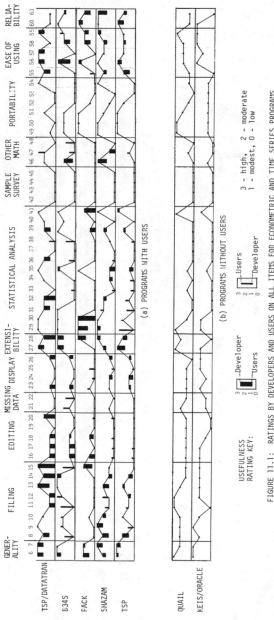

FIGURE 11.1: RATINGS BY DEVELOPERS AND USERS ON ALL ITEMS FOR ECONOMETRIC AND TIME SERIES PROGRAMS

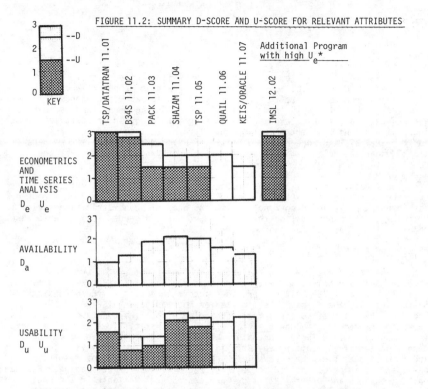

FIGURE 11.2: SUMMARY D-SCORE AND U-SCORE FOR RELEVANT ATTRIBUTES

Another chapter generally containing programs with strong Econometric Capabilities is Chapter 7 (see Figure 7.2).

*In addition, for this specific program, $U_e > 1.5$.

### 11.01 <u>TSP/DATATRAN</u>

INTRODUCTION

The TSP/DATATRAN system is a large, diversified collection of statistical procedures, which are available through an interactive computer language. Originally designed to operate in a batch system and to carry out econometric time-series processing (thus the name TSP), TSP/DATATRAN has been steadily revised and expanded in response to changing computer technology in order to meet the increasing demand from the academic community for a battery of research tools for the analysis and forecasting of time series. TSP/DATATRAN is now operational under the Multics time-sharing system (Honeywell) as a part of the Consistent System.

The first version of TSP was written at M.I.T. by Prof. Robert E. Hall. Over the years, TSP was taken to many different computer centers. New routines have been steadily added most notably in the area of econometric time-series processing. This orientation towards the methods and needs of time series has remained the principal function of TSP/DATATRAN.

PROCESSING AND DISPLAYING DATA

As described elsewhere in this book the Consistent System provides data storage, a data handling system, and a large variety of statistical and mathematical programs. TSP/DATATRAN provides automatic access to these Consistent System data storage files. The time series methodology of TSP/DATATRAN is backed up by a set of routines for rearranging and defining data. These transformation procedures include arithmetic operators, statistical as well as arithmetic functions, and logical operators which make it possible to write conditional statements subsetting any array of data needed for analysis. The amount and form of output can be controlled; graphics are available for displaying data; input can be controlled for either free or fixed format reading. Further backup routines and other statistical procedures are available in the Consistent System. A simple TSP/DATATRAN command makes data and results available to other programs in the Consistent System.

STATISTICAL ANALYSIS

Symbolic formula manipulation: both single equations and systems of equations (linear and nonlinear); Regression: ordinary, Bayesian, polynomial distributed lag, ARMA (Box-Jenkins), autoregressive, logit, ridge, two stage least squares or instrumental variables, three stage least squares, full information maximum likelihood, and non-linear, Residual analysis: Frequency domain analysis: spectral and cross-spectral analysis, fast Fourier transforms, complex arithmetic. Seasonal adjustment: moving average ratio method.

Analysis of variance; Principal components and eigenvalues; Factor analysis. Non-parametric tests.

Model building: symbolic manipulation of systems of equations (differentiation, division into linear blocks for recursive solution, determination and isolation of non-linearities); linear and non-linear solution of systems of equations.

Matrix operations:  inversion, generalized inversion, Cholesky decomposition, convolution, orthogonalization.  Macro facilities:  IF statements, DO
loops, logical and absolute branching.

THE DATATRAN LANGUAGE:

The DATATRAN language, for which this version of the TSP system has been
adapted, is designed to separate the user from the problems traditionally
inherent in moving back and forth from one program or system to another.  Any
transferring of control of data is done implicitly so that it is unnecessary to
know whether or not all of the functions being used are within the same package.
In DATATRAN, all statistical programs and matrix operations are considered as
simple or multi-valued functions which can be nested and/or included in an
arbitrary algebraic expression.

## TSP/DATATRAN: Time Series Processor

Developed by:                               Distributed by:

    John Brode                              Consistent System
    23 Berkeley Street                      Renaissance Computing, Inc.
    Cambridge, MA   02138                   P.O. Box 699
                                           Cambridge, MA   02139

Machine:                                    Operating Systems:

    Honeywell 6180                          Multics
    IBM 370/165 OS                          OS

Language:

    FORTRAN IV

Cost:

    Costs are negotiable.  See distributor above.

Documentation:

    A Manual is available from I.P.S. Publications Office, Building 39-484, M.I.T., Cambridge, Massachusettes, 02142.

## 11.02  B34S


### INTRODUCTION

B34S is an applications program which is typically used by Econometricians. The B34S program has been developed since 1972 by Houston H. Stokes with the help of others.  The aim has been to provide a vehicle whereby the best code developed by others in stand-alone programs could be combined into an integrated package.

### CAPABILITIES:  Processing and Displaying Data

The B34S program will read data from a sequential file, observation by observation for all variables, and from a Random Access Data Bank whose basic structure is compatable with Wharton Econometric software.  B34S will merge data from various sources, change the frequency of data and plot, graph and list data in various formats.

### CAPABILITIES:  Statistical Analysis

Basic statistics include means, variances, standard deviation, maximum and minimum.  Statistical techniques include ordinary least squares, weighted least analysis, Bayesian Analysis, Error Component Analysis, LIML, 2SLS, 3SLS, I3SLS. Limited Dependent Variable analysis routines include PROBIT, LOGIT and TOBIT Analysis.  Time Series Transfer Function and Intervention analysis and Vector Autoregressive and Moving Average models.  Extensive OLS specification tests are provided including Recursive Residual analysis, normality and heterosedasticity tests and dynamic specifications tests.  With the exception of routines developed by Jennings for Simultaneous Equations Estimation, all linear algebra is performed using LINPACK and all eigen analysis using EISPACK.

B34S provides the capability to estimate Markov Probability models and provides an extensive decompositions of the transition matrices as suggested by Theil.  B34S has the capability for non-linear estimation using the Meeter routine GAUSHAUS where the user has the option of dynamically branching to a library of subroutines which can be coded for the user's specific model.

### EXTENSIBILITY

Since B34S allows dynamic branching to external load modules, users can code their procedures in a FORTRAN subroutine and extend B34S.

### INTERFACES WITH OTHER SYSTEMS

B34S will read sequential files in F, E, Z, free and A8 format and files in unformated double precision.  B34S will also read a random access data bank. B34S users can input SPEAKEASY, BRAP, TSP, SAS and SPSS control cards for a branch to these packages for features not found in B34S.

### PROPOSED ADDITIONS IN NEXT YEAR

The MINIPACK nonlinear routines will be added as options to GAUSHAUS.

REFERENCES

Henry, Neil, John McDonald and H. Houston Stokes, "The Estimation of Dynamic Economic Relations from a Time Series of Cross Sections: A Programming Modification," Annals of Economic and Social Measurement. V(1), 1976, pp. 153-155.
Jennings, L., "Simultaneous Equations Estimation," in Journal of Econometrics, Annals of Applied Econometrics 1980-1, Vol. 12 (1980) pp. 23-39.
Kosobud, Richard and Houston H. Stokes, "Simulation of World Oil Shocks: A Markov Analysis of OPEC and Consumer Behavior," The Energy Journal, VI #2, 1980, pp. 55-84.
Metter, D., "Problems in the analysis of nonlinear models by least squares," Unpublished Ph.D. Thesis Univ. of Wisconsin 1964.
Stokes, Houston H., and Hugh Neuburger, "The Effect of Monetary Changes on Interest Rates: A Box-Jenkins Approach," Review of Economics and Statistics, LXI(4), November 1979, pp. 534-548.
Theil, H., "Social Mobility and Social Distance: A Markov Chain Approach," in H. Theil Statistical Decomposition Analysis, Amsterdam North Holland 1972.

SAMPLE JOB

```
EC     3   10 2 1     04020309
ERROR COMP TEST OUTPUT ON DATA NOT APPROPRIATE FOR EC ANALYSIS
```

## Sample Output

```
     EC ANALYSIS, IREAD SET TO     O ON EC CARD


   ERROR COMP TEST OUTPUT ON DATA NOT APPROPRIATE FOR EC ANALYSIS


   THE NUMBER OF INDIVIDUALS OR REGIONS IN THE CROSS SECTION IS     3
   THE NUMBER OF PERIODS IN THE TIME SERIES IS     1O
   THE NUMBER OF INDEPENDENT VARIABLES (EXCLUDING THE CONSTANT) IS     2

      2O8 WORDS OF  258OO AVAILABLE USED.
   ERROR COMP TEST OUTPUT ON DATA NOT APPROPRIATE FOR EC ANALYSIS

   OLS REGRESSION OF   Y(1)   ON LEVELS:


   COEFFICIENT OF DETERMINATION, R**2 = O.9148
   SUM OF SQUARED RESIDUALS =    2O.471
   STANDARD ERROR OF ESTIMATE =    .87O74
   1/COND OF MATRIX XPX    .94159305D-O1
```

| VARIABLE COEFFICIENT | | STANDARD ERROR | T-RATIO |
|---|---|---|---|
| X(1) | 2.225447 | .1916351 | 11.6129 |
| X2)**2 | .9680021 | .1007758 | 9.6055 |
| CONSTANT | 3.199276 | .2034959 | 15.7216 |

```
    RHOHAT CALCULATED FROM OLS EQUATION  O.136O

   ERROR COMP TEST OUTPUT ON DATA NOT APPROPRIATE FOR EC ANALYSIS

   OLS REGRESSION OF   Y(1)   ON TRANSFORMED VARIABLES:


   COEFFICIENT OF DETERMINATION, R**2 = O.9132
   SUM OF SQUARED RESIDUALS =    21.614
   STANDARD ERROR OF ESTIMATE =    .89471
   1/COND OF MATRIX XPX    .144O8851
```

| VARIABLE COEFFICIENT | | STANDARD ERROR | T-RATIO |
|---|---|---|---|
| X(1) | 2.167378 | .1865960 | 11.6154 |
| X2)**2 | .9547788 | .9782830D-O1 | 9.7597 |
| CONSTANT | 2.144968 | .1827808 | 11.7352 |

## B34S Data Analysis Program

Developed by:                                   Distributed by:

    Dr. Houston H. Stokes                            Same
    Department of Economics
    University of Illinois
    Box 4348
    Chicago, Illinois 60680
    (312)-996-2684

Computer Makes:                                 Operating Systems:

    IBM 360, 370                                     MVS, CMS
    DEC 20 version at U of
    Chicago not supported by
    Houston H. Stokes

                                            Interface Systems:

                                          TSP, SAS, SPSS, SPEAKEASY, BRAP

Source Languages:  Primarily FORTRAN, assembler for dynamic core allocation and dynamic branch to other load modules.

Cost:

    B34S is available to commercial organizations upon payment of an annual retainer of $1000.00 to Dr. Houston H. Stokes. The fee includes program load module (for IBM 370), test data, manual, subroutine libraries in load module form, installation instructions and telephone consultation to enable organization to handle a limited number of routine questions concerning the use and operation of B34S. The university and government fee is negotiable and in the area of $500.00.

Documentation:

Stokes, Houston H., "The B34S Data Analysis Program: A Short Writeup" Report FY 77-1, College of Business Administration Working Paper Series, University of Illinois at Chicago Circle, revised 14 July 1980, approximately 130 pages.
Thornber, Hudson, "Manual for (B34T, 8 Mar 66): A Stepwise Regression Program." Center for Mathematical Studies in Business and Economics Report 6603 (March 66) Univ. of Chicago, pp. 58 with supplements "BLUS Addendum to Technical Report 6603..." 1 August 1968, pp. 12 and "BAYES Addendum to Technical Report 6603..." 12 September 1967, pp. 14.

## 11.03  PACK

### INTRODUCTION

PACK is a stand-alone program which can be used for the analysis of time series models using the Box-Jenkins philosophy.  It is a batch program which requires the user to specify input parameters in the input file.  The program is quite general, correspondingly there are many options which must be explictly specified.

The program has been modified to run interactively and it has been installed in over 600 computer installations.

The program contains two main programs which read input parameters and call appropriate subroutines for computation and display of results.

### CAPABILITIES:  Statistical Analysis

The Box-Jenkins analyses performed by the program are:
1) Univariate time series model identification.
2) Univariate time series model estimation, diagnostic checking, forecasting.
3) Single input transfer function model identification.
4) Single input transfer function model estimation, diagnostic checking, forecasting.
5) Multiple input transfer function model identification (assuming independence of inputs).
6) Multiple input transfer function model estimation, diagnostic checking, forecasting.
7) Intervention model estimation, diagnostic checking, forecasting.

Programs dated April, 1972 from University of Wisconsin performed only Analyses 1 and 2 above.  Program dated February, 1974 from Ohio State added partial capability in Analyses 3 and 4.  Program dated December, 1974 from Ohio State performed Analyses 1, 2, 3, and 4 completely.  The present program is shipped in one and only one form, and performs all seven of the analyses listed above.

In addition, supplemental routines exist to:
*    Generate good starting values for Transfer Function Estimation
*    Express Univariate Box-Jenkins Models as weighted sum of past
*    Express Transfer Function and Noise model as ordinary regression equation
*    Compute variance of aggregated forecasting (Univariate and Multivariate)

REFERENCES

Pack, D.J., Goodman, M.L., and Miller, R.B., "Computer Programs for the Analy-
    sis of Univariate Time Series Using the Methods of Box and Jenkins",
    Technical Report #296, Dept. of Statistics, University of Wisconsin,
    Madison (April, 1972).

SAMPLE JOB

Input

```
      1
    157
(20 F4.3)
245 623 7223 1026 1326 5025 2723 3426 0628 5029 6130 0032 3032 1031 6531 4031 1330 4233 1631 6534 04
357 835 9133 3731 0225 9815 6213 5411 2610 4608 8810 9621 6862 4842 7932 7562 8142 8372 7122 8522 960
285 1324 7324 3335 8399 8411 7420 9457 2443 6395 4343 9324 4339 2264 1239 6228 6248 9242 6238 42272
230 2240 824 2023 2722 8823 5922 6824 0223 0423 5024 5826 1727 4627 5227 1927 3526 9427 1929 452837
279 2275 1280 3285 629 1429 1628 9729 0929 2029 9531 4 3332 0337 9345 3352 2352 2352 9353 2353 53484
348 2347 8347 9350 6352 7357 5362 4385 6382 8392 9394 2393 2389 5381 0383 1383 6391 2403 2408 24 362
459 646 704 6264 611 464 2453 9485 5493 2535 6538 7534 4500 7475 9455 4428 8385 2364 0348 0430 84275
445 1458 84 7625 0125 0814 9695 1445 3655 6215 5445 3825 0955 2025 3345 4925 9166 177
U S TREASURY BILLS INTEREST RATE, MONTHLY JANUARY 1956 THROUGH JANUARY 1969
      4
    .0000   .2500   .5000   .7500  1.0000
      1     1     0
      1     1
      2
      1     2
    .4000   .1000
      0
      1
    .0000   .0040
    100     1
      1
      1
     24     0    12    24     1     0
     24     2     3     4     1
     85    157
605059005750
THIS IS END OF THIS TEST DATA SET
```

Output

END OF ESTIMATION FOR MODEL  1
SUMMARY OF MODEL  1

\*\*\*\*\*\*\*\*\*\*\*\*\*\*\*\*\*\*\*\*\*\*\*\*\*\*\*\*\*\*\*\*\*\*\*\*\*\*\*\*\*\*\*\*\*\*\*\*\*\*\*\*\*\*\*\*\*\*\*\*\*\*\*\*\*\*\*\*\*\*\*\*\*\*\*\*\*\*\*\*\*\*\*\*\*\*\*\*\*\*\*\*

DATA  -  Z = U S TREASURY BILLS INTEREST RATE, MONTHLY JANUARY 1956 THROUGH JANUARY 1969          157 OBSERVATIONS

DIFFERENCING ON   Z - 1)  1 OF ORDER   1

\*\*\*\*\*\*\*\*\*\*\*\*\*\*\*\*\*\*\*\*\*\*\*\*\*\*\*\*\*\*\*\*\*\*\*\*\*\*\*\*\*\*\*\*\*\*\*\*\*\*\*\*\*\*\*\*\*\*\*\*\*\*\*\*\*\*\*\*\*\*\*\*\*\*\*\*\*\*\*\*\*\*\*\*\*\*\*\*\*\*\*\*

UNIVARIATE MODEL PARAMETERS

\*\*\*\*\*\*\*\*\*\*\*\*\*\*\*\*\*\*\*\*\*\*\*\*\*\*\*\*\*\*\*\*\*\*\*\*\*\*\*\*\*\*\*\*\*\*\*\*\*\*\*\*\*\*\*\*\*\*\*\*\*\*\*\*\*\*\*\*\*\*\*\*\*\*\*\*\*\*\*\*\*\*\*\*\*\*\*\*\*\*\*\*

| PARAMETER NUMBER | PARAMETER TYPE | PARAMETER ORDER | ESTIMATED VALUE | LOWER LIMIT 95 PER CENT | UPPER LIMIT |
|---|---|---|---|---|---|
| 1 | AUTOREGRESSIVE 1 | 1 | 0.37875E+00 | 0.21781E+00 | 0.53970E+00 |
| 2 | AUTOREGRESSIVE 1 | 2 | 0.63108E-01 | -.99614E-01 | 0.22583E+00 |

\*\*\*\*\*\*\*\*\*\*\*\*\*\*\*\*\*\*\*\*\*\*\*\*\*\*\*\*\*\*\*\*\*\*\*\*\*\*\*\*\*\*\*\*\*\*\*\*\*\*\*\*\*\*\*\*\*\*\*\*\*\*\*\*\*\*\*\*\*\*\*\*\*\*\*\*\*\*\*\*\*\*\*\*\*\*\*\*\*\*\*\*

OTHER INFORMATION AND RESULTS

\*\*\*\*\*\*\*\*\*\*\*\*\*\*\*\*\*\*\*\*\*\*\*\*\*\*\*\*\*\*\*\*\*\*\*\*\*\*\*\*\*\*\*\*\*\*\*\*\*\*\*\*\*\*\*\*\*\*\*\*\*\*\*\*\*\*\*\*\*\*\*\*\*\*\*\*\*\*\*\*\*\*\*\*\*\*\*\*\*\*\*\*

RESIDUAL SUM OF SQUARES     0.68820E+01     154 D.F.     RESIDUAL MEAN SQUARE     0.44688E-01

NUMBER OF RESIDUALS                    157                RESIDUAL STANDARD ERROR     0.21140E+00
AUTOCORRELATION FUNCTION

DATA  -  THE ESTIMATED RESIDUALS -  MODEL 1                                     157 OBSERVATIONS

ORIGINAL SERIES
MEAN OF THE SERIES =0.14064E-01
ST. DEV. OF SERIES =0.20889E+00
NUMBER OF OBSERVATIONS =  157

1- 12   -0.00   0.01  -0.02   0.05  -0.03   0.05  -0.14  -0.23  -0.05  -0.01   0.05  -0.04   0.10

PACK

Marketed by:

    Automatic Forecasting Systems, Inc.
    P.O. Box 563
    Hatboro, Pa. 19040
    215-675-0652

Computer Makes:                          Operating Systems:

    Burroughs B3700, B3800               IBM - OS, VS, TSO, VM-CMS
    CDC 6000 series                         DOS-VS
     CYBER series                        CDC - NOS, SCOPE 3.4
    DEC System 10, 20                     DEC - LINK 19 OVERLAY
    DEC PDP II                            XDS Sigma 7
    Honeywell 69/66/XX                    UNIVAC 1106/1108
     6000, 600                           BURROUGHS 6700
    IBM 360, 370                          HONEYWELL 6000
    Univac Series 1108

Source Languages: FORTRAN

Cost:

    There is a one-time charge of $200 for universities; $400 for all others.
There are supplementary routines available individually or at a **total package**
cost of $800.  There is an annual maintenance service available at a cost of
$50 with extended services available at a cost of $100 yearly.

Documentation:

    An extensive programmer and User's Guide is available at $25 per copy.
This contains sample output, detailed setup instructions and information
regarding program overlay and the implications of dimension statements.

    Advisories:  Error notices, amendments, modifications and other advisories.
Includes membership in Users Group with scheduled annual meetings.

    Extended Support:  Extensive telephone consultation to support your usage
or to answer questions related to Box-Jenkins modelling including model selection.  Includes all privileges of advisory subscription.

## 11.04  SHAZAM


### INTRODUCTION

SHAZAM is a computer program for general and specialized uses in econometrics written by Kenneth T. White.  The program can be run in batch mode or interactively at a computer terminal.  Computer core storage is dynamically allocated so that large problems are only limited by the size of the machine. SHAZAM is designed to grow so that new algorithms and procedures can easily be added by any programmer familiar with the internal structure of the program.

### CAPABILITIES:  Analysis of Statistical Data

Features of SHAZAM include ordinary least squares, two-stage least squares, seemingly unrelated regressions and iterative estimation of seemingly unrelated regressions, three-stage least squares and iterative three-stage least squares, models with first and second order autocorrelated disturbances, estimation of Box-Cox type nonlinear functional forms, principal components and factor analysis, regresson on principal components, ridge regression, regressions by matrix decompositions, random number generation for Monte Carlo samples, forecasting, and plotting.  Any set of linear restrictions or hypothesis tests can be used in the estimation.  A wide variety of output statistics are available with each procedure.

The autocorrelation section of SHAZAM includes maximum likelihood or least squares estimation by a grid search or iterative Cochrane-Orcutt procedure and inclusion or deletion of initial observations, exact and higher-order Durbin-Watson type tests, test based on Golub's uncorrelated residuals, Dhrymes corrections in a time series, Savin-White type simultaneous testing for functional form and autocorrelation, and forecasting using Goldberger's best linear unbiased predictor.

### PROPOSED ADDITIONS

The program is maintained by the developer who plans to improve the control language and add more econometric procedures.

### INTERFACES WITH OTHER SYSTEMS

SHAZAM can interface with CITIBASE DATA.

### REFERENCE

White, Kenneth J.  "A General Computer Program for Econometric Methods - SHAZAM" Econometrics, January 1978, 239-240.

SHAZAM:   A General Computer Program for Econometric Methods

Developed by:                              Distributed by:

    Dr. Kenneth J. White                   Same
    Department of Economics
    Rice University
    Houston, TX  77001

Machine:                                   Operating Systems:

    IBM 370                                OS, VS, VM, TSO, MTS
    Honeywell                              GCOS
    CDC                                    NOS

Language:

    FORTRAN IV

Cost:

    Universities -- $125; non-universities - $225.

## 11.05  TSP

### INTRODUCTION

TSP is a large econometric package which does estimation, variable transformation, plotting, and model simulation. It has been designed to be easy for
an economist or other non-programmer to learn to use, with such features as
free format input. It was developed originally by Robert E. Hall at MIT in
1966/1967 and has been continously developed since then by a variety of people.
Maintenance, distribution, and development was taken over by Bronwyn H. Hall in
1972. Since then over four hundred installations have acquired TSP.

### CAPABILITIES

TSP is designed for econometric estimation. In normal operation, no files
are used except the usual Fortran input and output so that the package is as
machine independent as possible. The program loads data in free or fixed format
from cards, tape or disk. Transformations may be performed on the input data,
either using algebraic-like expressions or certain commonly used techniques
such as seasonal adjustment or Divisia index calculations. The data may be
printed, punched, saved on disk, or plotted.

The estimation methods available in TSP include all those normally used in
econometric work, ordinary least squares, two and three stage least squares,
multi-equation least squares. In addition full information maximum likelihood
estimation of a nonlinear simultaneous equations model and solution or simulation of the same type of model is available. All nonlinear estimation methods
in TSP use analytic derivatives which are computed internally by the program.
A user-written procedure facility is available and there are a set of procedures for the usual matrix operations so that the user can program his own
estimators to run in TSP.

The latest release, version 3.5, includes:

* New auto-regressive procedure.
* Improved forecasting capabilities.
* New User's Manual
* Extensive improvements to the code to enhance efficiency and
  portability.
* New time series plotting facility.
* Extended character set to allow logical operators.
* Symbolic lags for time series or vectors.

### ADDITIONS NEXT 12 MONTHS

* Addition of Charles Nelson's (University of Washington) Box-Jenkins
  subroutines.
* Enhancement of data management capabilities.
* Rewritten result retrieval/output facility.
* Terminal width output (80 characters).
* Symbolic differentiation.

## REFERENCES

Velleman, P.F., J. R. Seaman, and I.E. Allen, "Evaluating Package Regression
    Routines," Proc. Stat. Comp. Sec., American Statistical Association, (1977)
Berndt, Ernst,K., Bronwyn H. Hall, Robert E. Hall and Jerry A. Hausman,
    "Estimation and Inference in Nonlinear Structural Models," Annals of Econo-
    mic and Social Measurement, 1974., pp. 653-665.

SAMPLE JOB: FIML

```
NAME ILUSFIML 'ILLUSTRATIVE MODEL FOR TSP VERSION 3.5' ;
SMPL 1 30 ; LOAD ;
?               MAKE AN ID VARIABLE STARTING IN 1946 AND A TIME VARIABLE.
MAKEID 1946 ; MAKEID TIME 1 ;
GENR P = P/100 ; GENR LP = LOG(P) ;
GENR G = GOVEXP+EXPORTS ; GENR R = RS ;
?               CHANGE SAMPLE FOR THE OPERATIONS ON LAGGED VARIABLES.
SMPL 3 30 ;
GENR M = (M+M(-1))/2 ;
SMPL 4 30 ;
GENR LM = LOG(M) ;
?
?           STARTING VALUES FOR THE PARAMETERS OF THE MODEL.
?
CONST DELTA 15 ;
PARAM A -21.5643 B .638235 LAMBDA .708188 ALPHA .893807
      D -6.57444 F 8.33842 PSI 1.12213 PHI -.0904265
      TREND .00351055 PO .543617 ;
?
?           SPECIFYING THE EQUATIONS FOR A SIMULTANEOUS EQUATIONS MODEL.
?
IDENT GNPID GNP-CONS-I-G ;
FRML CONSEQ CONS = A + B*GNP ;
FRML INVEQ I = LAMBDA*I(-1) + ALPHA*GNP/(DELTA+R) ;
FRML INTRSTEQ R = D + F*(LOG(GNP)+LP-LM) ;
FRML PRICEQ LP = LP(-1) + PSI*(LP(-1)-LP(-2)) + PHI*LOG(GNP)
      + TREND*TIME +PO ;
PAGE ;
FIML (ENDOG = (GNP,CONS,I,R,LP)) GNPID CONSEQ INVEQ INTRSTEQ PRICEQ ;
STOP ; END ;
```

## Output from Sample Job:  FIML

FULL INFORMATION MAXIMUM LIKELIHOOD RESULTS
*********************************************

```
    4 STOCHASTIC EQUATIONS
    1 IDENTITIES
   10 PARAMETERS
   27 OBSERVATIONS

            EQUATIONS:  CONSEQ    INVEQ    INTRSTEQ   PRICEQ    GNPID
ENDOGENOUS VARIABLES:  GNP       CONS     I          R         LP

           LOG OF LIKELIHOOD FUNCTION =   -191.797

           COVARIANCE MATRIX OF UNTRANSFORMED RESIDUALS
```

```
   ................................................................
 1  .    147.095      -64.2249     1.57023      .307606
 2  .    -64.2249     188.048     -.940456     -.296519
 3  .    1.57023      -.940456     1.12541      .900594E-02
 4  .    .307606      -.296519     .900594E-02  .969583E-03
            1            2            3            4
```

| RIGHT-HAND VARIABLE | ESTIMATED COEFFICIENT | STANDARD ERROR | T-STATISTIC |
|---|---|---|---|
| A | -15.3312 | 13.9740 | -1.09712 |
| B | .630865 | .138968E-01 | 45.3963 |
| LAMBDA | .735408 | .167058 | 4.40210 |
| ALPHA | .816494 | .514776 | 1.58611 |
| D | -6.76547 | 2.15664 | -3.13705 |
| F | 8.48422 | 1.47111 | 5.76721 |
| PSI | -.479733 | .205944 | -2.32944 |
| PHI | .635919 | .165646 | 3.83902 |
| TREND | -.195115E-01 | .504855E-02 | -3.86477 |
| P0 | -3.88537 | 1.03612 | -3.74991 |

<u>TSP, VERSION 3.5:  TIME SERIES PROCESSOR</u>

<u>Developed by</u>:

    Robert E. Hall
    Bronwyn H. Hall
    and numerous others

<u>Distributed by</u>:

    Bronwyn H. Hall
    204 Junipero Serra Blvd.
    Stanford, CA 94305
    U.S.A.

<u>Computer Makes</u>:

    IBM 360/370/3033
    Amdahl
    Itel AS series
    Univac 1100 series
    CDC Cyber
    CDC 6500/6600/6700
    Burroughs 6700/7700/7800
    DEC 10/20
    VAX 11/70 (V3.4)
    Prime (V3.4)
    Siemans, Tosbac, etc.
    Honeywell 66 series

<u>Operating Systems</u>:

    OS/MVT,MVS,TSO,VM/CMS,DOS
    "
    "
    EXEC
    NOS, SCOPE
    "
    MCP 2/3
    TOPS 10/20

    GCOS, Multics

<u>Source Languages</u>:  FORTRAN

    A few assembler routines on IBM and Univac for databank procedures.

<u>Cost</u>:

    There is a one-time fee of $900 for the first copy, $600 to universities and colleges.  This includes source on tape, load modules if IBM, one copy of the User's Manual and Programmer's Guide, an installation memo, and a copy of the testrun output.  Bug fixes are provided in printed form free of charge. New versions may be obtained by previous purchasers for a reduced price.

<u>Documentation</u>

Hall, B.H., and R.E. Hall, <u>Time Series Processor User's Manual</u>.  Stanford, CA:
    Bronwyn H. Hall.
Hall, B.H., and R.E. Hall, <u>Time Series Processor Programmer's Guide</u>.  Stanford,
    CA:  Bronwyn H. Hall.

11.06  <u>QUAIL</u>


## INTRODUCTION

QUAIL (QUAlitative, Intermittant, and Limited Dependent Variable Statistical Program) is a special-purpose computer system for analysis of statistical
models involving non-continuous dependent variables, and for manipulation and
storage of associated data arrays.  Its principal applications have been
econometric analysis of transportation mode choice and residential energy
demand.  QUAIL was developed by J. Berkman, D. Brownstone, D. McFadden, and
H. Wills between 1975 and 1978.  It can be used interactively or in batch and
is installed in over fifty computer installations.

## CAPABILITIES:  Processing and Displaying Data

The QUAIL system consists of separate compiler and interpreter programs, so
it is easy and inexpensive to check for syntax and keypunch errors.  The QUAIL
language is format-free and English-based.  There are extensive options for the
advanced user, but most of these options have default values which are sufficient for most users.

New variables can be created via general arithmetic and logical operations
on variables.  Variables can also be sorted and general arithmetic transformations within variables are possible.  Missing data are handled automatically,
and multiple missing data codes are permissable.  Data can be input or output
with user-specified formats or FORTRAN binary format.  There are only rudimen-
tary facilities for plotting data.

QUAIL has a permanent data storage system.  Up to two QUAIL tapes can be
used in a job, and the tape format minimizes data loss due to tape defects or
system crashes.  QUAIL tape directories allow for 40 character labels for each
variable stored on the tape, and selective tape-copying facilities are also
provided.

## CAPABILITIES:  Statistical Analysis

QUAIL's main statistical procedure is for estimating the Multinomial Logit
Model (see McFadden).  These estimations can be carried out with linear equality
restrictions on the parameters.  Possible estimation methods are maximum likelihood, nonlinear least squares, and weighted exogenous sample maximum likelihood.
The logit procedures can handle data sets which are too large to fit into the
computer's memory.

Other QUAIL statistical procedures include:  ordinary least squares, two-
stage least squares, lump it (see Berkman), and simple covariance matrices.
All QUAIL statistical procedures automatically handle missing data and allow
for general subsample selection.  Output from statistical procedures (i.e.,
residual vectors) can be saved as QUAIL variables for use in future computations and users can control printing of auxiliary statistics.

## PROPOSED ADDITIONS

QUAIL is not a supported program.  It is unlikely that any publicly available changes or additions will be made in the next twelve months.

REFERENCES

Berkman, J., and Brownstone, D., QUAIL 4.0 User's Manual, Computer Center,
     University of California Berkeley (1979).
McFadden, D., "Quantal Choice Analysis:  A Survey," Annals of Economic and
     Social Measurement 5, pp. 363-390 (1976).

SAMPLE JOB

    Input deck:

```
program na(logitsample)$
smpl in(1,52)$
list tty$

/* create the idcase and idalt vectors */
dimen ca(26) al(2)$

/* read and print the data */
read va(choice, time, cost) ft(f2.Ø,2f8.3)$
print va(idcase,idalt,choice,time,cost)$

/* now for the logit analysis */
logest dv(choice) iv(time,cost) al(1,2) du(1)$
end$
```

## Partial Output from Logest Procedure

log of iterations.

initial parameter values are-  0.          0.          0.

| itera-<br>tion | step<br>mode | max<br>size | mean square<br>adj | log<br>of gradient | percent increase<br>likelihood | in log likelihood |
|---|---|---|---|---|---|---|
| 1 | nr | 1.32 | 1.000 | 6.659 | -.1386183785669946e+02 | 30.019 |
| 2 | nr | .99 | .779 | .400 | -.1380725682900915e+02 | .396 |
| 3 | nr | 1.30 | .720e-02 | .322e-04 | -.1380723765533827e+02 | .139e-03 |
| 4 | nr | 1.01 | .399e-05 | .164e-06 | -.1380723766338608e+02 | .136e-10 |
| 5 | nr | 1.73 | .349e-07 | .120e-06 | -.1380723766339608e+02 | 0. |

summary of iterations.

convergence after   5 iterations. mean square of gradient=   .120e-06

  11 evaluations of the likelihood and gradients,   6 of the hessian.

logitsample  25 feb 79   19:45:54    page   4   logit estimation       1

```
         logit estimation results
         *************************
      the dependent variable is choice
```

| variable<br>name | logit<br>estimate | standard<br>error | t-<br>statistic |
|---|---|---|---|
| time | -.3034e-02 | .2306e-02 | -1.3155 |
| cost | .1685e-02 | .4795e-02 | .3514 |
| alt  1.000 | -.3920 | 1.0348 | -.3788 |

| auxiliary statistics. | at convergence | at zero |
|---|---|---|
| log likelihood | -13.8072 | -18.0218 |
| sum of squared residuals | 23.4692 | 26.0000 |
| degrees of freedom | 23.0000 | 26.0000 |
| percent correctly predicted | 69.2308 | 50.0000 |

| goodness of fit statistics. | about zero |
|---|---|
| likelihood ratio index | .2339 |
| likelihood ratio statistic | 8.4292 |

| 7 logest | * | .549 cp seconds | 524 words ecs, | 240 words common |
|---|---|---|---|---|
| 8 end | * | .011 cp seconds | 524 words ecs, | 240 words common |

  1.109 cp seconds   total execution
max blank common— 347 in stmt   6, max ecs— 524 in stmt   7

## QUAIL 4.0

Developed by:

    J. Berkman, D. Brownstone
    D. McFadden, and H. Wills
    Department of Economics
    University of California Berkeley
    Berkeley, CA  94720

Distributed by:

    David Brownstone
    Department of Economics
    Princeton University
    Princeton, NJ  08544

Computer Makes:

    CDC 6000-7000 series
      CYBER series
    IBM 360-370, 3033
    Amdahl 470

Operating Systems:

    All current CDC-SCOPE systems
    IBM OS, VS, SVS, MVS
      VM-CMS (with minor modifications)

Source Languages:  Primarily FORTRAN IV, with some assembler language routines.

Cost:

    $400.  Includes source and load modules, test decks, and sample JCL on magnetic tape.  Also includes one copy each of the QUAIL 4.0 User's and Programmer's Manuals.  Maintenance is not included.

Documentation:

Berkman, J., and Brownstone, D., QUAIL 4.0 User's Manual, Computer Center, University of California Berkeley (1979).
Berkman, J., and Brownstone, D., QUAIL 4.0 Programmer's Manual, Computer Center, University of California Berkeley (1979).

## 11.07  KEIS/ORACLE

### INTRODUCTION

The Kentucky Economic Information System (KEIS) is an economic data base
system.  It provides the user with the ability to display and analyze time
series data and it also provides data.  There is an extensive public data base,
accessible by any user, containing data pertaining primarily to Kentucky; however, users may also create and maintain private data bases for their own use.

This system is an implementation of the Online Retrieval and Computational
Language for Economists (the ORACLE), which began to be developed by Charles
Renfro in 1969 and which is designed to support the construction, maintenance
and use of large macro econometric models.  The state data base began to be
developed in 1973.  The KEIS extends the features of ORACLE, permitting the
appropriate use of the data by academics, government officials, businessmen,
and others.  The KEIS is available on two state-wide computer networks in
Kentucky, the Kentucky Educational Computing Network (KECNET) and the state
government computing network.  The system can be used in both interactive and
batch mode via a telecommunications terminal over an ordinary telephone line
from anywhere or from dedicated terminals throughout Kentucky.  Some features
of the system are available on a mini-computer network in Kentucky serving
agricultural extension agents and others.

### CAPABILITIES:  Processing and Displaying Data

The KEIS supports a wide range of data processing features invoked by a
verbal-mathematical free-format command language that allows a variety of operations on the data, including the generation of stochastic data for use in
Monte Carlo experiments.  It has such capabilities as seasonal adjustment
(X11 and X11-Q), frequency conversion, and interpolation, in addition to permitting a very wide range of data transformations using the standard arithmetic
operators and a variety of special functions.  However, the data management
focus of the system is upon the storage and processing of time series data.

The KEIS supports plots of series against time, or against other series;
the tabular presentation of data at various levels of user control, ranging
from "canned" tables to those such that the user controls both content and format; it allows the simultaneous use of multiple data banks; it permits the
creation of cross-section variables from a group of time series; data can be
displayed both as retrieved from a data bank and after any transformations.
The graphics facilities of the system depend upon the computer network used.

### CAPABILITIES:  Statistical Analysis

Excluding the interface capabilities, the statistical analysis features of
the KEIS are limited to those that are generally required in the estimation,
maintenance, and use of econometric models, including the testing and solution
of such models.  Furthermore, it generally provides the facilities that are
necessary for the construction and use of large econometric models (300+
equations) in a practical context--as opposed to small experimental models.
It allows such regression operations as autoregressive corrections (Cochrane-
Orcutt and search), linear restrictions on parameters, instrumental variables
(Two Stage Least Squares), various forms of distributed lags and other essentially linear techniques.  However, it does not now provide for full

information estimation methods (e.g., Three Stage Least Squares) and it is limited in its ability to perform non-linear parameter estimation. The usual plots of actual and predicted dependent variable values and associated regression statistics are provided.

A related, important feature of the system is that, once estimated, equations can be saved, either for subsequent printing (if a CRT is used) or in order to create and solve multi-equation models. For soft-coded models, using the ORACLE language, the limit is now 500 equations. However, large models (hard-coded in FORTRAN) can be solved using a well-integrated solution program; to this end, parameter estimates can be saved in a file mutually accessible by both estimation and solution programs.

## EXTENSIBILITY

The ORACLE provides <u>some</u> facility to program new statistical features: the data transformation options are extensive and the regression options mutually compatible (whenever computationally feasible), so as to allow for the simultaneous specification of multiple options. In addition, user-written programs can be added.

## INTERFACES

The KEIS interfaces with SAS in particular, in order to extend its range of statistical features. However, KEIS data and results can additionally be exported in both card image punched card and magnetic device form; user supplied data can similarly be read into the KEIS.

## PROPOSED ADDITIONS

New edition of user manual in preparation. New data editing-verification features. A program to translate ORACLE commands into FORTRAN in order to better integrate the large-model solution program.

## SAMPLE JOB

Batch setup:

```
SET PRINTWIDTH=80
JOBTITLE: MODLER SETUP - TWO REGRESSION EQUATIONS, THREE VARIABLES
INPUT BANK=ABANK
SET FREQUENCY=ANNUAL
SET DATES=1951-1979
FETCH:YWS$=YWS$(7)
FETCH:YOL$=YOL$(8)
XXA=YWS$(7)+YOL$(8)
TIME=TIME(1,1951)
SUPPRESS GRAPH
YOL$=F(YWS$,TIME)
SAVE EQUATION AS EQ#1
YWS$=F(TIME)
SAVE EQUATION AS EQ#2
DEFINE MODEL:NAME=MODEL1,DESC:SIMPLE THREE EQUATION MODEL
MODEL FREQUENCY=ANNUAL
XXA=YWS$+YOL$
INCLUDE:EQ#1
INCLUDE:EQ#2
MODEL DEFINED, SAVE, PRINT
READ ASSUMPTIONS
TIME,1980:30,31,33,33,34,35
PRINT ASSUMPTIONS
FORECAST 1980-1985
PRINT FORECAST VALUES
END
```

Output

```
***************************************
KENTUCKY ECONOMIC INFORMATION SYSTEM

MODLER - MODELING LANGUAGE FOR ECONOMETRIC RESEARCH

15134158 MONDAY,  29 DEC 1980
***************************************
SET PRINTWIDTH COMMAND RECEIVED
   PRINTWIDTH HAS BEEN SET AT  80 COLUMNS
JOBTITLE: MODLER SETUP - TWO REGRESSION EQUATIONS, THREE VARIABLES

CHARACTERISTICS OF INPUT BANK:
BANK DESCRIPTION: KENTUCKY ANNUAL ECONOMIC DATA BANK
NUMBER OF SERIES IN BANK: 3833
BANK LAST UPDATED AT 17:31:54 ON 23 DEC 1980

FREQUENCY SET.
PROGRAM NOW EXPECTS DATA THAT HAS A FREQUENCY OF  1 OBSERVATIONS PER YEAR

DATES SET.  PERIOD IS:  1951-  1979

MEMORY FILE CREATION DETAILS:

FETCH SERIES COMMAND:
   FETCH:YWS$=YWS$(7)
SERIES: YWS$    STORED IN MEMORY AS INTERNAL SERIES NUMBER   1
****
FETCH SERIES COMMAND:
   FETCH:YOL$=YOL$(8)
SERIES: YOL$    STORED IN MEMORY AS INTERNAL SERIES NUMBER   2
****
OPERATION PERFORMED:
SERIES: XXA=YWS$(7)+YOL$(8)
SERIES: XXA    STORED IN MEMORY AS INTERNAL SERIES NUMBER   3
****
OPERATION PERFORMED:
SERIES: TIME=TIME(1:1951)
SERIES: TIME    STORED IN MEMORY AS INTERNAL SERIES NUMBER   4
****
SUPPRESS GRAPH COMMAND RECEIVED
  GRAPH OF ACTUALS AND PREDICTED AGAINST TIME WILL NOT BE PRINTED.


EQUATION NUMBER:  1

SAMPLE PERIOD: 195101-197901
NUMBER OF OBSERVATIONS:  29
ORDINARY LEAST SQUARES

DEPENDENT VARIABLE     2-YOL$

C(  1-YWS$     )=      0.14397            T=  29.43626
          (       0.00557)
C(  4-TIME     )=    -16.60607            T=  -6.53395
          (       2.54150)
C(  5-CONSTANT )=   -231.99297            T= -12.48719
          (      18.57848)

VARIANCE= 0.23194805E 04
STANDARD DEVIATION= 48.16104126  (ADJUSTED FOR DEGREES OF FREEDOM)
R-SQUARED=  0.9912
F TEST( 2, 26)= 1583.4146
DURBIN-WATSON D STATISTIC=  0.31678

EQUATION SAVED AS EQ01
DEPENDENT VARIABLE: YOL$


EQUATION NUMBER:  2

SAMPLE PERIOD: 195101-197901
NUMBER OF OBSERVATIONS:  29
ORDINARY LEAST SQUARES

DEPENDENT VARIABLE     1-YWS$

C(  4-TIME     )=    413.93964            T=  11.20864
          (      36.93039)
C(  5-CONSTANT )=   -510.68278            T=  -0.80511
          (     634.30029)

VARIANCE= 0.27684250E 07
STANDARD DEVIATION= 1663.91845703  (ADJUSTED FOR DEGREES OF FREEDOM)
R-SQUARED=  0.8166
F TEST( 1, 27)= 125.6335
DURBIN-WATSON D STATISTIC=  0.08946

EQUATION SAVED AS EQ02
DEPENDENT VARIABLE: YWS$

MODEL DEFINED. NAME IS MODEL1
MODEL SAVED AS MODEL1


MODEL NUMBER 1

SIMPLE THREE EQUATION MODEL

EQUATION #  1

XXA=YWS$+YOL$

EQUATION #  2

YOL$( 2)=C( 1)*YWS$( 1)+C( 2)*TIME( 4)+C( 3)

EQUATION #  3

YWS$( 1)=C( 4)*TIME( 4)+C( 5)

PARAMETERS:

C(  1)=    0.144  C(  2)=  -16.606  C(  3)=  -231.993  C(  4)=  413.939
C(  5)=  -510.683


                              ASSUMPTIONS
              MODLER SETUP - SIMPLE THREE EQUATION MODEL
         SIMPLE THREE EQUATION MODEL - TWO REGRESSION EQUATIONS, THREE VARIABLES

                1980      1981      1982      1983      1984      1985
XXA             NA        NA        NA        NA        NA        NA
YOL$            NA        NA        NA        NA        NA        NA
YWS$            NA        NA        NA        NA        NA        NA
TIME            30.000    31.000    32.000    33.000    34.000    35.000


                             FORECAST VALUES
              MODLER SETUP - SIMPLE THREE EQUATION MODEL
         SIMPLE THREE EQUATION MODEL - TWO REGRESSION EQUATIONS, THREE VARIABLES

                1980       1981       1982       1983       1984       1985
YWS$          11907.500  12321.438  13149.316  13563.258  13977.195
YOL$           1222.299             1374.101               1478.336
XXA           13129.797  13595.000  14525.414  14990.625  15455.828
TIME             30.000     31.000     32.000     33.000     34.000     35.000
```

KEIS/ORACLE (Kentucky Information System/Online
Retrieval and Computational Language for Economists)

Developed by:                              Distributed by:

  ORACLE:  Charles G. Renfro              Director
          1421 Richmond Road              Center for Applied Economic Research
          Lexington, Kentucky 40502       451 Commerce Building
                              University of Kentucky
  KEIS  (including data banks and       Lexington, Kentucky 40506
          additional programs,            (606/258-4626
          procedures, etc.):

          Center for Applied Economic
          Research
          451 Commerce Building
          University of Kentucky
          Lexington, Kentucky 40506

Computer Makes:                            Operating Systems:

  IBM 360, 370                             OS, VM/CMS, MVS/TSO

Source Languages:

Primarily FORTRAN, but some stand-alone programs in PL/1 and some Assembly
language subroutines.  Procedures written in JCL, TSO Command Language and a
subset of CMS.

Cost:

The royalty fee to use the KEIS is $500 per year.  Fee is waived in the
case of governments, faculty doing non-funded research, and non-profit organizations.  However, in all cases, the user bears the cost of operating the
computer.

Documentation:

Charles G. Renfro and Paul A. Coomes., Kentucky Economic Information System
    User Manual. Lexington, KY:  Office of Research, College of Business and
    Economics, University of Kentucky, 1977.
Charles G. Renfro., "On the Development of A Comprehensive Public Data Base for
    Aggregate State and Local Economic Data" Review of Public Data Use,
    Volume 7, Number 5/6, December 1979, pp. 1-10.
Charles G. Renfro., "An Online Information System for Aggregate State and Local
    Area Economic Data" Journal of the American Society for Information
    Science,  Vol. 31, Number 5, September 1980, pp. 319-33 3.
Margaret O. Adams., "An Information System for State and Local Government
    Planning and Analysis:  The Example of the Kentucky Economic Information
    System" Center for Applied Economic Research Economic Studies Series,
    Number 2.  Lexington, KY:  Center for Applied Economic Research, University
    of Kentucky, June 1980, 14pp.
Charles G. Renfro., "ORACLE:  The Online Retrieval and Computational Language
    for Economists"  In preparation.

CHAPTER 12


MATHEMATICAL SUBROUTINE LIBRARIES


CONTENTS

INTRODUCTION

While Chapters 6 and 7 contain most general-purpose statistical systems, Chapters 8 to 12 contain special-purpose programs.  In particular Chapter 12 consists of mathematical or statistical program modules with which the user can fashion his own statistical analysis.

The ratings by the respective developers on selected items are displayed in Table 12.1.  Figure 12.1 compares developers' and users' ratings on all items, while Figure 12.2 summarizes developers' ratings (D-scores) and users' ratings (U-scores) on several relevant attributes.  These are explained in detail in Section 1.4 of Chapter 1.  Figure 12.2 also points to other chapters which describe programs with strong mathematical capabilities.

The two well-known libraries IMSL and NAG, which contain nearly five hundred subroutines each, include many subroutines specifically for statistical purposes.  The uses of EISPACK are not so specifically directed towards statistics but its eigen-routines are useful in many statistical applications.

DATAPAC also contains statistical subroutines, with simulation and graphics capabilities.  NMGS2 and its associate subroutines provide algorithms for the solution of linear equations on small computers.

REPOMAT is not a subroutine library, but rather an interactive program for matrix algebra computations which are useful for implementing many statistical procedures.

As can be seen from the taxonomy of Table 1.2 or from Figures 12.2 and 7.2, a number of other programs provide mathematical options for writing additional procedures.  And we could include some programming languages, such as APL.

In all uses of statistical software there should be a concern for the accuracy of computed results.  For users to be able to use software confidently requires high quality algorithms, high quality programming, and exhaustive

## TABLE 12.1:  RATINGS BY DEVELOPERS ON SELECTED ITEMS

### (i) Capabilities

Usefulness Rating Key:
3 - high
2 - moderate
1 - modest
. - low

| | | Complex Structures | File Management | Consistency Checks | Probabilistic Checks | Compute Tables | Print Tables | Multiple Regression | Anova/Linear Model | Linear Multivariate | Multi-way Tables | Nonparametric | Exploratory | Robust | Non-linear | Time Series | Econometric |
|---|---|---|---|---|---|---|---|---|---|---|---|---|---|---|---|---|---|
| | | 11 | 14 | 18 | 19 | 24 | 25 | 30 | 31 | 32 | 33 | 36 | 37 | 38 | 39 | 35 | 41 |
| 12.01 | DATAPAC | 1 | 2 | 2 | 2 | 1 | 1 | 1 | 1 | 1 | . | 1 | 2 | 1 | 1 | 2 | . |
| 12.02 | IMSL | . | . | . | . | 1 | 1 | 3 | 3 | 3 | 3 | 3 | 1 | 2 | 3 | 3 | 3 |
| 12.03 | REPOMAT | 1 | 1 | . | . | . | . | 2 | 2 | . | . | . | . | 1 | 1 | . | 3 |
| 12.04 | NMGS2 | 2 | . | . | . | . | . | 3 | 2 | . | . | . | . | . | . | . | . |
| 12.05 | NAG LIBRARY | . | . | . | . | . | . | 1 | 1 | . | . | . | . | . | 2 | . | . |
| 12.06 | EISPACK | . | . | . | . | . | . | . | 1 | . | . | . | . | . | . | . | . |

## FOR MATHEMATICAL SUBROUTINE LIBRARIES

### (ii) User Interface

| | Survey Estimates | Survey Variances | Simulation | Math Functions | Operations Research | Availability | Installations | Computer Makes | Mini Version | Core Requirements | Batch/Interactive | Stat. Training | Computer Training | Language Simplicity | Documentation | User Convenience | Maintenance | Tested for Accuracy |
|---|---|---|---|---|---|---|---|---|---|---|---|---|---|---|---|---|---|---|
| | 43 | 44 | 46 | 47 | 48 | 49 | 50 | 51 | 52 | 53 | 54 | 55 | 56 | 57 | 58 | 59 | 60 | 61 |
| DATAPAC | . | . | 3 | 1 | . | 2 | 2 | 3 | 3 | 2 | 3 | 2 | 2 | . | 1 | 2 | 3 | 1 |
| IMSL | 2 | 2 | 3 | 3 | 2 | 3 | 3 | 3 | 3 | 3 | 3 | 1 | 1 | . | 2 | 2 | 3 | 1 |
| REPOMAT | . | . | 2 | 2 | . | . | . | 1 | 1 | 2 | 3 | 2 | 1 | 2 | 1 | 3 | 1 | 1 |
| NMGS2 | . | . | . | 3 | 2 | 1 | 3 | 2 | 3 | 1 | . | 3 | 3 | . | 3 | 3 | 3 | 3 |
| NAG LIBRARY | . | . | 2 | 2 | 1 | 2 | 3 | 3 | 3 | . | . | 1 | 1 | . | 2 | 1 | 3 | 2 |
| EISPACK | . | . | . | 3 | . | 3 | 3 | 3 | 3 | 3 | . | 3 | 1 | . | 3 | 3 | 3 | 3 |

testing.  Few programs in this book have undergone testing enough to warrant a rating of "3" on item 61 of our questionnaire.  One exception is EISPACK.

If a statistical package has been tested exhaustively, a user, as well as editors and readers of journals in which his results are published, will be confident in the accuracy of his results.  However if the user has fashioned his own program from subroutines, even though they may be of the highest quality, there exists the possibility of programming errors, or of the inappropriate combination of certain algorithms.  Francis and Sedransk (1979) observed such programming and logical errors in an experiment which required some participants to program some calculations and others to use packages.  Some of the participants who used packages were seen to obtain incorrect results by carelessly using an omnibus option for a problem which differed slightly from the standard problem. It is very easy to make mistakes with a package that is easy to use.

The remainder of this chapter contains descriptions of these mathematical and statistical subroutines in the format described in Table 1.5.

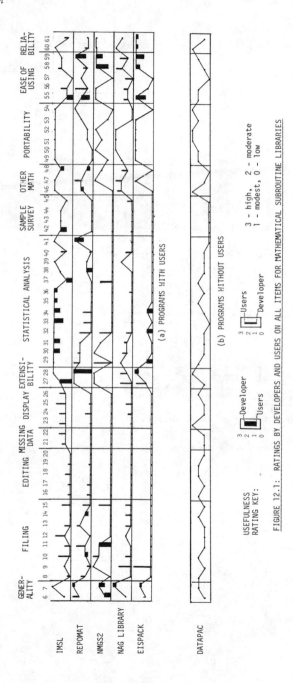

FIGURE 12.1: RATINGS BY DEVELOPERS AND USERS ON ALL ITEMS FOR MATHEMATICAL SUBROUTINE LIBRARIES

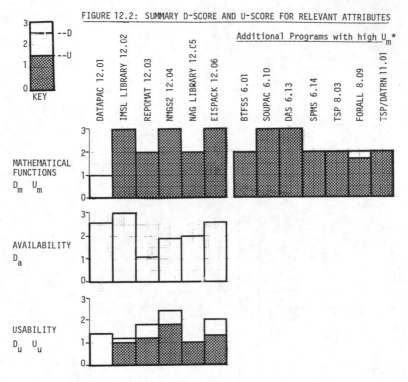

FIGURE 12.2: SUMMARY D-SCORE AND U-SCORE FOR RELEVANT ATTRIBUTES

Another chapter generally containing programs with strong
Mathematical Capabilities is Chapter 7 (see Figure 7.2).

*In addition, for these specific programs, $U_m > 1.5$.

## 12.01  DATAPAC


### INTRODUCTION

DATAPAC (an acronym for data analysis package) is a homogeneous, self-contained set of 175 FORTRAN subroutines for data analysis and for associated probability calculations. It was developed originally by James J. Filliben in 1971 in response to statistical consulting problems encountered at the National Bureau of Standards. The DATAPAC subroutines are grouped into 14 categories: 1) cum. dist. functions (21 dists.); 2) prob. density functions (7 dists.); 3) percent point functions (20 dists,); 4) sparsity functions (7 dists.); 5) random number generators (23 dists.); 6 probability plots (18 dists.); 7) elementary statistics (20 dists.); 8) general analyses (10 analyses); 9) time series analyses; 10) polynomial fitting; 11) printer (wide-carriage) graphics (12 types); 12) terminal (narrow-width) graphics (6 types); 13) format-free input/output (4 routines); 14) data manipulation (17 routines). Ease of use is promoted by a simple argument structure (typically only 2 or 3 arguments) and consistent subroutine nomenclature (e.g., NORCDF for the normal cum. dist. function; NORPDF for the normal prob. dens. function; NORPPF for normal percent point function; NORRAN for normal random number generator; NORPLT for normal probability plot; CAUCDF for Cauchy cum. dist. function; CAUPDF for Cauchy density function, and so forth). Portability is promoted by underlying ANSI FORTRAN (PFORT) code. Since 1971, DATAPAC has been installed and run on a wide variety of computers.

### CAPABILITIES:  Processing and Displaying Data

Subroutines are provided for format-free I/O. The various classical and graphical analysis subroutines discussed below serve as tools for the processing and editing of data. Sorting, ranking, deleting, and subset extraction subroutines are provided.

Graphics capabilities include 12 subroutines for wide-carriage discrete devices (such as the high-speed printer) and 6 subroutines for narrow-width discrete devices (such as terminals). Such graphics capabilities include scatter plots, lag-1 plots, multi-trace plots, character plots, plots restricted to subsets, and a single-page 4-plot analysis (run sequence plot, lag-1 plot, histogram and normal probability plot).

### CAPABILITIES:  Statistical Analysis

Analysis capabilities include frequency tabulation; runs analysis for testing randomness; location estimation; scale estimation; normal outlier analysis; tolerance limits analysis; probability plot correlation coefficient analysis for symmetric distributional family (which includes a battery of test statistics for normality), a probability plot correlation coefficient analysis for the Weibull distributional family, a similar analysis for the extreme value family, weighted/unweighted polynomial fitting, and time series analysis.

Graphical data analysis capabilities include scatter plots; lag-1 plots; histograms; probability plots (21 distributions); probability plot correlation coefficient distributional anlaysis (symmetric family, Weibull family, and extreme value family); 4-plot per page univariate analysis; autocorrelation plots. Many of the above analysis and graphical techniques, though nominally

univariate in nature, have routine application to residuals from any general
model.

Classical statistical capabilities include elementary statistics (20 statistics), cum. dist. functions (21 dist.), prob. density functions (7 dist.),
percent point functions (20 dist.), sparsity functions (7 dist.), and random
number generation (23 dist.).

## EXTENSIBILITY

The various subroutines in the DATAPAC library have, by construction, been
coded to be as stand-alone as possible so as to minimize any interactions
between subroutines.  Over 100 of the subroutines are completely stand-alone.
An additional 37 call only 1 other subroutine.  DATAPAC is entirely self-
contained -- all calls are within the library.  This stand-alone nature was
implemented to simplify implementation and to encourage extensibility.  Capabilities in the form of new subroutines may be added without perturbing existing
subroutines within the DATAPAC library.

## INTERFACES WITH OTHER SYSTEMS

Due to the nature of subroutine libraries, and due to the stand-alone
nature of the DATAPAC library, DATAPAC may be easily merged with other subroutine libraries, and/or the library may serve as an underlying basis for
incorporating DATAPAC capabilities in higher-level languages/systems.

## PROPOSED ADDITIONS IN NEXT YEAR

None--all code developments have been absorbed and directed into DATAPLOT--
the interactive high-level graphics language described elsewhere in this software compendium.

## SAMPLE JOB

Carry out a univariate analysis.  Read in data; plot x(i) versus i; lag-1
plot to test for autocorrelation; generate a histogram to check general distributional shape; perform a runs analysis to check for randomness; generate a
normal probability plot to test for normality; perform a normal outlier analysis; generate an autocorrelation and spectral plot to test for frequency structure; carry out a general symmetric distribution analysis; carry out a general
Weibull analysis; compute tolerance limits (both normal and distribution-free).

```
    DIMENSION X(1000)
    CALL READ(1,80,X,N)
    CALL PLOTX(X,N)          (output shown)
    CALL PLOTXX(X,N)         (output shown)
    CALL HIST(X,N)           (output shown)
    CALL RUNS(X,N)
    CALL NORPLT(X,N)         (output shown)
    CALL NOROUT(X,N)
    CALL TIME(X,N)
    CALL TAIL(X,N)
    CALL WEIB(X,N)
    CALL TOL(X,N)
    STOP
    END
```

Graphical Output

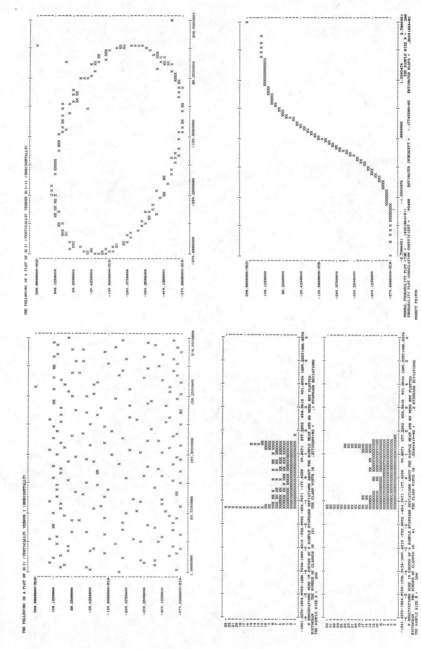

DATAPAC:  Data Analysis Package

Developed by:                              Distributed by:

James J. Filliben                          Same
National Bureau of Standards
Statistical Engineering Laboratory
Administration Building A-337
Washington, D.C.  20234
Phone:  301-921-3651

Computer Makes:                            Operating Systems:

Univac 1108                                Exec 8
Univac 1110                                Exec 8
IBM 360/65                                 DOS/MVT
IBM 370/145
IBM 370/148                                DOS/VS
IBM 370/158
IBM 370/168
IBM 370/169
IBM 3033                                   OS/MVS2
CDC 6600                                   CRONOS
CDC 172                                    NOS
CDC CYBER 175
DEC 10
PRIME 400                                  PRIMOS 4
INTERDATA

Source Languages:  ANSI FORTRAN/PFORT

Cost:

    DATAPAC has been distributed since 1971; there is no cost.  To obtain DATA-
PAC, send a 2400-foot magnetic tape to the above address, and include computer
information such as manufacturer/model/operating system, and tape information
such as tracks, density, parity, and mode (ASCII, BCD, EBCDIC, etc.).  Along
with the extensive internal documentation included in each subroutine, there is
also external documentation which will be provided with the tape.

Documentation:

Filliben, James J., (1976).  DATAPAC- A Data Analysis Package, in Proceedings of
    the Ninth Interface Symposium on Computer Science and Statistics.  Prindle,
    Weber & Schmidt, Boston.

## 12.02  IMSL

### INTRODUCTION

The IMSL Library is a unified, supported set of Fortran subroutines spanning the statistical and mathematical areas. Its elements constitute computational kernels for application software to solve problems in science and engineering. Some routines are written in single precision, some in double precision, and some are available in both versions. Special routines are provided for extended precision arithmetic, and accumulations in most routines are done in a higher precision than the working precision. The first edition of the Library was released by IMSL, Inc. in 1970. Edition 8 of the Library, released in 1980, contains 495 subroutines. It is currently installed at over 1100 sites in more than thirty-six countries.

### CAPABILITIES:  Processing and Displaying Data

Data to be processed by IMSL must be entered in a Fortran array. Several utility subroutines are available in the Library for partitioning of Fortran matrices, transposition and algebraic operations on matrices, and modification of the mode of storage of matrices. Other utility routines provide for transformation of variables. Some of the routines are designed for "out-of-core" processing of the data in order to handle large data sets with a limited available memory.

Except for some special utility subroutines, the subroutines are input/output free.

Capabilities are available in the Library for two-dimensional line-printer plots of up to ten functions on a single set of axes, vertical or horizontal histograms, stem-and-leaf plots, and boxplots.

### CAPABILITIES:  Statistical Analysis

Major divisions (chapters) of the Library exist for basic statistics, analysis of variance, regression analysis, categorized data analysis, nonparametric statistics, time series analysis and forecasting, multivariate analysis, generation of random deviates from various distributions, inferences based on various sampling designs, and probability distribution function evaluation. Within each of these chapters there are routines for most of the standard statistical procedures in the respective areas, as well as routines for less commonly employed methods such as probability density function estimation by a penalized maximum likelihood procedure or by the kernel method, regression analysis using the least absolute values criterion or using the minimax criterion, nonlinear regression analysis, and so on.

The chapter of subroutines for generation of random numbers includes three different uniform distribution generators, giving the user a choice of multipliers and an option for shuffling of the sequence. The uniform generators produce identical sequences on all computers on which the IMSL Library is supported. Other subroutines in this chapter generate deviates from most of the standard distributions, as well as from general discrete distributions using the alias or the table-lookup method, and from general continuous distributions

using a quasi-cubic spline interpolation method. In addition there are subroutines for generation of order statistics, random permutations, random
samples, and deviates in a time series.

Other chapters of the Library contain routines for differential equations;
linear algebra; transforms; eigensystem analysis; optimization, including linear
programming; interpolation, approximation, and smoothing; solution of nonlinear
equations; and special mathematical functions.

PROPOSED ADDITIONS IN NEXT YEAR

New editions of the IMSL Library are released on a one to two year cycle,
and maintenance updates are issued between new edition releases. Enhancements
result from several sources including in-house personnel, the IMSL Advisory
Board, and current users, who are encouraged to provide feedback using standard
Program Attention Request forms or Request for Ability Inclusion forms. New
routines that have been requested and that are tentatively scheduled for inclusion in the next edition are in the areas of nonparametric analysis of variance, multidimensional scaling, multivariate analysis, and regression analysis,
with particular emphasis on regression diagnostics. Also, more routines are
planned that will tie existing routines together in a natural way for a given
application. In these cases, existing modules will remain accessible to the
user who wants to follow or to control the intermediate computations.

SAMPLE JOB

```
      DOUBLE PRECISION DSEED
      INTEGER NR,NI(5),K,MAXL,IER
      REAL X(501)
      DATA NI/5*100/
      NR=100
      DSEED=123457.DO
      CALL GGUBS (DSEED,NR,X)
      CALL GGNPM (DSEED,NR,X(101))
      CALL GGEXN (DSEED,1.,NR,X(201))
      CALL GGTRA (DSEED,NR,X(301))
      CALL GGNLG (DSEED,NR,0.,1.,X(401))
      K=5
      MAXL=80
      CALL USBOX (X,K,NI,MAXL,IER)
      WRITE(6,10)
   10 FORMAT(40H BOX PLOTS OF SAMPLES OF SIZE 100 FROM -,//,40X,
     * 26H UNIFORM(0,1) DISTRIBUTION,/,40X,
     * 25H NORMAL(0,1) DISTRIBUTION,/,40X,
     * 28H EXPONENTIAL(1) DISTRIBUTION,/,40X,
     * 29H TRIANGULAR(0,1) DISTRIBUTION,/,40X,
     * 29H LOG-NORMAL(0,1) DISTRIBUTION
      STOP
      END
```

Partial Output

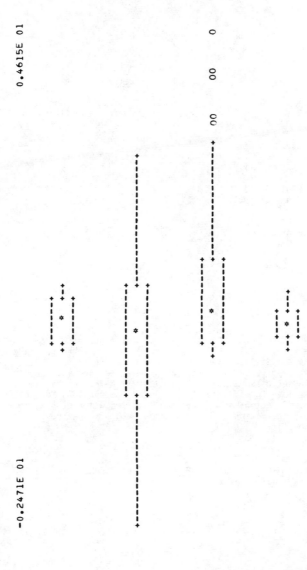

-0.2471E 01

0.4615E 01

BOX PLOTS OF SAMPLES OF SIZE 100 FROM -

UNIFORM(0,1) DISTRIBUTION
NORMAL(0,1) DISTRIBUTION
EXPONENTIAL(1) DISTRIBUTION
TRIANGULAR(0,1) DISTRIBUTION

IMSL Library - Edition 8:  Mathematical and Statistical Library Package

Developed by:

    IMSL, Inc.
    Sixth Floor, NBC Building
    7500 Bellaire Boulevard
    Houston, Texas, U.S.A. 77036

Distributed by:

    Same

Computer Makes:

    Amdahl 470
    Burroughs 6700/7700
    Control Data 6000/7000
      Cyber 70/170
    Data General (Eclipse/Nova with
      Fortran 5)
    Digital Equipment VAX, VAX-11/780,
      Dec 10/20, PDP11
    Hewlett Packard 3000-II & III
    Honeywell and Cii Honeywell Bull
      DPS 8/66/6000
    IBM 360/370/303X, 4300 Series,
      IBM compatible mainframes
    ITEL AS/3/4/5/6
    Telefunken TR 440
    UNIVAC 1100
    XEROX Sigma 6, 7, 9, 11, 560

Operating Systems:

    Operating Systems Independent
    (Standard Fortran Compilers

Interfaced Systems:

    Does not apply

Source Languages:  FORTRAN

Cost:

    The IMSL Library leases for $1600 per year for large scale computing systems and $1400 per year for small scale computing systems.  Degree granting institutions receive a 25% discount.  A subscription includes:  the IMSL Library in Fortran on magnetic tape; Library accessing, generation, and maintenance utility programs on magnetic tape; a maintained Library Reference Manual, in both printed and microfiche forms; the IMSL Numerical Computations Newsletter; a maintained General Information Manual; automatic distribution of new Library editions and updates; consultation on Library usage; and a mechanism for requesting new routines.

Documentation

IMSL General Information Manual.  Edition 8.  IMSL, Inc. 1980.
IMSL Library Reference Manual.  Edition 8.  IMSL, Inc. 1980.

## 12.03  REPOMAT

### INTRODUCTION

REPOMAT is an interactive program that simplifies matrix algebra computation. Reverse Polish Logic is used as the organizing principle for both the
computational flow of the program and the input instructions to run the program.
There are a number of utility commands that input and output matrices to and
from disk, the user's terminal, the system's line printer, and the program's
in-core storage. The program contains a fairly complete list of matrix algebra
operators, including eigen-structure analysis. The commands for these operators, for the most part, correspond to standard matrix algebra symbols. A
number of operators allow the user to manipulate matrices (e.g. partition or
join). In addition to the standard matrix operators, input-output commands,
and matrix manipulation operators, the program has several convenience features.
The program generates random matrices (uniform, Bernoulli, normal), contains
logical operators, and has a loop command that facilitates simulation and inter-
ative search calculations.

REPOMAT is designed to minimize the core required to run it. Thus only
diagonal elements of diagonal matrices are stored, and only upper triangular
elements of symmetric matrices are stored. All elements of general matrices are
stored and scalars are treated as (1x1) matrices. REPOMAT can be configured to
the size of the system in a number of ways. At the University of Pittsburgh,
the most commonly used version has 10,800 in-core storage elements, and will
accomodate up to 30 matrices, 20 stack arguments in RPL, 9 nested levels of
interaction loops, and 9 nested levels of externally supplied instruction files.
The syntax of the command language follows standard Reverse Polish Logic convention. A 38 page HELP file is available with the program.

### EXTENSIBILITY

Most any statistical calculation that is representable by a matrix algebra
expression is feasible. Limitations stem from sample size and, perhaps, some
new matrix algebra operator that is not included in the current version. User
written operators can be added to the program.

### INTERFACES WITH OTHER SYSTEMS

REPOMAT can read ASCII card image matrix files.

### PROPOSED ADDITIONS IN NEXT YEAR

None anticipated.

### SAMPLE JOB

Lines in capital letters following > are user input lines, and lines in
capital letters without the > are REPOMAT requests or file information. Lines
in () are explanatory comments.

```
.RUN REPOS
REPOMAT VERSION 4
BEGIN                     (REPOMAT awaits first instruction line.)
>READ GEN XG              (Read file GEN.REP stored on disk and name it XG.)
EXAMPLE:  GEN. REP        (Header line of file GEN.REP)
OK
>TYPE XG                  (Type matrix XG at terminal.)
ROW     1   COL    1    (Row and Col of first element in line.)
            0.1000E+01    0.2000E+01
ROW     2   COL    1
            0.3000E+01    0.4000E+01
>SET X            (Create a new matrix and name it X.)
 ENTER NUMBER OF ROWS AND COL'S OF X
>2 2
 ENTER ELEMENTS BY ROWS
 ROW 1
>1 2
 ROW 2
>2 3
 OK
>STORE X SYM      (Change storage mode of X to symmetric mode.)
>TYPE X
 ROW     1  COL    1
            0.1000E+01
 ROW     2  COL    1
            0.2000E+01    0.4000E+01
>X XG + TYPE      (Add X and XG and type result at terminal.)
 CURRENT STACK RESULT
 ROW     1  COL    1
            0.2000E+01    0.4000E+01
 ROW     2  COL    1
            0.5000E+01    0.8000E+01
>X EIGEN          (Compute the eigenstructure of X.)
>=L REV = V       (Name eigen values L stored in first stack position, reverse
                   first and second stack positions, name new first stack
                   position V for eigen vectors.)
>V L V INV * * TYPE    (Compute X from its eigenstructure expressed in RPL, and
                        type result.)
 CURRENT STACK RESULT
 ROW     1  COL    1
            0.1000E+01    0.2000E+01
 ROW     2  COL    1
            0.3000E+01    0.4000E+01
>END       (End of run.)
```

(Given data matrices X and Y, each with n observations/rows, and defining U as
a column vector of 1's, then OLS regression coefficients can be computed by
the following expression.)
>U X JOINR Y RLDWN DUP DUP '* INV REV RLUP ' * *
(The latter part of the line can be stored in an instruction file and given an
operator name such as OLS.  Then the REPOMAT input line would be:)
>U X JOINR OLS

REPOMAT

Developed by:                                    Distributed by:

    Norman P. Hummon                            Same
    2J28 Forbes Quadrangle
    University of Pittsburgh
    Pittsburgh, PA 15260

Computer Make:                                   Operating System:

    DEC System 10                               DEC Monitor

Source Language:

    FORTRAN-10 (DEC version of Fortran IV)

Cost:

    There is a one time charge for universities, governments and non-profit
organizations, and commercial use, of tape copying charges at the University of
Pittsburgh.

12.04  <u>NMGS2</u>

## INTRODUCTION

Although the Gram-Schmidt process has long been noted for its numerical stability, a difficulty has been with core requirements, especially for small computers. NMGS2 (Longley,1975) and NMGS3 carry two vectors and a single subscripted upper triangular inverse matrix in main core. Additional core is required for various side statistics, such as weights, the regressand which must be saved, and an extra column for the sum of the squares of the orthonormal vectors used to produce the standard deviations of the conditional values of $y_i$, the standard errors of forecast, and the standard deviations of the residuals. Dimension statements are variable. Core required for a (500 x 31) order matrix is 128 K, where K = 1024 bytes. Both programs use a LOGICAL*1 string of bytes to prevent underflow and overflow. Both programs will process data scaled $10^{-70}$.

SUBROUTINE SKINNY(Longley,1978) requires from two to N columns to be carried in main core. There is a reduced version which will process a (1000 x 1000) order matrix with a minimum of 124 K in long precision, Longley and Dash (1980).

SUBROUTINE STINGY (Longley, 1979a) blocks the data by rows, core required being from two to M rows. With blocks of 100 rows and 13 columns, core required is about 62 K, with maximum limit of disk space of about 600,000,000 bytes on an IBM 360/65 Computer.

Although SUBROUTINES SKINNY and STINGY will run on small computers, no matter how large the computer there is contention for resources, so that at some point there may be a trade-off between core and disk I/Os.

All four versions are fully documented in the comment statements, so that anyone receiving a tape and a listing can run the programs.

## CAPABILITIES

NMGS2 contains nineteen options with minimum output being about seven pages. This particular program may not run on the H Level Compiler because of "table size". NMGS3 will run on the H Level Compiler. There is also a version which will run on the H Level Compiler Extended, about 32 digits. NMGS3 computes the trace and determinant of the design matrix which is compared with the trace and determinant computed from the singular values. In addition, NMGS3 can compute Stewart's Sensitivity Index, Stewart (October 1979).

Both SUBROUTINES STINGY and SKINNY will process the Leontief Input-Output matrix. Output is minimal, being confined to factored correlation matrix, length of vectors used in computing F-ratios and partial correlations, t-ratios, tolerance values, etc., including plots of standardized residuals. All will compute weighted least squares. And all will process a square unsymmetric inverse matrix.

PROPOSED ADDITIONS IN NEXT YEAR

The above programs have been written with two primary aims in mind: (1) numerical stability, and (2) reduction in core and cost of processing a problem.  All programs except Longley and Dash (1980) reduce the columns of the design matrix to unit length which introduces considerable savings in computation of side statistics and which serves as a means of preconditioning the matrix.  There are some problems, however, for which this normalizing process will not necessarily increase average decimal digit accuracy of the solution. Changes in the programs can detect some of these particular types of problems. The singular value decomposition subroutine works well for problems for which N is small and for which the singular values are large and dispariate.  An attempt will be made to overcome the difficulty in cases where the singular values are in clusters.

Longley (1979b) indicated that with modification in the program, Classical Gram-Schmidt Process possesses the same numerical stability as Modified Gram-Schmidt.  Longley (1980) has demonstrated conclusively that this is true.  An attempt will be made to rewrite SUBROUTINE SKINNY to take advantage of Classical Gram-Schmidt which forms the upper triangular matrix by columns in an effort to save disk I/O's.  (Modified Gram-Schmidt forms the upper triangular matrix by rows).

SAMPLE JOB

NPROB Card (FORMAT (I5)), 1, 99,999 problems

N M Card (N columns and M rows) with about nineteen options:  If a particular option is wanted, user must ask for it.  Otherwise NMGS2 and NMGS3 will produce input data, two tables of F-ratios, solution, beta coefficients, t-ratios, standard deviations of regression coefficients, table of Actual, Computed, S. D. y, SEF, S. D. Residuals, standardized residuals, etc., and plot of standardized residuals.

TITLE CARD, Column 1-72.

NAME CARDS
If user supplied format is wanted, card must be placed between M N card and title card.

Partial output of NMGS2 with Joan R. Rosenblatt's third degree ill-conditioned polynomial with data scaled $1.0 \times 10^{-70}$ appears on next page.

PROBLEM 14

DR. ROSENBLATT'S 3D DEG. POLYNOMIAL WITH INPUTS IN 1.0D-70

OBSERVED AND CALCULATED REGRESSAND, STANDARD DEVIATIONS AND STANDARD ERROR OF FORECAST (SEF) OF COMPUTED VALUES, RESIDUALS, AND STANDARD DEVIATIONS OF RESIDUALS FOR Y=LOG Z

| | OBSERVED | CALCULATED | STD. DEVIATIONS CALCULATED | STD. ERROR OF FORECAST | RESIDUALS | STD. DEVIATIONS RESIDUALS |
|---|---|---|---|---|---|---|
| 1 | 0.1672097857935711D-69 | 0.1614788288386826D-69 | 0.2406935OD-70 | 0.3420536OD-70 | 0.573O9569548890970-71 | 0.3367411 2D-71 |
| 2 | 0.1378397900948138D-69 | 0.1592586882161833D-69 | 0.2072685OD-70 | 0.31941776D-70 | -0.21418658121369500-70 | 0.12691300D-70 |
| 3 | 0.471291711105893860-70 | 0.23121024354451970-70 | 0.1456483 4D-70 | 0.2833385 7D-70 | 0.24008146711441890-70 | 0.194560 70D-70 |
| 4 | -0.749579997691106OD-70 | -0.863697858768970-70 | 0.1404791 1D-70 | 0.280716380-70 | 0.114117861076783 7D-70 | 0.1983253 1D-70 |
| 5 | -0.2362510270487489D-69 | -0.2187834655407074D-69 | 0.1512527 1D-70 | 0.2862598 3D-70 | -0.1746756150804146D-70 | 0.1902365OD-70 |
| 6 | -0.3832682652518240-69 | -0.37060117746156680-69 | 0.1421876 9D-70 | 0.2815752 9D-70 | -0.1266708906361553D-70 | 0.1971039 5D-70 |
| 7 | -0.5279014255846261D-69 | -0.5383040845036432D-69 | 0.2265677 2D-70 | 0.3322653 1D-70 | -0.104026589190171 3D-70 | 0.8794529 9D-71 |
| FORECAST VALUES | | | | | | |
| 8 | 0.1061701281374208D-69 | 0.10617012813742080-69 | 0.1834931 6D-70 | 0.304527580-70 | | |

INDICATOR OF ACCURACY OF COMPUTED VALUES BY SOLUTION VECTOR IS     11.761 DIGITS.

STANDARD ERROR OF RESIDUALS IS,     0.24303766D-70

MEAN DEVIATION OF RESIDUALS IS,     0.14729585D-70

NMGS2, NMGS3, SUBROUTINES SKINNY AND STINGY: Normalized, Modified
Gram-Schmidt Subroutines

Developed by:                                      Distributed by:

    James W. Longley and                           James W. Longley
    John B. Dash
    Bureau of Labor Statistics
    Washington, D.C. 20212

Computer Makes:                                    Operating Systems:

    IBM 360, 370                                  IBM - TSO

Source Languages:  FORTRAN IV

Cost:  Cost of delivery

Documentation and References

Longley, James W. (1975).  Normalized Modified Gram-Schmidt Algorithm for the
    Solution of Linear Least Squares Equations.  Proc. of the Computer Sci.
    and Stat.:  Eighth An. Symposium on the Interface, Los Angeles ·
    Univ. of Ca.  158-67
Longley, James Wildon (1978).  Core Saving Techniques for Modified Gram-Schmidt
    Orthogonalization.  1978 Stat. Comp. Sec., Proc. of the Amer. Stat. Assoc.
    San Diego, 152-9.
Longley, James Wildon (1979a).  Pre-conditioning, Ill-conditioning in the
    Matrix Process, in Wang, Peter C. C., Information Linkage between Ap. Math.
    and Ind.  New York:  Academic Press, 451-58.
Longley, James W. (1979b).  Out-of-Core Gram-Schmidt Orthogonalization by
    Blocked Rows.  1979 Stat. Comp. Sec., Proc. of the Amer. Stat. Assoc.,
    Washington, D. C., 72-81.
Longley, James W. and Dash, John B., (1980).  Out of Core Gram-Schmidt Ortho-
    gonalization for Computers with Limited Capacity, in Schoenstadt, Arthur,
    L., Information Linkage between Ap. Math. and Ind. II.  New York:
    Academic Press, 119-123.
Longley, James W., (October 6, 1980).  Modified Gram-Schmidt Process vs.
    Classical Gram-Schmidt.  Conference on Applications of Numerical Analysis
    and Special Functions in Statistics.  University of Maryland, October 2-8,
    1980.
Stewart, G.W., (October 1979).  Assessing the Effects of Variable Error in
    Linear Regression.  Tech. Report 818, Computer Science Technical Report
    Series, University of Maryland.

12.05  <u>NAG</u>

## INTRODUCTION

The NAG Library is a structured subroutine library which is designed to solve mathematical and statistical problems in science, engineering, medicine, commerce, government, and education. NAG endeavours to keep the library contents abreast of advances in numerical analysis and numerical software techniques. New editions (Marks) are produced annually. The Fortran library (Mark 8) contains 466 primary routines. Versions of the library in Algol 60 and Algol 68 are also available. The library is installed at over 400 user sites and is available on over 30 computer systems. The library and documentation have a structured design which reflects the sub-divisions in the areas covered.

### CAPABILITIES: Processing and Displaying Data

The main emphasis in the library is on algorithms. Data management, editing, and results tabulation are not significant design goals. Some line-printer scatter plotting, and a forthcoming graphical supplement are the main facilities under this heading.

### CAPABILITIES: Statistical and Numerical Analysis

Statistics: moment-based measures, scatter plots, distribution functions and inverses, Normal scores calculation and plotting, product-moment and non-parametric correlations, multiple linear regression, analysis of variance, pseudo-random number generators, non-parametric tests.

Numerical analysis: Includes these items of special statistical interest- singular value decomposition, $L_1$, $L_2$, $L_\infty$ curve and surface fitting, constrained and unconstrained nonlinear optimisation, matrix factorisations, calculation of eigenvalues and eigenvectors.

### PROPOSED ADDITIONS IN NEXT YEAR

A graphical supplement will be added to the Fortran library. This will include line, curve, and function drawing, contouring, and surface view drawing. It will interface with several plotting packages.

Routines for univariate time series analysis based on Box-Jenkins methods will be added. Differencing, calculation of sample autocorrelations and partial autocorrelations, model estimation, and forecasting from an estimated model can be performed.

A series of routines for solving systems of non-linear equations will be added.

### REFERENCES

Whelan, J.P., "Using the NAG Library in Industrial Research", <u>Numerical Software - Needs and Availability,</u> D.A.H. Jacobs (Ed.), Academic Press (1978).
McLain, D.H., "Interpolation Methods for Erroneous Data", <u>Mathematical Methods in Computer Graphics and Design</u>, K.W. Brodlie (Ed.), Academic Press (1980).

## SAMPLE JOB

The following program reads in eight observations on each of two variables, and then performs a simple linear regression with the first variable as the independent variable, and the second variable as the dependent variable. Finally the results are printed.

## Input

```
      C     G02CAF EXAMPLE PROGRAM TEXT
      C     MARK 4 RELEASE NAG COPYRIGHT 1974.
      C     MARK 4.5 REVISED
            REAL TITLE(7), X(8), Y(8), RESULT(20)
            INTEGER NIN, NOUT, I, IFAIL
            DATA NIN /5/, NOUT /6/
            READ (NIN,99999) TITLE
            WRITE (NOUT,99998) (TITLE(I),I=1,6)
            READ (NIN,99997) (X(I),Y(I),I=1,8)
            WRITE (NOUT,99996) (I,X(I),Y(I),I=1,8)
            IFAIL = 1
            CALL G02CAF(8, X, Y, RESULT, IFAIL)
            IF (IFAIL) 20, 40, 20
         20 WRITE (NOUT,99995) IFAIL
            GO TO 60
         40 WRITE (NOUT,99994) (RESULT(I),I=1,5)
            WRITE (NOUT,99993) RESULT(6), RESULT(8), RESULT(10)
            WRITE (NOUT,99992) RESULT(7), RESULT(9), RESULT(11)
            WRITE (NOUT,99991) (RESULT(I),I=12,20)
         60 CONTINUE
            STOP
      99999 FORMAT (6A4, 1A3)
      99998 FORMAT (4(1X/), 1H , 5A4, 1A3, 7HRESULTS/1X)
      99997 FORMAT (2F10.5)
      99996 FORMAT (36H0 CASE      INDEPENDENT     DEPENDENT/10H NUMBER  ,
           * 25H VARIABLE           VARIABLE//(1H , I4, 2F15.4))
      99995 FORMAT (22H0ROUTINE FAILS, IFAIL=, I2/)
      99994 FORMAT (46H0MEAN OF INDEPENDENT VARIABLE             = ,
           * F8.4/46H MEAN OF   DEPENDENT VARIABLE              = ,
           * F8.4/46H STANDARD DEVIATION OF INDEPENDENT VARIABLE = ,
           * F8.4/46H STANDARD DEVIATION OF   DEPENDENT VARIABLE = ,
           * F8.4/46H CORRELATION COEFFICIENT                   = , F8.4)
      99993 FORMAT (46H0REGRESSION COEFFICIENT                   = ,
           * F8.4/46H STANDARD ERROR OF COEFFICIENT             = ,
           * F8.4/46H T-VALUE FOR COEFFICIENT                   = , F8.4)
      99992 FORMAT (46H0REGRESSION CONSTANT                      = ,
           * F8.4/46H STANDARD ERROR OF CONSTANT                = ,
           * F8.4/46H T-VALUE FOR CONSTANT                      = , F8.4)
      99991 FORMAT (32H0ANALYSIS OF REGRESSION TABLE :-//13H       SOURCE,
           * 55H        SUM OF SQUARES  D.F.   MEAN SQUARE     F-VALUE//
           * 18H DUE TO REGRESSION, F14.4, F8.0, 2F14.4/14H ABOUT  REGRES,
           * 4HSION, F14.4, F8.0, F14.4/18H TOTAL              , F14.4,
           * F8.0)
            END
```

Output

```
GO2CAF EXAMPLE PROGRAM RESULTS


     CASE      INDEPENDENT      DEPENDENT
    NUMBER       VARIABLE        VARIABLE

      1          1.0000          20.0000
      2          0.0000          15.5000
      3          4.0000          28.3000
      4          7.5000          45.0000
      5          2.5000          24.5000
      6          0.0000          10.0000
      7         10.0000          99.0000
      8          5.0000          31.2000


MEAN OF INDEPENDENT VARIABLE                =    3.7500
MEAN OF   DEPENDENT VARIABLE                =   34.1875
STANDARD DEVIATION OF INDEPENDENT VARIABLE  =    3.6253
STANDARD DEVIATION OF   DEPENDENT VARIABLE  =   28.2604
CORRELATION COEFFICIENT                     =    0.9096

REGRESSION COEFFICIENT                      =    7.0905
STANDARD ERROR OF COEFFICIENT               =    1.3224
T-VALUE FOR COEFFICIENT                     =    5.3620

REGRESSION CONSTANT                         =    7.5982
STANDARD ERROR OF CONSTANT                  =    6.6858
T-VALUE FOR CONSTANT                        =    1.1365

ANALYSIS OF REGRESSION TABLE :-


        SOURCE       SUM OF SQUARES   D.F.   MEAN SQUARE   F-VALUE

DUE TO REGRESSION       4625.3033      1.    4625.3033    28.7511
ABOUT   REGRESSION       965.2454      6.     160.8742
TOTAL                   5590.5487      7.
```

## NAG Library, Mark 8:  Numerical Algorithms Group Ltd.

Developed by:

 NAG, Ltd.
 NAG Central Office
 7, Banbury Road
 Oxford      OX2 6NN
 England

Computer Makes

 Burroughs B5700
 Burroughs B6700
 CDC 3000L
 CDC 7600/600/Cyber
 Cray-1
 DEC PDP-11
 DEC System 10
 DEC System 20
 DEC VAX-11
 GEC 4000
 Harris Vulcan
 Hewlett Packard 3000
 Honeywell (GCOS, MULTICS)
 IBM 360/370
 ICL 1900
 ICL 2900
 ICL 4100
 ICL System 4
 NORD 10/50
 P.E. Interdata 32
 Philips 14/1800
 PRIME 300/400
 Rank Xerox 530
 Siemens BS 2000
 Telefunken TR440
 Univac 1100
 Xerox Sigma 6-7

Distributed by:

 The NAG Library Service Co-ordinator
 Numerical Algorithms Group Limited
 NAG Central Office
 7, Banbury Road
 Oxford      OX2 6NN
 United Kingdom

Tel:  National            0865 511245
      International  +44 865 511245

Telex:  83147 (ref:  NAG)

North American readers may find it
more convenient to contact:

The Company Secretary
Numerical Algorithms Group (USA) Inc.
1250 Grace Court
Downers Grove
Illinois 60516
U.S.A.

(Tel:  (312) 971 2337)

Languages:

 FORTRAN
 ALGOL 60
 ALGOL 68

Cost:

 The annual fee for one language version is £700 sterling for universities;
£858 sterling for government and non-profit organizations; and £858 sterling
for commercial use.  Includes source, load modules, test programs, installation
instructions, 1 manual, and maintenance.

Documentation

NAG Fortran Library Manual, Mark 8 (5 volumes), 1980.
NAG Algol 60 Library Manual, Mark 7 (4 volumes), 1979.
NAG Algol 68 Library Manual, Mark 3 (4 volumes), 1980.
NAG Mini-Manual, Mark 7 (1 volume), 1979.

12.06  EISPACK

## CAPABILITIES

EISPACK is a collection of Fortran subroutines to solve the standard eigen-problem for any one of the following classes of matrices: real general, certain real tridiagonal, real symmetric, real symmetric band, real symmetric tri-diagonal, complex general, and complex Hermitian. It will also solve the generalized eigenproblem for real symmetric positive-definite and real general matrix systems. EISPACK subroutines perform the singular value decomposition of a real retangular matrix and solve associated least squares problems.

REFERENCES (where user guide was reviewed)

Zentralblatt für Mathematek, Band 289
Mathematics of Computation, January 1978
Computing Reviews, March 1976
Documentation (Frankreich), No. 214, Sept. 1976 (Vol. 6)
Documentation (Frankreich), Oct. 1977 (Vol. 51)
Telecommunication Journal, Dec. 1974 (Vo. 6)
Telecommunication Journal, (Schweiz), Apr. 1978 (Vol. 51)
Australian Computer Journal, Vol. 10, May 1978

EISPACK R.2

Distributed by

National Energy Software Center
Building 221
Argonne National Laboratory
Argonne, IL  60439

and

IMSL
Sixth Floor - NBC Building
7500 Bellaire Blvd.
Houston, TX  77036

Machine:

IBM 360/370
CDC 6000/7000
Univac 1108/1110
Honeywell 6070
DEC PDP 10
Burroughs B6700

Operating Systems:

OS/360, MTS, MCP
SCOPE
EXEC-8
GECOS
TOPS-10

Language:

ANS Fortran

Cost:

One-time charge (machine-dependent)

Documentation:

Smith, B.T., Boyle, J.M., Dongarra, J.J., Garbow, B.S., Ikebe, Y., Klema, V.C., Moler, C.B., Matrix Eigensystem Routines-EISPACK Guide, 2nd ed. Lecture Notes in Computer Science, Vol. 6, Heidelberg:  Springer-Verlag 1976.
Garbow, B.S., Boyle, J.M., Dongarra, J.J., Moler, C.B., Matrix Eigensystem Routines-EISPACK Guide Extension, Lecture Notes in Computer Science, Vol.51 Heidelberg:  Springer-Verlag 1977.

## SAMPLE PROGRAM

```
      C       SAMPLE PROGRAM ILLUSTRATING THE USE OF THE EISPACK SUBROUTINES.
      C
      C       THIS PROGRAM READS A REAL SYMMETRIC MATRIX  A   FROM FORTRAN
      C       LOGICAL UNIT 5 AND COMPUTES EIGENVALUES  W  IN THE
      C       INTERVAL (0,3) AND THE ASSOCIATED ORTHONORMAL EIGENVECTORS  Z.
      C       SEE SECTION 2.1.13.
      C
      C       REAL SYMMETRIC MATRIX  A, NO LARGER THAN ORDER  20.
      C       REAL EIGENVALUES  W, AT MOST  3  OF THEM.
      C       ORTHONORMAL EIGENVECTORS  Z, AT MOST  3  OF THEM.
      C
              REAL      A(20,20),W(3),Z(20,3)
              REAL      RLB,RUB,EPS1
      C
      C       TEMPORARY STORAGE ARRAYS.
      C
              REAL      FV1(20),FV2(20),FV3(20),FV4(20),FV5(20),FV6(20),
             +          FV7(20),FV8(20)
              INTEGER IV1(3)
      C
      C       ROW AND COLUMN DIMENSION PARAMETERS ASSIGNED.
      C
              NM = 20
              MM = 3
      C
      C       READ IN THE REAL SYMMETRIC MATRIX OF ORDER  N  ROW-WISE.
      C
        10 READ(5,20) N
        20 FORMAT(I4)
              IF (N .LE. 0) STOP
      C
              DO 40 I = 1, N
                READ(5,30) (A(I,J),J=1,N)
        30      FORMAT(4E16.8)
        40 CONTINUE
      C
              WRITE(6,50) N
        50 FORMAT(///23H ORDER OF THE MATRIX IS,I4//16H MATRIX ELEMENTS)
      C
              DO 70 I = 1, N
                WRITE(6,60) (A(I,J),J=1,N)
        60      FORMAT(1X,1P4E16.8)
        70 CONTINUE
      C
      C       INITIALIZE THE INTERVAL (RLB,RUB).
      C
              RLB = 0.0
              RUB = 3.0
```

```
C
C        THE FOLLOWING PATH IS COPIED FROM SECTION 2.1.13.
C
         CALL TRED1(NM,N,A,FV1,FV2,FV3)
         EPS1 = 0.0
         CALL BISECT(N,EPS1,FV1,FV2,FV3,RLB,RUB,MM,M,W,IV1,IERR,FV4,FV5)
         IF (IERR .NE. 0) GO TO 99999
         CALL TINVIT(NM,N,FV1,FV2,FV3,M,W,IV1,Z,IERR,FV4,FV5,FV6,FV7,FV8)
         IF (IERR .NE. 0) GO TO 99999
         CALL TRBAK1(NM,N,A,FV2,M,Z)
C
C        PRINT THE M EIGENVALUES AND CORRESPONDING EIGENVECTORS.
C
         IF (M .EQ. 0) GO TO 88888
         WRITE(6,90) M
  90 FORMAT(/47H NUMBER OF EIGENVALUES IN THE INTERVAL (0,3) IS,I4)
C
         DO 110 I = 1, M
            WRITE(6,100) W(I),(Z(J,I),J=1,N)
 100     FORMAT(/11H EIGENVALUE/1X,1PE16.8//14H CORRESPONDING,
     +             12H EIGENVECTOR/(1X,3E16.8))
 110 CONTINUE
C
         GO TO 10
C
C        THERE ARE NO EIGENVALUES IN (0,3).
C
88888 WRITE(6,88100)
88100 FORMAT(43H0NO EIGENVALUES OF A IN THE INTERVAL (0,3).)
         GO TO 10
C
C        HANDLING OF IERR PARAMETER
C
99999 IF (IERR .GT. 0) GO TO 99200
         IERR = -IERR
         WRITE(6,99100) IERR
99100 FORMAT(52H0AT LEAST ONE EIGENVECTOR FAILED TO CONVERGE, NAMELY,I5)
         GO TO 10
C
99200 WRITE(6,99300) M
99300 FORMAT(/35H NOT ENOUGH SPACE ALLOCATED FOR THE,I4,
     +          35H EIGENVALUES IN THE INTERVAL (0,3).)
         GO TO 10
         END
```

## Data Read from Unit 5

```
    3
 -1.0              1.0           -1.0
  1.0              1.0           -1.0
 -1.0             -1.0            1.0
    4
  1.875           -0.5            0.375        -0.25
 -0.5             2.25           -0.25         -0.375
  0.375           -0.25           2.125        -0.125
 -0.25            -0.375         -0.125         2.125
    0
```

## Output

```
ORDER OF THE MATRIX IS    3

MATRIX ELEMENTS
 -1.00000000E 00   1.00000000E 00 -1.00000000E 00
  1.00000000E 00   1.00000000E 00 -1.00000000E 00
 -1.00000000E 00 -1.00000000E 00   1.00000000E 00

NUMBER OF EIGENVALUES IN THE INTERVAL (0,3) IS    1

EIGENVALUE
  2.56154823E 00

CORRESPONDING EIGENVECTOR
 -3.69047642E-01 -6.57192111E-01  6.57192290E-01


ORDER OF THE MATRIX IS    4

MATRIX ELEMENTS
  1.87500000E 00 -5.00000000E-01  3.75000000E-01 -2.50000000E-01
 -5.00000000E-01  2.25000000E 00 -2.50000000E-01 -3.75000000E-01
  3.75000000E-01 -2.50000000E-01  2.12500000E 00 -1.25000000E-01
 -2.50000000E-01 -3.75000000E-01 -1.25000000E-01  2.12500000E 00

NOT ENOUGH SPACE ALLOCATED FOR THE    4 EIGENVALUES IN THE INTERVAL (0,3).
```

### CHAPTER 13

# AN OVERVIEW OF THE SOFTWARE AND THE ASSOCIATED HARDWARE

## CONTENTS

## 13.1  A SOFTWARE-HARDWARE REFERENCE TABLE

The developers of all programs were asked to list the makes and models of all computers on which a version of their programs had been installed. Figure 13.1 is a compilation of these computer models with references to all programs which run on them. The first thirteen computer models are plug-compatible with IBM.

Note that this is a compilation of machine types on which the programs have been installed. Some of these programs, particularly the smaller ones that have been written in a portable language, may be portable to other machines with some modifications.

## 13.2  TWO OVERALL RANKINGS OF SOFTWARE

Tables 13.1 and 13.2 provide an overview of the usefulness of all the programs in this book from two points of view. First, in Table 13.1 all the programs for which we have users are ranked by their $\Omega$-scores (defined in detail in Section 1.5): a program's $\Omega$-score can be viewed as an average user-rating on those features which the developer claims to be strengths of his program. Thus SIR, which its developer claimed to have strengths in file management, editing, and ease of use, and was rated very high on these same features by its users, was ranked first with its $\Omega$-score of 2.8. (Note that 2.8 is recorded as $2^8$ on Tables 13.1, 13.2, and A.1.)

Second, in Table 13.2 all programs are ranked by their overall D-scores (defined in detail in Section 1.5): a program's overall D-score is the average of its developer's ratings on twenty-seven items which cover the entire range of statistical and usability features. Thus a program's D-score is a measure of its usefulness as a total statistical analysis system as claimed by its developer. It is not surprising that the general-purpose statistical packages from Chapters 6 and 7 dominate the leading positions on this list.

The two other overall ratings, the U-score and $\Delta$-score, on Tables 13.1 and 13.2 are also defined in Section 1.5: the U-score is the user's version of the D-score, while the $\Delta$-score is the developer's version of the $\Omega$-score.

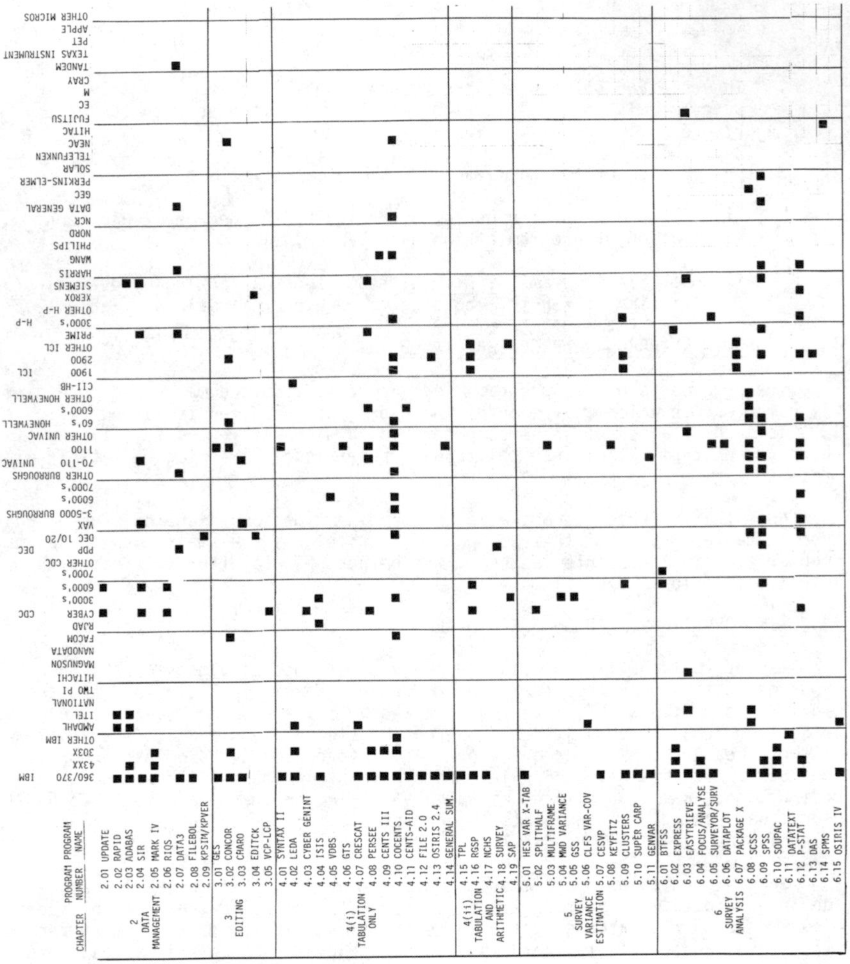

FIGURE 13.1: THE RELATION BETWEEN PROGRAMS AND THE COMPUTER MODELS ON WHICH THE PROGRAMS HAVE BEEN INSTALLED

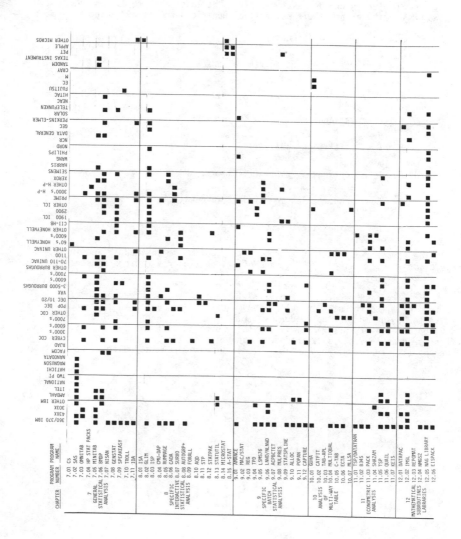

On the right-hand side of both Tables 13.1 and 13.2 is printed a simple profile of each program in terms of the eight statistical attributes, as defined in Table 1.6, for which the developer claims at least some modest capability. The eight attributes and their respective specific D-scores reflected here are:

1. $D_f$ : file management,

2. $D_d$ : detecting and editing errors,

3. $D_t$ : tabulation,

4. $D_v$ : variance estimation for surveys

5. $D_s$ : statistical analysis

6. $D_c$ : contingency table (multi-way) analysis

7. $D_e$ : econometrics and time series analysis

8. $D_m$ : mathematical functions

The specific-attribute D-scores, $D_f$, $D_d$, etc. are defined in Section 1.4.3. For example, $D_f$ is simply the average of its developer's ratings on seven file management items.

Opposite each program in Tables 13.1 and 13.2 are printed the initial letters of all attributes for which the specific D-score for that program is equal to 1.5 or more. The reader can tell at a glance what capabilities the developers of every program claim for their programs. For example, SAS, the first program in Table 13.2, has some capabilities for all eight attributes, whereas SIR, the first program in Table 13.1, has capabilities only for the first two.

TABLE 13.1: <u>A RANKING OF THOSE PROGRAMS WITH USERS BY THEIR $\Omega$-SCORE, WITH</u> <u>INDICATIONS OF SIGNIFICANT SPECIFIC D-SCORES (>1.5). See text.</u>

| PROGRAM NUMBER | PROGRAM NAME | U | D | $\Omega$ | $\Delta$ | $D_f$ | $D_d$ | $D_t$ | $D_v$ | $D_s$ | $D_c$ | $D_e$ | $D_m$ |
|---|---|---|---|---|---|---|---|---|---|---|---|---|---|
| | | | | | | \multicolumn{8}{c}{D-Scores >1.5} | | | | | | | |
| 2.04 | SIR | $1^3$ | $1^3$ | $2^8$ | $2^9$ | f | d | | | | | | |
| 7.01 | CS | $2^2$ | $2^4$ | $2^6$ | $2^8$ | f | d | t | v | s | | e | m |
| 7.02 | SAS | $2^5$ | $2^7$ | $2^6$ | $2^8$ | f | d | t | v | s | c | e | m |
| 6.13 | DAS | $1^8$ | $1^9$ | $2^6$ | $2^7$ | f | d | t | | s | c | | m |
| 11.02 | B34S | $1^1$ | $1^2$ | $2^4$ | $2^6$ | f | | | | s | | e | |
| 4.07 | CRESCAT | $1^0$ | $1^0$ | $2^4$ | $2^5$ | f | | t | | | | | |
| 4.02 | LEDA | $1^2$ | $1^3$ | $2^4$ | $2^6$ | f | d | t | | | | | |
| 12.06 | EISPACK | $0^3$ | $0^4$ | $2^3$ | $3^0$ | | | | | | | | m |
| 12.02 | IMSL | $1^4$ | $1^6$ | $2^3$ | $2^7$ | | | | v | s | c | e | m |
| 9.11 | POPAN | $1^0$ | $1^0$ | $2^2$ | $2^4$ | f | | t | | | | | |
| 6.01 | BTFSS | $1^2$ | $1^3$ | $2^2$ | $2^3$ | f | | t | | | | | m |
| 6.14 | SPMS | $1^6$ | $1^7$ | $2^2$ | $2^3$ | f | d | | | s | | | m |
| 6.02 | EXPRESS | $1^5$ | $2^0$ | $2^2$ | $2^8$ | f | d | t | | | | | |
| 2.02 | RAPID | $0^7$ | $0^7$ | $2^2$ | $2^4$ | f | | | | | | | |
| 7.08 | GENSTAT 4.02 | $1^5$ | $1^9$ | $2^2$ | $2^6$ | f | | t | | s | c | | m |
| 6.15 | OSIRIS IV | $1^7$ | $2^2$ | $2^2$ | $2^7$ | f | d | t | v | s | | | |
| 6.09 | SPSS | $1^2$ | $1^3$ | $2^2$ | $2^5$ | f | | t | | s | | | |
| 4.10 | COCENTS | $0^8$ | $1^0$ | $2^1$ | $2^8$ | f | | t | | | | | |
| 4.06 | GTS | $0^8$ | $1^1$ | $2^1$ | $2^9$ | f | | t | | | | | |
| 9.06 | LNWD/NLNWD | $0^5$ | $0^7$ | $2^1$ | $2^8$ | | | | | | | | |
| 6.10 | SOUPAC | $1^2$ | $1^2$ | $2^1$ | $2^1$ | f | | | | | | | m |
| 4.04 | ISIS | $1^0$ | $1^2$ | $2^1$ | $2^4$ | f | d | t | | | | | |
| 7.06 | BMDP | $1^3$ | $1^8$ | $2^0$ | $2^5$ | f | d | t | | s | c | | |
| 12.04 | NMGS2 | $0^6$ | $0^8$ | $2^0$ | $2^6$ | | | | | | | | m |
| 6.12 | P-STAT 78 | $1^4$ | $1^9$ | $2^0$ | $2^6$ | f | d | t | | | | | m |
| 4.16 | RGSP | $1^0$ | $1^3$ | $2^0$ | $2^5$ | f | d | t | v | | | | |
| 11.05 | TSP | $0^7$ | $0^8$ | $2^0$ | $2^1$ | | | | | | | e | |
| 11.04 | SHAZAM | $0^9$ | $1^2$ | $2^0$ | $2^4$ | f | | | | s | | e | |
| 4.08 | PERSEE | $0^9$ | $1^2$ | $1^9$ | $2^7$ | f | | t | | | | | |
| 4.11 | CENTS-AID | $0^7$ | $1^0$ | $1^9$ | $2^4$ | f | | t | | | | | |
| 8.04 | CMU-DAP | $0^8$ | $0^9$ | $1^9$ | $2^1$ | | | | | | c | | |
| 2.03 | ADABAS | $0^6$ | $0^8$ | $1^9$ | $2^4$ | f | | | | | | | |
| 8.05 | RUMMAGE | $0^8$ | $1^0$ | $1^9$ | $2^2$ | | | | | | c | | |
| 2.08 | FILEBOL | $0^5$ | $0^7$ | $1^8$ | $2^7$ | | d | | | | | | |
| 7.10 | TROLL | $1^4$ | $2^0$ | $1^8$ | $2^4$ | f | d | t | | s | c | e | m |
| 8.09 | FORALL | $1^1$ | $1^3$ | $1^8$ | $2^1$ | | | t | | s | | | m |
| 12.05 | NAG LIBRARY | $0^4$ | $0^4$ | $1^8$ | $1^8$ | | | | | | | | m |
| 6.11 | DATATEXT | $1^2$ | $1^8$ | $1^7$ | $2^6$ | f | | t | | s | c | | |
| 6.07 | PACKAGE X | $1^1$ | $1^4$ | $1^7$ | $2^3$ | f | d | | | | | | |
| 7.05 | MINITAB II | $1^2$ | $1^9$ | $1^7$ | $2^6$ | f | d | t | | s | | e | m |
| 8.02 | GLIM | $0^8$ | $1^2$ | $1^7$ | $2^3$ | | | t | | s | c | | |
| 8.08 | AUTOGRP+ | $0^8$ | $1^5$ | $1^7$ | $2^7$ | f | | t | | | | | |
| 12.03 | REPOMAT | $0^6$ | $0^7$ | $1^7$ | $2^1$ | | | | | | | e | m |
| 9.01 | AMANCE | $1^0$ | $1^3$ | $1^7$ | $2^2$ | | | t | | s | | | |
| 5.09 | CLUSTERS | $0^4$ | $0^5$ | $1^6$ | $2^2$ | | | | v | | | | |
| 7.11 | IDA | $1^0$ | $1^7$ | $1^6$ | $2^4$ | f | | | | s | c | e | m |
| 4.15 | TPL | $0^9$ | $1^4$ | $1^6$ | $2^6$ | f | | t | v | | | | |
| 10.01 | GUHA | $0^4$ | $0^6$ | $1^6$ | $2^0$ | | | | | | c | | |
| 9.03 | REG | $0^8$ | $1^1$ | $1^6$ | $2^0$ | | | | | | | | |
| 2.06 | RIQS | $0^6$ | $1^0$ | $1^6$ | $2^6$ | f | | | | | | | |

TABLE 13.1 Contintued

| PROGRAM NUMBER | PROGRAM NAME | $U$ | $D$ | $\Omega$ | $\Delta$ | $D_f$ | $D_d$ | $D_t$ | $D_v$ | $D_s$ | $D_c$ | $D_e$ | $D_m$ |
|---|---|---|---|---|---|---|---|---|---|---|---|---|---|
| 9.12 | CAPTURE | $0^2$ | $0^3$ | $1^5$ | $2^3$ | | | | | | | | |
| 11.01 | TSP/DATATRAN | $0^9$ | $1^6$ | $1^5$ | $2^4$ | f | | | | s | | e | m |
| 10.06 | ECTA | $0^2$ | $0^3$ | $1^5$ | $2^0$ | | | | | | c | | |
| 9.05 | LSML76 | $0^6$ | $0^8$ | $1^5$ | $2^2$ | | | | v | | | | |
| 10.04 | MULTIQUAL | $0^4$ | $0^6$ | $1^4$ | $2^1$ | | | | | | c | | |
| 4.01 | SYNTAX II | $0^7$ | $1^4$ | $1^4$ | $2^6$ | f | d | t | | | | | |
| 5.10 | SUPERCARP | $0^4$ | $0^6$ | $1^4$ | $2^2$ | | | | v | | | | |
| 4.19 | SAP | $0^3$ | $0^5$ | $1^4$ | $1^6$ | | | | | | | | |
| 8.11 | STP | $0^5$ | $0^9$ | $1^3$ | $2^3$ | | | | | | | | |
| 6.03 | EASYTRIEVE | $0^9$ | $2^2$ | $1^3$ | $2^6$ | f | d | t | v | | | | |
| 5.02 | SPLITHALF | $0^4$ | $0^7$ | $1^3$ | $1^9$ | | | | v | | | | |
| 8.06 | CADA | $0^3$ | $0^5$ | $1^2$ | $2^0$ | | | | | | | | |
| 7.04 | HP STAT PACKS | $0^7$ | $1^6$ | $1^2$ | $2^5$ | | | t | | s | c | e | m |
| 6.05 | SURVEYOR/SURV | $1^0$ | $2^0$ | $1^2$ | $2^4$ | f | d | t | | s | c | | |
| 10.07 | MLLSA | $0^2$ | $0^3$ | $1^2$ | $1^7$ | | | | | | c | | |
| 8.03 | ISP | $0^6$ | $1^1$ | $1^2$ | $2^1$ | f | | | | | | | m |
| 5.04 | MWD VARIANCE | $0^4$ | $0^6$ | $1^2$ | $2^0$ | | | | | | | | |
| 10.03 | TAB-APL | $0^3$ | $0^3$ | $1^2$ | $1^3$ | | | | | | c | | |
| 5.05 | GSS EST. | $0^2$ | $0^5$ | $1^0$ | $1^8$ | | | | v | | | | |
| 11.03 | PACK | $0^3$ | $0^6$ | $1^0$ | $2^3$ | | | | | | | e | |
| 7.03 | OMNITAB 80 | $0^6$ | $2^0$ | $0^8$ | $2^6$ | f | d | t | v | s | c | | m |
| 2.09 | KPSIM/KPVER | $0^2$ | $0^5$ | $0^7$ | $1^7$ | | | | | | | | |
| 6.06 | DATAPLOT | $0^5$ | $1^8$ | $0^6$ | $2^3$ | f | d | t | v | | | e | m |
| 9.07 | ACPBCTET | $0^2$ | $0^4$ | $0^5$ | $1^8$ | | | | | | | | |

TABLE 13.2:  A RANKING OF ALL PROGRAMS BY THEIR OVERALL D-SCORES, WITH
INDICATIONS OF SIGNIFICANT SPECIFIC D-SCORES (>1.5). See Text.

| PROGRAM NUMBER | PROGRAM NAME | $D$ | $\Delta$ | $D_f$ | $D_d$ | $D_t$ | $D_v$ | $D_s$ | $D_c$ | $D_e$ | $D_m$ |
|---|---|---|---|---|---|---|---|---|---|---|---|
| | | | | D-Scores >1.5 | | | | | | | |
| 7.02 | SAS | $2^7$ | $2^8$ | f | d | t | v | s | c | e | m |
| 7.01 | CS | $2^4$ | $2^8$ | f | d | t | v | s | | e | m |
| 7.09 | SPEAKEASY III | $2^3$ | $2^8$ | f | d | t | v | s | | e | m |
| 6.15 | OSIRIS IV | $2^2$ | $2^7$ | f | d | t | v | s | | | |
| 6.03 | EASYTRIEVE | $2^2$ | $2^6$ | f | d | t | v | | | | |
| 6.04 | FOCUS | $2^1$ | $2^7$ | f | d | t | v | s | | | |
| 6.02 | EXPRESS | $2^0$ | $2^8$ | f | d | t | | | | | |
| 7.10 | TROLL | $2^0$ | $2^4$ | f | d | t | | s | c | e | m |
| 7.03 | OMNITAB 80 | $2^0$ | $2^6$ | f | d | t | v | s | c | | m |
| 6.05 | SURVEYOR/SURV | $2^0$ | $2^4$ | f | d | t | | s | c | | |
| 6.13 | DAS | $1^9$ | $2^7$ | f | d | t | | s | c | | m |
| 7.08 | GENSTAT 4.02 | $1^9$ | $2^6$ | f | | t | | s | c | | m |
| 6.12 | P-STAT 78 | $1^9$ | $2^6$ | f | d | t | | | | | m |
| 7.05 | MINITAB II | $1^9$ | $2^6$ | f | d | t | | s | | e | m |
| 7.06 | BMDP | $1^8$ | $2^5$ | f | d | t | | s | c | | |
| 6.11 | DATATEXT | $1^8$ | $2^6$ | f | | t | | s | c | | |
| 6.06 | DATAPLOT | $1^8$ | $2^3$ | f | d | t | v | | | e | m |
| 7.11 | IDA | $1^7$ | $2^4$ | f | | t | | s | c | e | m |
| 6.08 | SCSS | $1^7$ | $2^5$ | f | d | t | | | | | |
| 6.14 | SPMS | $1^7$ | $2^3$ | f | d | | | s | | | |
| 8.07 | SURVO | $1^7$ | $2^4$ | f | d | t | | | | e | |
| 7.07 | NISAN | $1^7$ | $2^1$ | f | | t | v | s | c | e | m |
| 11.01 | TSP/DATATRAN | $1^6$ | $2^4$ | f | | | | s | | e | m |
| 12.02 | IMSL LIBRARY | $1^6$ | $2^7$ | | | | v | s | c | e | m |
| 7.04 | HPSTAT PACKS | $1^6$ | $2^5$ | | | t | | s | c | e | m |
| 9.02 | MAC/STAT | $1^5$ | $2^4$ | f | | t | | | c | | m |
| 8.01 | ISA | $1^5$ | $2^0$ | f | | t | | s | c | | m |
| 8.08 | AUTOGRP+ | $1^5$ | $2^7$ | f | | t | | | | | |
| 6.07 | PACKAGE X | $1^4$ | $2^3$ | f | d | | | | | | |
| 4.01 | SYNTAX II | $1^4$ | $2^6$ | f | d | t | | | | | |
| 4.15 | TPL | $1^4$ | $2^6$ | f | | t | v | | | | |
| 2.04 | SIR | $1^3$ | $2^9$ | f | d | | | | | | |
| 6.09 | SPSS | $1^3$ | $2^5$ | f | | t | | s | | | |
| 4.02 | LEDA | $1^3$ | $2^6$ | f | d | | | | | | |
| 8.09 | FORALL | $1^3$ | $2^1$ | | | t | | s | | | m |
| 6.01 | BTFSS | $1^3$ | $2^3$ | f | | t | | | | | m |
| 4.16 | RGSP | $1^3$ | $2^5$ | f | d | t | v | | | | |
| 9.01 | AMANCE | $1^3$ | $2^2$ | | | t | | s | | | |
| 11.02 | B34S | $1^2$ | $2^6$ | f | | | | s | | e | |
| 6.10 | SOUPAC | $1^2$ | $2^1$ | f | | | | | | | m |
| 4.08 | PERSEE | $1^2$ | $2^7$ | f | | t | | | | | |
| 11.06 | QUAIL | $1^2$ | $2^5$ | f | | | | | | e | |
| 2.05 | MARK IV | $1^2$ | $2^6$ | f | d | | | | | | |
| 4.04 | ISIS | $1^2$ | $2^4$ | f | d | t | | | | | |
| 11.04 | SHAZAM | $1^2$ | $2^4$ | f | | | | s | | e | |
| 8.10 | AQD | $1^2$ | $2^2$ | f | | | | s | | | |
| 8.02 | GLIM | $1^2$ | $2^3$ | | | t | | s | c | | |
| 11.07 | KEIS/ORACLE | $1^1$ | $2^2$ | f | | t | | | | e | |

TABLE 13.2 Continued

| PROGRAM NUMBER | PROGRAM NAME | $D$ | $\Delta$ | $D_f$ | $D_d$ | $D_t$ | $D_v$ | $D_s$ | $D_c$ | $D_e$ | $D_m$ |
|---|---|---|---|---|---|---|---|---|---|---|---|
| 9.04 | TPD-3 | $1^1$ | $1^9$ | | | | | s | | | |
| 4.09 | CENTS III | $1^1$ | $2^8$ | f | | t | | | | | |
| 4.05 | VDBS | $1^1$ | $2^6$ | f | | t | | | | | |
| 4.06 | GTS | $1^1$ | $2^9$ | f | | t | | | | | |
| 9.03 | REG | $1^1$ | $2^0$ | | | | | | | | |
| 8.03 | ISP | $1^1$ | $2^1$ | f | | | | | | | m |
| 12.01 | DATAPAC | $1^1$ | $1^9$ | | d | | | | | | |
| 4.18 | SURVEY | $1^0$ | $2^2$ | | d | t | | | | | |
| 4.07 | CRESCAT | $1^0$ | $2^5$ | f | | t | | | | | |
| 9.11 | POPAN | $1^0$ | $2^4$ | f | | t | | | | | |
| 4.10 | COCENTS | $1^0$ | $2^8$ | f | | t | | | | | |
| 8.05 | RUMMAGE | $1^0$ | $2^2$ | | | | | | c | | |
| 2.06 | RIQS | $1^0$ | $2^6$ | f | | | | | | | |
| 4.11 | CENTS-AID | $1^0$ | $2^4$ | f | | t | | | | | |
| 8.13 | STATUTIL | $0^9$ | $2^0$ | f | | | | | | | m |
| 4.03 | CYBER GENINT | $0^9$ | $2^2$ | | d | t | | | | | |
| 9.10 | ALLOC | $0^9$ | $2^3$ | | | t | | | | | |
| 8.04 | CMU-DAP | $0^9$ | $2^1$ | | | | | | c | | |
| 8.11 | STP | $0^9$ | $2^3$ | | | | | | | | |
| 8.12 | STATPAK | $0^8$ | $2^0$ | | | | | | | | |
| 9.05 | LSML76 | $0^8$ | $2^2$ | | | | v | | | | |
| 5.01 | HES VAR-X TAB | $0^8$ | $2^1$ | f | | t | v | | | | |
| 4.12 | FILE 2.0 | $0^8$ | $2^4$ | | d | t | | | | | |
| 3.02 | CONCOR | $0^8$ | $2^5$ | | d | | | | | | |
| 3.01 | GES | $0^8$ | $2^3$ | f | d | | | | | | |
| 11.05 | TSP | $0^8$ | $2^1$ | | | | | | | e | |
| 2.03 | ADABAS | $0^8$ | $2^4$ | f | | | | | | | |
| 5.03 | MULTI-FRAME | $0^8$ | $2^0$ | f | | t | v | | | | |
| 2.07 | DATA3 | $0^8$ | $2^1$ | f | | | | | | | |
| 12.04 | NMGS2 | $0^8$ | $2^6$ | | | | | | | | m |
| 2.02 | RAPID | $0^7$ | $2^4$ | f | | | | | | | |
| 8.14 | MICROSTAT | $0^7$ | $2^0$ | | | | | | | | |
| 4.17 | NCHS X-TAB | $0^7$ | $2^0$ | | | t | | | | | |
| 4.13 | OSIRIS 2.4 | $0^7$ | $2^4$ | | | t | | | | | |
| 12.03 | REPOMAT | $0^7$ | $2^1$ | | | | | | | e | m |
| 5.02 | SPLITHALF | $0^7$ | $1^9$ | | | | v | | | | |
| 9.06 | LNWD/NLNWD | $0^7$ | $2^8$ | | | | | | | | |
| 2.08 | FILEBOL | $0^7$ | $2^7$ | | d | | | | | | |
| 10.04 | MULTIQUAL | $0^6$ | $2^1$ | | | | | | c | | |
| 11.03 | PACK | $0^6$ | $2^3$ | | | | | | | e | |
| 3.04 | EDITCK | $0^6$ | $2^1$ | | d | | | | | | |
| 10.01 | GUHA | $0^6$ | $2^0$ | | | | | | c | | |
| 5.04 | MWD VARIANCE | $0^6$ | $2^0$ | | | | | | | | |
| 5.10 | SUPER CARP | $0^6$ | $2^2$ | | | | v | | | | |
| 8.15 | A-STAT 79 | $0^5$ | $2^0$ | | | t | | | | | |
| 3.05 | VCP-LCP | $0^5$ | $2^0$ | | | | | | | | |
| 4.19 | SAP | $0^5$ | $1^6$ | | | | | | | | |
| 8.06 | CADA | $0^5$ | $2^0$ | | | | | | | | |
| 5.05 | GSS EST. | $0^5$ | $1^8$ | | | | v | | | | |
| 2.09 | KPSIM/KPVER | $0^5$ | $1^7$ | | | | | | | | |
| 10.02 | CATFIT | $0^5$ | $2^2$ | | | t | | c | | | |

TABLE 13.2 Continued

| PROGRAM NUMBER | PROGRAM NAME | $D$ | $\Delta$ | $D_f$ | $D_d$ | $D_t$ | $D_v$ | $D_s$ | $D_c$ | $D_e$ | $D_m$ |
|---|---|---|---|---|---|---|---|---|---|---|---|
| 5.06 | CLFS VAR-COV | $0^5$ | $2^6$ | | | | v | | | | |
| 5.09 | CLUSTERS | $0^5$ | $2^2$ | | | | v | | | | |
| 9.09 | STATSPLINE | $0^4$ | $1^5$ | | | | | | | | |
| 4.14 | GEN. SUMMARY | $0^4$ | $1^8$ | | | t | | | | | |
| 3.03 | CHARO | $0^4$ | $2^4$ | | | | | | | | |
| 2.01 | UPDATE | $0^4$ | $2^4$ | | | | | | | | |
| 5.11 | GENVAR | $0^4$ | $2^0$ | | | | v | | | | |
| 5.07 | CESVP | $0^4$ | $2^0$ | | | | v | | | | |
| 12.06 | EISPACK | $0^4$ | $3^0$ | | | | | | | | m |
| 12.05 | NAG LIBRARY | $0^4$ | $1^8$ | | | | | | | | m |
| 9.07 | ACPBCTET | $0^4$ | $1^8$ | | | | | | | | |
| 10.05 | C-TAB | $0^3$ | $1^3$ | | | | | | c | | |
| 5.08 | KEYFITZ | $0^3$ | $1^8$ | | | | v | | | | |
| 10.03 | TAB-APL | $0^3$ | $1^3$ | | | | | | c | | |
| 9.12 | CAPTURE | $0^3$ | $2^3$ | | | | | | | | |
| 10.06 | ECTA | $0^3$ | $2^0$ | | | | | | c | | |
| 10.07 | MLLSA | $0^3$ | $1^7$ | | | | | | c | | |
| 9.08 | MULPRES | $0^1$ | $1^0$ | | | | | | | | |

APPENDIX

CONTENTS

## QUESTIONNAIRE - Q2b

### FEATURES OF A STATISTICAL PROGRAM OR PACKAGE

*To be filled out by the developer or distributor of the program*

*   This questionnaire was designed to help a user answer the following questions:

    i) <u>Capabilities</u>:  Was the program designed to help solve problems like mine?

    ii) <u>Portability</u>:  Can the program be transported conveniently to my computer?

    iii) <u>Ease of learning and using</u>:  Is the program sufficiently easy to learn and use that it will actually be useful in solving problems?

    iv) <u>Reliability</u>:  Is the program maintained by some reliable organization, and has it been extensively tested for accuracy?

*   The structure of the questionnaire is designed to address these four questions in turn:

    I.   <u>General Information</u>:  questions 1 to 5

    II.  <u>Program Capabilities</u>:  questions 6 to 48

        A.  Architecture and Generality of Purpose

        B.  Processing and Displaying Data

            B1  File building and manipulation

            B2  Editing - Error detection and reporting

            B3  Handling of erroneous or missing data

            B4  Tabulation, display, and graphics

        C.  Mathematical Analysis of Statistical Data

            C1  User language and extensibility

            C2  Capabilities of procedures for statistical analysis

            C3  Selection and analysis of complex sample surveys

            C4  Capabilities for other mathematical procedures

    III. <u>Portability</u>:  questions 49 to 54

    IV.  <u>Ease of Learning and Using</u>:  questions 55 to 59

    V.   <u>Reliability and Maintenance</u>:  questions 60 to 61

1

## QUESTIONNAIRE

*RATING INSTRUCTIONS*:  *Throughout this questionnaire, please use the 0-3 or a-d rating schemes beside each question. For the 0-3 scheme please refer to the ratings below. Please circle the* number *or* letter *of your answer. Please circle* one *and* only one *answer per question, but you may add written comments to any answer on the back of that page.*

*0 to 3 Ratings*:

0   *No facilities in this area, or not an intended purpose.*

1   *A few functions in this area are present, or a minor purpose or byproduct.*

2   *Moderate capabilities, or a significant purpose.*

3   *Complete coverage in all aspects of the area, or one of the principal purposes. (In the statement of all of the questions with an 0-3 rating scheme, we have defined what we consider to be complete coverage.)*

*a to d Ratings*:

*These describe a program's generality of use to people other than the developer, ranging from "a" which is "low", to "d" which is "high". For precise meaning, reference must be made to the specific question.*

1)   NAME OF PROGRAM OR PACKAGE AND VERSION NUMBER (include both acronym and complete name):

2)   NAME AND ADDRESS OF PROGRAM DEVELOPER:

3)   NAME AND ADDRESS OF CURRENT PROGRAM DISTRIBUTOR (if different from above):

4)   TODAY'S DATE:

5)   NAME AND ADDRESS OF PERSON FILLING OUT THIS QUESTIONNAIRE:

2

## II.  PROGRAM CAPABILITIES

### A.  Architecture and Generality of Purpose

6)  How would you categorize your program's architecture?

Special or single program or subroutine...................... a

A collection of distinct programs or subroutines............. b

Integrated package or program limited in some respect
(e.g. lack of filing system or high level language)......... c

Unified and extensible filing and computing system........... d

7)  Concerning the generality of purpose, this program or package is

An educational tool only..................................... a

Special purpose program or algorithms........................ b

A package useful for a specific set of tasks (e.g. analysis
of repeated experiments or surveys, or tabulation, or data
management).................................................. c

A general and extensible system for scientific and
statistical data exploration and analysis.................... d

### B.  Processing and Displaying Data

#### B1.  *File building and manipulation*

8)  Summary question for this section:  The purpose of
this program is file building and manipulation          0   1   2   3

9)  What is the largest sized data set that your program will
handle efficiently?

Small data sets - under 1000 raw data records................ a

1000 - 10,000 raw data records............................... b

10,000 - 200,000 raw data records............................ c

200,000 and up raw data records.............................. d

10)  Flexibility of data input:  alphabetic characters, free
format, flexible format, variety of sources and data
types, multiple record lengths and types.................0   1   2   3

11)  Ability to define and file complex data structures:
hierarchical, matrices, vectors, tables, variable by
case or case by variable.................................0   1   2   3

3

12) <u>Flexibility</u> <u>in</u> <u>handling</u> <u>missing</u> <u>data</u>:  many possible
    codes, information stored with file, automatic ex-
    clusion from operations with manual override, choice
    of algorithms for deletion of cases with missing data... 0   1   2   3

13) <u>Power</u> <u>of</u> <u>file</u> <u>storage</u> <u>and</u> <u>retrieval</u> <u>system</u>:  storage,
    retrieval, and portability of all file structures and
    intermediate results of computations, optional for-
    mats for transportability, optional binary storage,
    storage of labels and codebook information............. 0   1   2   3

14) <u>Power</u> <u>in</u> <u>file</u> <u>manipulation</u>:  rearranging structures
    (e.g. hierarchical to rectangular), merging, sorting,
    updating, subfile processing........................... 0   1   2   3

15) <u>Flexibility</u> <u>of</u> <u>labelling</u> <u>facilities</u>:  storage of
    user-defined labels of arbitrary length for struc-
    tures, variables and values, choice of outputting
    labels, names, or both................................. 0   1   2   3

### B2.  *Editing--Error detection and reporting*

16) <u>Summary</u> <u>question</u> <u>for</u> <u>this</u> <u>section</u>:  a purpose of this
    program is editing *(If you anser "0" to this ques-
    tion you may skip to section B3.)*..................... 0   1   2   3

17) <u>Power</u> <u>and</u> <u>flexibility</u> <u>of</u> <u>specialized</u> <u>editing</u> <u>capa-</u>
    <u>bilities</u>:  filing system for storing intermediate
    results, high level editing language as opposed to
    a set of conditional source-code-like commands,
    ability to flag and recall errors, adaptable to wide
    variety of data types.................................. 0   1   2   3

18) <u>Power</u> <u>of</u> <u>consistency</u> <u>checking</u> <u>facilities</u>:  range
    checks, wild code checks, logical checks between
    items, arithmetic consistency checks, file struc-
    ture and skip checks................................... 0   1   2   3

19) <u>Variety</u> <u>of</u> <u>probabilistic</u> <u>checks</u>:  univariate checks,
    probability plots, multivariate checks, e.g. gener-
    alized distance and leverage for individual points,
    detection of isolated clusters......................... 0   1   2   3

20) <u>Ease</u> <u>of</u> <u>handling</u> <u>records</u> <u>in</u> <u>error</u>:  informative and
    automatic cataloguing and display of records and vari-
    ables in error, user choice and convenience in record
    deletion or recoding and correcting old file, collec-
    tion of error statistics by error type and by variable. 0   1   2   3

4

B3.   *Handling of erroneous or missing data*

21)   Summary question for this section:  a purpose of this
program is to handle a variety of kinds of erroneous
or missing data. *(If you answer "0" to this question
you may skip to Section B4.)*........................... 0   1   2   3

22)   Flexibility and power of procedures for imputation
(using relationships between variates, or other
techniques to automatically replace faulty or missing
items):  user choice of imputation options, such as
hot deck, cold deck, regression or user supplied
formula; automatic flagging of imputed values......... 0   1   2   3

B4.   *Tabulation, display, graphics*

23)   Summary question for this section:  a purpose of this
program is tabulation, display or graphics. *(If you
answer "0" to this question you may skip to section C.).* 0   1   2   3

24)   Power and flexibility in computing tables:  ability to
recode variables, produce multi-way hierarchical tables,
weighted summary tables, permit user control of arithmetic operations on the table, e.g. percentages, totals,
means, standard deviations, quantiles, skewness, for
any dimension........................................ 0   1   2   3

25)   Flexibility in printing tables:  automatic placement,
according to any user-specified format, of numbers,
labels and titles, including printing of multiway tables
with nesting and concatenation in several dimensions... 0   1   2   3

26)   Power of graphics capabilities:  user control of variables to be displayed, of scales and plot characters,
choice of line printer, video, or special graphics hardware; multiple, three-dimensional and contour plots.... 0   1   2   3

C.   Mathematical Analysis of Statistical Data

C1.   *User language and extensibility*

27)   Level of user language for mathematics and statistics:
high level language that will facilitate the writing
of transformations, new statistical procedures, macros,
and operating on matrices............................. 0   1   2   3

28) How accessible is the source code for user modification?

    Source code private; not supplied with program........                              a

    Source code supplied, but is not internally commented
    or documented - modifications by a user would be very
    difficult...............................................                             b

    Source code supplied, the code is internally commented -
    a systems level programmer would be able to make modifi-
    cations without undue troubles.........................                             c

    Source code is supplied, the code is structured, well
    documented and commented - a user familiar with the
    source language would have little trouble in writing
    modifications and extensions..........................                              d

## C2.  *Capabilities of procedures for statistical analysis*

29) Summary question for this section:  a purpose of this
    program is analysis of statistical data.  *(If you answer
    "0" to this question you may skip to section C3.)*......  0  1  2  3

30) Multiple regression:  stepwise regression, all possible
    regressions, ridge regression, wide variety of residual
    plots available, summary statistics (Durbin-Watson, etc.)
    regression through origin.............................  0  1  2  3

31) ANOVA and the general linear model:  Tests of hypotheses
    and weighted least squares estimates of arbitrary speci-
    fied estimable functions when the design matrix has less
    than full rank, and the covariance matrix of the resid-
    uals is assumed to be known but not scalar; estimated
    covariance matrix of such estimates, recovery of intra-
    block information, user-specified transformations, probit
    and logit analysis....................................  0  1  2  3

32) Linear multivariate analysis:  Principal component analyses
    and plots, discriminant analysis with estimated probabili-
    ties of misclassification, Hotelling's $T^2$, multivariate
    analysis of variance, canonical regression............  0  1  2  3

33) Multi-way contingency tables:  Fitting of log-linear
    models, estimates of multiplicative and linear para-
    meters, printing of multi-way and marginal tables,
    tests for association, orthogonal contrasts, likeli-
    hood ratio and Pearsonian $\chi^2$, partitioned $\chi^2$, model
    selection procedures, user control of dependent vari-
    able and constant added to cells......................  0  1  2  3

34) Other multidimensional techniques:  Factor analysis,
    cluster analysis, multidimensional scaling, AID, MCA,
    etc...................................................  0  1  2  3

35) <u>Time series analysis</u>: Box-Jenkins model building
and forecasting techniques, cross-spectral analysis,
multiple model building, simultaneous time and frequency domain analysis................................ 0  1  2  3

36) <u>Nonparametric techniques</u>: Significance levels (or many
critical values) for test statistics corresponding to a
wide variety of commonly used nonparametric techniques
(e.g., Wilcoxon matched-pairs signed-ranks, Fisher
exact, Mann-Whitney, Kolmogorov-Smirnov, one and two
sample, Kruskal-Wallis analysis of variance, Wilcoxon
two-sample for censored data, etc.).................... 0  1  2  3

37) <u>Exploratory methods</u>: Stem and leaf display, letter values,
schematic plots, alias box plots, spread vs. level plots,
resistant smoothing, median polish, diagnostic plots,
product-ratio plot..................................... 0  1  2  3

38) <u>Robust methods</u>: Measures of location: median, trimmed
mean, M-estimators (Huber, sine, bi-weight), outmean,
adaptive location estimator. Measures of scale: interquartile range, median absolute deviation from the median.
Linear and non-linear models: fit by least absolute
residuals and iteratively reweighted least squares..... 0  1  2  3

39) <u>General curve fitting</u>: e.g., including non-linear regression, splines, etc..................................... 0  1  2  3

40) <u>Bayesian methods</u>: Ability to determine probabilities
corresponding to common posterior and preposterior
distributions (e.g., beta, beta-binomial, normal-gamma,
gamma-2, Student, inverted-beta-2, etc.), utility and
probability assessment................................. 0  1  2  3

41) <u>Econometric methods</u>: Two and three stage least
squares estimation of simultaneous linear and non-
linear equations, distributed lag models, auto-
correlated error terms, pooling of time series and
cross-sectional data, qualitative dependent variables, time-varying parameter models.................. 0  1  2  3

*Please describe other statistical procedures not mentioned in
section C2 on a separate sheet and attach to the questionnaire.*

## C3.  *Selection and analysis of complex sample surveys*

42) <u>Summary question for this section</u>: a purpose of this
program is the selection or analysis of data for complex sample surveys. *(If you answer "0" to this
question you may skip to section C4.)*................. 0  1  2  3

43) <u>Facilities</u> <u>for</u> <u>computing</u> <u>point</u> <u>estimates</u>: ability
to compute weighted linear (means, totals) and
nonlinear (ratios, correlations, regression coefficients) point estimates for a wide variety of complex
probability sample designs............................ 0  1  2  3

44) <u>Facilities</u> <u>for</u> <u>computing</u> <u>estimates</u> <u>of</u> <u>variance</u> <u>for</u>
<u>complex</u> <u>survey</u> <u>designs</u>: ability to compute estimates
of variance of weighted linear and non-linear estima-
tors with user supplied formula, Taylor approximation, replicated sampling, etc........................ 0  1  2  3

45) <u>Facilities</u> <u>for</u> <u>selecting</u> <u>probability</u> <u>samples</u>:
automatic selection of any specified complex proba-
bility sample from a machine-readable frame, stratified, clustered, etc., random-digit-dial sample........ 0  1  2  3

### C4. *Capabilities for other mathematical procedures*

46) <u>Simulation</u>: Generation of pseudo-random observations from (i) continuous distributions: uniform
(contaminated) Gaussian, exponential, gamma, $\chi^2$, t,
F, beta, lambda family, symmetric stable laws, multivariate Gaussian; (ii) discrete distributions:
uniform, binomial, multinomial, Poisson, hypergeometric, negative binomial, other user-specified
distributions. User choice of starting value.
Algorithm documented and tested....................... 0  1  2  3

47) <u>Mathematical</u> <u>functions</u>: Matrix arithmetic, eigenanalysis, orthogonal and singular value decompo-
sition, quadrature, values and integrals of probability distribution functions......................... 0  1  2  3

48) <u>Operations</u> <u>research</u>: Linear, integer, and non-
linear programming, dynamic programming, input-
output analysis, queueing and Markov models........... 0  1  2  3

*Please describe other mathematical procedures not mentioned in
section C4 on a separate sheet and attach to the questionnaire.*

8

## III. PORTABILITY

49) The program is

Not exported at present, but developer
might be willing to make it available................... a

Currently available via a time-sharing network
or service bureau (but not (c))........................ b

Currently available for export to other installations (but not (b))................................. c

Both (b) and (c)...................................... d

50) At how many institutions (computer centers) is this
program currently available?

One institution only.................................. a

Two to ten............................................ b

Eleven to fifty....................................... c

Fifty-one or more..................................... d

51) On how many makes of computer (brand name/series, e.g. PDP/11)
has a version of this program been successfully installed and
run? (Several plug-compatible series, such as IBM 360 and IBM
370 and Amdahl 470, should be considered as one make.)

Only one, and very machine dependent.................. a

One, but easily modified for other machines........... b

Two to four........................................... c

Five or more.......................................... d

52) Can this program be run on a mini, or small computer?

Not possible.......................................... a

A small version might be possible but would require
major modifications which would limit the program's
flexibility........................................... b

A mini version is under development................... c

An existing version of this program will run on a
small or mini computer................................ d

9

53) What are the physical core requirements for a typical problem? For programs which run on a variety of operating systems or computers please respond according to the most widely used configuration.

    191 K characters (bytes) or more........................          a

    111 K to 190 K characters (bytes)......................          b

    49 K to 110 K characters (bytes).......................          c

    48 K or less characters (bytes)........................          d

54) The operating environment of your program is

    Batch execution only...................................          a

    Partially interactive execution (e.g., interactive syntactic checking of the user program with batch execution).............................................          b

    Interactive execution only (not including RJE, TSO, etc.)..................................................          c

    Either batch or interactive execution (not including RJE, TSO, etc.)........................................          d

## IV. EASE OF LEARNING AND USING

55) Concerning the intended users of this program, what amount of training in <u>statistics</u> should the typical user have had in order to be an <u>effective</u> user?

    A bachelor's degree in statistics, or equivalent professional training and experience...................          a

    A year's study or experience in applied statistics.....          b

    An elementary, college-level course in statistics, or equivalent.........................................          c

    Virtually no training in statistics....................          d

56) Concerning the intended users again, what amount of
training in <u>computing</u> or in <u>computer</u> <u>science</u> should the
typical user have had in order to be an <u>effective</u> user?

A bachelor's degree in computing, or equivalent professional training and experience......................       a

A year's study or experience in computing applications.     b

An elementary, college-level course in computing or
equivalent.............................................     c

Virtually no training in computing.....................     d

57) User command language consists mainly of

Fixed position alpha or numeric codes..................     a

Codes in fixed order with punctuation to indicate
omitted codes..........................................     b

Free field alphanumeric commands with specified syntax.     c

English-like (or other natural language) verbs and nouns
or sentences...........................................     d

58) The following documentation is available for the program:

No documentation.......................................     a

General user instructions..............................     b

User manual and systems guide..........................     c

User manual, systems guide, novice or special use
manuals, details of computational methods..............     d

59) <u>Level</u> <u>of</u> <u>other</u> <u>user</u> <u>conveniences</u>:  informative error
messages, user protection and warning system, clearly
labelled and readable.output..........................  0  1  2  3

### V.  RELIABILITY AND MAINTENANCE

60) The program

Is not being maintained................................     a

Has limited maintenance................................     b

Has full organizational maintenance, but is not being
further developed......................................     c

Has full organizational maintenance, and is under active
development, evolution and updating....................     d

11

61) Has the program been tested for statistical accuracy?

Testing still underway................................                a

Yes, tested internally on unpublished data............                b

Yes, all procedures tested with published data, so
that user can reproduce test results..................                c

Yes, extensive external tests by an independent
group, or extensive comparative tests using published data with known solutions, have been done
and published.........................................                d

TABLE A.1:  RATINGS OF A STATISTICAL PROGRAM:  PART I
CAPABILITIES:  PROCESSING AND DISPLAYING DATA

Usefulness Ratings
0=low
1=modest
2=moderate
3=high
-=no data

Tabular Scheme

| α |
| β |
| γ |

Note: $2^7$=2.7

α=Developer rating
β=Average user rating
γ=Number of users

Column groups: GENERALITY (6 Subroutines or Package, 7 Generality of Purpose); FILING (8 Filing Summary, 9 Data Set Size, 10 Flexible Data Input, 11 Complex Structures, 12 Filing Missing Data, 13 Storage/Retrieval, 14 File Manipulation, 15 Flexible Output); EDITING (16 Editing Summary, 17 Editing Language, 18 Consistency Checks, 19 Probability Checks, 20 Error Handling); MISSING DATA (21 Missing Data Summary, 22 Imputation); DISPLAY (23 Display Summary, 24 Compute Tables, 25 Print Tables, 26 Graphics)

| Chapter | Program | 6 | 7 | 8 | 9 | 10 | 11 | 12 | 13 | 14 | 15 | 16 | 17 | 18 | 19 | 20 | 21 | 22 | 23 | 24 | 25 | 26 |
|---|---|---|---|---|---|---|---|---|---|---|---|---|---|---|---|---|---|---|---|---|---|---|
| Chapter 2 | UPDATE | 0 | 1 | 3 | 3 | 2 | 0 | 0 | 0 | 3 | 0 | 0 | 0 | 0 | 0 | 0 | 0 | 0 | 0 | 0 | 0 | 0 |
| | RAPID | 3 | 3 | 3 | 3 | 2 | 2 | 1 | 3 | 2 | 2 | 0 | 0 | 0 | 0 | 0 | 0 | 0 | 0 | 0 | 0 | 0 |
| | | $2^3$ | $2^3$ | $2^7$ | 3 | 2 | $1^7$ | $0^7$ | $2^7$ | $2^3$ | 2 | $0^7$ | 0 | 0 | 0 | $0^3$ | 0 | 0 | $0^3$ | 0 | $0^3$ | 0 |
| | | 3 | 3 | 3 | 3 | 3 | 3 | 3 | 3 | 3 | 3 | 3 | 3 | 3 | 3 | 3 | 3 | 2 | 3 | 2 | 3 | 2 |
| | ADABAS | 3 | 3 | 3 | 3 | 3 | 2 | 0 | 2 | 2 | 2 | 0 | 0 | 0 | 0 | 0 | 0 | 0 | 2 | 0 | 2 | 0 |
| | | 3 | 2 | 3 | 3 | 3 | 2 | 3 | 3 | 3 | 1 | 0 | 0 | 0 | 0 | 0 | - | 0 | 0 | 0 | 0 | 0 |
| | | 1 | 1 | 1 | 1 | 1 | 1 | 1 | 1 | 1 | 1 | 1 | 1 | 1 | 1 | 1 | 0 | 1 | 1 | 1 | 1 | 1 |
| | SIR | 3 | 2 | 3 | 3 | 3 | 3 | 3 | 3 | 3 | 3 | 3 | 3 | 3 | 0 | 3 | 1 | 1 | 1 | 1 | 0 | 2 |
| | | 3 | 2 | 3 | 3 | 3 | $2^7$ | $2^3$ | 3 | 3 | 3 | $2^7$ | $2^7$ | 3 | 0 | 3 | $2^3$ | $0^7$ | 2 | $1^3$ | $1^7$ | $1^3$ |
| | | 3 | 3 | 3 | 3 | 2 | 3 | 3 | 3 | 3 | 3 | 3 | 3 | 3 | 2 | 3 | 3 | 3 | 3 | 3 | 3 | 3 |
| | MARK IV | 3 | 2 | 3 | 3 | 3 | 3 | 1 | 3 | 3 | 2 | 2 | 2 | 2 | 0 | 2 | 0 | 0 | 1 | 1 | 1 | 1 |
| | RIQS | 3 | 3 | 3 | 2 | 3 | 1 | 1 | 3 | 3 | 3 | 1 | 2 | 1 | 0 | 1 | 0 | 0 | 2 | 0 | 2 | 3 |
| | | 2 | 2 | - | 2 | 2 | 1 | 1 | 2 | 2 | 2 | 0 | 0 | 0 | 0 | 0 | 0 | 0 | 0 | 0 | 0 | 2 |
| | | 1 | 1 | 0 | 1 | 1 | 1 | 1 | 1 | 1 | 1 | 1 | 1 | 1 | 1 | 1 | 1 | 1 | 1 | 1 | 1 | 1 |
| | DATA3 | 2 | 2 | 3 | 1 | 3 | 2 | 2 | 1 | 2 | 2 | 3 | 2 | 1 | 1 | 0 | 1 | 0 | 1 | 0 | 0 | 0 |
| | FILEBOL | 0 | 2 | 3 | 3 | 3 | 0 | 0 | 0 | 1 | 0 | 3 | 2 | 1 | 0 | 3 | 1 | 0 | 0 | 0 | 0 | 0 |
| | | 0 | 2 | 3 | $2^5$ | 2 | 1 | $1^5$ | $1^5$ | $2^5$ | 0 | $1^5$ | $1^5$ | 2 | 0 | $2^5$ | $1^5$ | 1 | 1 | $0^5$ | $0^5$ | 0 |
| | | 1 | 2 | 2 | 2 | 2 | 2 | 2 | 2 | 2 | 2 | 2 | 2 | 2 | 1 | 2 | 2 | 1 | 2 | 2 | 2 | 2 |
| | KPSIM/KPVER | 0 | 2 | 2 | 1 | 1 | 2 | 1 | 0 | 0 | 0 | 1 | 1 | 2 | 0 | 2 | 0 | 0 | 0 | 0 | 0 | 0 |
| | | 0 | 2 | 3 | - | 2 | 2 | 0 | 0 | 0 | - | 0 | 0 | 0 | 0 | 0 | 3 | 3 | 0 | 0 | 0 | 0 |
| | | 1 | 1 | 1 | 0 | 1 | 1 | 1 | 1 | 1 | 0 | 1 | 1 | 1 | 1 | 1 | 1 | 1 | 1 | 1 | 1 | 1 |
| Chapter 3 | GES | 3 | 2 | 2 | 3 | 1 | 2 | 1 | 1 | 2 | 1 | 3 | 1 | 3 | 1 | 3 | 1 | 1 | 0 | 0 | 0 | 0 |
| | CONCOR | 2 | 2 | 0 | 3 | 2 | 1 | 0 | 1 | 2 | 0 | 3 | 3 | 3 | 0 | 3 | 3 | 2 | 0 | 0 | 0 | 0 |
| | CHARO | 0 | 2 | 0 | 3 | 0 | 0 | 0 | 0 | 0 | 0 | 3 | 2 | 3 | 0 | 0 | 0 | 0 | 3 | 0 | 0 | 0 |
| | EDITCK | 0 | 1 | 0 | 1 | 2 | 0 | 0 | 0 | 0 | 2 | 3 | 2 | 3 | 0 | 3 | 0 | 0 | 0 | 0 | 0 | 0 |
| | VCP-LCP | 0 | 1 | 2 | 3 | 2 | 0 | 1 | 1 | 0 | 0 | 3 | 0 | 3 | 0 | 2 | 0 | 0 | 0 | 0 | 0 | 0 |
| Chapter 4 (part i) | SYNTAX II | 3 | 3 | 3 | 3 | 3 | 3 | 3 | 3 | 2 | 3 | 3 | 2 | 3 | 0 | 2 | 1 | 1 | 3 | 3 | 3 | 0 |
| | | 2 | 2 | 2 | 3 | 3 | 1 | 2 | 0 | 1 | 3 | 0 | 0 | 0 | 0 | 0 | 0 | 0 | 3 | 2 | 1 | 0 |
| | | 1 | 1 | 1 | 1 | 1 | 1 | 1 | 1 | 1 | 1 | 1 | 1 | 1 | 1 | 1 | 1 | 1 | 1 | 1 | 1 | 1 |
| | LEDA | 2 | 2 | 3 | 3 | 1 | 2 | 2 | 3 | 3 | 3 | 3 | 3 | 2 | 0 | 3 | 2 | 2 | 3 | 3 | 2 | 0 |
| | | $2^5$ | 2 | 3 | 3 | 2 | 3 | 3 | 3 | 3 | 2 | $2^5$ | 2 | $2^5$ | 0 | $2^5$ | 3 | 3 | 3 | $2^5$ | 3 | 0 |
| | | 2 | 2 | 2 | 2 | 2 | 2 | 1 | 2 | 2 | 2 | 2 | 1 | 2 | 2 | 2 | 2 | 1 | 2 | 2 | 2 | 2 |
| | CYBER GENINT | 3 | 2 | 2 | 2 | 1 | 2 | 1 | 1 | 2 | 0 | 3 | 2 | 2 | 0 | 2 | 1 | 2 | 3 | 3 | 3 | 0 |
| | ISIS | 3 | 2 | 3 | 2 | 3 | 2 | 2 | 2 | 2 | 0 | 2 | 3 | 3 | 0 | 2 | 0 | 1 | 2 | 2 | 3 | 0 |
| | | - | - | - | - | - | - | - | - | - | - | - | 2 | 2 | 0 | 1 | - | 0 | - | - | - | - |
| | | 0 | 0 | 0 | 0 | 0 | 0 | 0 | 0 | 0 | 0 | 0 | 1 | 1 | 1 | 0 | 0 | 1 | 0 | 0 | 0 | 0 |
| | VDBS | 2 | 2 | 3 | 3 | 1 | 3 | 2 | 2 | 3 | 2 | 2 | 2 | 0 | 0 | 0 | 0 | 0 | 3 | 3 | 3 | 2 |
| | GTS | 3 | 2 | 3 | 3 | 3 | 3 | 2 | 1 | 3 | 3 | 0 | 0 | 0 | 0 | 0 | 0 | 0 | 3 | 3 | 3 | 0 |
| | | 2 | 2 | $1^3$ | $2^7$ | 3 | 2 | $1^3$ | $1^7$ | $0^7$ | $1^7$ | 0 | 0 | 0 | 0 | 0 | 0 | 0 | 3 | 2 | $2^3$ | 0 |
| | | 3 | 3 | 3 | 3 | 3 | 3 | 3 | 3 | 3 | 3 | 3 | 3 | 3 | 3 | 3 | 2 | 3 | 3 | 3 | 3 | 3 |
| | CRESCAT | 3 | 3 | 3 | 2 | 3 | 3 | 3 | 1 | 3 | 1 | 3 | 2 | 0 | 0 | 0 | 0 | 0 | 3 | 3 | 2 | 2 |
| | | 0 | 1 | 3 | - | 3 | 3 | - | - | 3 | 3 | 1 | - | - | - | - | - | - | 3 | - | - | - |
| | | 1 | 1 | 1 | 0 | 1 | 1 | 0 | 0 | 1 | 1 | 1 | 0 | 0 | 0 | 0 | 0 | 0 | 1 | 0 | 0 | 0 |

TABLE A.1:  RATINGS OF A STATISTICAL PROGRAM: PART I cont.
CAPABILITIES:  PROCESSING AND DISPLAYING DATA

Usefulness Ratings
0=low
1=modest
2=moderate
3=high
-=no data

Tabular Scheme

| α |
|---|
| β |
| γ |

Note: $2^7$=2.7

α=Developer rating
β=Average user rating
γ=Number of users

| | GENERALITY | | FILING | | | | | | | | EDITING | | | | | MISSING DATA | | DISPLAY | | | |
|---|---|---|---|---|---|---|---|---|---|---|---|---|---|---|---|---|---|---|---|---|---|
| | Subroutines or Package | Generality of Purpose | Filing Summary | Data Set Size | Flexible Data Input | Complex Structures | Filing Missing Data | Storage/Retrieval | File Manipulation | Flexible Output | Editing Summary | Editing Language | Consistency | Probability Checks | Error Handling | Missing Data Summary | Imputation | Display Summary | Compute Tables | Print Tables | Graphics |
| | 6 | 7 | 8 | 9 | 10 | 11 | 12 | 13 | 14 | 15 | 16 | 17 | 18 | 19 | 20 | 21 | 22 | 23 | 24 | 25 | 26 |
| PERSEE | 1 | 3 | 3 | 3 | 3 | 3 | 3 | 2 | 2 | 3 | 3 | 3 | 1 | 0 | 0 | 0 | 0 | 3 | 3 | 2 | 0 |
| | 1 | 2 | 2 | 1 | 3 | 2 | 3 | 3 | 2 | 1 | 1 | 1 | 2 | 0 | 2 | 2 | 0 | 3 | 3 | 1 | 0 |
| | 1 | 1 | 1 | 1 | 1 | 1 | 1 | 1 | 1 | 1 | 1 | 1 | 1 | 1 | 1 | 1 | 1 | 1 | 1 | 1 | 1 |
| CENTS III | 2 | 2 | 1 | 3 | 3 | 3 | 3 | 2 | 1 | 1 | 1 | 1 | 1 | 0 | 1 | 2 | 1 | 3 | 3 | 3 | 0 |
| COCENTS | 2 | 3 | 1 | 3 | 3 | 3 | 3 | 2 | 1 | 1 | 0 | 0 | 0 | 0 | 0 | 2 | 1 | 3 | 3 | 3 | 0 |
| | $1^3$ | 2 | $0^3$ | $2^7$ | $2^3$ | $1^3$ | $2^7$ | 0 | 0 | $2^7$ | 1 | $0^5$ | 1 | 0 | 0 | $0^3$ | $0^7$ | 3 | 3 | 3 | 0 |
| | 3 | 3 | 3 | 3 | 3 | 3 | 3 | 3 | 3 | 3 | 3 | 2 | 3 | 3 | 3 | 3 | 3 | 3 | 3 | 2 | 3 |
| CENTS-AID | 3 | 2 | 1 | 3 | 3 | 2 | 2 | 2 | 1 | 3 | 0 | 0 | 0 | 0 | 0 | 2 | 1 | 3 | 2 | 2 | 0 |
| | - | 2 | 0 | 3 | $2^5$ | $2^5$ | 0 | $0^5$ | $0^5$ | $2^5$ | 0 | 0 | 0 | 0 | 0 | 0 | 0 | 3 | 3 | $2^5$ | 0 |
| | 0 | 2 | 1 | 2 | 2 | 2 | 1 | 2 | 2 | 2 | 1 | 1 | 1 | 1 | 1 | 1 | 2 | 2 | 2 | 2 | 2 |
| FILE 2.0 | 2 | 2 | 2 | 3 | 2 | 0 | 1 | 0 | 1 | 0 | 3 | 2 | 1 | 0 | 3 | 0 | 0 | 3 | 3 | 3 | 0 |
| OSIRIS 2.4 | 2 | 2 | 2 | 2 | 2 | 0 | 0 | 0 | 0 | 3 | 1 | 1 | 1 | 0 | 0 | 0 | 0 | 3 | 3 | 3 | 0 |
| GEN. SUMMARY | 3 | 2 | 1 | 3 | 1 | 1 | 1 | 1 | 0 | 0 | 0 | 0 | 0 | 0 | 0 | 0 | 0 | 2 | 1 | 2 | 0 |
| TPL | 2 | 2 | 1 | 3 | 2 | 3 | 3 | 2 | 2 | 3 | 1 | 1 | 1 | 0 | 1 | 2 | 1 | 3 | 3 | 3 | 0 |
| | 2 | 2 | $1^7$ | $2^3$ | $0^3$ | $1^3$ | $1^7$ | $1^7$ | 1 | $2^3$ | $0^7$ | $0^7$ | $1^3$ | $0^3$ | 1 | $0^3$ | 0 | $2^7$ | $2^7$ | $2^7$ | $0^7$ |
| | 3 | 3 | 3 | 3 | 3 | 3 | 3 | 3 | 3 | 3 | 3 | 3 | 3 | 3 | 3 | 3 | 3 | 3 | 3 | 3 | 3 |
| RGSP | 2 | 2 | 2 | 3 | 3 | 2 | 2 | 1 | 0 | 0 | 1 | 1 | 3 | 2 | 2 | 1 | 1 | 3 | 3 | 3 | 0 |
| | $1^8$ | 2 | $1^5$ | $2^8$ | $2^5$ | $2^7$ | $2^8$ | 2 | 2 | $2^8$ | 2 | 2 | $2^3$ | 0 | 2 | $1^5$ | $1^7$ | $2^8$ | 3 | 3 | $0^5$ |
| | 4 | 4 | 4 | 4 | 4 | 3 | 4 | 3 | 4 | 4 | 4 | 4 | 4 | 4 | 4 | 4 | 3 | 4 | 4 | 4 | 4 |
| NCHS-XTAB | 2 | 2 | 1 | 2 | 1 | 1 | 2 | 1 | 0 | 2 | 0 | 0 | 0 | 0 | 0 | 1 | 0 | 2 | 2 | 2 | 0 |
| SURVEY | 2 | 2 | 2 | 1 | 1 | 0 | 1 | 2 | 1 | 1 | 2 | 2 | 3 | 1 | 2 | 1 | 1 | 3 | 2 | 2 | 2 |
| SAP | 1 | 2 | 0 | 2 | 2 | 1 | 2 | 0 | 0 | 0 | 0 | 0 | 1 | 0 | 1 | 0 | 0 | 1 | 1 | 0 | 0 |
| | - | 2 | 0 | - | 1 | - | - | - | - | - | - | 0 | 0 | 0 | 0 | - | 0 | 2 | 2 | - | - |
| | 0 | 1 | 1 | 0 | 1 | 0 | 0 | 0 | 0 | 0 | 0 | 1 | 1 | 1 | 1 | 0 | 1 | 1 | 1 | 0 | 0 |
| HES VAR-X TAB | 2 | 2 | 1 | 2 | 2 | 0 | 1 | 2 | 1 | 3 | 0 | 0 | 0 | 0 | 0 | 1 | 0 | 1 | 2 | 1 | 0 |
| SPLITHALF | 0 | 2 | 0 | 2 | 1 | 1 | 2 | 1 | 2 | 1 | 0 | 0 | 0 | 0 | 0 | 0 | 0 | 1 | 1 | 1 | 0 |
| | 0 | 2 | $0^5$ | 2 | 0 | 0 | 0 | 0 | 0 | 0 | 0 | 0 | 0 | 0 | 0 | $0^5$ | 0 | $1^5$ | $1^5$ | 1 | 0 |
| | 1 | 2 | 2 | 2 | 2 | 2 | 2 | 1 | 2 | 2 | 2 | 2 | 2 | 2 | 2 | 2 | 2 | 2 | 2 | 2 | 2 |
| MULTI-FRAME | 3 | 2 | 3 | 3 | 1 | 1 | 1 | 1 | 2 | 2 | 0 | 0 | 0 | 0 | 0 | 0 | 0 | 3 | 2 | 1 | 0 |
| MWD VARIANCE | 2 | 2 | 2 | 3 | 2 | 1 | 0 | 0 | 2 | 0 | 0 | 0 | 0 | 0 | 0 | 0 | 0 | 3 | 2 | 0 | 0 |
| | 1 | $1^7$ | 1 | $2^5$ | $0^7$ | 0 | $0^7$ | $0^3$ | 1 | 0 | 0 | 0 | 0 | 0 | 0 | $0^7$ | 0 | $0^3$ | $0^3$ | 0 | 0 |
| | 1 | 3 | 2 | 2 | 3 | 3 | 3 | 3 | 3 | 2 | 3 | 3 | 3 | 3 | 3 | 3 | 3 | 2 | 2 | 3 | 2 |
| GSS EST. | 2 | 2 | 2 | 1 | 0 | 0 | 1 | 1 | 1 | 0 | 0 | 0 | 0 | 0 | 1 | 0 | 0 | 1 | 1 | 1 | 0 |
| | 2 | 2 | 0 | 2 | 1 | 0 | 0 | 0 | 0 | 0 | 0 | 0 | 0 | 0 | 0 | 0 | 0 | 2 | 1 | 0 | 0 |
| | 1 | 1 | 1 | 1 | 1 | 1 | 1 | 1 | 1 | 1 | 1 | 1 | 1 | 1 | 1 | 1 | 1 | 1 | 1 | 1 | 1 |
| CLFS VAR-COV | 3 | 2 | 0 | 3 | 0 | 0 | 0 | 0 | 0 | 0 | 0 | 0 | 0 | 0 | 0 | 0 | 0 | 0 | 0 | 0 | 0 |
| CESVP | 0 | 1 | 1 | 3 | 0 | 0 | 0 | 1 | 0 | 0 | 0 | 0 | 0 | 0 | 0 | 0 | 0 | 1 | 0 | 0 | 0 |
| KEYFITZ | 1 | 1 | 1 | 2 | 0 | 0 | 0 | 0 | 0 | 0 | 0 | 0 | 0 | 0 | 0 | 0 | 0 | 0 | 0 | 0 | 0 |
| CLUSTERS | 3 | 1 | 0 | 3 | 1 | 0 | 1 | 0 | 0 | 0 | 0 | 0 | 0 | 0 | 0 | 0 | 0 | 0 | 0 | 0 | 0 |
| | 1 | 2 | 1 | 3 | 2 | 0 | 2 | 0 | 1 | 1 | 0 | 0 | 0 | 0 | 0 | 0 | 0 | 2 | 2 | 0 | 0 |
| | 1 | 1 | 1 | 1 | 1 | 1 | 1 | 1 | 1 | 1 | 1 | 1 | 1 | 1 | 1 | 1 | 1 | 1 | 1 | 1 | 1 |

Chapter 4 (part ii) — TPL

Chapter 5 — HES VAR-X TAB

TABLE A.1:   RATINGS OF A STATISTICAL PROGRAM: PART I cont.
CAPABILITIES:   PROCESSING AND DISPLAYING DATA

**Usefulness Ratings**
0=low
1=modest
2=moderate
3=high
-=no data

**Tabular Scheme**

| α |
|---|
| β |
| γ |

Note: $2^7$=2.7

α=Developer rating
β=Average user rating
γ=Number of users

Column key:

| # | Category | Label |
|---|----------|-------|
| 6 | GENERALITY | Subroutines or Package |
| 7 | GENERALITY | Generality of Purpose |
| 8 | FILING | Filing Summary |
| 9 | FILING | Data Set Size |
| 10 | FILING | Flexible Data Input |
| 11 | FILING | Complex Structures |
| 12 | FILING | Filing Missing Data |
| 13 | FILING | Storage/Retrieval |
| 14 | FILING | File Manipulation |
| 15 | FILING | Flexible Output |
| 16 | EDITING | Editing Summary |
| 17 | EDITING | Editing Language |
| 18 | EDITING | Consistency |
| 19 | EDITING | Probability Checks |
| 20 | EDITING | Error Handling |
| 21 | MISSING DATA | Missing Data Summary |
| 22 | MISSING DATA | Imputation |
| 23 | DISPLAY | Display Summary |
| 24 | DISPLAY | Compute Tables |
| 25 | DISPLAY | Print Tables |
| 26 | DISPLAY | Graphics |

| Program | 6 | 7 | 8 | 9 | 10 | 11 | 12 | 13 | 14 | 15 | 16 | 17 | 18 | 19 | 20 | 21 | 22 | 23 | 24 | 25 | 26 |
|---|---|---|---|---|---|---|---|---|---|---|---|---|---|---|---|---|---|---|---|---|---|
| SUPER CARP | 2 | 2 | 0 | 3 | 1 | 0 | 2 | 0 | 0 | 1 | 0 | 0 | 0 | 0 | 0 | 1 | 0 | 0 | 0 | 0 | 0 |
|  | 2 | 2 | 0 | 2 | 0 | 0 | 1 | 0 | 0 | 1 | 0 | 0 | 0 | 0 | 0 | 1 | 0 | 0 | 0 | 0 | 0 |
|  | 1 | 1 | 1 | 1 | 1 | 1 | 1 | 1 | 1 | 1 | 1 | 1 | 1 | 1 | 1 | 1 | 1 | 1 | 1 | 1 | 1 |
| GENVAR | 1 | 2 | 0 | 1 | 1 | 1 | 0 | 0 | 0 | 0 | 0 | 0 | 0 | 0 | 0 | 0 | 0 | 0 | 0 | 0 | 0 |
| BTFSS | 3 | 3 | 3 | 2 | 2 | 3 | 2 | 3 | 3 | 2 | 1 | 1 | 1 | 0 | 0 | 1 | 0 | 3 | 3 | 1 | 0 |
|  | 3 | 2 | 2 | 2 | 2 | 3 | 3 | 2 | 2 | 2 | 2 | 0 | 2 | 1 | 2 | 2 | 1 | 2 | 3 | 2 | - |
|  | 1 | 1 | 1 | 1 | 1 | 1 | 1 | 1 | 1 | 1 | 1 | 1 | 1 | 1 | 1 | 1 | 1 | 1 | 1 | 1 | 0 |
| EXPRESS | 3 | 3 | 3 | 3 | 3 | 3 | 2 | 3 | 1 | 2 | 3 | 3 | 3 | 1 | 2 | 1 | 1 | 3 | 3 | 3 | 3 |
|  | 3 | 3 | 3 | $2^5$ | 3 | 3 | 2 | $2^5$ | 3 | 3 | 2 | 2 | 2 | 0 | $0^5$ | 2 | $0^5$ | 3 | 2 | 2 | $2^5$ |
|  | 2 | 2 | 2 | 2 | 2 | 2 | 2 | 2 | 2 | 1 | 2 | 1 | 1 | 1 | 2 | 2 | 2 | 2 | 2 | 2 | 2 |
| EASYTRIEVE | 3 | 3 | 3 | 3 | 3 | 3 | 3 | 3 | 3 | 3 | 3 | 2 | 3 | 2 | 3 | 3 | 1 | 3 | 3 | 2 | 2 |
|  | 3 | $2^5$ | $2^5$ | 2 | $2^5$ | 0 | - | $2^5$ | $2^5$ | 3 | $2^5$ | 1 | $2^5$ | 0 | $0^5$ | 2 | 2 | 0 | 0 | 0 | 0 |
|  | 2 | 2 | 2 | 2 | 2 | 2 | 0 | 2 | 2 | 2 | 2 | 2 | 2 | 1 | 2 | 2 | 2 | 0 | 0 | 0 | 0 |
| FOCUS | 3 | 3 | 3 | 3 | 3 | 3 | 3 | 3 | 3 | 3 | 2 | 2 | 2 | 2 | 3 | 2 | 2 | 3 | 3 | 3 | 3 |
| SURVEYOR/SURV | 1 | 3 | 3 | 3 | 2 | 2 | 3 | 2 | 2 | 2 | 3 | 3 | 1 | 3 |  | 0 | 0 | 3 | 3 | 3 | 2 |
|  | $2^5$ | 2 | $1^5$ | 3 | 2 | $0^5$ | 2 | $1^5$ | $2^5$ | $0^5$ | 3 | 3 | 0 | 2 |  | 1 | 0 | 3 | $2^5$ | $2^5$ | 0 |
|  | 2 | 2 | 2 | 2 | 2 | 2 | 1 | 2 | 2 | 2 | 2 | 2 | 2 | 2 |  | 2 | 2 | 2 | 2 | 2 | 2 |
| DATAPLOT | 3 | 3 | 2 | 1 | 3 | 2 | 2 | 2 | 2 | 2 | 2 | 1 | 2 | 3 | 2 | 3 | 3 | 3 | 3 | 1 | 3 |
|  | 1 | 3 | 0 | - | 0 | 0 | - | 0 | 0 | 0 | 0 | 0 | 0 | 0 | 0 | 0 | 0 | 3 | 1 | 1 | 3 |
|  | 1 | 1 | 1 | 0 | 1 | 0 | 1 | 0 | 1 | 1 | 1 | 1 | 1 | 1 | 1 | 1 | 1 | 1 | 1 | 1 | 1 |
| PACKAGE X | 3 | 3 | 3 | 2 | 3 | 2 | 2 | 2 | 2 | 2 | 3 | 3 | 2 | 1 | 2 | 2 | 1 | 3 | 2 | 0 | 2 |
|  | 3 | 3 | 2 | 2 | 2 | 2 | $2^5$ | 2 | $1^5$ | 2 | 1 | $0^5$ | 1 | $0^5$ | $0^5$ | 2 | 1 | 3 | 2 | $1^5$ | $1^5$ |
|  | 2 | 2 | 2 | 1 | 2 | 2 | 2 | 2 | 2 | 2 | 2 | 2 | 2 | 2 | 2 | 2 | 2 | 2 | 2 | 2 | 2 |
| SCSS | 3 | 3 | 2 | 3 | 2 | 1 | 3 | 2 | 1 | 3 | 2 | 2 | 2 | 3 | 2 | 3 | 2 | 3 | 3 | 2 | 2 |
| SPSS | 3 | 3 | 2 | 3 | 2 | 1 | 3 | 2 | 2 | 3 | 0 | 0 | 0 | 0 | 0 | 2 | 0 | 2 | 2 | 1 | 0 |
|  | $1^7$ | $2^7$ | $2^3$ | $2^5$ | 2 | 1 | $2^7$ | 2 | $1^7$ | $2^7$ | $1^3$ | $1^3$ | $1^3$ | $0^7$ | $1^3$ | $1^3$ | 1 | $2^7$ | $2^3$ | 1 | $1^3$ |
|  | 3 | 3 | 3 | 2 | 3 | 3 | 3 | 3 | 3 | 3 | 3 | 3 | 3 | 3 | 3 | 3 | 3 | 3 | 3 | 3 | 3 |
| SOUPAC | 3 | 3 | 2 | 2 | 2 | 1 | 2 | 2 | 2 | 1 | 1 | 1 | 1 | 0 | 1 | 1 | 1 | 1 | 1 | 0 | 1 |
|  | $2^5$ | 3 | 1 | 3 | $2^5$ | $1^5$ | $2^5$ | 2 | 2 | 1 | $1^5$ | $1^5$ | $0^5$ | $1^5$ | $1^5$ | $1^5$ | 3 | $2^5$ | 3 | 2 | $1^5$ |
|  | 2 | 2 | 2 | 2 | 2 | 2 | 2 | 2 | 2 | 2 | 2 | 2 | 2 | 2 | 2 | 2 | 1 | 2 | 2 | 2 | 2 |
| DATATEXT | 3 | 3 | 3 | 3 | 3 | 1 | 3 | 3 | 2 | 3 | 2 | 2 | 2 | 0 | 1 | 3 | 0 | 3 | 3 | 1 | 1 |
|  | 2 | 2 | 1 | 2 | 2 | $1^5$ | 3 | $2^5$ | 1 | 2 | 2 | 2 | 3 | 1 | 2 | 2 | 2 | $2^5$ | 2 | $1^5$ | $1^5$ |
|  | 2 | 2 | 2 | 1 | 2 | 2 | 2 | 2 | 1 | 2 | 2 | 2 | 2 | 2 | 2 | 2 | 2 | 2 | 2 | 2 | 2 |
| P-STAT 78 | 3 | 3 | 3 | 3 | 3 | 3 | 3 | 3 | 3 | 3 | 2 | 2 | 3 | 1 | 2 | 2 | 1 | 3 | 3 | 3 | 2 |
|  | 3 | 3 | 3 | 3 | 2 | 2 | 3 | 2 | 3 | 2 | 2 | $1^5$ | 2 | $0^5$ | 2 | 2 | $1^5$ | $2^5$ | $2^5$ | 2 | 2 |
|  | 2 | 2 | 2 | 2 | 2 | 2 | 2 | 2 | 2 | 2 | 2 | 2 | 2 | 2 | 2 | 2 | 2 | 2 | 2 | 2 | 2 |
| DAS | 3 | 3 | 3 | 2 | 3 | 3 | 3 | 2 | 3 | 2 | 3 | 3 | 2 | 2 | 1 | 3 | 3 | 3 | 3 | 0 | 3 |
|  | 3 | 3 | 3 | 2 | 3 | 2 | 3 | 3 | 3 | 2 | - | - | - | - | - | 3 | 3 | 2 | 3 | 3 | 2 |
|  | 1 | 1 | 1 | 1 | 1 | 1 | 1 | 1 | 1 | 1 | 0 | 0 | 0 | 0 | 0 | 1 | 1 | 1 | 1 | 1 | 1 |
| SPMS | 3 | 3 | 3 | 3 | 3 | 2 | 2 | 3 | 1 |  | 2 | 2 | 2 | 2 | 1 | 1 | 1 | 2 | 1 | 0 | 2 |
|  | - | 2 | - | - | - | 2 | 2 | - | - | - | 2 | - | 2 | - | - | 2 | 2 | - | - | - | 2 |
|  | 0 | 1 | 0 | 0 | 0 | 1 | 1 | 0 | 0 | 0 | 1 | 0 | 1 | 0 | 0 | 1 | 1 | 0 | 0 | 0 | 1 |
| OSIRIS IV | 3 | 3 | 3 | 2 | 3 | 3 | 3 | 3 | 3 | 3 | 3 | 3 | 3 | 1 | 2 | 3 | 0 | 2 | 3 | 2 | 1 |
|  | 3 | 3 | 3 | 3 | $2^3$ | $1^7$ | 3 | $2^3$ | $2^7$ | $1^7$ | $2^3$ | $2^7$ | $2^7$ | $1^7$ | $2^3$ | 2 | $0^3$ | 2 | $2^7$ | $0^7$ | 1 |
|  | 3 | 3 | 3 | 3 | 3 | 3 | 3 | 3 | 3 | 3 | 3 | 3 | 3 | 3 | 3 | 3 | 3 | 3 | 3 | 3 | 3 |

Chapter 6 (associated with BTFSS row)

Appendix 2

TABLE A.1: RATINGS OF A STATISTICAL PROGRAM: PART I cont.
CAPABILITIES: PROCESSING AND DISPLAYING DATA

Usefulness Ratings
0=low
1=low
2=moderate
3=high
-=no data

Tabular Scheme

| α |
|---|
| β |
| γ |

Note: $2^7$=2.7

α=Developer rating
β=Average user ratio
γ=Number of users

| | | GENERALITY | | FILING | | | | | | | | EDITING | | | | | MISSING DATA | | DISPLAY | | | |
|---|---|---|---|---|---|---|---|---|---|---|---|---|---|---|---|---|---|---|---|---|---|---|
| | | Subroutines or Package | Generality of Purpose | Filing Summary | Data Set Size | Flexible Data Input | Complex Structures | Filing Missing Data | Storage/Retrieval | File Manipulation | Flexible Output | Editing Summary | Editing Language | Consistency | Probability Checks | Error Handling | Missing Data Summary | Imputation | Display Summary | Compute Tables | Print Tables | Graphics |
| | | 6 | 7 | 8 | 9 | 10 | 11 | 12 | 13 | 14 | 15 | 16 | 17 | 18 | 19 | 20 | 21 | 22 | 23 | 24 | 25 | 26 |
| Chapter 7 | CS | 3 | 3 | 3 | 3 | 3 | 3 | 3 | 3 | 3 | 3 | 3 | 3 | 3 | 1 | 3 | 2 | 2 | 3 | 3 | 3 | 3 |
| | | 3 | 3 | 3 | 2 | 2 | 3 | 3 | 3 | 3 | 3 | 3 | 3 | 2 | 3 | 3 | 3 | - | 3 | 3 | 2 | - |
| | | 1 | 1 | 1 | 1 | 1 | 1 | 1 | 1 | 1 | 1 | 1 | 1 | 1 | 1 | 1 | 1 | 0 | 1 | 1 | 1 | 0 |
| | SAS | 3 | 3 | 3 | 3 | 3 | 3 | 3 | 3 | 3 | 3 | 3 | 3 | 3 | 3 | 3 | 3 | 1 | 3 | 3 | 2 | 2 |
| | | 3 | 3 | 3 | 3 | 3 | $2^7$ | 3 | 3 | 3 | 2 | 2 | 2 | $2^3$ | 3 | $2^5$ | 2 | 2 | $2^7$ | 3 | $2^5$ | $2^5$ |
| | | 3 | 3 | 3 | 2 | 3 | 3 | 3 | 3 | 3 | 3 | 2 | 3 | 3 | 2 | 2 | 2 | 2 | 3 | 3 | 2 | 2 |
| | OMNITAB 80 | 3 | 3 | 2 | 1 | 3 | 3 | 2 | 1 | 3 | 2 | 2 | 2 | 1 | 3 | 1 | 1 | 1 | 3 | 3 | 1 | 3 |
| | | 2 | 1 | 0 | 0 | 1 | 0 | 0 | 0 | 1 | 0 | 0 | 0 | 0 | 0 | 0 | 0 | 0 | 2 | 2 | 2 | 0 |
| | | 1 | 1 | 1 | 1 | 1 | 1 | 1 | 1 | 1 | 1 | 1 | 1 | 1 | 1 | 1 | 1 | 1 | 1 | 1 | 1 | 1 |
| | HP STAT PACKS | 3 | 3 | 2 | 1 | 1 | 1 | 1 | 1 | 2 | 0 | 2 | 2 | 0 | 1 | 0 | 1 | 1 | 3 | 3 | 1 | 3 |
| | | 1 | 3 | 1 | 1 | 0 | 1 | 0 | 0 | 1 | 0 | 1 | 1 | 0 | 0 | 0 | 0 | 0 | 2 | 1 | 0 | 2 |
| | | 1 | 1 | 1 | 1 | 1 | 1 | 1 | 1 | 1 | 1 | 1 | 1 | 1 | 1 | 1 | 1 | 1 | 1 | 1 | 1 | 1 |
| | MINITAB II | 3 | 3 | 2 | 1 | 3 | 3 | 2 | 2 | 2 | 1 | 3 | 1 | 1 | 2 | 2 | 3 | 1 | 3 | 3 | 3 | 3 |
| | | $2^5$ | 3 | $1^5$ | $0^5$ | $2^5$ | 2 | $1^5$ | 2 | 1 | 1 | 1 | $0^5$ | $0^5$ | $0^5$ | $0^5$ | 2 | 1 | $2^5$ | 1 | 2 | $2^5$ |
| | | 2 | 2 | 2 | 2 | 2 | 1 | 2 | 2 | 2 | 2 | 2 | 2 | 2 | 2 | 2 | 2 | 1 | 2 | 2 | 2 | 2 |
| | BMDP | 2 | 3 | 2 | 1 | 2 | 1 | 3 | 2 | 1 | 1 | 1 | 1 | 2 | 3 | 2 | 3 | 2 | 3 | 2 | 1 | 2 |
| | | 2 | $2^7$ | $1^3$ | $2^3$ | $1^3$ | $0^3$ | $2^7$ | 2 | $0^7$ | 1 | 1 | $0^3$ | 1 | 2 | 1 | $1^3$ | 2 | $1^7$ | $1^3$ | $0^3$ | 1 |
| | | 3 | 3 | 3 | 3 | 3 | 3 | 3 | 3 | 3 | 3 | 3 | 3 | 3 | 3 | 3 | 3 | 3 | 3 | 3 | 3 | 3 |
| | NISAN | 3 | 3 | 2 | 1 | 2 | 2 | 1 | 2 | 2 | 2 | 0 | 0 | 0 | 0 | 0 | 1 | 1 | 2 | 2 | 2 | 2 |
| | GENSTAT 4.02 | 3 | 3 | 1 | 1 | 3 | 3 | 2 | 3 | 2 | 3 | 1 | 1 | 1 | 1 | 1 | 1 | 1 | 3 | 3 | 3 | 2 |
| | | $2^5$ | 3 | 1 | $1^5$ | 2 | 3 | 2 | 2 | 2 | 3 | $0^5$ | 0 | $0^5$ | 0 | $0^5$ | $0^5$ | $0^5$ | $2^5$ | $2^5$ | $2^5$ | 2 |
| | | 2 | 2 | 2 | 2 | 2 | 2 | 2 | 2 | 2 | 2 | 2 | 2 | 2 | 2 | 2 | 2 | 2 | 2 | 2 | 2 | 2 |
| | SPEAKEASY III | 3 | 3 | 2 | 1 | 3 | 3 | 1 | 3 | 2 | 3 | 2 | 3 | 2 | 1 | 1 | 2 | 1 | 3 | 3 | 3 | 3 |
| | TROLL | 3 | 3 | 2 | 2 | 2 | 2 | 2 | 3 | 3 | 2 | 2 | 2 | 0 | 3 | 1 | 1 | 0 | 2 | 2 | 2 | 3 |
| | | 3 | 3 | $1^7$ | $1^5$ | $1^3$ | $1^7$ | $1^7$ | $2^3$ | $1^3$ | 1 | $0^7$ | $0^5$ | 0 | 0 | 0 | $0^5$ | $0^5$ | $1^3$ | 1 | 1 | $2^7$ |
| | | 3 | 3 | 3 | 2 | 3 | 3 | 3 | 3 | 3 | 3 | 3 | 2 | 2 | 1 | 1 | 2 | 2 | 3 | 3 | 3 | 1 |
| | IDA | 3 | 3 | 2 | 0 | 2 | 1 | 2 | 2 | 2 | 2 | 2 | 1 | 0 | 2 | 1 | 2 | 1 | 3 | 2 | 3 | 1 |
| | | 2 | 0 | 1 | 0 | 3 | 3 | 2 | 0 | 2 | 2 | 0 | 0 | 0 | 0 | 0 | 2 | 0 | 2 | 1 | 1 | 1 |
| | | 1 | 1 | 1 | 1 | 1 | 1 | 1 | 1 | 1 | 1 | 1 | 1 | 1 | 1 | 1 | 1 | 1 | 1 | 1 | 1 | 1 |
| Chapter 8 | ISA | 1 | 3 | 1 | 1 | 2 | 2 | 2 | 1 | 2 | 1 | 1 | 1 | 0 | 1 | 1 | 1 | 0 | 2 | 2 | 2 | 2 |
| | GLIM | 2 | 3 | 0 | 1 | 2 | 1 | 1 | 1 | 1 | 2 | 0 | 0 | 0 | 0 | 0 | 0 | 0 | 1 | 2 | 1 | 1 |
| | | $1^3$ | $2^7$ | 1 | $0^7$ | $1^3$ | $0^5$ | $1^3$ | 1 | $0^5$ | 1 | $0^3$ | 1 | 0 | $0^5$ | $0^5$ | $0^3$ | $0^3$ | $1^3$ | $0^7$ | $0^3$ | $1^3$ |
| | | 3 | 3 | 3 | 3 | 3 | 2 | 3 | 2 | 2 | 3 | 3 | 2 | 2 | 2 | 2 | 3 | 3 | 3 | 3 | 3 | 3 |
| | ISP | 3 | 3 | 2 | 0 | 3 | 2 | 2 | 2 | 2 | 0 | 1 | 1 | 0 | 1 | 1 | 2 | 0 | 2 | 2 | 0 | 2 |
| | | 3 | 3 | 2 | 0 | 0 | 0 | 2 | 2 | 2 | 0 | 0 | 0 | 0 | 0 | 0 | 1 | 0 | 0 | 0 | 0 | 0 |
| | | 1 | 1 | 1 | 1 | 1 | 1 | 1 | 1 | 1 | 1 | 1 | 1 | 1 | 1 | 1 | 1 | 1 | 1 | 1 | 1 | 1 |
| | CMU-DAP | 3 | 2 | 2 | 0 | 2 | 1 | 2 | 1 | 2 | 2 | 2 | 0 | 1 | 0 | 0 | 1 | 0 | 2 | 1 | 0 | 2 |
| | | 2 | $1^5$ | $0^5$ | 0 | $2^5$ | $1^5$ | 2 | $1^5$ | 2 | 2 | 0 | 0 | 0 | $0^5$ | 0 | 0 | 0 | 3 | $1^5$ | $1^5$ | 2 |
| | | 1 | 2 | 2 | 2 | 2 | 2 | 2 | 2 | 2 | 2 | 2 | 2 | 2 | 2 | 2 | 2 | 2 | 2 | 2 | 2 | 2 |
| | RUMMAGE | 2 | 3 | 1 | 3 | 2 | 1 | 0 | 2 | 0 | 1 | 1 | 0 | 0 | 1 | 1 | 1 | 0 | 2 | 1 | 1 | 2 |
| | | 2 | $2^3$ | $0^3$ | - | $1^7$ | $1^5$ | 2 | $0^5$ | $0^3$ | $0^7$ | $0^3$ | $0^3$ | 0 | $0^3$ | 0 | $1^7$ | $1^5$ | 2 | 1 | 1 | $1^3$ |
| | | 1 | 3 | 3 | 0 | 3 | 2 | 3 | 2 | 3 | 3 | 3 | 3 | 3 | 3 | 3 | 3 | 2 | 2 | 2 | 1 | 3 |
| | CADA | 3 | 3 | 1 | 0 | 0 | 0 | 0 | 1 | 0 | 1 | 1 | 1 | 1 | 0 | 0 | 0 | 0 | 1 | 1 | 0 | 1 |
| | | 2 | 2 | 0 | 0 | 1 | 0 | 0 | 1 | 1 | 1 | 0 | 0 | 0 | 0 | 0 | 0 | 0 | 2 | 1 | 1 | 1 |
| | | 1 | 1 | 1 | 1 | 1 | 1 | 1 | 1 | 1 | 1 | 1 | 1 | 1 | 1 | 1 | 1 | 1 | 1 | 1 | 1 | 1 |

TABLE A.1: RATINGS OF A STATISTICAL PROGRAM: PART I cont.

CAPABILITIES: PROCESSING AND DISPLAYING DATA

Usefulness Ratings
0=low
1=modest
2=moderate
3=high
-=no data

Tabular Scheme

| α |
| β |   Note: $2^7$=2.7
| γ |

α=Developer rating
β=Average user rating
γ=Number of users

Column groups: GENERALITY | FILING | EDITING | MISSING DATA | DISPLAY

| Program | 6 Subroutines or Package | 7 Generality of Purpose | 8 Filing Summary | 9 Data Set Size | 10 Flexible Data Input | 11 Complex Structures | 12 Filing Missing Data | 13 Storage/Retrieval | 14 File Manipulation | 15 Flexible Output | 16 Editing Summary | 17 Editing Language | 18 Consistency | 19 Probability Checks | 20 Error Handling | 21 Missing Data Summary | 22 Imputation | 23 Display Summary | 24 Compute Tables | 25 Print Tables | 26 Graphics |
|---|---|---|---|---|---|---|---|---|---|---|---|---|---|---|---|---|---|---|---|---|---|
| SURVO | 2 | 3 | 0 | 3 | 2 | 2 | 2 | 2 | 2 | 3 | 1 | 2 | 1 | 0 | 3 | 1 | 2 | 1 | 1 | 3 | 0 |
| AUTOGRP+ | 3 | 3 | 3 | 3 | 3 | 2 | 2 | 2 | 3 | 3 | 2 | 1 | 1 | 0 | 0 | 1 | 0 | 3 | 3 | 2 | 2 |
| | 3 | 3 | $1^5$ | $1^5$ | 2 | $1^5$ | 2 | $1^5$ | $1^5$ | 2 | 0 | 0 | 0 | 0 | 0 | 1 | $1^5$ | $2^5$ | $2^5$ | 1 | $1^5$ |
| | 2 | 2 | 2 | 2 | 2 | 2 | 2 | 2 | 2 | 2 | 2 | 2 | 2 | 2 | 2 | 2 | 2 | 2 | 2 | 2 | 2 |
| FORALL | 2 | 3 | 1 | 1 | 2 | 2 | 1 | 2 | 1 | 1 | 1 | 1 | 1 | 1 | 1 | 1 | 1 | 2 | 2 | 2 | 2 |
| | $2^7$ | 3 | 2 | 1 | $2^3$ | $1^5$ | $1^5$ | $1^5$ | $1^5$ | 1 | 1 | 1 | 2 | $0^5$ | 1 | 0 | 0 | 2 | $1^5$ | $1^5$ | $1^5$ |
| | 3 | 3 | 3 | 3 | 3 | 2 | 2 | 2 | 2 | 2 | 3 | 2 | 1 | 2 | 1 | 2 | 2 | 3 | 2 | 2 | 2 |
| AQD | 2 | 3 | 2 | 3 | 2 | 1 | 1 | 2 | 2 | 1 | 0 | 0 | 0 | 0 | 0 | 0 | 0 | 1 | 2 | 0 | 1 |
| STP | 2 | 2 | 1 | 1 | 3 | 1 | 1 | 1 | 1 | 1 | 1 | 1 | 0 | 0 | 0 | 1 | 0 | 1 | 2 | 0 | 0 |
| | $1^7$ | $2^3$ | $0^3$ | 0 | $0^7$ | $0^3$ | 0 | $1^3$ | 1 | $0^3$ | $0^7$ | $0^7$ | 0 | 0 | 0 | 0 | 0 | $1^7$ | 1 | 0 | $0^7$ |
| | 3 | 3 | 3 | 2 | 3 | 3 | 3 | 3 | 3 | 3 | 3 | 3 | 3 | 3 | 3 | 3 | 3 | 3 | 3 | 3 | 3 |
| STATPAK | 3 | 2 | 1 | 2 | 2 | 1 | 0 | 1 | 1 | 0 | 1 | 1 | 0 | 0 | 1 | 0 | 0 | 1 | 1 | 0 | 1 |
| STATUTIL | 1 | 2 | 2 | 1 | 3 | 2 | 1 | 2 | 2 | 2 | 2 | 2 | 0 | 0 | 0 | 0 | 0 | 0 | 1 | 1 | 1 |
| MICROSTAT | 1 | 3 | 2 | 0 | 1 | 0 | 0 | 2 | 2 | 2 | 2 | 1 | 0 | 0 | 0 | 0 | 0 | 1 | 0 | 0 | 1 |
| A-STAT 79 | 3 | 3 | 1 | 1 | 1 | 0 | 1 | 1 | 1 | 0 | 0 | 0 | 0 | 0 | 0 | 1 | 1 | 2 | 2 | 1 | 0 |
| AMANCE | 3 | 3 | 0 | 2 | 1 | 1 | 2 | 1 | 2 | 1 | 1 | 1 | 1 | 2 | 0 | 0 | 0 | 2 | 2 | 2 | 2 |
| | 2 | 3 | 2 | 1 | 2 | 2 | 2 | 2 | 0 | $1^5$ | $1^5$ | $1^5$ | 0 | 2 | 1 | 1 | 0 | $1^5$ | 2 | 1 | $1^5$ |
| | 2 | 2 | 2 | 1 | 2 | 2 | 1 | 1 | 1 | 2 | 2 | 2 | 1 | 1 | 1 | 1 | 1 | 2 | 1 | 1 | 2 |
| MAC/STAT | 3 | 3 | 3 | 3 | 3 | 2 | 3 | 3 | 2 | 2 | 0 | 0 | 0 | 0 | 0 | 0 | 0 | 3 | 3 | 3 | 3 |
| REG | 2 | 2 | 1 | 2 | 2 | 1 | 1 | 1 | 1 | 2 | 1 | 0 | 1 | 1 | 1 | 1 | 0 | 1 | 1 | 1 | 0 |
| | 2 | $2^5$ | 1 | 2 | $1^5$ | 2 | 2 | $2^5$ | $0^5$ | 0 | 1 | 1 | $1^5$ | $0^5$ | $0^5$ | $0^5$ | - | 1 | $1^5$ | 0 | 0 |
| | 2 | 2 | 2 | 1 | 2 | 2 | 2 | 2 | 2 | 1 | 2 | 2 | 2 | 2 | 2 | 2 | 0 | 2 | 2 | 1 | 2 |
| TPD-3 | 3 | 3 | 1 | 2 | 2 | 2 | 1 | 1 | 1 | 0 | 2 | 2 | 0 | 1 | 0 | 1 | 1 | 2 | 1 | 0 | 1 |
| LSML76 | 2 | 3 | 0 | 2 | 2 | 1 | 2 | 0 | 0 | 3 | 0 | 0 | 0 | 0 | 0 | 0 | 0 | 0 | 0 | 0 | 0 |
| | $1^3$ | 2 | $0^3$ | 2 | $0^3$ | $0^3$ | $1^3$ | $0^7$ | 0 | $0^7$ | $0^3$ | $0^7$ | 1 | $0^3$ | $0^3$ | $0^7$ | $0^3$ | $0^3$ | $0^3$ | 0 | 0 |
| | 3 | 3 | 3 | 3 | 3 | 3 | 3 | 3 | 3 | 3 | 3 | 3 | 3 | 3 | 3 | 3 | 3 | 3 | 3 | 3 | 3 |
| LNWD/NLNWD | 2 | 3 | 0 | 0 | 1 | 0 | 1 | 0 | 0 | 0 | 0 | 0 | 0 | 0 | 0 | 0 | 0 | 2 | 2 | 0 | 0 |
| | 2 | $1^5$ | 0 | 0 | $1^5$ | 0 | 1 | 1 | $0^5$ | $1^5$ | 0 | 0 | $0^5$ | $1^5$ | $0^5$ | $1^5$ | 0 | 2 | 1 | 0 | $0^5$ |
| | 2 | 2 | 2 | 1 | 2 | 2 | 2 | 2 | 2 | 2 | 1 | 2 | 2 | 2 | 2 | 2 | 2 | 2 | 2 | 2 | 2 |
| ACPBCTET | 0 | 1 | 1 | 0 | 1 | 0 | 0 | 1 | 0 | 0 | 0 | 0 | 0 | 0 | 0 | 0 | 0 | 0 | 0 | 0 | 0 |
| | 0 | 1 | 0 | - | - | 0 | 1 | 0 | - | | 0 | 0 | 0 | 0 | 0 | 0 | 0 | 0 | 0 | 0 | 0 |
| | 1 | 1 | 1 | 0 | 0 | 0 | 1 | 1 | 1 | 0 | 1 | 1 | 1 | 1 | 1 | 1 | 1 | 1 | 1 | 1 | 1 |
| MULPRES | 0 | 2 | 0 | 0 | 0 | 0 | 0 | 0 | 0 | 0 | 0 | 0 | 0 | 0 | 0 | 0 | 0 | 2 | 0 | 0 | 1 |
| STATSPLINE | 3 | 2 | 1 | 1 | 2 | 0 | 0 | 0 | 1 | 1 | 0 | 0 | 1 | 0 | 1 | 0 | 0 | 2 | 1 | 0 | 2 |
| ALLOC | 1 | 2 | 0 | 0 | 2 | 1 | 0 | 0 | 0 | 2 | 0 | 0 | 2 | 1 | 1 | 0 | 0 | 2 | 2 | 2 | 0 |
| POPAN | 3 | 2 | 2 | 2 | 3 | 1 | 1 | 3 | 3 | 2 | 1 | 1 | 1 | 0 | 2 | 0 | 0 | 2 | 2 | 2 | 0 |
| | 3 | 2 | 3 | 2 | 2 | 3 | 3 | 3 | 3 | 2 | 3 | 2 | 3 | 0 | 3 | 3 | 3 | 3 | 3 | 1 | 1 |
| | 1 | 1 | 1 | 1 | 1 | 1 | 1 | 1 | 1 | 1 | 1 | 1 | 1 | 1 | 1 | 1 | 1 | 1 | 1 | 1 | 1 |
| CAPTURE | 2 | 2 | 0 | 1 | 2 | 0 | 0 | 0 | 0 | 0 | 0 | 0 | 0 | 0 | 0 | 0 | 0 | 1 | 0 | 0 | 1 |
| | $0^5$ | 1 | 0 | 1 | $1^5$ | 0 | 0 | 0 | 0 | 0 | 0 | 0 | 0 | 0 | 0 | 0 | 0 | $0^5$ | 0 | 0 | 0 |
| | 2 | 2 | 2 | 2 | 2 | 1 | 2 | 2 | 2 | | 2 | 2 | 2 | 2 | 2 | 2 | 2 | 2 | 2 | 2 | 2 |

Chapter 9 — AMANCE

TABLE A.1:   RATINGS OF A STATISTICAL PROGRAM: PART I cont.
CAPABILITIES:   PROCESSING AND DISPLAYING DATA

Usefulness Ratings
0=low
1=modest
2=moderate
3=high
-=no data

Tabular Scheme

| α |
|---|
| β |
| γ |

Note: $2^7$=2.7

α=Developer rating
β=Average user rating
γ=Number of users

| | | GENERALITY | | FILING | | | | | | | | EDITING | | | | | MISSING DATA | | DISPLAY | | | |
|---|---|---|---|---|---|---|---|---|---|---|---|---|---|---|---|---|---|---|---|---|---|---|
| | | Subroutines or Package | Generality of Purpose | Filing Summary | Data Set Size | Flexible Data Input | Complex Structures | Filing Missing Data | Storage/Retrieval | File Manipulation | Flexible Output | Editing Summary | Editing Language | Consistency | Probability Checks | Error Handling | Missing Data Summary | Imputation | Display Summary | Compute Tables | Print Tables | Graphics |
| | | 6 | 7 | 8 | 9 | 10 | 11 | 12 | 13 | 14 | 15 | 16 | 17 | 18 | 19 | 20 | 21 | 22 | 23 | 24 | 25 | 26 |
| Chapter 10 | GUHA | 2 | 2 | 1 | 1 | 1 | 0 | 2 | 0 | 0 | 1 | 2 | 1 | 1 | 0 | 0 | 2 | 2 | 0 | 0 | 0 | 0 |
| | | 2 | $2^7$ | $0^3$ | $1^3$ | $0^7$ | $0^5$ | $1^7$ | 0 | 1 | $0^5$ | $0^3$ | 0 | $0^3$ | 0 | $0^7$ | $1^7$ | $0^5$ | $0^3$ | $0^3$ | $0^3$ | 0 |
| | | 3 | 3 | 3 | 3 | 3 | 3 | 3 | 3 | 3 | 2 | 3 | 3 | 3 | 3 | 3 | 3 | 2 | 3 | 3 | 3 | 3 |
| | CATFIT | 1 | 2 | 1 | 0 | 0 | 0 | 1 | 0 | 1 | 1 | 0 | 0 | 0 | 0 | 0 | 0 | 0 | 3 | 2 | 1 | 0 |
| | TAB-APL | 1 | 2 | 1 | 0 | 0 | 1 | 0 | 1 | 0 | 1 | 0 | 0 | 0 | 0 | 0 | 0 | 0 | 1 | 1 | 1 | 0 |
| | | $0^5$ | 2 | 0 | 0 | 3 | 2 | 0 | 0 | 0 | $0^5$ | 0 | 0 | 0 | 0 | 0 | 0 | 0 | 2 | $1^5$ | 1 | 0 |
| | | 2 | 2 | 2 | 1 | 1 | 2 | 1 | 2 | 2 | 2 | 2 | 2 | 2 | 2 | 2 | 2 | 2 | 2 | 2 | 2 | 2 |
| | MULTIQUAL | 2 | 2 | 0 | 1 | 2 | 2 | 2 | 1 | 0 | 2 | 0 | 0 | 0 | 0 | 0 | 0 | 0 | $0^7$ | $0^7$ | $0^7$ | 0 |
| | | 1 | 2 | 0 | 0 | $1^3$ | 1 | $0^7$ | $0^3$ | 0 | 1 | 0 | 0 | 0 | 0 | 0 | 0 | 0 | $0^7$ | $0^7$ | $0^7$ | 0 |
| | | 3 | 3 | 3 | 1 | 3 | 3 | 3 | 3 | 3 | 3 | 3 | 3 | 3 | 3 | 3 | 2 | 2 | 3 | 3 | 3 | 3 |
| | C-TAB | 0 | 1 | 0 | 0 | 1 | 0 | 0 | 0 | 0 | 0 | 0 | 1 | 1 | 0 | 0 | 0 | 0 | 1 | 1 | 1 | 0 |
| | ECTA | 0 | 2 | 0 | 0 | 0 | 0 | 0 | 0 | 0 | 0 | 0 | 0 | 0 | 0 | 0 | 0 | 0 | 1 | 0 | 1 | 0 |
| | | $0^7$ | $1^3$ | 0 | $0^3$ | $0^3$ | $0^7$ | 0 | $0^3$ | 0 | 1 | 0 | 0 | 0 | 0 | 0 | 0 | 0 | $0^7$ | $0^7$ | $0^7$ | 0 |
| | | 3 | 3 | 3 | 3 | 3 | 3 | 3 | 3 | 3 | 3 | 3 | 3 | 3 | 3 | 3 | 3 | 3 | 3 | 3 | 3 | 3 |
| | MLLSA | 0 | 1 | 0 | 0 | 0 | 0 | 0 | 0 | 0 | 0 | 0 | 0 | 0 | 0 | 0 | 0 | 0 | 0 | 0 | 0 | 0 |
| | | 1 | $1^3$ | $0^7$ | 2 | $1^3$ | 1 | $0^3$ | 0 | $0^7$ | $1^7$ | $0^7$ | $0^7$ | $0^7$ | $0^7$ | $0^3$ | 0 | 0 | $0^7$ | $0^7$ | $0^7$ | $0^3$ |
| | | 3 | 3 | 3 | 2 | 3 | 3 | 3 | 2 | 3 | 3 | 3 | 3 | 3 | 3 | 3 | 3 | 3 | 3 | 3 | 3 | 3 |
| Chapter 11 | TSP/DATATRAN | 3 | 3 | 2 | 3 | 2 | 1 | 1 | 3 | 2 | 3 | 1 | 1 | 1 | 0 | 2 | 0 | 0 | 1 | 0 | 0 | 1 |
| | | 2 | 2 | 1 | - | - | 0 | 0 | 1 | 0 | 0 | 0 | 0 | 0 | 0 | 0 | 0 | 0 | 0 | 0 | 0 | 0 |
| | | 1 | 1 | 1 | 0 | 0 | 1 | 1 | 1 | 1 | 1 | 1 | 1 | 1 | 1 | 1 | 1 | 1 | 1 | 1 | 1 | 1 |
| | B34S | 3 | 3 | 0 | 2 | 2 | 0 | 2 | 3 | 3 | 0 | 1 | 1 | 1 | 0 | 0 | 1 | 0 | 0 | 0 | 0 | 1 |
| | | $2^3$ | 3 | $1^7$ | $2^3$ | $2^3$ | $2^3$ | 3 | 3 | $2^3$ | 2 | 3 | 2 | 3 | 2 | 2 | 2 | $1^5$ | $0^3$ | $0^3$ | $0^3$ | $0^3$ |
| | | 3 | 3 | 3 | 3 | 3 | 3 | 2 | 3 | 3 | 2 | 3 | 2 | 3 | 2 | 2 | 2 | 2 | 3 | 3 | 3 | 3 |
| | PACK | 2 | 2 | 0 | 0 | 1 | 0 | 0 | 0 | 0 | 2 | 1 | 0 | 1 | 0 | 0 | 0 | 0 | 1 | 0 | 0 | 0 |
| | | 1 | $1^7$ | 0 | 0 | 1 | 0 | 0 | $0^3$ | 0 | $0^5$ | 0 | 0 | $0^3$ | 0 | $0^3$ | 0 | 0 | $1^3$ | $0^5$ | 1 | 0 |
| | | 3 | 3 | 2 | 1 | 3 | 2 | 3 | 1 | 3 | 2 | 3 | 3 | 3 | 3 | 3 | 3 | 3 | 3 | 2 | 3 | 2 |
| | SHAZAM | 2 | 3 | 2 | 3 | 2 | 1 | 1 | 2 | 1 | 1 | 0 | 0 | 0 | 0 | 0 | 1 | 0 | 1 | 1 | 0 | 1 |
| | | $1^5$ | 2 | 1 | 2 | $1^5$ | $1^5$ | 2 | 1 | $0^5$ | 1 | $0^5$ | $0^5$ | $0^5$ | 0 | $0^5$ | 1 | 0 | 1 | $0^5$ | $0^5$ | $0^5$ |
| | | 2 | 2 | 2 | 2 | 2 | 2 | 2 | 2 | 2 | 2 | 2 | 2 | 2 | 2 | 2 | 2 | 2 | 2 | 2 | 2 | 2 |
| | TSP | 3 | 3 | 2 | 0 | 2 | 1 | 1 | 2 | 1 | 0 | 0 | 0 | 0 | 0 | 0 | 0 | 0 | 1 | 1 | 0 | 1 |
| | | $2^7$ | $2^3$ | 1 | 1 | 2 | $1^3$ | $0^7$ | $1^7$ | $0^3$ | 0 | $0^3$ | $0^3$ | 0 | 0 | 0 | 0 | 0 | $0^7$ | $0^3$ | $0^3$ | $0^7$ |
| | | 3 | 3 | 3 | 2 | 3 | 3 | 3 | 3 | 3 | 2 | 3 | 3 | 3 | 3 | 3 | 3 | 3 | 3 | 3 | 3 | 3 |
| | QUAIL | 3 | 2 | 2 | 2 | 2 | 3 | 3 | 2 | 1 | 1 | 1 | 1 | 1 | 1 | 1 | 2 | 1 | 1 | 1 | 1 | 0 |
| | KEIS/ORACLE | 3 | 3 | 2 | 3 | 2 | 0 | 2 | 2 | 2 | 2 | 2 | 2 | 1 | 0 | 0 | 1 | 1 | 2 | 1 | 2 | 2 |
| Chapter 12 | DATAPAC | 1 | 2 | 2 | 1 | 2 | 1 | 1 | 1 | 2 | 1 | 2 | 1 | 2 | 2 | 1 | 1 | 1 | 3 | 1 | 1 | 2 |
| | IMSL LIBRARY | 1 | 3 | 0 | 1 | 0 | 0 | 1 | 0 | 0 | 0 | 0 | 0 | 0 | 0 | 0 | 1 | 1 | 1 | 1 | 1 | 1 |
| | | 1 | 3 | $0^5$ | - | 2 | 1 | $2^5$ | 1 | 0 | $1^5$ | $0^5$ | $0^5$ | 1 | $0^5$ | 0 | $1^5$ | 2 | 2 | 2 | $2^5$ | 2 |
| | | 2 | 2 | 2 | 0 | 2 | 1 | 2 | 1 | 2 | 2 | 2 | 2 | 2 | 2 | 2 | 2 | 1 | 2 | 2 | 2 | 2 |
| | REPOMAT | 2 | 3 | 1 | 0 | 1 | 0 | 2 | 1 | 0 | 0 | 0 | 0 | 0 | 0 | 0 | 0 | 0 | $0^3$ | 1 | 0 | 0 |
| | | 3 | $2^7$ | $1^5$ | $0^5$ | $0^5$ | $2^3$ | 0 | $2^3$ | $0^5$ | $1^3$ | $0^7$ | $0^7$ | 0 | $0^3$ | 0 | 0 | 0 | $0^3$ | 1 | $0^3$ | $0^3$ |
| | | 3 | 3 | 2 | 2 | 2 | 3 | 2 | 3 | 2 | 3 | 3 | 3 | 3 | 3 | 3 | 2 | 3 | 3 | 3 | 3 | 3 |
| | NMGS2 | 1 | 2 | 0 | 3 | 2 | 2 | 0 | 0 | 0 | 0 | 0 | 0 | 0 | 0 | 0 | 0 | 0 | 0 | 0 | 0 | 0 |
| | | 0 | 1 | 0 | - | 3 | 0 | 0 | 0 | 0 | 2 | 0 | 0 | 0 | 0 | 0 | 0 | 0 | 0 | 0 | 0 | 0 |
| | | 1 | 1 | 1 | 0 | 1 | 1 | 1 | 1 | 1 | 1 | 1 | 1 | 1 | 1 | 1 | 1 | 1 | 1 | 1 | 1 | 1 |

TABLE A.1:  RATINGS OF A STATISTICAL PROGRAM: PART I cont.

CAPABILITIES:  PROCESSING AND DISPLAYING DATA

Usefulness Ratings
0=low
1=modest
2=moderate
3=high
-=no data

Tabular Scheme

| γ |
| β |
| γ |

Note: $2^7$=2.7

α=Developer rating
β=Average user rating
γ=Number of users

| | GENERALITY | | FILING | | | | | | | | EDITING | | | | | MISSING DATA | | DISPLAY | | | |
|---|---|---|---|---|---|---|---|---|---|---|---|---|---|---|---|---|---|---|---|---|---|
| | Subroutines or Package | Generality of Purpose | Filing Summary | Data Set Size | Flexible Data Input | Complex Structures | Filing Missing Data | Storage/Retrieval | File Manipulation | Flexible Output | Editing Summary | Editing Language | Consistency | Probability Checks | Error Handling | Missing Data Summary | Imputation | Display Summary | Compute Tables | Print Tables | Graphics |
| | 6 | 7 | 8 | 9 | 10 | 11 | 12 | 13 | 14 | 15 | 16 | 17 | 18 | 19 | 20 | 21 | 22 | 23 | 24 | 25 | 26 |
| NAG LIBRARY | 1 | 3 | 0 | 0 | 0 | 0 | 1 | 0 | 0 | 0 | 0 | 0 | 0 | 0 | 0 | 0 | 0 | 0 | 0 | 0 | 0 |
| | 1 | $2^5$ | 0 | 2 | 1 | 0 | 1 | 1 | 1 | 1 | 0 | 0 | 0 | 0 | 0 | 0 | 0 | $0^3$ | $0^7$ | 1 | 0 |
| | 3 | 2 | 3 | 2 | 3 | 3 | 3 | 3 | 3 | 3 | 3 | 3 | 3 | 3 | 3 | 3 | 3 | 3 | 3 | 3 | 3 |
| EISPACK | 1 | 2 | 0 | 0 | 0 | 0 | 0 | 0 | 0 | 0 | 0 | 0 | 0 | 0 | 0 | 0 | 0 | 0 | 0 | 0 | 0 |
| | $1^3$ | $1^7$ | 0 | 3 | 0 | 0 | 0 | 0 | 0 | 0 | 0 | 0 | 0 | 0 | 0 | 0 | 0 | 0 | 0 | 0 | 0 |
| | 3 | 3 | 3 | 1 | 2 | 2 | 2 | 2 | 2 | 2 | 3 | 3 | 3 | 3 | 3 | 3 | 3 | 3 | 3 | 3 | 3 |

TABLE A.1:  RATINGS OF A STATISTICAL PROGRAM: PART II

CAPABILITIES:  MATHEMATICAL ANALYSIS OF STATISTICAL DATA

Usefulness Ratings
0=low
1=modest
2=moderate
3=high
-=no data

Tabular Scheme

$2^7 = 2.7$

α=Developer rating
β=Average user rating
γ=Number of users

| | | EXTENSIBILITY | | STATISTICAL ANALYSIS | | | | | | | | | | | | | SAMPLE SURVEY | | | | OTHER MATH | | |
|---|---|---|---|---|---|---|---|---|---|---|---|---|---|---|---|---|---|---|---|---|---|---|---|
| | | Language Math. Power | Code Modifiability | Stat. Analysis Summary | Multiple Regression | Anova/Linear Models | Linear Multivariate | Multi-way Tables | Other Multivariate | Time Series | Non-Parametric | Exploratory | Robust | Non-linear | Bayesian | Econometric | Sample Survey Summary | Compute Estimates | Compute Variances | Select Sample | Simulation | Math. Functions | Operations Research |
| | | 27 | 28 | 29 | 30 | 31 | 32 | 33 | 34 | 35 | 36 | 37 | 38 | 39 | 40 | 41 | 42 | 43 | 44 | 45 | 46 | 47 | 48 |
| **Chapter 2** | UPDATE | 3 | 1 | 0 | 0 | 0 | 0 | 0 | 0 | 0 | 0 | 0 | 0 | 0 | 0 | 0 | 0 | 0 | 0 | 0 | 0 | 0 | 0 |
| | RAPID | 0 | 2 | 0 | 0 | 0 | 0 | 0 | 0 | 0 | 0 | 0 | 0 | 0 | 0 | 0 | 0 | 0 | 0 | 0 | 0 | 0 | 0 |
| | | $0^3$ | $2^7$ | 0 | 0 | 0 | 0 | 0 | 0 | 0 | 0 | 0 | 0 | 0 | 0 | 0 | 0 | 0 | 0 | 0 | 0 | 0 | 0 |
| | | 3 | 3 | 3 | 3 | 3 | 3 | 3 | 3 | 3 | 3 | 3 | 3 | 3 | 3 | 3 | 2 | 3 | 3 | 3 | 3 | 3 | 3 |
| | ADABAS | 2 | 0 | 0 | 0 | 0 | 0 | 0 | 0 | 0 | 0 | 0 | 0 | 0 | 0 | 0 | 0 | 0 | 0 | 0 | 0 | 0 | 0 |
| | | 1 | 2 | 0 | 0 | 0 | 0 | 0 | 0 | 0 | 0 | 0 | 0 | 0 | 0 | 0 | 0 | 0 | 0 | 0 | 0 | 0 | 0 |
| | | 1 | 1 | 1 | 1 | 1 | 1 | 1 | 1 | 1 | 1 | 1 | 1 | 1 | 1 | 1 | 1 | 1 | 1 | 1 | 1 | 1 | 1 |
| | SIR | 1 | 1 | 0 | 0 | 0 | 0 | 0 | 0 | 0 | 0 | 0 | 0 | 0 | 0 | 0 | 0 | 0 | 0 | 0 | 0 | 0 | 0 |
| | | $0^7$ | 1 | $0^3$ | 0 | 0 | 0 | 0 | 0 | 0 | 0 | 0 | 0 | 0 | 0 | 0 | 0 | 0 | 0 | 0 | 0 | 0 | 0 |
| | | 3 | 3 | 3 | 3 | 3 | 3 | 3 | 3 | 3 | 3 | 3 | 3 | 3 | 3 | 3 | 3 | 3 | 3 | 3 | 3 | 3 | 3 |
| | MARK IV | 1 | 0 | 0 | 0 | 0 | 0 | 0 | 0 | 0 | 0 | 0 | 0 | 0 | 0 | 0 | 0 | 0 | 0 | 0 | 0 | 0 | 0 |
| | RIQS | 2 | 0 | 1 | 0 | 0 | 0 | 0 | 0 | 0 | 0 | 0 | 0 | 0 | 0 | 0 | 0 | 0 | 0 | 0 | 0 | 0 | 0 |
| | | 3 | 2 | 0 | 0 | 0 | 0 | 0 | 0 | 0 | 0 | 0 | 0 | 0 | 0 | 0 | 0 | 0 | 0 | 0 | - | 0 | 0 |
| | | 1 | 1 | 1 | 1 | 1 | 1 | 1 | 1 | 1 | 1 | 1 | 1 | 1 | 1 | 1 | 1 | 1 | 1 | 1 | 0 | 1 | 1 |
| | DATA3 | 0 | 0 | 0 | 0 | 0 | 0 | 0 | 0 | 0 | 0 | 0 | 0 | 0 | 0 | 0 | 0 | 0 | 0 | 0 | 0 | 0 | 0 |
| | FILEBOL | 0 | 2 | 0 | 0 | 0 | 0 | 0 | 0 | 0 | 0 | 0 | 0 | 0 | 0 | 0 | 0 | 0 | 0 | 0 | 0 | 0 | 0 |
| | | 0 | 1 | 0 | 0 | 0 | 0 | 0 | 0 | 0 | 0 | 0 | 0 | 0 | 0 | 0 | 0 | 0 | 0 | 0 | 0 | 0 | 0 |
| | | 2 | 2 | 2 | 1 | 1 | 2 | 2 | 2 | 2 | 2 | 2 | 2 | 2 | 2 | 2 | 2 | 2 | 2 | 2 | 2 | 2 | 2 |
| | KPSIM/KPVER | 0 | 3 | 0 | 0 | 0 | 0 | 0 | 0 | 0 | 0 | 0 | 0 | 0 | 0 | 0 | 0 | 0 | 0 | 0 | 0 | 0 | 0 |
| | | 0 | 3 | 0 | 0 | 0 | 0 | 0 | 0 | 0 | 0 | 0 | 0 | 0 | 0 | 0 | 0 | 0 | 0 | 0 | 0 | 0 | 0 |
| | | 1 | 1 | 1 | 1 | 1 | 1 | 1 | 1 | 1 | 1 | 1 | 1 | 1 | 1 | 1 | 1 | 1 | 1 | 1 | 1 | 1 | 1 |
| **Chapter 3** | GES | 0 | 3 | 0 | 0 | 0 | 0 | 0 | 0 | 0 | 0 | 0 | 0 | 0 | 0 | 0 | 0 | 0 | 0 | 0 | 0 | 0 | 0 |
| | CONCOR | 2 | 1 | 0 | 0 | 0 | 0 | 0 | 0 | 0 | 0 | 0 | 0 | 0 | 0 | 0 | 0 | 0 | 0 | 0 | 0 | 0 | 0 |
| | CHARO | 0 | 0 | 0 | 0 | 0 | 0 | 0 | 0 | 0 | 0 | 0 | 0 | 0 | 0 | 0 | 0 | 0 | 0 | 0 | 0 | 0 | 0 |
| | EDITCK | 0 | 3 | 0 | 0 | 0 | 0 | 0 | 0 | 0 | 0 | 0 | 0 | 0 | 0 | 0 | 0 | 0 | 0 | 0 | 0 | 0 | 0 |
| | VCP-LCP | 3 | 1 | 0 | 0 | 0 | 0 | 0 | 0 | 0 | 0 | 0 | 0 | 0 | 0 | 0 | 0 | 0 | 0 | 0 | 0 | 0 | 0 |
| **Chapter 4** (part i) | SYNTAX II | 3 | 2 | 1 | 0 | 0 | 0 | 0 | 0 | 0 | 0 | 0 | 0 | 0 | 0 | 0 | 3 | 1 | 0 | 1 | 1 | 1 | 1 |
| | | - | 1 | 0 | 0 | 0 | 0 | 0 | 0 | 0 | 0 | 0 | 0 | 0 | 0 | 0 | 0 | 0 | 0 | 0 | 0 | 0 | 0 |
| | | 0 | 1 | 1 | 1 | 1 | 1 | 1 | 1 | 1 | 1 | 1 | 1 | 1 | 1 | 1 | 1 | 1 | 1 | 1 | 1 | 1 | 1 |
| | LEDA | 0 | 0 | 0 | 0 | 0 | 0 | 0 | 0 | 0 | 0 | 0 | 0 | 0 | 0 | 0 | 0 | 0 | 0 | 0 | 0 | 0 | 0 |
| | | $0^5$ | 0 | 0 | 0 | 0 | 0 | 0 | 0 | 0 | 0 | 0 | 0 | 0 | 0 | 0 | 0 | 0 | 0 | 0 | 0 | 0 | 0 | 0 |
| | | 2 | 2 | 2 | 2 | 2 | 2 | 2 | 2 | 2 | 2 | 2 | 2 | 2 | 2 | 2 | 2 | 2 | 2 | 2 | 2 | 1 | 1 |
| | CYBER GENINT | 1 | 2 | 0 | 0 | 0 | 0 | 0 | 0 | 0 | 0 | 0 | 0 | 0 | 0 | 0 | 0 | 0 | 0 | 0 | 0 | 0 | 0 |
| | ISIS | 2 | 2 | 0 | 0 | 0 | 0 | 0 | 0 | 0 | 0 | 0 | 0 | 0 | 0 | 0 | 0 | 0 | 0 | 0 | 0 | 1 | 0 |
| | | - | - | - | 0 | 0 | 0 | 0 | 0 | 0 | 0 | 0 | 0 | 0 | 0 | 0 | - | 0 | 0 | 0 | - | - | - |
| | | 0 | 0 | 0 | 1 | 1 | 1 | 1 | 1 | 1 | 1 | 1 | 1 | 1 | 1 | 1 | 0 | 1 | 1 | 1 | 0 | 0 | 0 |
| | VDBS | 1 | 1 | 0 | 0 | 0 | 0 | 0 | 0 | 0 | 0 | 0 | 0 | 0 | 0 | 0 | 0 | 0 | 0 | 0 | 0 | 0 | 0 |
| | GTS | 3 | 3 | 0 | 0 | 0 | 0 | 0 | 0 | 0 | 0 | 0 | 0 | 0 | 0 | 0 | 0 | 0 | 0 | 0 | 0 | 0 | 0 |
| | | 1 | 0 | 0 | 0 | 0 | 0 | 0 | 0 | 0 | 0 | 0 | 0 | 0 | 0 | 0 | 0 | 0 | 0 | 0 | 0 | 0 | 0 |
| | | 2 | 1 | 3 | 3 | 3 | 3 | 3 | 3 | 3 | 3 | 3 | 3 | 3 | 3 | 3 | 3 | 3 | 3 | 3 | 3 | 3 | 3 |
| | CRESCAT | 3 | 0 | 0 | 0 | 0 | 0 | 0 | 0 | 0 | 0 | 0 | 0 | 0 | 0 | 0 | 0 | 0 | 0 | 0 | 1 | 1 | 1 |
| | | - | - | 1 | 0 | 0 | 0 | 0 | 0 | 0 | 1 | 0 | 0 | 0 | 0 | 0 | 0 | 0 | 0 | 0 | 0 | 0 | 0 |
| | | 0 | 0 | 1 | 1 | 1 | 1 | 1 | 1 | 1 | 1 | 1 | 1 | 1 | 1 | 1 | 1 | 1 | 1 | 1 | 0 | 0 | 0 |

## TABLE A.1: RATINGS OF A STATISTICAL PROGRAM: PART II cont.
### CAPABILITIES: MATHEMATICAL ANALYSIS OF STATISTICAL DATA

Usefulness Ratings
0=low
1=modest
2=moderate
3=high
-=no data

Tabular Scheme

| α |
|---|
| β |
| γ |

Note: $2^7$=2.7

α=Developer rating
β=Average user rating
γ=Number of users

| Program | EXTENSIBILITY | | STATISTICAL ANALYSIS | | | | | | | | | | | | | SAMPLE SURVEY | | | | OTHER MATH | | |
|---|---|---|---|---|---|---|---|---|---|---|---|---|---|---|---|---|---|---|---|---|---|---|
| | Language Math. Power | Code Modifiability | Stat. Analysis Summary | Multiple Regression | Anova/Linear Models | Linear Multivariate | Multi-way Tables | Other Multivariate | Time Series | Non-Parametric | Exploratory | Robust | Non-linear | Bayesian | Econometric | Sample Survey Summary | Compute Estimates | Compute Variances | Select Sample | Simulation | Math. Functions | Operations Research |
| | 27 | 28 | 29 | 30 | 31 | 32 | 33 | 34 | 35 | 36 | 37 | 38 | 39 | 40 | 41 | 42 | 43 | 44 | 45 | 46 | 47 | 48 |
| PERSEE | 0 | 0 | 0 | 0 | 0 | 0 | 0 | 0 | 0 | 0 | 0 | 0 | 0 | 0 | 0 | 0 | 0 | 0 | 0 | 0 | 0 | 0 |
| | 0 | 1 | 1 | 2 | 0 | 0 | 1 | 2 | 0 | 0 | 0 | 0 | 0 | 0 | 0 | 2 | 1 | 0 | 1 | 0 | 0 | 0 |
| | 1 | 1 | 1 | 1 | 1 | 1 | 1 | 1 | 1 | 1 | 1 | 1 | 1 | 1 | 1 | 1 | 1 | 1 | 1 | 1 | 1 | 1 |
| CENTS III | 1 | 1 | 0 | 0 | 0 | 0 | 0 | 0 | 0 | 0 | 0 | 0 | 0 | 0 | 0 | 0 | 0 | 0 | 0 | 0 | 0 | 0 |
| COCENTS 1.3 | 1 | 1 | 0 | 0 | 0 | 0 | 0 | 0 | 0 | 0 | 0 | 0 | 0 | 0 | 0 | 0 | 0 | 0 | 0 | 0 | 0 | 0 |
| | 1 | $2^7$ | 0 | 0 | 0 | 0 | 0 | 0 | 0 | 0 | 0 | 0 | 0 | 0 | 0 | 0 | 0 | 0 | 0 | 0 | 0 | 0 |
| | 3 | 3 | 3 | 2 | 3 | 3 | 3 | 3 | 3 | 3 | 3 | 3 | 3 | 3 | 3 | 3 | 3 | 3 | 3 | 3 | 3 | 3 |
| CENTS-AID | 0 | 3 | 1 | 0 | 0 | 0 | 1 | 0 | 0 | 0 | 0 | 0 | 0 | 0 | 0 | 0 | 0 | 0 | 0 | 0 | 0 | 0 |
| | 1 | - | 0 | 0 | 0 | 0 | 0 | 0 | 0 | 0 | 0 | 0 | 0 | 0 | 0 | 0 | 0 | 0 | 0 | 0 | 0 | 0 |
| | 1 | 0 | 2 | 2 | 2 | 2 | 2 | 2 | 2 | 2 | 2 | 2 | 2 | 2 | 2 | 1 | 2 | 2 | 2 | 2 | 1 | 1 |
| FILE 2.0 | 0 | 2 | 0 | 0 | 0 | 0 | 0 | 0 | 0 | 0 | 0 | 0 | 0 | 0 | 0 | 0 | 0 | 0 | 0 | 0 | 0 | 0 |
| OSIRIS 2.4 | 1 | 1 | 0 | 0 | 0 | 0 | 0 | 0 | 0 | 0 | 0 | 0 | 0 | 0 | 0 | 0 | 0 | 0 | 0 | 0 | 0 | 0 |
| GEN. SUMMARY | 0 | 3 | 0 | 0 | 0 | 0 | 0 | 0 | 0 | 0 | 0 | 0 | 0 | 0 | 0 | 0 | 0 | 0 | 0 | 0 | 0 | 0 |
| TPL *(Chapter 4 part ii)* | 1 | 2 | 1 | 0 | 0 | 0 | 0 | 0 | 0 | 1 | 0 | 2 | 0 | 0 | 0 | 2 | 2 | 2 | 0 | 0 | 0 | 0 |
| | $0^7$ | 1 | 0 | 0 | 0 | 0 | 0 | 0 | 0 | 0 | 0 | 0 | 0 | 0 | 0 | 1 | 0 | 0 | 0 | 0 | 0 | 0 |
| | 3 | 3 | 3 | 3 | 3 | 3 | 3 | 3 | 3 | 3 | 3 | 3 | 3 | 3 | 3 | 3 | 3 | 3 | 3 | 3 | 3 | 3 |
| RGSP | 1 | 1 | 0 | 0 | 0 | 0 | 0 | 0 | 0 | 0 | 0 | 0 | 0 | 0 | 0 | 2 | 2 | 2 | 0 | 0 | 0 | 0 |
| | $1^7$ | $0^8$ | $0^5$ | 0 | 0 | 0 | 0 | 0 | $0^3$ | 0 | 0 | 0 | 0 | 0 | 0 | $1^8$ | $1^5$ | $1^3$ | 0 | 0 | 0 | 0 |
| | 3 | 4 | 4 | 4 | 4 | 4 | 4 | 4 | 4 | 4 | 4 | 4 | 4 | 4 | 4 | 4 | 4 | 4 | 4 | 4 | 4 | 4 |
| NCHS-XTAB | 3 | 2 | 0 | 0 | 0 | 0 | 0 | 0 | 0 | 0 | 0 | 0 | 0 | 0 | 0 | 2 | 2 | 0 | 0 | 0 | 0 | 0 |
| SURVEY | 1 | 0 | 2 | 2 | 0 | 1 | 0 | 0 | 0 | 0 | 0 | 0 | 0 | 0 | 0 | 1 | 1 | 0 | 0 | 0 | 0 | 0 |
| SAP | 0 | 2 | 1 | 0 | 0 | 0 | 0 | 0 | 0 | 0 | 0 | 0 | 0 | 0 | 0 | 1 | 1 | 1 | 0 | 0 | 0 | 0 |
| | - | - | 2 | 0 | 0 | 0 | - | 0 | - | - | 0 | 0 | 0 | 0 | 0 | 0 | 0 | 0 | 0 | - | - | - |
| | 0 | 0 | 1 | 0 | 1 | 1 | 0 | 1 | 0 | 0 | 1 | 1 | 1 | 1 | 1 | 1 | 1 | 1 | 1 | 0 | 0 | 0 |
| HES VAR X-TAB *(Chapter 5)* | 0 | 2 | 0 | 0 | 0 | 0 | 0 | 0 | 0 | 0 | 0 | 0 | 0 | 0 | 0 | 3 | 2 | 2 | 0 | 0 | 0 | 0 |
| SPLITHALF | 1 | 1 | 0 | 0 | 0 | 0 | 0 | 0 | 0 | 0 | 0 | 0 | 0 | 0 | 0 | 2 | 2 | 2 | 0 | 0 | 0 | 0 |
| | 0 | - | $2^5$ | 0 | 0 | 0 | 0 | 0 | 0 | 0 | 0 | 0 | 0 | 0 | 0 | $2^5$ | $2^5$ | $2^5$ | 0 | 0 | 0 | 0 |
| | 2 | 0 | 2 | 2 | 2 | 2 | 2 | 2 | 2 | 2 | 2 | 2 | 2 | 2 | 2 | 2 | 2 | 2 | 2 | 2 | 2 | 2 |
| MULTI-FRAME | 2 | 3 | 0 | 0 | 0 | 0 | 0 | 0 | 0 | 0 | 0 | 0 | 0 | 0 | 0 | 2 | 2 | 2 | 0 | 0 | 0 | 0 |
| MWD VARIANCE | 0 | 2 | 3 | 1 | 0 | 0 | 0 | 0 | 0 | 0 | 0 | 0 | 0 | 0 | 0 | 1 | 1 | 1 | 0 | 0 | 0 | 0 |
| | $0^7$ | $1^7$ | $1^5$ | $0^7$ | $0^7$ | 0 | 0 | 0 | 0 | 0 | 0 | 0 | 0 | 0 | 0 | 2 | $1^5$ | 2 | 0 | 0 | 0 | 0 |
| | 3 | 3 | 2 | 3 | 3 | 3 | 3 | 3 | 3 | 3 | 3 | 3 | 3 | 3 | 3 | 2 | 2 | 2 | 2 | 2 | 2 | 3 |
| GSS EST. | 1 | 1 | 0 | 0 | 0 | 0 | 0 | 0 | 0 | 0 | 0 | 0 | 0 | 0 | 0 | 3 | 3 | 2 | 0 | 0 | 0 | 0 |
| | 1 | 1 | 0 | 0 | 0 | 0 | 0 | 0 | 0 | 0 | 0 | 0 | 0 | 0 | 0 | 2 | 1 | 1 | 2 | 0 | 0 | 0 |
| | 1 | 1 | 1 | 1 | 1 | 1 | 1 | 1 | 1 | 1 | 1 | 1 | 1 | 1 | 1 | 1 | 1 | 1 | 1 | 1 | 1 | 1 |
| CLFS VAR-COV | 0 | 1 | 0 | 0 | 0 | 0 | 0 | 0 | 0 | 0 | 0 | 0 | 0 | 0 | 0 | 3 | 3 | 3 | 0 | 0 | 0 | 0 |
| CESVP | 1 | 1 | 0 | 0 | 0 | 0 | 0 | 0 | 0 | 0 | 0 | 0 | 0 | 0 | 0 | 3 | 3 | 3 | 0 | 0 | 0 | 0 |
| KEYFITZ | 0 | 1 | 0 | 0 | 0 | 0 | 0 | 0 | 0 | 0 | 0 | 0 | 0 | 0 | 0 | 3 | 2 | 3 | 0 | 0 | 0 | 0 |
| CLUSTERS | 1 | 2 | 0 | 0 | 0 | 0 | 0 | 0 | 0 | 0 | 0 | 0 | 0 | 0 | 0 | 3 | 2 | 2 | 0 | 0 | 0 | 0 |
| | 2 | 2 | 3 | 0 | 0 | 0 | 0 | 0 | 0 | 0 | 0 | 0 | 0 | 0 | 0 | 3 | 1 | 2 | 0 | 0 | 0 | 0 |
| | 1 | 1 | 1 | 1 | 1 | 1 | 1 | 1 | 1 | 1 | 1 | 1 | 1 | 1 | 1 | 1 | 1 | 1 | 1 | 1 | 1 | 1 |

TABLE A.1:　RATINGS OF A STATISTICAL PROGRAM: PART II cont.

CAPABILITIES:　MATHEMATICAL ANALYSIS OF STATISTICAL DATA

Usefulness Ratings
0=low
1=modest
2=moderate
3=high
-=no data

Tabular Scheme

| α |
|---|
| β |
| γ |

Note: $2^7 = 2.7$

α=Developer rating
β=Average user rating
γ=Number of users

Column legend — EXTENSIBILITY: 27 Language Math. Power, 28 Code Modifiability. STATISTICAL ANALYSIS: 29 Stat. Analysis Summary, 30 Multiple Regression, 31 Anova/Linear Models, 32 Linear Multivariate, 33 Multi-way Tables, 34 Other Multivariate, 35 Time Series, 36 Non-Parametric, 37 Exploratory, 38 Robust, 39 Non-linear, 40 Bayesian, 41 Econometric. SAMPLE SURVEY: 42 Sample Survey Summary, 43 Compute Estimates, 44 Compute Variances, 45 Select Sample. OTHER MATH: 46 Simulation, 47 Math. Functions, 48 Operations Research.

| Program | | 27 | 28 | 29 | 30 | 31 | 32 | 33 | 34 | 35 | 36 | 37 | 38 | 39 | 40 | 41 | 42 | 43 | 44 | 45 | 46 | 47 | 48 |
|---|---|---|---|---|---|---|---|---|---|---|---|---|---|---|---|---|---|---|---|---|---|---|---|
| SUPER CARP | α | 0 | 1 | 2 | 1 | 0 | 0 | 0 | 0 | 0 | 0 | 0 | 0 | 0 | 0 | 0 | 2 | 2 | 1 | 0 | 0 | 0 | 0 |
| | β | 0 | 2 | 3 | 2 | 0 | 0 | 0 | 0 | 0 | 0 | 0 | 0 | 0 | 0 | 1 | 3 | 2 | 2 | 3 | 0 | 0 | 0 |
| | γ | 1 | 1 | 1 | 1 | 1 | 1 | 1 | 1 | 1 | 1 | 1 | 1 | 1 | 1 | 1 | 1 | 1 | 1 | 1 | 1 | 1 | 1 |
| GENVAR | | 1 | 1 | 3 | 0 | 0 | 0 | 0 | 0 | 0 | 0 | 0 | 0 | 0 | 0 | 0 | 3 | 2 | 3 | 0 | 0 | 0 | 0 |
| BTFSS | α | 1 | 0 | 2 | 1 | 1 | 0 | 1 | 0 | 1 | 0 | 0 | 0 | 0 | 0 | 0 | 2 | 1 | 1 | 0 | 1 | 2 | 0 |
| | β | - | - | 2 | 2 | 2 | 1 | 2 | 2 | - | 2 | 0 | 1 | 0 | 0 | - | 3 | 2 | 2 | 1 | - | - | 0 |
| | γ | 0 | 0 | 1 | 1 | 1 | 1 | 1 | 1 | 0 | 1 | 1 | 1 | 1 | 1 | 0 | 1 | 1 | 1 | 1 | 0 | 0 | 1 |
| EXPRESS | α | 3 | 0 | 3 | 2 | 1 | 1 | 0 | 3 | 2 | 0 | 0 | 0 | 1 | 0 | 0 | 2 | 1 | 1 | 0 | 1 | 1 | 1 |
| | β | 2 | 0 | $2^5$ | 2 | 1 | 1 | $0^5$ | 2 | $1^5$ | $0^5$ | 0 | 0 | $0^5$ | 0 | 0 | 1 | $0^5$ | $0^5$ | $0^5$ | 0 | $0^5$ | 0 |
| | γ | 2 | 2 | 2 | 2 | 1 | 2 | 2 | 2 | 2 | 2 | 2 | 2 | 2 | 2 | 1 | 2 | 2 | 2 | 2 | 2 | 2 | 2 |
| EASYTRIEVE | α | 2 | 0 | 3 | 2 | 1 | 1 | 1 | 1 | 1 | 1 | 1 | 1 | 2 | 1 | 1 | 3 | 2 | 2 | 1 | 1 | 1 | 1 |
| | β | 0 | 0 | 0 | 0 | 0 | 0 | 0 | 0 | 0 | 0 | 0 | 0 | 0 | 0 | 0 | 1 | $0^5$ | $0^5$ | $0^5$ | 0 | $0^5$ | 0 |
| | γ | 2 | 2 | 2 | 2 | 2 | 2 | 2 | 2 | 2 | 2 | 2 | 2 | 2 | 2 | 2 | 2 | 2 | 2 | 2 | 1 | 2 | 2 |
| FOCUS | | 2 | 0 | 3 | 3 | 2 | 2 | 1 | 1 | 1 | 1 | 1 | 1 | 1 | 1 | 0 | 2 | 2 | 1 | 1 | 1 | 0 | 0 |
| SURVEYOR/SURV | α | 2 | 0 | 2 | 2 | 2 | 2 | 2 | 2 | 0 | 2 | 2 | 2 | 2 | 0 | 0 | 2 | 2 | 0 | 2 | 0 | 0 | 0 |
| | β | 0 | 0 | 1 | 0 | $0^5$ | 0 | 0 | 0 | 0 | 0 | 0 | $0^5$ | 0 | 0 | 0 | $0^5$ | $0^5$ | 0 | 0 | 2 | 2 | 2 |
| | γ | 2 | 2 | 2 | 2 | 2 | 2 | 2 | 2 | 2 | 2 | 2 | 2 | 2 | 2 | 2 | 2 | 2 | 2 | 2 | 2 | 2 | 2 |
| DATAPLOT | α | 3 | 3 | 3 | 2 | 1 | 1 | 0 | 0 | 3 | 1 | 3 | 2 | 3 | 2 | 1 | 1 | 2 | 2 | 1 | 2 | 2 | 1 |
| | β | 3 | 2 | 2 | 2 | 1 | 1 | 1 | 1 | 0 | 0 | 0 | 1 | 2 | 1 | 0 | 0 | 0 | 0 | 0 | 0 | 1 | 1 |
| | γ | 1 | 1 | 1 | 1 | 1 | 1 | 1 | 1 | 1 | 1 | 1 | 1 | 1 | 1 | 1 | 1 | 1 | 1 | 1 | 1 | 1 | 1 |
| PACKAGE X | α | 3 | 0 | 2 | 2 | 1 | 0 | 1 | 0 | 0 | 3 | 0 | 0 | 0 | 0 | 1 | 0 | 0 | 0 | 0 | 1 | 0 | 0 |
| | β | $2^5$ | 2 | 3 | 2 | $1^5$ | 1 | 1 | 1 | $0^5$ | 2 | 0 | 1 | 1 | 0 | 0 | 1 | $0^5$ | 0 | 1 | 2 | 0 | 0 |
| | γ | 2 | 1 | 2 | 2 | 2 | 2 | 2 | 2 | 2 | 1 | 1 | 1 | 1 | 2 | 2 | 2 | 2 | 2 | 2 | 1 | 2 | 2 |
| SCSS | | 2 | 2 | 3 | 3 | 1 | 1 | 1 | 2 | 0 | 1 | 2 | 1 | 0 | 0 | 0 | 1 | 1 | 0 | 1 | 1 | 1 | 0 |
| SPSS | α | 1 | 2 | 2 | 2 | 1 | 2 | 0 | 2 | 0 | 3 | 0 | 0 | 0 | 0 | 0 | 1 | 1 | 0 | 0 | 2 | 0 | 0 |
| | β | $0^7$ | - | $2^3$ | $1^3$ | 1 | $1^7$ | 1 | 2 | $0^3$ | $2^3$ | 0 | 0 | 0 | 0 | 0 | $0^7$ | $0^7$ | 0 | $0^5$ | $0^5$ | 0 | 0 |
| | γ | 3 | 0 | 3 | 3 | 3 | 3 | 3 | 3 | 3 | 3 | 2 | 3 | 3 | 3 | 3 | 3 | 2 | 2 | 2 | 2 | 3 | 3 |
| SOUPAC | α | 2 | 2 | 1 | 2 | 1 | 2 | 1 | 1 | 1 | 1 | 0 | 0 | 0 | 0 | 1 | 0 | 0 | 0 | 0 | 1 | 3 | 1 |
| | β | 3 | 1 | 3 | 3 | $2^5$ | 3 | 1 | $2^5$ | $1^5$ | $1^5$ | 0 | 1 | $2^5$ | 0 | 3 | 3 | 3 | $2^5$ | $1^5$ | 2 | 3 | $1^5$ |
| | γ | 2 | 1 | 2 | 2 | 2 | 2 | 2 | 2 | 2 | 2 | 2 | 2 | 2 | 2 | 2 | 2 | 2 | 2 | 2 | 2 | 2 | 2 |
| DATATEXT | α | 2 | 2 | 3 | 3 | 3 | 2 | 3 | 2 | 1 | 2 | 0 | 2 | 0 | 0 | 0 | 0 | 0 | 0 | 0 | 0 | 0 | 0 |
| | β | $1^5$ | 0 | $2^5$ | 2 | 2 | 1 | $0^5$ | $0^5$ | 0 | 0 | 0 | 0 | 0 | 0 | 0 | 2 | $2^5$ | 0 | $0^5$ | $0^5$ | 0 | 0 |
| | γ | 2 | 1 | 2 | 1 | 2 | 2 | 2 | 2 | 2 | 2 | 2 | 2 | 2 | 2 | 2 | 2 | 2 | 1 | 2 | 2 | 2 | 2 |
| P-STAT 78 | α | 3 | 3 | 2 | 2 | 1 | 2 | 1 | 2 | 0 | 1 | 0 | 0 | 0 | 0 | 0 | 0 | 0 | 0 | 0 | 0 | 2 | 0 |
| | β | 3 | 2 | 2 | $1^5$ | $0^5$ | $1^5$ | 0 | 1 | 0 | 0 | 0 | 0 | 0 | 0 | 0 | $2^5$ | 1 | $0^5$ | $0^5$ | 0 | 2 | 0 |
| | γ | 2 | 2 | 2 | 2 | 2 | 2 | 2 | 2 | 2 | 2 | 2 | 2 | 2 | 2 | 2 | 2 | 2 | 2 | 2 | 2 | 2 | 2 |
| DAS | α | 3 | 0 | 3 | 3 | 1 | 3 | 2 | 3 | 0 | 0 | 0 | 0 | 1 | 0 | 1 | 0 | 0 | 0 | 0 | 2 | 3 | 0 |
| | β | 3 | 0 | 3 | 3 | 2 | 3 | - | 3 | 0 | - | - | - | 1 | 0 | 0 | - | - | - | - | 3 | - | 0 |
| | γ | 1 | 1 | 1 | 1 | 1 | 1 | 1 | 0 | 1 | 1 | 0 | 0 | 1 | 1 | 1 | 0 | 0 | 0 | 0 | 1 | 0 | 1 |
| SPMS | α | 3 | 1 | 3 | 2 | 1 | 2 | 1 | 2 | 0 | 3 | 0 | 0 | 2 | 0 | 0 | 1 | 1 | 1 | 1 | 3 | 2 | 0 |
| | β | - | - | 1 | 2 | - | 2 | - | - | - | - | - | - | - | - | - | 1 | 2 | - | - | - | - | - |
| | γ | 0 | 0 | 1 | 1 | 0 | 1 | 0 | 0 | 0 | 0 | 0 | 0 | 0 | 0 | 0 | 1 | 1 | 0 | 0 | 0 | 0 | 0 |
| OSIRIS IV | α | 2 | 3 | 3 | 2 | 2 | 2 | 0 | 3 | 0 | 3 | 0 | 1 | 0 | 0 | 1 | 3 | 3 | 3 | 0 | 0 | 0 | 0 |
| | β | $1^7$ | 1 | $2^7$ | $1^7$ | 2 | $2^3$ | $1^3$ | 3 | 0 | $1^7$ | $0^7$ | $0^3$ | 0 | 0 | $0^3$ | $2^7$ | $1^7$ | $1^5$ | $1^5$ | $0^7$ | 0 | 0 |
| | γ | 3 | 2 | 3 | 3 | 3 | 3 | 3 | 3 | 3 | 3 | 3 | 3 | 3 | 3 | 3 | 3 | 3 | 2 | 2 | 3 | 3 | 3 |

(Chapter 6 — BTFSS)

TABLE A.1:  RATINGS OF A STATISTICAL PROGRAM: PART II cont.

CAPABILITIES:  MATHEMATICAL ANALYSIS OF STATISTICAL DATA

Usefulness Ratings
0=low
1=modest
2=moderate
3=high
-=no data

Tabular Scheme

| α |
|---|
| β |
| γ |

Note: $2^7$=2.7

α=Developer rating
β=Average user rating
γ=Number of users

| | | EXTENSIBILITY | | STATISTICAL ANALYSIS | | | | | | | | | | | | | SAMPLE SURVEY | | | | OTHER MATH | | |
|---|---|---|---|---|---|---|---|---|---|---|---|---|---|---|---|---|---|---|---|---|---|---|---|---|
| | | Language Math. Power | Code Modifiability | Stat. Analysis Summary | Multiple Regression | Ancva/Linear Models | Linear Multivariate | Multi-way Tables | Other Multivariate | Time Series | Non-Parametric | Exploratory | Robust | Non-linear | Bayesian | Econometric | Sample Survey Summary | Compute Estimates | Compute Variances | Select Sample | Simulation | Math. Functions | Operations Research |
| | | 27 | 28 | 29 | 30 | 31 | 32 | 33 | 34 | 35 | 36 | 37 | 38 | 39 | 40 | 41 | 42 | 43 | 44 | 45 | 46 | 47 | 48 |
| Chapter 7 | CS | 3 | 3 | 3 | 1 | 1 | 1 | 1 | 2 | 3 | 2 | 1 | 2 | 1 | 1 | | 3 | 2 | 1 | 3 | 2 | 3 | 2 |
| | | 3 | - | 3 | 3 | 3 | 3 | 3 | 2 | - | 3 | 3 | 2 | 2 | - | - | 3 | 3 | 2 | 2 | 2 | 2 | - |
| | | 1 | 0 | 1 | 1 | 1 | 1 | 1 | 1 | 0 | 1 | 1 | 1 | 1 | 0 | 0 | 1 | 1 | 1 | 1 | 1 | 1 | 0 |
| | SAS | 3 | 2 | 3 | 3 | 3 | 3 | 2 | 2 | 2 | 2 | 2 | 3 | 0 | 2 | | 1 | 1 | 2 | 1 | 1 | 3 | 0 |
| | | 3 | 3 | 3 | 3 | $2^7$ | $2^5$ | 3 | 2 | $2^5$ | 2 | $1^5$ | $1^5$ | $2^5$ | 0 | 3 | 2 | $1^5$ | $1^5$ | $1^5$ | 2 | 3 | 1 |
| | | 3 | 3 | 2 | 3 | 2 | 2 | 2 | 2 | 2 | 2 | 2 | 2 | 1 | 1 | | 3 | 2 | 2 | 2 | 2 | 2 | 1 |
| | OMNITAB 80 | 3 | 3 | 3 | 3 | 2 | 2 | 1 | 0 | 2 | 3 | 3 | 1 | 0 | 1 | | 1 | 3 | 2 | 2 | 3 | 3 | 1 |
| | | 0 | 2 | 1 | 2 | - | 0 | 0 | 0 | 0 | 0 | 0 | 0 | 0 | 0 | | 0 | 0 | 0 | 0 | 0 | 2 | 0 |
| | | 1 | 1 | 1 | 1 | 0 | 1 | 1 | 1 | 1 | 1 | 1 | 1 | 1 | 1 | | 1 | 1 | 1 | 1 | 1 | 1 | 1 |
| | HP STAT PACKS | 2 | 3 | 3 | 3 | 2 | 2 | 2 | 1 | 2 | 3 | 2 | 0 | 3 | 0 | 1 | 2 | 1 | 0 | 1 | 3 | 3 | 2 |
| | | 0 | - | 2 | 2 | 1 | 0 | 0 | 0 | 1 | 2 | 0 | 1 | 2 | 0 | 0 | 2 | 2 | 1 | 0 | 0 | 2 | 1 |
| | | 1 | 0 | 1 | 1 | 1 | 1 | 1 | 1 | 1 | 1 | 1 | 1 | 1 | 1 | | 1 | 1 | 1 | 1 | 1 | 1 | 1 |
| | MINITAB II | 3 | 3 | 3 | 3 | 2 | 1 | 0 | 1 | 3 | 2 | 3 | 2 | 1 | 0 | 1 | 1 | 1 | 1 | 1 | 3 | 3 | 0 |
| | | 2 | 3 | 3 | $2^5$ | 2 | 1 | $1^5$ | $0^5$ | $1^5$ | 2 | 1 | 0 | 0 | 0 | 0 | 0 | 0 | 0 | 0 | 3 | 1 | 0 |
| | | 2 | 2 | 2 | 2 | 2 | 2 | 2 | 2 | 1 | 1 | 2 | 2 | 2 | 2 | | 2 | 2 | 2 | 2 | 2 | 2 | 2 |
| | BMDP | 1 | 0 | 3 | 3 | 2 | 3 | 3 | 3 | 0 | 2 | 1 | 2 | 3 | 0 | 0 | 1 | 1 | 1 | 1 | 1 | 1 | 0 |
| | | $1^3$ | $1^7$ | 3 | 3 | 2 | 3 | $2^7$ | $2^3$ | $0^3$ | $1^7$ | 1 | $2^3$ | $1^7$ | $0^3$ | 0 | 0 | 0 | 0 | 0 | $0^7$ | $0^3$ | $0^3$ |
| | | 3 | 3 | 3 | 3 | 3 | 3 | 3 | 3 | 3 | 3 | 3 | 3 | 3 | 3 | | 3 | 3 | 3 | 3 | 3 | 3 | 3 |
| | NISAN | 2 | 3 | 3 | 3 | 3 | 3 | 2 | 3 | 1 | 3 | 2 | 2 | 2 | 3 | 2 | 2 | 2 | 2 | 1 | 2 | 2 | 1 |
| | GENSTAT 4.02 | 3 | 0 | 3 | 2 | 3 | 3 | 3 | 2 | 0 | 2 | 0 | 2 | 2 | 0 | 1 | 0 | 0 | 0 | 0 | 1 | 2 | 0 |
| | | 3 | - | 3 | 2 | 2 | $2^5$ | $2^5$ | $2^5$ | $0^5$ | $0^5$ | 0 | 1 | 2 | 0 | 0 | 1 | 1 | 2 | 0 | 1 | 2 | 0 |
| | | 2 | 0 | 2 | 2 | 2 | 2 | 2 | 2 | 2 | 2 | 1 | 2 | 2 | 1 | 1 | 1 | 1 | 1 | 1 | 1 | 2 | 1 |
| | SPEAKEASY III | 3 | 3 | 3 | 3 | 2 | 3 | 1 | 1 | 3 | 1 | 1 | 2 | 3 | 0 | 3 | 3 | 1 | 3 | 2 | 3 | 3 | 3 |
| | TROLL | 3 | 0 | 3 | 3 | 3 | 2 | 2 | 2 | 0 | 3 | 2 | 3 | 0 | 3 | 0 | 0 | 0 | 0 | 0 | 2 | 2 | 1 |
| | | $2^7$ | 1 | $2^7$ | $2^7$ | $2^5$ | 3 | 2 | - | $2^7$ | 0 | 2 | $2^5$ | $2^3$ | - | $2^7$ | - | 0 | 0 | 0 | $2^7$ | $2^7$ | 2 |
| | | 3 | 2 | 3 | 3 | 2 | 1 | 1 | 0 | 3 | 1 | 2 | 2 | 3 | 0 | 3 | 0 | 1 | 1 | 1 | 3 | 3 | 1 |
| | IDA | 2 | 2 | 3 | 3 | 2 | 1 | 2 | 2 | 3 | 3 | 0 | 1 | 1 | 3 | 1 | 0 | 0 | 0 | 0 | 3 | 3 | 0 |
| | | 2 | 1 | 3 | 3 | 2 | 0 | 3 | 0 | 2 | 1 | 0 | 0 | 0 | 0 | 0 | 0 | 0 | 0 | 0 | 1 | - | - |
| | | 1 | 1 | 1 | 1 | 1 | 1 | 1 | 1 | 1 | 1 | 1 | 1 | 1 | 1 | | 1 | 1 | 1 | 1 | 1 | 0 | 0 |
| Chapter 8 | ISA | 3 | 2 | 3 | 2 | 3 | 3 | 2 | 2 | 2 | 1 | 3 | 2 | 1 | 0 | 0 | 1 | 1 | 1 | 1 | 2 | 2 | 2 |
| | GLIM | 3 | 3 | 3 | 3 | 2 | 0 | 3 | 1 | 0 | 1 | 2 | 3 | 2 | 0 | 1 | 0 | 0 | 0 | 0 | 2 | 1 | 0 |
| | | $2^7$ | 1 | 3 | 2 | $2^7$ | 0 | 3 | 0 | 0 | 0 | 0 | $1^3$ | $1^7$ | 0 | 3 | 0 | 0 | 0 | 0 | 1 | 1 | 0 |
| | | 3 | 2 | 3 | 3 | 3 | 3 | 3 | 3 | 3 | 3 | 3 | 3 | 3 | 3 | | 3 | 3 | 3 | 3 | 3 | 3 | 3 |
| | ISP | 3 | 2 | 3 | 2 | 1 | 1 | 1 | 1 | 0 | 3 | 3 | 1 | 0 | 0 | | 0 | 0 | 0 | 0 | 1 | 2 | 0 |
| | | 3 | 3 | 3 | 3 | 1 | 2 | 1 | 0 | 2 | 0 | 3 | 3 | 0 | 0 | 0 | 0 | 0 | 0 | 0 | 2 | 3 | 0 |
| | | 1 | 1 | 1 | 1 | 1 | 1 | 1 | 1 | 1 | 1 | 1 | 1 | 1 | 1 | | 1 | 1 | 1 | 1 | 1 | 1 | 1 |
| | CMU-DAP | 2 | 2 | 3 | 3 | 1 | 0 | 2 | 0 | 0 | 0 | 3 | 2 | 2 | 0 | 0 | 0 | 0 | 0 | 0 | 2 | 0 | 0 |
| | | 1 | 2 | $2^5$ | $2^5$ | $0^5$ | 0 | $1^5$ | 0 | 0 | 0 | 3 | 2 | 1 | 0 | 0 | 0 | 0 | 0 | 0 | $1^5$ | 0 | 0 |
| | | 2 | 1 | 2 | 2 | 2 | 2 | 2 | 2 | 2 | 2 | 2 | 2 | 2 | 2 | 1 | 2 | 2 | 2 | 2 | 2 | 2 | 2 |
| | RUMMAGE | 2 | 3 | 2 | 2 | 3 | 2 | 3 | 0 | 0 | 0 | 1 | 2 | 0 | 0 | 0 | 0 | 0 | 0 | 0 | 0 | 1 | 0 |
| | | 2 | $1^3$ | 3 | $2^3$ | $2^7$ | $1^5$ | 3 | 0 | 0 | 0 | 0 | $0^3$ | 0 | 0 | 0 | 0 | 0 | 0 | 0 | 0 | 0 | 0 |
| | | 3 | 3 | 3 | 3 | 3 | 2 | 1 | 2 | 3 | 3 | 3 | 3 | 3 | 3 | 3 | 3 | 3 | 3 | 3 | 3 | 3 | 3 |
| | CADA | 0 | 1 | 3 | 2 | 2 | 0 | 1 | 0 | 0 | 0 | 2 | 1 | 0 | 3 | 0 | 0 | 0 | 0 | 0 | 0 | 0 | 0 |
| | | 0 | 3 | 2 | 2 | 0 | 0 | 0 | 1 | 0 | 0 | 3 | 1 | 0 | 3 | 0 | 0 | 0 | 0 | 0 | 0 | 1 | 0 |
| | | 1 | 1 | 1 | 1 | 1 | 1 | 1 | 1 | 1 | 1 | 1 | 1 | 1 | 1 | | 1 | 1 | 1 | 1 | 1 | 1 | 1 |

TABLE A.1:   RATINGS OF A STATISTICAL PROGRAM: PART II cont.
CAPABILITIES:  MATHEMATICAL ANALYSIS OF STATISTICAL DATA

**Usefulness Ratings**
0=low
1=modest
2=moderate
3=high
-=no data

**Tabular Scheme**

| α |
|---|
| β |
| γ |

Note: $2^7 = 2.7$

α=Developer rating
β=Average user rating
γ=Number of users

Column groups: EXTENSIBILITY — Language Math. Power (27), Code Modifiability (28); STATISTICAL ANALYSIS — Stat. Analysis Summary (29), Multiple Regression (30), Anova/Linear Models (31), Linear Multivariate (32), Multi-way Tables (33), Other Multivariate (34), Time Series (35), Non-Parametric (36), Exploratory (37), Robust (38), Non-linear (39), Bayesian (40), Econometric (41); SAMPLE SURVEY — Sample Survey Summary (42), Compute Estimates (43), Compute Variances (44), Select Sample (45); OTHER MATH — Simulation (46), Math. Functions (47), Operations Research (48).

| Program | 27 | 28 | 29 | 30 | 31 | 32 | 33 | 34 | 35 | 36 | 37 | 38 | 39 | 40 | 41 | 42 | 43 | 44 | 45 | 46 | 47 | 48 |
|---|---|---|---|---|---|---|---|---|---|---|---|---|---|---|---|---|---|---|---|---|---|---|
| SURVO | 3 | 2 | 3 | 2 | 1 | 2 | 1 | 2 | 3 | 1 | 0 | 0 | 0 | 0 | 1 | 0 | 0 | 0 | 0 | 0 | 1 | 0 |
| AUTOGRP+ | 3 | 1 | 2 | 0 | 1 | 0 | 1 | 3 | 0 | 3 | 1 | 2 | 0 | 0 | 0 | 0 | 0 | 0 | 0 | 3 | 1 | 2 |
|  | $2^5$ | $0^5$ | $2^5$ | 0 | $0^5$ | 0 | $0^5$ | 1 | 0 | $1^5$ | $0^5$ | $0^5$ | 0 | $0^5$ | 0 | $0^5$ | 0 | $0^5$ | $0^5$ | 2 | 0 | $0^5$ |
|  | 2 | 2 | 2 | 2 | 2 | 2 | 2 | 2 | 2 | 2 | ? | 2 | 2 | 2 | 2 | 2 | 1 | 2 | 2 | 2 | 1 | 2 |
| FORALL | 3 | 3 | 2 | 2 | 1 | 1 | 1 | 1 | 1 | 3 | 0 | 1 | 1 | 0 | 0 | 0 | 0 | 0 | 0 | 1 | 2 | 0 |
|  | $2^3$ | 2 | 3 | 2 | 2 | 1 | $0^7$ | $0^3$ | 1 | $2^3$ | 0 | 0 | $1^5$ | 0 | 0 | 0 | 0 | 0 | 0 | $1^5$ | $1^7$ | $0^5$ |
|  | 3 | 1 | 3 | 3 | 3 | 3 | 3 | 3 | 3 | 3 | 3 | 3 | 2 | 3 | 2 | 2 | 3 | 3 | 3 | 2 | 3 | 2 |
| AQD | 0 | 0 | 2 | 2 | 1 | 2 | 1 | 2 | 0 | 0 | 1 | 2 | 2 | 0 | 2 | 0 | 0 | 0 | 0 | 0 | 0 | 0 |
| STP | 2 | 2 | 2 | 2 | 1 | 0 | 0 | 1 | 1 | 2 | 0 | 0 | 0 | 0 | 0 | 0 | 0 | 0 | 0 | 1 | 0 | 0 |
|  | $0^3$ | 1 | $2^7$ | $1^3$ | 1 | 0 | $0^3$ | $0^3$ | 0 | $1^7$ | $0^3$ | 0 | $0^3$ | 0 | 0 | $0^7$ | $0^3$ | 0 | 0 | 0 | 0 | 0 |
|  | 3 | 1 | 3 | 3 | 3 | 3 | 3 | 3 | 3 | 3 | 3 | 3 | 3 | 3 | 3 | 3 | 3 | 3 | 3 | 3 | 3 | 3 |
| STATPAK | 2 | 1 | 3 | 2 | 1 | 1 | 1 | 1 | 1 | 2 | 0 | 0 | 0 | 0 | 0 | 0 | 0 | 0 | 0 | 0 | 0 | 1 |
| STATUTIL | 3 | 1 | 2 | 1 | 1 | 0 | 0 | 0 | 0 | 1 | 1 | 1 | 0 | 1 | 0 | 0 | 0 | 0 | 0 | 2 | 2 | 0 |
| MICROSTAT | 2 | 1 | 2 | 2 | 1 | 0 | 0 | 0 | 1 | 2 | 0 | 0 | 1 | 0 | 0 | 0 | 0 | 0 | 0 | 0 | 0 | 0 |
| A-STAT 79 | 1 | 1 | 1 | 1 | 1 | 0 | 0 | 0 | 0 | 0 | 0 | 0 | 0 | 0 | 0 | 0 | 0 | 0 | 0 | 0 | 0 | 0 |
| AMANCE | 0 | 2 | 3 | 3 | 3 | 3 | 0 | 2 | 0 | 0 | 0 | 0 | 0 | 0 | 0 | 1 | 1 | 0 | 0 | 0 | 0 | 0 |
|  | 0 | 2 | 3 | 2 | 2 | 3 | 1 | 1 | - | 0 | 0 | - | 2 | 0 | - | 1 | - | - | - | - | 1 | - |
|  | 1 | 1 | 2 | 2 | 2 | 2 | 1 | 1 | 0 | 1 | 1 | 0 | 1 | 1 | 0 | 1 | 0 | 0 | 0 | 0 | 1 | 0 |
| MAC/STAT | 2 | 3 | 3 | 2 | 2 | 2 | 2 | 1 | 0 | 0 | 0 | 0 | 0 | 0 | 0 | 0 | 0 | 0 | 0 | 3 | 2 | 0 |
| REG | 1 | 1 | 3 | 2 | 3 | 2 | 1 | 0 | 0 | 0 | 0 | 1 | 1 | 0 | 0 | 1 | 2 | 0 | 0 | 0 | 1 | 0 |
|  | $1^5$ | 2 | 3 | $2^5$ | 3 | $2^5$ | $2^5$ | $0^5$ | 0 | 0 | 0 | 1 | 2 | 0 | $0^5$ | $1^5$ | $0^5$ | 0 | 0 | 0 | $0^5$ | 0 |
|  | 2 | 2 | 2 | 2 | 2 | 2 | 2 | 2 | 1 | 2 | 1 | 2 | 1 | 1 | 2 | 2 | 2 | 2 | 1 | 1 | 2 | 2 |
| TPD-3 | 2 | 1 | 3 | 2 | 2 | 2 | 1 | 1 | 0 | 1 | 0 | 0 | 2 | 0 | 1 | 1 | 1 | 1 | 0 | 2 | 0 | 0 |
| LSML76 | 2 | 2 | 3 | 2 | 3 | 0 | 0 | 0 | 0 | 0 | 0 | 0 | 1 | 0 | 0 | 2 | 2 | 1 | 0 | 0 | 0 | 0 |
|  | 2 | 2 | $2^7$ | $0^7$ | $2^7$ | $0^3$ | $0^7$ | 0 | $0^3$ | 0 | 0 | $0^7$ | 2 | 0 | $1^7$ | $0^3$ | $1^7$ | 1 | $0^7$ | 0 | $0^3$ | 0 |
|  | 3 | 2 | 3 | 3 | 3 | 3 | 3 | 2 | 3 | 3 | 3 | 3 | 3 | 3 | 3 | 3 | 3 | 3 | 3 | 3 | 3 | 2 |
| LNWD/NLNWD | 2 | 3 | 3 | 3 | 0 | 0 | 0 | 0 | 0 | 0 | 2 | 0 | 3 | 0 | 0 | 0 | 0 | 0 | 0 | 0 | 1 | 0 |
|  | 0 | 2 | $2^5$ | 3 | 2 | 1 | 1 | 0 | 0 | 0 | 0 | 0 | 1 | 0 | 0 | 1 | 1 | $0^5$ | 0 | 0 | 0 | 0 |
|  | 2 | 2 | 2 | 2 | 2 | 2 | 2 | 2 | 2 | 2 | 2 | 2 | 2 | 2 | 2 | 2 | 2 | 2 | 2 | 2 | 2 | 2 |
| ACPBCTET | 2 | 1 | 3 | 3 | 1 | 0 | 0 | 0 | 0 | 0 | 0 | 0 | 3 | 0 | 0 | 0 | 0 | 0 | 0 | 0 | 0 | 0 |
|  | - | 3 | 2 | 1 | 0 | 0 | 1 | 0 | 0 | 0 | 1 | 1 | 0 | 0 | 1 | 0 | 0 | 0 | 0 | 0 | 0 | 1 |
|  | 0 | 1 | 1 | 1 | 1 | 1 | 1 | 1 | 1 | 1 | 1 | 1 | 1 | 1 | 1 | 1 | 1 | 1 | 1 | 1 | 1 | 1 |
| MULPRES | 3 | 2 | 2 | 1 | 0 | 0 | 0 | 0 | 0 | 0 | 0 | 0 | 0 | 0 | 0 | 0 | 0 | 0 | 0 | 0 | 0 | 0 |
| STATSPLINE | 2 | 2 | 2 | 0 | 0 | 0 | 0 | 0 | 0 | 0 | 0 | 0 | 2 | 0 | 0 | 0 | 0 | 0 | 0 | 0 | 0 | 0 |
| ALLOC | 0 | 3 | 3 | 0 | 0 | 0 | 0 | 3 | 0 | 3 | 0 | 2 | 0 | 2 | 0 | 0 | 0 | 0 | 0 | 0 | 0 | 0 |
| POPAN | 3 | 3 | 0 | 0 | 0 | 0 | 0 | 0 | 0 | 0 | 0 | 0 | 0 | 0 | 0 | 0 | 0 | 0 | 0 | 0 | 0 | 0 |
|  | 3 | 3 | 1 | 0 | 0 | 0 | 1 | 0 | 0 | 0 | 0 | 0 | 0 | 0 | 0 | 0 | 0 | 0 | 1 | 2 | 0 | 0 |
|  | 1 | 1 | 1 | 1 | 1 | 1 | 1 | 1 | 1 | 1 | 1 | 1 | 1 | 1 | 1 | 1 | 1 | 1 | 1 | 1 | 1 | 1 |
| CAPTURE | 0 | 3 | 3 | 0 | 0 | 0 | 0 | 0 | 0 | 0 | 0 | 0 | 0 | 0 | 0 | 0 | 0 | 0 | 0 | 1 | 0 | 0 |
|  | 0 | 3 | 0 | 0 | 0 | 0 | 0 | 0 | 0 | 0 | 0 | 0 | 0 | 0 | 0 | 0 | 0 | 0 | 0 | 0 | 0 | 0 |
|  | 2 | 1 | 2 | 2 | 2 | 2 | 2 | 2 | 2 | 2 | 2 | 2 | 2 | 2 | 2 | 2 | 2 | 2 | 2 | 2 | 2 | 1 |

Chapter 9

TABLE A.1: <u>RATINGS OF A STATISTICAL PROGRAM:   PART III cont.</u>

Usefulness Ratings
0=low
1=modest
2=moderate
3=high
-=no data

Tabular Scheme

| α |
|---|
| β |
| γ |

Note: $2^7=2.7$

α=Developer rating
β=Average user rating
γ=Number of users

| Chapter | Program | PORTABILITY | | | | | | EASE OF LEARNING AND USING | | | | | RELIABILITY | | AVERAGE RATINGS FOR ATTRIBUTES (See Section 1.4.3) | | | | | | | | | |
|---|---|---|---|---|---|---|---|---|---|---|---|---|---|---|---|---|---|---|---|---|---|---|---|---|
| | | Availability (49) | Installations (50) | Computer Makes (51) | Mini Version (52) | Core Requirements (53) | Batch/Interactive (54) | Stat. Training (55) | Computer Training (56) | Language Simplicity (57) | Documentation (58) | User Convenience (59) | Maintenance (60) | Tested for Accuracy (61) | Filing $D_f/U_f$ | Detecting Errors $D_d/U_d$ | Tabulation $D_t/U_t$ | Variance Est. $D_v/U_v$ | Statistical Anal. $D_s/U_s$ | Contingency Tab. $D_c/U_c$ | Econometrics $D_e/U_e$ | Mathematics $D_m/U_m$ | Availability $D_a$ | Usability $D_u/U_u$ |
| Chapter 10 | GUHA | 2 | 2 | 2 | 2 | 0 | 0 | 1 / $1^7$ / 3 | 2 / 1 / 3 | 0 / 1 / 3 | 2 / $2^3$ / 3 | 2 / $1^7$ / 3 | 3 / 3 / 3 | 1 / 1 / 2 | $0^7$ / $0^6$ | $0^5$ / $0^1$ | 0 / 0 | $0^5$ / 0 | $0^3$ / $0^2$ | 2 / $1^5$ | 0 / 0 | 0 / 0 | $1^6$ | $1^4$ / $1^1$ |
| | CATFIT | 3 | 1 | 0 | 3 | 3 | 2 | 3 | 3 | 2 | 1 | 3 | 1 | 1 | $0^4$ | 0 | $1^5$ | 0 | 0 | 3 | 0 | 0 | $1^9$ | $2^4$ |
| | TAB-APL | 2 | 3 | 2 | 3 | 3 | 2 | 1 / $1^5$ / 2 | 3 / $2^5$ / 2 | 0 / 0 / 1 | 1 / $0^5$ / 2 | 1 / $1^5$ / 2 | 1 / 0 / 1 | 2 / 1 / 1 | $0^4$ / $0^3$ | 0 / 0 | 1 / 1 | 0 / 0 | 0 / 0 | 2 / 2 | 0 / 0 | 0 / 0 | $2^3$ | $1^2$ / $1^0$ |
| | MULTI-QUAL | 2 | 3 | 3 | 1 | 1 | 0 | 1 / $0^5$ / 2 | 3 / $2^3$ / 3 | 2 / 2 / 3 | 1 / 2 / 3 | 3 / 2 / 3 | 3 / $1^3$ | 2 / $1^5$ | $1^4$ / $0^6$ | 0 / 0 | 0 / 0 | 0 / 0 | 0 / 0 | 3 / 3 | 0 / 0 | 0 / 0 | $1^9$ | 2 / $1^6$ |
| | C-TAB | 2 | 1 | 2 | 0 | 2 | 0 | 1 | 3 | 0 | 1 | 1 | 1 | 2 | $0^1$ / $0^5$ | $0^5$ | 1 | 0 | 0 | 2 | 0 | 0 | $1^1$ | $1^2$ |
| | ECTA | 2 | 3 | 3 | 1 | 2 | 0 | 1 / $1^3$ / 3 | 3 / 3 / 3 | 0 / 0 / 3 | 1 / $1^3$ / 3 | 2 / $1^7$ / 3 | 1 / 2 / 2 | 2 / $2^3$ / 3 | 0 / 0 | 0 / 0 | $0^5$ / $0^3$ | 0 / 0 | 0 / 0 | 3 / $1^7$ | 0 / 0 | 0 / 0 | $1^7$ | $1^4$ / $1^3$ |
| | MLLSA | 2 | 2 | 2 | 0 | 0 | 0 | 1 / $0^7$ / 3 | 2 / 2 / 3 | 0 / $0^3$ / 3 | 2 / $1^3$ / 3 | 1 / $1^3$ / 3 | 3 / $1^5$ / 2 | 2 / $1^7$ / 3 | 0 / 0 | 0 / 0 | 0 / 0 | 0 / 0 | $0^2$ / $0^2$ | 2 / $1^3$ | $0^5$ / 0 | 0 / 0 | $1^3$ | $1^2$ / 1 |
| Chapter 11 | TSP/DATATRAN | 1 | 0 | 2 | 0 | 0 | 3 | 2 / 1 / 1 | 3 / 2 / 1 | 3 / 2 / 1 | 1 / 1 / 1 | 3 / 2 / 1 | 1 / 1 / 1 | 1 / - / 0 | $2^1$ / $0^9$ | 1 / 0 | 0 / 0 | 0 / 0 | $2^2$ / $1^3$ | 0 / 0 | 3 / 3 | 2 / 2 | 1 | $2^4$ / $1^6$ |
| | B34S | 2 | 2 | 2 | 0 | 0 | 0 | 0 / $0^3$ / 3 | 3 / 1 / 3 | 0 / 0 / 3 | 2 / 1 / 3 | 2 / $2^7$ / 3 | 3 / 3 / 3 | 2 / $1^7$ / 3 | $1^7$ / $1^6$ | $0^5$ / $0^2$ | 0 / 0 | 0 / 0 | $1^5$ / $1^5$ | 0 / 0 | 3 / $2^8$ | 0 / 0 | $1^3$ | $1^4$ / $0^8$ |
| | PACK | 3 | 3 | 3 | 1 | 0 | 0 | 0 / 1 / 2 | 3 / $2^5$ / 2 | 0 / $0^3$ / 3 | 2 / 1 / 3 | 2 / $1^3$ / 3 | 3 / $2^3$ / 3 | 2 / 1 / 2 | $0^4$ / $0^2$ | $0^3$ / $0^1$ | 0 / 0 | 0 / 0 | $0^7$ / 0 | 0 / 0 | $2^5$ / $1^5$ | 0 / 0 | $1^9$ | $1^4$ / 1 |
| | SHAZAM | 3 | 3 | 2 | 1 | 0 | 3 | 2 / $1^5$ / 2 | 3 / 3 / 2 | 2 / 2 / 2 | 2 / $1^5$ / 2 | 3 / $2^5$ / 2 | 3 / 3 / 2 | 3 / $1^5$ / 2 | $1^6$ / $1^1$ | 0 / 0 | $0^5$ / $0^3$ | 0 / 0 | $1^7$ / $1^4$ | 0 / 0 | 2 / $1^5$ | 1 / 0 | $2^1$ | $2^4$ / $2^1$ |
| | TSP | 3 | 3 | 3 | 2 | 0 | 0 | 2 / 1 / 3 | 3 / $2^3$ / 3 | 2 / 2 / 3 | 2 / $1^7$ / 3 | 2 / 2 / 3 | 3 / $2^7$ / 3 | 2 / 1 / 1 | 1 / $0^8$ | 0 / 0 | $0^5$ / $0^1$ | 0 / 0 | $0^8$ / $0^8$ | 0 / 0 | 2 / $1^5$ | 1 / 1 | 2 | $2^2$ / $1^8$ |
| | QUAIL | 3 | 3 | 2 | 1 | 0 | 1 | 1 | 2 | 2 | 2 | 3 | 1 | 1 | 2 | 1 | 1 | 0 | $0^2$ | 1 | 2 | 1 | $1^6$ | 2 |
| | KEIS/ORACLE | 1 | 1 | 1 | 1 | 1 | 1 | 0 | 3 | 2 | 3 | 3 | 3 | 0 | $1^9$ | $0^8$ | $1^5$ | 0 | $0^3$ | 0 | $1^5$ | 0 | $1^3$ | $2^2$ |
| Chapter 12 | DATAPAC | 2 | 2 | 3 | 3 | 2 | 3 | 2 | 2 | 0 | 1 | 2 | 3 | 1 | $1^3$ | $1^5$ | 1 | 0 | 1 | 0 | 1 | 1 | $2^6$ | $1^4$ |
| | IMSL LIBRARY | 3 | 3 | 3 | 3 | 3 | 3 | 1 / 0 / 1 | 1 / 2 / 2 | 0 / - / 0 | 2 / $2^5$ / 2 | 2 / 2 / 2 | 3 / 3 / 2 | 1 / $1^5$ / 2 | $0^3$ / $0^3$ | 0 / 0 | 1 / 1 | 2 / 2 | 3 / $2^4$ | 3 / $1^5$ | 3 / $2^8$ | 3 / 3 | 3 | $1^2$ / 1 |
| | REPOMAT | 0 | 0 | 1 | 1 | 2 | 3 | 2 / $0^7$ / 3 | 1 / 3 / 3 | 2 / 2 / 3 | 1 / $1^7$ / 3 | 3 / $1^3$ / 3 | 1 / $1^5$ / 2 | 1 / 1 / 1 | $0^7$ / $0^6$ | 0 / 0 | 0 / 0 | 0 / 0 | $0^8$ / $0^8$ | 0 / 0 | $1^5$ / $0^8$ | 2 / 2 | $1^1$ | $1^8$ / $1^2$ |
| | NMGS2 | 1 | 3 | 2 | 3 | 1 | 0 | 3 / - / 0 | 3 / 3 / 1 | 0 / 0 / 1 | 3 / 1 / 1 | 3 / 2 / 1 | 3 / 3 / 1 | 3 / 3 / 1 | $1^0$ / $0^7$ | 0 / 0 | 0 / 0 | 0 / 0 | $0^8$ / $0^8$ | 0 / 0 | 0 / 0 | 3 / 3 | $1^9$ | $2^4$ / $1^8$ |

TABLE A.1:  RATINGS OF A STATISTICAL PROGRAM: PART II cont.

CAPABILITIES:  MATHEMATICAL ANALYSIS OF STATISTICAL DATA

Usefulness Ratings
0=low
1=modest
2=moderate
3=high
-=no data

Tabular Scheme

| α |
|---|
| β |
| γ |

Note: $2^7 = 2.7$

α=Developer rating
β=Average user rating
γ=Number of users

| | EXTENSIBILITY | | STATISTICAL ANALYSIS | | | | | | | | | | | | | SAMPLE SURVEY | | | | OTHER MATH | | |
|---|---|---|---|---|---|---|---|---|---|---|---|---|---|---|---|---|---|---|---|---|---|---|
| | Language Math. Power | Code Modifiability | Stat. Analysis Summary | Multiple Regression | Anova/Linear Models | Linear Multivariate | Multi-way Tables | Other Multivariate | Time Series | Non-Parametric | Exploratory | Robust | Non-linear | Bayesian | Econometric | Sample Survey Summary | Compute Estimates | Compute Variances | Select Sample | Simulation | Math. Functions | Operations Research |
| | 27 | 28 | 29 | 30 | 31 | 32 | 33 | 34 | 35 | 36 | 37 | 38 | 39 | 40 | 41 | 42 | 43 | 44 | 45 | 46 | 47 | 48 |
| NAG LIBRARY | 0 | 1 | 2 | 1 | 1 | 0 | 0 | 0 | 0 | 0 | 0 | 0 | 2 | 0 | 0 | 0 | 0 | 0 | 0 | 2 | 2 | 1 |
| | 2 | 2 | 2 | $1^7$ | $1^3$ | 1 | $0^3$ | 0 | 0 | $0^7$ | 0 | 0 | 2 | $1^3$ | $0^3$ | $0^3$ | $0^3$ | 0 | 1 | $2^7$ | 3 | 3 |
| | 3 | 3 | 3 | 3 | 3 | 3 | 3 | 3 | 3 | 3 | 3 | 3 | 3 | 3 | 3 | 3 | 3 | 3 | 3 | 3 | 3 | 3 |
| EISPACK | 1 | 3 | 1 | 0 | 0 | 1 | 0 | 1 | 0 | 0 | 0 | 0 | 0 | 0 | 0 | 0 | 0 | 0 | 0 | 0 | 3 | 0 |
| | $2^7$ | $2^3$ | 0 | 0 | 0 | 0 | 0 | 0 | 0 | 0 | 0 | 0 | 0 | 0 | 0 | 0 | 0 | 0 | 0 | 0 | 3 | 0 |
| | 3 | 3 | 3 | 3 | 3 | 3 | 3 | 3 | 3 | 3 | 3 | 3 | 3 | 3 | 3 | 3 | 3 | 3 | 3 | 3 | 3 | 2 |

TABLE A.1:   RATINGS OF A STATISTICAL PROGRAM:   PART III

Usefulness Ratings
0=low
1=modest
2=moderate
3=high
-=no data

Tabular Scheme

| α |
|---|
| β |
| γ |

Note: $2^7 = 2.7$

α=Developer rating
β=Average user rating
γ=Number of users

| Chapter | Program | \[PORTABILITY\] 49 Availability | 50 Installations | 51 Computer Makes | 52 Mini Version | 53 Core Requirements | 54 Batch/Interactive | \[EASE OF LEARNING AND USING\] 55 Stat. Training | 56 Computer Training | 57 Language Simplicity | 58 Documentation | 59 User Convenience | \[RELIABILITY\] 60 Maintenance | 61 Tested for Accuracy | Filing $D_f/U_f$ | Detecting Errors $D_d/U_d$ | Tabulation $D_t/U_t$ | Variance Est. $D_v/U_v$ | Statistical Anal. $D_s/U_s$ | Contingency Tab. $D_c/U_c$ | Econometrics $D_e/U_e$ | Mathematics $D_m/U_m$ | Availability $D_a/U_a$ | Usability $D_u/U_u$ |
|---|---|---|---|---|---|---|---|---|---|---|---|---|---|---|---|---|---|---|---|---|---|---|---|---|
| Chapter 2 | UPDATE | 0 | 0 | 1 | 0 | 2 | 0 | 3 | 3 | 0 | 1 | 3 | 2 | 0 | $1^1$ | 0 | 0 | 0 | 0 | 0 | 0 | 0 | $0^7$ | 2 |
| | RAPID | 2 | 1 | 0 | 0 | 0 | 1 | 3 | 2 | 2 | 2 | 3 | 3 | 0 | $2^1$ | 0 | 0 | 0 | 0 | 0 | 0 | 0 | 1 | $2^4$ |
| | | | | | | | | 3 | 1 | 2 | $2^7$ | 2 | 3 | 0 | 2 | 0 | 0 | 0 | 0 | 0 | 0 | 0 | | 2 |
| | | | | | | | | 3 | 3 | 3 | 3 | 3 | 3 | 1 | | | | | | | | | | |
| | ADABAS | 3 | 3 | 2 | 3 | 0 | 3 | 3 | 3 | 2 | 3 | 3 | 3 | 0 | 2 | 0 | 1 | 0 | 0 | 0 | 0 | 0 | $2^4$ | $2^8$ |
| | | | | | | | | 3 | 3 | 2 | 2 | 3 | 3 | 2 | $1^9$ | 0 | 0 | 0 | 0 | 0 | 0 | 0 | | $2^4$ |
| | | | | | | | | 1 | 1 | 1 | 1 | 1 | 1 | 1 | | | | | | | | | | |
| | SIR | 3 | 2 | 2 | 2 | 1 | 3 | 3 | 2 | 3 | 2 | 3 | 3 | 1 | 3 | $2^3$ | $0^5$ | 0 | 0 | 0 | 0 | 0 | $2^3$ | $2^6$ |
| | | | | | | | | 3 | $2^3$ | 2 | 2 | $2^3$ | 3 | 1 | $2^9$ | $2^2$ | $0^5$ | 0 | 0 | 0 | 0 | 0 | | $2^3$ |
| | | | | | | | | 3 | 3 | 3 | 3 | 3 | 3 | 1 | | | | | | | | | | |
| | MARK IV | 3 | 3 | 2 | 0 | 0 | 0 | 3 | 3 | 0 | 3 | 3 | 3 | 1 | $2^6$ | $1^5$ | 1 | 0 | 0 | 0 | 0 | 0 | $1^6$ | $2^4$ |
| | RIQS | 2 | 2 | 0 | 0 | 0 | 3 | 3 | 3 | 3 | 2 | 3 | 3 | 0 | $2^3$ | 1 | 1 | 0 | 0 | 0 | 0 | 0 | $1^4$ | $2^8$ |
| | | | | | | | | 3 | 2 | 3 | 2 | 2 | - | - | $1^7$ | 0 | 0 | 0 | 0 | 0 | 0 | 0 | | $2^4$ |
| | | | | | | | | 1 | 1 | 1 | 1 | 1 | 0 | 0 | | | | | | | | | | |
| | DATA3 | 1 | 0 | 1 | 3 | 3 | 2 | 2 | 3 | 3 | 1 | 3 | 3 | 0 | $1^9$ | 1 | 0 | 0 | 0 | 0 | 0 | 0 | $1^9$ | $2^4$ |
| | FILEBOL | 2 | 1 | 0 | 1 | 2 | 0 | 3 | 3 | 2 | 2 | 3 | 2 | 0 | 1 | $1^5$ | 0 | 0 | 0 | 0 | 0 | 0 | $1^1$ | $2^6$ |
| | | | | | | | | 3 | $1^5$ | $2^5$ | 1 | $1^5$ | 0 | - | $0^8$ | $1^3$ | 0 | 0 | 0 | 0 | 0 | 0 | | $1^8$ |
| | | | | | | | | 2 | 2 | 2 | 2 | 2 | 1 | 0 | | | | | | | | | | |
| | KPSIM/KPVER | 2 | 0 | 0 | 0 | 3 | 2 | 3 | 3 | 2 | 1 | 1 | 2 | 1 | 1 | $1^3$ | 0 | 0 | 0 | 0 | 0 | 0 | $1^3$ | 2 |
| | | | | | | | | 3 | 2 | 0 | 1 | 3 | 1 | - | $0^6$ | 0 | 0 | 0 | 0 | 0 | 0 | 0 | | $1^4$ |
| | | | | | | | | 1 | 1 | 1 | 1 | 1 | 1 | 0 | | | | | | | | | | |
| Chapter 3 | GES | 2 | 0 | 2 | 1 | 1 | 3 | 2 | 1 | 0 | 1 | 2 | 3 | 1 | $1^6$ | 2 | 0 | 0 | 0 | 0 | 0 | 0 | $1^7$ | $1^2$ |
| | CONCOR | 2 | 2 | 0 | 0 | 2 | 0 | 3 | 1 | 2 | 1 | 3 | 3 | 1 | $1^3$ | $2^3$ | 0 | 0 | 0 | 0 | 0 | 0 | $1^3$ | 2 |
| | CHARO | 2 | 1 | 2 | 3 | 2 | 0 | 3 | 3 | 0 | 1 | 3 | 2 | 1 | $0^4$ | $1^3$ | 0 | 0 | 0 | 0 | 0 | 0 | $1^7$ | 2 |
| | EDITCK | 0 | 0 | 1 | 3 | 2 | 0 | 3 | 1 | 0 | 1 | 2 | 3 | 1 | $0^7$ | 2 | 0 | 0 | 0 | 0 | 0 | 0 | $1^3$ | $1^4$ |
| | VCP-LCP | 0 | 0 | 1 | 0 | 1 | 0 | 3 | 3 | 1 | 1 | 3 | 3 | 0 | $1^0$ | $1^3$ | 0 | 0 | 0 | 0 | 0 | 0 | $0^7$ | $1^8$ |
| Chapter 4 (part i) | SYNTAX II | 2 | 1 | 0 | 2 | 1 | 0 | 3 | 3 | 3 | 2 | 2 | 1 | 2 | $2^9$ | $1^8$ | 3 | $0^5$ | 0 | 0 | 0 | 1 | 1 | $2^6$ |
| | | | | | | | | 3 | 3 | 3 | 2 | 2 | 0 | - | $1^9$ | 0 | $1^5$ | 0 | 0 | 0 | 0 | 0 | | $2^6$ |
| | | | | | | | | 1 | 1 | 1 | 1 | 1 | 1 | 0 | | | | | | | | | | |
| | LEDA | 2 | 2 | 2 | 0 | 0 | 1 | 3 | 2 | 3 | 3 | 3 | 3 | 1 | $2^3$ | 2 | $2^5$ | 0 | 0 | 0 | 0 | 0 | $1^4$ | $2^8$ |
| | | | | | | | | $2^5$ | $2^5$ | $2^5$ | 3 | $2^5$ | 3 | 2 | $2^1$ | $1^6$ | $2^3$ | 0 | 0 | 0 | 0 | 0 | | $2^5$ |
| | | | | | | | | 2 | 2 | 2 | 2 | 2 | 2 | 1 | | | | | | | | | | |
| | CYBER GENINT | 0 | 0 | 0 | 0 | 0 | 0 | 3 | 2 | 2 | 2 | 2 | 2 | 1 | $1^3$ | $1^5$ | 3 | 0 | 0 | 0 | 0 | 0 | $0^3$ | $2^2$ |
| | ISIS | 2 | 1 | 2 | 0 | 1 | 0 | 3 | 1 | 2 | 2 | 3 | 2 | 2 | $1^9$ | 2 | $2^5$ | 0 | 0 | 0 | 0 | 1 | $1^1$ | $2^2$ |
| | | | | | | | | 3 | - | 3 | 1 | - | 0 | - | $1^9$ | $1^3$ | $2^5$ | 0 | 0 | 0 | 0 | 1 | | 2 |
| | | | | | | | | 1 | 0 | 1 | 1 | 0 | 1 | 0 | | | | | | | | | | |
| | VDBS | 0 | 0 | 0 | 3 | 1 | 3 | 3 | 2 | 2 | 3 | 3 | 3 | 2 | $2^3$ | $0^5$ | 3 | 0 | 0 | 0 | 0 | 0 | $1^4$ | $2^6$ |
| | GTS | 0 | 0 | 1 | 1 | 0 | 1 | 3 | 3 | 3 | 3 | 3 | 3 | 2 | $2^6$ | 0 | 3 | 0 | 0 | 0 | 0 | 0 | $1^9$ | 3 |
| | | | | | | | | 3 | $2^3$ | 2 | $2^7$ | $2^7$ | $2^7$ | 2 | $1^8$ | 0 | $2^1$ | 0 | 0 | 0 | 0 | 0 | | $2^5$ |
| | | | | | | | | 3 | 3 | 3 | 3 | 3 | 3 | 2 | | | | | | | | | | |
| | CRESCAT | 2 | 1 | 0 | 0 | 0 | 0 | 3 | 3 | 2 | 2 | 2 | 1 | 1 | $2^3$ | $0^5$ | $2^5$ | 0 | 0 | 0 | 0 | 1 | $0^6$ | $2^4$ |
| | | | | | | | | 3 | 1 | 2 | 1 | 1 | 1 | - | $2^3$ | $0^5$ | $2^5$ | 0 | 0 | 0 | 1 | | | $1^8$ |
| | | | | | | | | 1 | 1 | 1 | 1 | 1 | 1 | 0 | | | | | | | | | | |

AVERAGE RATINGS FOR ATTRIBUTES
See Section 1.4.3

TABLE A.1: RATINGS OF A STATISTICAL PROGRAM: PART III cont.

Usefulness Ratings
0=low
1=modest
2=moderate
3=high
-=no data

Tabular Scheme

$\begin{array}{|c|} \hline \alpha \\ \beta \\ \gamma \\ \hline \end{array}$   Note: $2^7 = 2.7$

α=Developer rating
β=Average user rating
γ=Number of users

| | PORTABILITY | | | | | | EASE OF LEARNING AND USING | | | | | RELIABILITY | | AVERAGE RATINGS FOR ATTRIBUTES (See Section 1.4.3) | | | | | | | | | |
|---|---|---|---|---|---|---|---|---|---|---|---|---|---|---|---|---|---|---|---|---|---|---|---|
| | Availability (49) | Installations (50) | Computer Makes (51) | Mini Version (52) | Core Requirements (53) | Batch/Interactive (54) | Stat. Training (55) | Computer Training (56) | Language Simplicity (57) | Documentation (58) | User Convenience (59) | Maintenance (60) | Tested for Accuracy (61) | Filing $D_f/U_f$ | Detecting Errors $D_d/U_d$ | Tabulation $D_t/U_t$ | Variance Est. $D_v/U_v$ | Statistical Anal. $D_s/U_s$ | Contingency Tab. $D_c/U_c$ | Econometrics $D_e/U_e$ | Mathematics $D_m/U_m$ | Availability $D_a$ | Usability $D_u/U_u$ |
| PERSEE | 3 | 2 | 3 | 2 | 1 | 0 | 3 | 3 | 3 | 2 | 3 | 3 | 1 | $2^7$ | 1 | $2^5$ | 0 | 0 | 0 | 0 | 0 | 2 | $2^8$ |
| | | | | | | | 2 | 2 | 2 | 1 | 3 | 3 | 1 | 2 | $0^5$ | 2 | 0 | 0 | 0 | 0 | 0 | | 2 |
| | | | | | | | 1 | 1 | 1 | 1 | 1 | 1 | 1 | | | | | | | | | | |
| CENTS III | 2 | 3 | 0 | 3 | 3 | 0 | 3 | 1 | 2 | 2 | 3 | 3 | 2 | $2^3$ | $0^8$ | 3 | 0 | 0 | 0 | 0 | 0 | 2 | $2^2$ |
| COCENTS 1.3 | 2 | 3 | 3 | 3 | 2 | 0 | 3 | 1 | 2 | 2 | 3 | 2 | 2 | $2^3$ | 0 | 3 | 0 | 0 | 0 | 0 | 0 | $2^1$ | $2^2$ |
| | | | | | | | 2 | $2^3$ | 2 | $2^3$ | $2^3$ | 2 | $2^3$ | $1^4$ | 0 | 3 | 0 | 0 | 0 | 0 | 0 | | $1^9$ |
| | | | | | | | 2 | 3 | 3 | 3 | 3 | 2 | 3 | | | | | | | | | | |
| CENTS-AID | 3 | 3 | 2 | 1 | 2 | 0 | 3 | 3 | 3 | 2 | 3 | 3 | 2 | $2^3$ | 0 | 2 | 0 | 0 | 1 | 0 | 0 | $0^7$ | $2^8$ |
| | | | | | | | 3 | $2^5$ | 3 | 2 | $2^5$ | $2^5$ | 1 | $1^6$ | 0 | 2 | 0 | 0 | 0 | 0 | 0 | | $2^6$ |
| | | | | | | | 2 | 2 | 2 | 2 | 2 | 2 | 1 | | | | | | | | | | |
| FILE 2.0 | 0 | 0 | 0 | 0 | 0 | 0 | 3 | 3 | 3 | 1 | 2 | 1 | 1 | 1 | $1^5$ | 3 | 0 | 0 | 0 | 0 | 0 | $0^1$ | $2^4$ |
| OSIRIS 2.4 | 2 | 1 | 2 | 0 | 0 | 0 | 3 | 2 | 2 | 2 | 2 | 2 | 1 | 1 | $0^5$ | 3 | 0 | 0 | 0 | 0 | 0 | 1 | $2^2$ |
| GEN. SUMMARY | 0 | 0 | 2 | 1 | 0 | 3 | 2 | 2 | 0 | 1 | 1 | 2 | 1 | 1 | 0 | $1^5$ | 0 | 0 | 0 | 0 | 0 | $1^1$ | $1^2$ |
| **Chapter 4 (part ii)** TPL | 3 | 3 | 1 | 0 | 0 | 0 | 3 | 3 | 3 | 3 | 3 | 3 | 2 | $2^6$ | $0^8$ | 3 | 2 | $0^2$ | 0 | 0 | 0 | $1^4$ | 3 |
| | | | | | | | $2^7$ | $1^3$ | 2 | $2^3$ | $2^3$ | 3 | 2 | $1^5$ | $0^7$ | $2^7$ | 0 | 0 | 0 | 0 | 0 | | $2^1$ |
| | | | | | | | 3 | 3 | 3 | 3 | 3 | 3 | 2 | | | | | | | | | | |
| RGSP | 2 | 2 | 2 | 2 | 1 | 0 | 2 | 3 | 2 | 2 | 3 | 3 | 2 | $1^6$ | 2 | 3 | 2 | 0 | 0 | 0 | 0 | $1^7$ | $2^4$ |
| | | | | | | | $1^7$ | 1 | 2 | $2^3$ | $1^3$ | 3 | $2^5$ | $1^5$ | $1^3$ | 3 | $1^4$ | 0 | 0 | 0 | 0 | | $1^6$ |
| | | | | | | | 3 | 4 | 4 | 3 | 3 | 4 | 2 | | | | | | | | | | |
| NCHS-XTAB | 0 | 0 | 1 | 1 | 0 | 0 | 3 | 3 | 0 | 1 | 3 | 2 | 2 | $1^3$ | 0 | 2 | 1 | 0 | 0 | 0 | 0 | $0^6$ | 2 |
| SURVEY | 1 | 3 | 0 | 0 | 0 | 2 | 3 | 3 | 3 | 3 | 2 | 1 | 2 | 1 | 2 | 2 | $0^5$ | $0^5$ | 0 | 0 | 0 | 1 | $2^8$ |
| SAP | 1 | 1 | 2 | 1 | 2 | 0 | 3 | 2 | 0 | 1 | 1 | 0 | 1 | 1 | $0^5$ | $0^5$ | 1 | 0 | 0 | 0 | 0 | 1 | $1^4$ |
| | | | | | | | 1 | 2 | 0 | - | - | - | - | $0^9$ | 0 | $0^5$ | 0 | 0 | 0 | 0 | 0 | | 1 |
| | | | | | | | 1 | 1 | 1 | 0 | 0 | 0 | 0 | | | | | | | | | | |
| **Chapter 5** HES VAR-X | 2 | 1 | 1 | 1 | 0 | 0 | 3 | 3 | 0 | 1 | 3 | 2 | 1 | $1^6$ | 0 | $1^5$ | 2 | 0 | 0 | 0 | 0 | 1 | 2 |
| SPLITHALF | 0 | 0 | 1 | 1 | 2 | 0 | 1 | 2 | 0 | 1 | 2 | 2 | 1 | $1^4$ | 0 | 1 | 2 | 0 | 0 | 0 | 0 | $0^9$ | $1^2$ |
| | | | | | | | $1^5$ | $2^5$ | 2 | 1 | 2 | $1^5$ | 1 | $0^3$ | 0 | 1 | 2 | 0 | 0 | 0 | 0 | | $1^2$ |
| | | | | | | | 2 | 2 | 2 | 1 | 2 | 2 | 1 | | | | | | | | | | |
| MULTI-FRAME | 0 | 0 | 0 | 0 | 0 | 3 | 0 | 0 | 0 | 1 | 2 | 3 | 1 | $1^6$ | 0 | $1^5$ | 2 | 0 | 0 | 0 | 0 | $0^9$ | $0^6$ |
| MWD VARIANCE | 0 | 0 | 1 | 0 | 2 | 0 | 0 | 1 | 0 | 1 | 2 | 2 | 1 | $1^1$ | 0 | 1 | 1 | $0^2$ | 0 | 0 | 0 | $0^7$ | $0^8$ |
| | | | | | | | 1 | $1^3$ | $0^7$ | $1^7$ | $1^7$ | $1^7$ | 1 | $0^6$ | 0 | $0^1$ | 1 | $0^1$ | 0 | 0 | 0 | | $0^7$ |
| | | | | | | | 3 | 3 | 3 | 3 | 3 | 3 | 2 | | | | | | | | | | |
| GSS EST. | 0 | 0 | 0 | 0 | 1 | 0 | 1 | 2 | 0 | 1 | 1 | 1 | 1 | $0^6$ | $0^3$ | 1 | $2^5$ | 0 | 0 | 0 | 0 | $0^3$ | 1 |
| | | | | | | | 3 | 3 | 0 | 3 | 1 | 1 | 1 | $0^1$ | 0 | $0^5$ | 1 | 0 | 0 | 0 | 0 | | 1 |
| | | | | | | | 1 | 1 | 1 | 1 | 1 | 1 | 1 | | | | | | | | | | |
| CLFS VAR-COV | 0 | 0 | 0 | 0 | 0 | 0 | 0 | 0 | 0 | 1 | 3 | 2 | 1 | $0^4$ | 0 | 0 | 3 | 0 | 0 | 0 | 0 | $0^3$ | $0^8$ |
| CESVP | 0 | 0 | 0 | 0 | 0 | 0 | 0 | 0 | 3 | 0 | 1 | 1 | 2 | $0^6$ | 0 | 0 | 3 | 0 | 0 | 0 | 0 | $0^1$ | $0^8$ |
| KEYFITZ | 0 | 0 | 0 | 0 | 0 | 0 | 1 | 2 | 1 | 1 | 1 | 2 | 2 | $0^3$ | 0 | 0 | $2^5$ | 0 | 0 | 0 | 0 | $0^4$ | $1^2$ |
| CLUSTERS | 2 | 2 | 2 | 1 | 2 | 0 | 0 | 3 | 0 | 3 | 1 | 1 | 1 | $0^7$ | 0 | 0 | 2 | 0 | 0 | 0 | 0 | $1^4$ | $1^4$ |
| | | | | | | | 0 | 2 | 0 | 1 | 2 | 0 | 2 | $0^7$ | 0 | 0 | $1^5$ | 0 | 0 | 0 | 0 | | $0^8$ |
| | | | | | | | 1 | 1 | 1 | 1 | 1 | 1 | 1 | | | | | | | | | | |

TABLE A.1: RATINGS OF A STATISTICAL PROGRAM: PART III cont

Usefulness Ratings
0=low
1=modest
2=moderate
3=high
-=no data

Tabular Scheme

| α |
|---|
| β |
| γ |

Note: $2^7$=2.7

α=Developer rating
β=Average user rating
γ=Number of users

| | | PORTABILITY | | | | | | EASE OF LEARNING AND USING | | | | | RELIABILITY | | AVERAGE RATINGS FOR ATTRIBUTES See Section 1.4.3 | | | | | | | | | |
|---|---|---|---|---|---|---|---|---|---|---|---|---|---|---|---|---|---|---|---|---|---|---|---|---|
| | | Availability 49 | Installations 50 | Computer Makes 51 | Mini Version 52 | Core Requirements 53 | Batch/Interactive 54 | Stat. Training 55 | Computer Training 56 | Language Simplicity 57 | Documentation 58 | User Convenience 59 | Maintenance 60 | Tested for Accuracy 61 | Filing $D_f$ $U_f$ | Detecting Errors $D_d$ $U_d$ | Tabulation $D_t$ $U_t$ | Variance Est. $D_v$ $U_v$ | Statistical Anal. $D_s$ $U_s$ | Contingency Tab. $D_c$ $U_c$ | Econometrics $D_e$ $U_e$ | Mathematics $D_m$ $U_m$ | Availability $D_a$ | Usability $D_u$ $U_u$ |
| | SUPER CARP | 2 | 2 | 2 | 1 | 0 | 0 | 0 | 3 | 0 | 2 | 2 | 3 | 2 | 1 | 0 | 0 | $1^5$ | $0^2$ | 0 | 0 | 0 | $1^4$ | $1^4$ |
| | | | | | | | | 0 | 1 | 0 | 2 | 0 | 1 | 1 | $0^6$ | 0 | 0 | $1^5$ | $0^2$ | 0 | 0 | 0 | | $0^6$ |
| | | | | | | | | 1 | 1 | 1 | 1 | 1 | 1 | 1 | | | | | | | | | | |
| | GENVAR | 2 | 1 | 2 | 1 | 3 | 0 | 0 | 2 | 1 | 1 | 2 | 0 | 1 | $0^4$ | 0 | 0 | $2^5$ | 0 | 0 | 0 | 0 | $1^3$ | $1^2$ |
| Chapter 6 | BTFSS | 0 | 1 | 0 | 0 | 0 | 0 | 3 | 3 | 2 | 1 | 2 | 1 | 1 | $2^4$ | $0^5$ | 2 | 1 | $0^7$ | 0 | 0 | 2 | $0^3$ | $2^2$ |
| | | | | | | | | 2 | 2 | 2 | 1 | 2 | 2 | 1 | $2^3$ | $0^3$ | 2 | 1 | $0^7$ | 0 | 0 | 2 | | $1^8$ |
| | | | | | | | | 1 | 1 | 1 | 1 | 1 | 1 | 1 | | | | | | | | | | |
| | EXPRESS | 3 | 2 | 2 | 3 | 0 | 3 | 3 | 3 | 3 | 3 | 3 | 3 | 1 | $2^9$ | $2^3$ | 3 | 1 | $1^3$ | 0 | 1 | 1 | $2^3$ | 3 |
| | | | | | | | | $2^5$ | 3 | $2^5$ | $2^5$ | 2 | 3 | - | $2^6$ | $1^1$ | 2 | $0^5$ | $1^1$ | 0 | $0^8$ | $0^5$ | | $2^5$ |
| | | | | | | | | 2 | 2 | 2 | 2 | 2 | 2 | 0 | | | | | | | | | | |
| | EASYTRIEVE | 3 | 3 | 3 | 0 | 2 | 0 | 1 | 3 | 3 | 3 | 3 | 3 | 1 | 3 | $2^5$ | $2^5$ | 2 | $1^3$ | 1 | 1 | 1 | 2 | $2^6$ |
| | | | | | | | | 3 | $1^5$ | $2^5$ | 2 | $2^5$ | 3 | 2 | $2^2$ | 1 | 0 | $0^5$ | 0 | 0 | 0 | $0^5$ | | 9 |
| | | | | | | | | 2 | 2 | 2 | 2 | 2 | 2 | 1 | | | | | | | | | | |
| | FOCUS | 3 | 3 | 0 | 0 | 0 | 3 | 3 | 3 | 3 | 3 | 3 | 3 | 2 | 3 | $2^3$ | 3 | $1^5$ | $1^7$ | 1 | $0^5$ | 0 | $1^7$ | 3 |
| | SURVEYOR/SURV | 3 | 2 | 2 | 3 | 1 | 3 | 3 | 3 | 2 | 2 | 3 | 3 | 1 | $2^3$ | $2^5$ | 3 | 1 | 2 | 2 | 0 | 0 | $2^4$ | $2^6$ |
| | | | | | | | | $2^5$ | $2^5$ | 0 | $0^5$ | $1^5$ | 2 | 1 | $1^6$ | $1^8$ | $2^5$ | $0^3$ | $0^1$ | 0 | 0 | 0 | | $1^4$ |
| | | | | | | | | 2 | 2 | 2 | 2 | 2 | 2 | 2 | | | | | | | | | | |
| | DATAPLOT | 0 | 0 | 1 | 1 | 0 | 3 | 2 | 3 | 3 | 2 | 3 | 3 | 2 | 2 | 2 | 2 | 2 | $1^3$ | 0 | 2 | 2 | $1^1$ | $2^6$ |
| | | | | | | | | - | - | - | 0 | 2 | 0 | 1 | $0^4$ | 0 | 1 | 0 | $1^0$ | 0 | 0 | 1 | | 2 |
| | | | | | | | | 0 | 0 | 0 | 1 | 1 | 1 | 1 | | | | | | | | | | |
| | PACKAGE X | 2 | 2 | 1 | 3 | 2 | 3 | 2 | 3 | 3 | 3 | 3 | 3 | 3 | $2^1$ | 2 | 1 | 0 | 1 | 1 | $0^5$ | 0 | $2^3$ | $2^8$ |
| | | | | | | | | 1 | 3 | $2^5$ | $1^5$ | 3 | 2 | $0^5$ | $1^9$ | $0^6$ | 1 | 0 | $0^8$ | 1 | 0 | 0 | | $2^2$ |
| | | | | | | | | 2 | 2 | 2 | 2 | 2 | 2 | 2 | | | | | | | | | | |
| | SCSS | 3 | 3 | 3 | 0 | 1 | 2 | 2 | 3 | 3 | 3 | 3 | 3 | 2 | $2^1$ | $2^3$ | $2^5$ | $0^5$ | $1^3$ | 1 | 0 | 1 | $2^1$ | $2^8$ |
| | SPSS | 3 | 3 | 3 | 3 | 1 | 0 | 3 | 3 | 2 | 3 | 3 | 3 | 1 | $2^3$ | 0 | $1^5$ | $0^5$ | $1^7$ | 0 | 0 | 0 | $2^3$ | $2^8$ |
| | | | | | | | | $0^3$ | $1^7$ | $1^7$ | $2^7$ | $2^3$ | 3 | 3 | $2^1$ | 0 | $1^5$ | $0^3$ | $1^4$ | 0 | 0 | 0 | | $1^7$ |
| | | | | | | | | 3 | 3 | 3 | 3 | 3 | 3 | 2 | | | | | | | | | | |
| | SOUPAC | 2 | 3 | 2 | 1 | 1 | 0 | 1 | 2 | 1 | 2 | 2 | 2 | 1 | $1^7$ | $0^8$ | $0^5$ | 0 | $1^2$ | 1 | 1 | 3 | $1^6$ | $1^6$ |
| | | | | | | | | 0 | 3 | 1 | 3 | $2^5$ | $2^5$ | 3 | $1^7$ | $0^6$ | $0^5$ | 0 | $1^2$ | 1 | 1 | 3 | | $1^4$ |
| | | | | | | | | 1 | 2 | 2 | 2 | 2 | 2 | 2 | | | | | | | | | | |
| | DATATEXT | 3 | 3 | 1 | 1 | 0 | 1 | 3 | 3 | 3 | 3 | 3 | 2 | 2 | $2^6$ | $1^3$ | 2 | 0 | 2 | 3 | $0^5$ | 0 | $1^6$ | 3 |
| | | | | | | | | 2 | $2^5$ | $2^5$ | 2 | 2 | $1^5$ | - | $1^9$ | $1^3$ | 2 | 0 | $0^9$ | $0^5$ | 0 | 0 | | $2^2$ |
| | | | | | | | | 2 | 2 | 2 | 2 | 1 | 2 | 0 | | | | | | | | | | |
| | P-STAT 78 | 3 | 3 | 3 | 3 | 1 | 3 | 2 | 3 | 2 | 2 | 3 | 3 | 2 | 3 | 2 | 3 | 0 | $1^3$ | 1 | 0 | 2 | $2^7$ | $2^4$ |
| | | | | | | | | $2^5$ | 2 | $2^5$ | $2^5$ | 3 | 3 | 2 | $2^4$ | $1^5$ | $2^3$ | 0 | $0^8$ | 0 | 0 | 0 | | $2^2$ |
| | | | | | | | | 2 | 2 | 2 | 2 | 2 | 2 | 2 | | | | | | | | | | |
| | DAS | 2 | 1 | 1 | 0 | 0 | 3 | 2 | 2 | 3 | 3 | 3 | 3 | 2 | $2^6$ | 2 | $1^5$ | 0 | $1^8$ | 2 | $0^5$ | 3 | $1^4$ | $2^6$ |
| | | | | | | | | 1 | 3 | 3 | 2 | 3 | 3 | - | $2^4$ | 2 | $1^5$ | 0 | $1^8$ | 2 | 0 | 3 | | $2^2$ |
| | | | | | | | | 1 | 1 | 1 | 1 | 1 | 1 | 0 | | | | | | | | | | |
| | SPMS | 0 | 0 | 1 | 3 | 2 | 0 | 1 | 2 | 3 | 1 | 3 | 2 | 1 | $2^4$ | $1^8$ | $0^5$ | 1 | 2 | 1 | 0 | 2 | $1^1$ | 2 |
| | | | | | | | | 2 | 2 | 3 | 2 | 2 | - | - | $2^3$ | $1^8$ | $0^5$ | 1 | 2 | 1 | 0 | 2 | | $1^8$ |
| | | | | | | | | 1 | 1 | 1 | 1 | 1 | 0 | 0 | | | | | | | | | | |
| | OSIRIS IV | 3 | 1 | 0 | 1 | 1 | 3 | 2 | 3 | 2 | 2 | 3 | 3 | 1 | $2^9$ | $2^3$ | $2^5$ | 3 | 2 | 0 | $0^5$ | 0 | $1^7$ | $2^4$ |
| | | | | | | | | 1 | $1^7$ | 2 | $2^3$ | $2^3$ | $2^7$ | 2 | $2^3$ | $2^1$ | $1^7$ | $1^6$ | $1^7$ | 0 | $0^1$ | 0 | | $1^8$ |
| | | | | | | | | 2 | 3 | 3 | 3 | 3 | 3 | 1 | | | | | | | | | | |

## TABLE A.1: RATINGS OF A STATISTICAL PROGRAM: PART III cont.

**Usefulness Ratings**
0=low
1=modest
2=moderate
3=high
-=no data

**Tabular Scheme**

| α |
|---|
| β |
| γ |

Note: $2^7 = 2.7$

α=Developer rating
β=Average user rating
γ=Number of users

Legend for attribute columns (See Section 1.4.3): Filing $D_f/U_f$, Detecting Errors $D_d/U_d$, Tabulation $D_t/U_t$, Variance Est. $D_v/U_v$, Statistical Anal. $D_s/U_s$, Contingency Tab. $D_c/U_c$, Econometrics $D_e/U_e$, Mathematics $D_m/U_m$, Availability $D_a$, Usability $D_u/U_u$

PORTABILITY: 49 Availability, 50 Installations, 51 Computer Makes, 52 Mini Version, 53 Core Requirements, 54 Batch/Interactive
EASE OF LEARNING AND USING: 55 Stat. Training, 56 Computer Training, 57 Language Simplicity, 58 Documentation, 59 User Convenience
RELIABILITY: 60 Maintenance, 61 Tested for Accuracy

### Chapter 7

| Program | 49 | 50 | 51 | 52 | 53 | 54 | 55 | 56 | 57 | 58 | 59 | 60 | 61 | Filing | Det.Err | Tab | Var.Est | Stat.Anal | Conting | Econ | Math | Avail | Usab |
|---|---|---|---|---|---|---|---|---|---|---|---|---|---|---|---|---|---|---|---|---|---|---|---|
| **CS** | 3 | 1 | 1 | 1 | 0 | 3 | 2 | 3 | 3 | 3 | 3 | 3 | 1 | 3 | $2^5$ | 3 | $1^5$ | $1^8$ | 1 | $1^5$ | 3 | $1^7$ | $2^8$ |
|  |  |  |  |  |  |  | 2 | 3 | 3 | 3 | 3 | 3 | - | $2^7$ | $2^3$ | $2^5$ | $1^5$ | $1^8$ | 1 | 1 | $1^5$ | 2 | $2^8$ |
|  |  |  |  |  |  |  | 1 | 1 | 1 | 1 | 1 | 1 | 0 |  |  |  |  |  |  |  |  |  |  |
| **SAS** | 3 | 3 | 0 | 1 | 1 | 3 | 3 | 3 | 2 | 3 | 3 | 3 | 3 | 3 | 3 | $2^5$ | $1^5$ | $2^7$ | 3 | 2 | 3 | 2 | $2^8$ |
|  |  |  |  |  |  |  | 3 | $2^7$ | $2^3$ | 3 | 3 | 3 | $2^7$ | $2^8$ | $2^4$ | $2^5$ | $1^3$ | $2^4$ | 3 | 2 | 3 |  | $2^7$ |
|  |  |  |  |  |  |  | 3 | 3 | 3 | 3 | 3 | 3 | 3 |  |  |  |  |  |  |  |  |  |  |
| **OMNITAB 80** | 3 | 3 | 3 | 1 | 0 | 3 | 3 | 3 | 3 | 3 | 3 | 3 | 3 | $2^1$ | $1^8$ | 2 | $2^5$ | $1^8$ | 2 | $0^5$ | 3 | $2^3$ | 3 |
|  |  |  |  |  |  |  | 0 | 2 | 2 | 3 | 2 | 3 | 2 | $0^3$ | 0 | $1^5$ | 0 | $0^7$ | 0 | 0 | 2 |  | $1^8$ |
|  |  |  |  |  |  |  | 1 | 1 | 1 | 1 | 1 | 1 | 1 |  |  |  |  |  |  |  |  |  |  |
| **HP STAT PACKS** | 2 | 3 | 0 | 3 | 3 | 2 | 2 | 3 | 0 | 2 | 3 | 3 | 2 | 1 | $0^8$ | 2 | $0^5$ | $2^3$ | 2 | $1^5$ | 3 | $2^3$ | 2 |
|  |  |  |  |  |  |  | 2 | 2 | 3 | 1 | 2 | 1 | - | $0^4$ | $0^3$ | $0^5$ | $0^5$ | $1^2$ | 0 | $0^5$ | 2 |  | $1^4$ |
|  |  |  |  |  |  |  | 1 | 1 | 1 | 1 | 1 | 1 | 0 |  |  |  |  |  |  |  |  |  |  |
| **MINITAB II** | 3 | 3 | 3 | 3 | 2 | 3 | 3 | 3 | 3 | 3 | 3 | 3 | 3 | 2 | $1^5$ | 3 | 1 | $1^7$ | 0 | 2 | 3 | $2^9$ | 3 |
|  |  |  |  |  |  |  | 2 | 3 | 3 | 3 | $2^5$ | 3 | $2^5$ | $1^5$ | $0^5$ | $1^5$ | 0 | $1^3$ | 0 | $0^8$ | 1 |  | $2^7$ |
|  |  |  |  |  |  |  | 2 | 2 | 2 | 2 | 2 | 2 | 2 |  |  |  |  |  |  |  |  |  |  |
| **BMDP** | 3 | 3 | 3 | 2 | 1 | 0 | 2 | 2 | 3 | 3 | 2 | 3 | 2 | $1^6$ | 2 | $1^5$ | 1 | $2^7$ | 3 | 0 | 1 | $2^1$ | $2^4$ |
|  |  |  |  |  |  |  | $0^7$ | $2^7$ | $2^3$ | $2^7$ | 2 | 3 | $2^7$ | $1^3$ | $1^1$ | $0^8$ | 0 | $2^3$ | $2^7$ | 0 | $0^3$ |  | $1^9$ |
|  |  |  |  |  |  |  | 3 | 3 | 3 | 3 | 3 | 3 | 3 |  |  |  |  |  |  |  |  |  |  |
| **NISAN** | 0 | 2 | 1 | 1 | 1 | 2 | 2 | 2 | 3 | 0 | 1 | 1 | 1 | $1^7$ | 0 | 2 | 2 | $2^8$ | 2 | $1^5$ | 2 | $1^1$ | $1^6$ |
| **GENSTAT** | 3 | 3 | 3 | 3 | 1 | 3 | 2 | 2 | 2 | 2 | 3 | 3 | 3 | $2^4$ | 1 | 3 | 0 | $2^3$ | 3 | $0^5$ | 2 | $2^7$ | $2^2$ |
|  |  |  |  |  |  |  | 1 | 3 | 2 | 2 | $2^5$ | 3 | - | $2^1$ | $0^3$ | $2^5$ | 0 | $1^8$ | $2^5$ | 0 | 2 |  | $1^9$ |
|  |  |  |  |  |  |  | 1 | 1 | 2 | 2 | 2 | 2 | 0 |  |  |  |  |  |  |  |  |  |  |
| **SPEAKEASY III** | 3 | 3 | 2 | 3 | 0 | 3 | 3 | 3 | 2 | 2 | 3 | 3 | 1 | $2^3$ | $1^8$ | 3 | 2 | $2^2$ | 1 | 3 | 3 | $2^4$ | $2^6$ |
| **TROLL** | 3 | 2 | 0 | 0 | 0 | 3 | 3 | 3 | 2 | 3 | 3 | 3 | 1 | $2^3$ | $1^5$ | 2 | 0 | $2^2$ | 2 | $2^5$ | 2 | $1^6$ | $2^8$ |
|  |  |  |  |  |  |  | $1^3$ | 3 | 2 | 3 | $2^5$ | 3 | 3 | $1^5$ | $0^1$ | 1 | 0 | $1^9$ | 2 | $2^3$ | 2 |  | $2^4$ |
|  |  |  |  |  |  |  | 3 | 3 | 3 | 3 | 2 | 3 | 1 |  |  |  |  |  |  |  |  |  |  |
| **IDA** | 3 | 3 | 3 | 3 | 0 | 3 | 1 | 3 | 3 | 3 | 3 | 3 | 1 | $1^6$ | 1 | $2^5$ | 0 | 2 | 2 | 2 | 3 | $2^6$ | $2^6$ |
|  |  |  |  |  |  |  | 3 | 3 | 2 | 1 | 3 | 3 | 1 | $1^3$ | 0 | 1 | 0 | 1 | 2 | 1 | 3 |  | $2^0$ |
|  |  |  |  |  |  |  | 1 | 1 | 1 | 1 | 1 | 1 | 1 |  |  |  |  |  |  |  |  |  |  |

### Chapter 8

| Program | 49 | 50 | 51 | 52 | 53 | 54 | 55 | 56 | 57 | 58 | 59 | 60 | 61 | Filing | Det.Err | Tab | Var.Est | Stat.Anal | Conting | Econ | Math | Avail | Usab |
|---|---|---|---|---|---|---|---|---|---|---|---|---|---|---|---|---|---|---|---|---|---|---|---|
| **ISA** | 0 | 1 | 2 | 3 | 3 | 2 | 1 | 2 | 3 | 1 | 1 | 1 | 1 | $1^6$ | $0^8$ | 2 | 1 | 2 | 2 | 1 | 2 | $1^7$ | $1^6$ |
| **GLIM** | 3 | 3 | 3 | 3 | 2 | 3 | 1 | 3 | 2 | 2 | 3 | 3 | 3 | $1^3$ | 0 | $1^5$ | 0 | $1^5$ | 3 | $0^5$ | 1 | $2^9$ | $2^2$ |
|  |  |  |  |  |  |  | 0 | $2^7$ | 2 | $1^7$ | 2 | 3 | - | $0^9$ | 0 | $0^5$ | 0 | $0^9$ | 3 | 0 | 1 |  | $1^7$ |
|  |  |  |  |  |  |  | 3 | 3 | 3 | 3 | 3 | 3 | 0 |  |  |  |  |  |  |  |  |  |  |
| **ISP** | 2 | 2 | 1 | 3 | 2 | 2 | 2 | 2 | 2 | 1 | 3 | 3 | 1 | $1^6$ | $0^8$ | 1 | 0 | 1 | 1 | $0^5$ | 2 | $2^1$ | 2 |
|  |  |  |  |  |  |  | 0 | 2 | 2 | 3 | 1 | 0 | - | $0^9$ | 0 | 0 | 0 | $0^7$ | 1 | $0^5$ | 2 |  | $1^2$ |
|  |  |  |  |  |  |  | 1 | 1 | 1 | 1 | 1 | 1 | 0 |  |  |  |  |  |  |  |  |  |  |
| **CMU-DAP** | 3 | 2 | 3 | 1 | 2 | 2 | 2 | 3 | 2 | 2 | 2 | 1 | 2 | $1^4$ | $0^3$ | $0^5$ | 0 | 1 | 2 | 0 | 0 | 2 | $2^2$ |
|  |  |  |  |  |  |  | 3 | 3 | 2 | 3 | 3 | 3 | 2 | $1^4$ | 0 | $0^5$ | 0 | $0^7$ | $1^5$ | 0 | 0 |  | $2^2$ |
|  |  |  |  |  |  |  | 2 | 2 | 2 | 2 | 2 | 2 | 1 |  |  |  |  |  |  |  |  |  |  |
| **RUMMAGE** | 3 | 2 | 3 | 2 | 1 | 3 | 1 | 3 | 2 | 1 | 2 | 3 | 2 | $1^3$ | $0^5$ | 1 | 0 | $1^2$ | 3 | 0 | 1 | $2^4$ | $1^8$ |
|  |  |  |  |  |  |  | 0 | 2 | 2 | 1 | $1^3$ | $2^7$ | 2 | 1 | $0^1$ | 1 | 0 | 1 | 3 | 0 | 0 |  | $1^3$ |
|  |  |  |  |  |  |  | 3 | 3 | 3 | 2 | 3 | 3 | 2 |  |  |  |  |  |  |  |  |  |  |
| **CADA** | 2 | 2 | 3 | 3 | 3 | 2 | 1 | 3 | 0 | 2 | 2 | 3 | 1 | $0^3$ | $0^5$ | $0^5$ | 0 | $0^7$ | 1 | 0 | 0 | $2^6$ | $1^6$ |
|  |  |  |  |  |  |  | 3 | 3 | 3 | 2 | 1 | 2 | 0 | $0^3$ | 0 | $0^5$ | 0 | $0^3$ | 0 | 0 | 0 |  | $1^4$ |
|  |  |  |  |  |  |  | 1 | 1 | 1 | 1 | 1 | 1 | 1 |  |  |  |  |  |  |  |  |  |  |

TABLE A.1: RATINGS OF A STATISTICAL PROGRAM: PART III cont.

Usefulness Ratings
0=low
1=modest
2=moderate
3=high
-=no data

Tabular Scheme

| α |
|---|
| β |
| γ |

Note: $2^7$=2.7

α=Developer rating
β=Average user rating
γ=Number of users

| | | PORTABILITY | | | | | | EASE OF LEARNING AND USING | | | | | RELIABILITY | | AVERAGE RATINGS FOR ATTRIBUTES See Section 1.4.3 | | | | | | | | | |
|---|---|---|---|---|---|---|---|---|---|---|---|---|---|---|---|---|---|---|---|---|---|---|---|---|
| | | Availability | Installations | Computer Makes | Mini Version | Core Requirements | Batch/Interactive | Stat. Training | Computer Training | Language Simplicity | Documentation | User Convenience | Maintenance | Tested for Accuracy | Filing $D_f$ $U_f$ | Detecting Errors $D_d$ $U_d$ | Tabulation $D_t$ $U_t$ | Variance Est. $D_v$ $U_v$ | Statistical Anal. $D_s$ $U_s$ | Contingency Tab. $D_c$ $U_c$ | Econometrics $D_e$ $U_e$ | Mathematics $D_m$ $U_m$ | Availability $D_a$ | Usability $D_u$ $U_u$ |
| | | 49 | 50 | 51 | 52 | 53 | 54 | 55 | 56 | 57 | 58 | 59 | 60 | 61 | | | | | | | | | | |
| | SURVO | 3 | 2 | 2 | 3 | 3 | 3 | 3 | 3 | 3 | 2 | 3 | 3 | 2 | $2^3$ | $1^5$ | 2 | 0 | $1^3$ | 1 | 2 | 1 | $2^7$ | $2^8$ |
| | AUTOGRP+ | 3 | 1 | 0 | 2 | 0 | 3 | 3 | 3 | 2 | 3 | 3 | 3 | 0 | $2^6$ | $0^5$ | $2^5$ | 0 | $1^2$ | 1 | 0 | 1 | $1^7$ | $2^8$ |
| | | | | | | | | 2 | 3 | 2 | $2^5$ | $1^5$ | 3 | - | $1^7$ | 0 | $1^8$ | 0 | $0^5$ | $0^5$ | 0 | 0 | | $2^2$ |
| | | | | | | | | 1 | 1 | 2 | 2 | 2 | 2 | 0 | | | | | | | | | | |
| | FORALL | 1 | 0 | 1 | 1 | 1 | 3 | 0 | 2 | 2 | 2 | 2 | 1 | 1 | $1^4$ | 1 | 2 | 0 | $1^5$ | 1 | $0^5$ | 2 | $1^1$ | $1^6$ |
| | | | | | | | | 0 | $0^5$ | 2 | $1^7$ | 2 | $1^3$ | $1^5$ | $1^3$ | $0^9$ | $1^5$ | 0 | $1^3$ | $0^7$ | $0^5$ | $1^7$ | | $1^2$ |
| | | | | | | | | 2 | 2 | 3 | 3 | 2 | 3 | 2 | | | | | | | | | | |
| | AQD | 3 | 2 | 2 | 2 | 2 | 3 | 3 | 3 | 3 | 2 | 3 | 3 | 1 | $1^7$ | 0 | 0 | 0 | $1^5$ | 1 | 0 | 0 | $1^4$ | $2^8$ |
| | STP | 3 | 3 | 0 | 1 | 2 | 3 | 3 | 3 | 2 | 2 | 3 | 3 | 2 | $1^3$ | $0^3$ | 1 | 0 | 1 | 0 | $0^5$ | 0 | $2^1$ | $2^6$ |
| | | | | | | | | 2 | 3 | 0 | $1^5$ | $1^7$ | 1 | 1 | $0^5$ | $0^2$ | $0^5$ | 0 | $0^7$ | 0 | $0^0$ | 0 | | $1^6$ |
| | | | | | | | | 3 | 3 | 2 | 2 | 3 | 1 | 1 | | | | | | | | | | |
| | STATPAK | 1 | 1 | 2 | 1 | 2 | 2 | 2 | 3 | 2 | 2 | 2 | .3 | 2 | 1 | $0^5$ | $0^5$ | 0 | $1^2$ | 1 | $0^5$ | 0 | $1^7$ | $2^2$ |
| | STATUTIL | 0 | 0 | 2 | 3 | 0 | 3 | 2 | 3 | 0 | 1 | 2 | 3 | 2 | $1^9$ | $0^5$ | 1 | 0 | $0^5$ | 0 | 0 | 2 | $1^6$ | $1^6$ |
| | MICROSTAT | 0 | 2 | 2 | 3 | 3 | 2 | 2 | 3 | 3 | 2 | 2 | 3 | 2 | 1 | $0^3$ | 0 | 0 | 1 | 0 | $0^5$ | 0 | $2^1$ | $2^4$ |
| | A-STAT 79 | 2 | 3 | 1 | 3 | 3 | 3 | 3 | 3 | 3 | 2 | 2 | 3 | 1 | $0^7$ | 0 | $1^5$ | 0 | $0^3$ | 0 | 0 | 0 | $2^6$ | $2^6$ |
| Chapter 9 | AMANCE | 0 | 1 | 2 | 3 | 3 | 0 | 1 | 2 | 0 | 3 | 1 | 1 | 1 | $1^4$ | 1 | 2 | $0^5$ | $1^8$ | 0 | 0 | 0 | $1^4$ | $1^4$ |
| | | | | | | | | $0^5$ | $1^5$ | 0 | 3 | 1 | $2^5$ | $1^5$ | 1 | $0^8$ | $1^5$ | $0^5$ | $1^3$ | 0 | 0 | 0 | | $1^2$ |
| | | | | | | | | 2 | 2 | 2 | 2 | 1 | 2 | 2 | | | | | | | | | | |
| | MAC/STAT | 0 | 0 | 1 | 1 | 2 | 0 | 0 | 2 | 0 | 2 | 3 | 1 | 1 | $2^6$ | 0 | 3 | 0 | $1^2$ | 2 | 0 | 2 | $0^7$ | $1^4$ |
| | REG | 3 | 1 | 2 | 0 | 0 | 0 | 1 | 2 | 2 | 1 | 2 | 2 | 1 | $1^4$ | $0^8$ | 1 | 1 | $1^3$ | 1 | 0 | 1 | $1^1$ | $1^6$ |
| | | | | | | | | $0^5$ | 3 | 2 | $1^5$ | 3 | $1^5$ | 2 | 1 | $0^5$ | $0^5$ | $0^3$ | $1^3$ | 1 | 0 | $0^5$ | | $1^5$ |
| | | | | | | | | 2 | 2 | 2 | 2 | 2 | 2 | 2 | | | | | | | | | | |
| | TPD-3 | 2 | 1 | 2 | 3 | 1 | 0 | 1 | 2 | 0 | 1 | 2 | 2 | 1 | $1^3$ | $0^8$ | $0^5$ | 1 | $1^7$ | 1 | $0^5$ | 0 | $1^6$ | $1^2$ |
| | LSML76 | 2 | 3 | 3 | 0 | 0 | 0 | 1 | 3 | 0 | 1 | 3 | 3 | 2 | $1^4$ | 0 | 0 | $1^5$ | 1 | 0 | 0 | 0 | $1^6$ | $1^6$ |
| | | | | | | | | $0^7$ | 2 | 0 | 1 | $2^7$ | $1^7$ | $1^7$ | $0^7$ | 0 | 0 | $1^3$ | $0^7$ | 0 | 0 | 0 | | $1^3$ |
| | | | | | | | | 3 | 3 | 3 | 3 | 3 | 3 | 3 | | | | | | | | | | |
| | LNWD/NLNWD | 3 | 3 | 3 | 1 | 2 | 1 | 3 | 3 | 0 | 3 | 3 | 3 | 3 | $0^3$ | 0 | 1 | 0 | 1 | 1 | 0 | 1 | $2^3$ | $2^4$ |
| | | | | | | | | 1 | 3 | 0 | 3 | $2^5$ | 3 | 3 | $0^3$ | 0 | $0^5$ | 0 | $0^7$ | 1 | 0 | 0 | | $1^9$ |
| | | | | | | | | 2 | 2 | 2 | 2 | 2 | 2 | 2 | | | | | | | | | | |
| | ACPBCTET | 2 | 0 | 0 | 0 | 0 | 0 | 1 | 3 | 0 | 1 | 0 | 1 | 2 | $0^3$ | 0 | 0 | 0 | $1^2$ | 0 | 0 | 0 | $0^4$ | 1 |
| | | | | | | | | 0 | 1 | 0 | 0 | 2 | - | 1 | $0^3$ | 0 | 0 | 0 | $0^3$ | 0 | 0 | 0 | | $0^2$ |
| | | | | | | | | 1 | 1 | 1 | 1 | 1 | 0 | 1 | | | | | | | | | | |
| | MULPRES | 0 | 0 | 2 | 0 | 3 | 0 | 2 | 3 | 0 | 1 | 1 | 2 | 1 | 0 | 0 | 0 | 0 | $0^2$ | 0 | 0 | 0 | 1 | $1^4$ |
| | STATSPLINE | 2 | 1 | 2 | 3 | 1 | 0 | 1 | 3 | 2 | 2 | 0 | 1 | 2 | $0^7$ | $0^5$ | $0^5$ | 0 | $0^3$ | 0 | 0 | 0 | $1^4$ | $1^6$ |
| | ALLOC | 3 | 2 | 2 | 1 | 1 | 0 | 1 | 0 | 0 | 3 | 2 | 3 | 2 | $0^7$ | 1 | 2 | 0 | 1 | 0 | 0 | 0 | $1^7$ | $1^2$ |
| | POPAN | 2 | 1 | 1 | 1 | 0 | 0 | 2 | 2 | 3 | 2 | 3 | 1 | 1 | $2^1$ | 1 | 2 | 0 | 0 | 0 | 0 | 0 | $0^9$ | $2^4$ |
| | | | | | | | | 1 | 1 | 3 | 3 | 3 | 3 | 3 | 2 | 1 | $1^5$ | 0 | 0 | 0 | 0 | 0 | | $2^0$ |
| | | | | | | | | 1 | 1 | 1 | 1 | 1 | 1 | 1 | | | | | | | | | | |
| | CAPTURE | 2 | 2 | 2 | 1 | 1 | 3 | 2 | 3 | 3 | 2 | 3 | 3 | 2 | $0^4$ | 0 | 0 | 0 | 0 | 0 | 0 | 0 | 2 | $2^6$ |
| | | | | | | | | 2 | $1^5$ | 3 | 1 | 2 | 2 | 2 | $0^4$ | 0 | 0 | 0 | 0 | 0 | 0 | 0 | | $1^9$ |
| | | | | | | | | 2 | 2 | 2 | 2 | 2 | 1 | 2 | | | | | | | | | | |

TABLE A.1: RATINGS OF A STATISTICAL PROGRAM: PART II cont.
CAPABILITIES: MATHEMATICAL ANALYSIS OF STATISTICAL DATA

Usefulness Ratings
0=low
1=modest
2=moderate
3=high
-=no data

Tabular Scheme

| α |
| β |
| γ |

Note: $2^7$=2.7

α=Developer rating
β=Average user rating
γ=Number of users

Column headings:

- EXTENSIBILITY: 27 = Langauge Math. Power, 28 = Code Modifiability
- STATISTICAL ANALYSIS: 29 = Stat. Analysis Summary, 30 = Multiple Regression, 31 = Anova/Linear Models, 32 = Linear Multivariate, 33 = Multi-way Tables, 34 = Other Multivariate, 35 = Time Series, 36 = Non-Parametric, 37 = Exploratory, 38 = Robust, 39 = Non-linear, 40 = Bayesian, 41 = Econometric
- SAMPLE SURVEY: 42 = Sample Survey Summary, 43 = Compute Estimates, 44 = Compute Variances, 45 = Select Sample
- OTHER MATH: 46 = Simulation, 47 = Math. Functions, 48 = Operations Research

| Chapter | Program | 27 | 28 | 29 | 30 | 31 | 32 | 33 | 34 | 35 | 36 | 37 | 38 | 39 | 40 | 41 | 42 | 43 | 44 | 45 | 46 | 47 | 48 |
|---|---|---|---|---|---|---|---|---|---|---|---|---|---|---|---|---|---|---|---|---|---|---|---|
| Chapter 10 | GUHA | 0 | 2 | 2 | 0 | 0 | 0 | 2 | 2 | 0 | 0 | 0 | 0 | 0 | 0 | 0 | 2 | 1 | 0 | 0 | 0 | 0 | 0 |
|  |  | $0^3$ | 1 | 2 | 0 | 0 | 0 | $1^5$ | 1 | 0 | 1 | $0^5$ | 0 | 0 | 0 | $0^3$ | 0 | 0 | 0 | 0 | 0 | 0 | 0 |
|  |  | 3 | 3 | 3 | 2 | 2 | 2 | 2 | 2 | 3 | 3 | 2 | 2 | 3 | 3 | 3 | 1 | 2 | 2 | 2 | 3 | 3 | 3 |
|  | CATFIT | 2 | 2 | 3 | 0 | 0 | 0 | 3 | 0 | 0 | 0 | 0 | 0 | 0 | 0 | 0 | 0 | 0 | 0 | 0 | 0 | 0 | 0 |
|  | TAB-APL | 0 | 2 | 3 | 0 | 0 | 0 | 2 | 0 | 0 | 0 | 0 | 0 | 0 | 0 | 0 | 0 | 0 | 0 | 0 | 0 | 0 | 0 |
|  |  | 3 | $2^5$ | 3 | 0 | 0 | 0 | 3 | 0 | 0 | 0 | 0 | 0 | 0 | 0 | 0 | 0 | 0 | 0 | 0 | 0 | 0 | 0 |
|  |  | 2 | 2 | 2 | 2 | 2 | 2 | 2 | 2 | 2 | 2 | 2 | 2 | 2 | 2 | 2 | 1 | 2 | 2 | 2 | 2 | 2 | 2 |
|  | MULTIQUAL | 3 | 1 | 3 | 0 | 0 | 0 | 3 | 0 | 0 | 0 | 0 | 0 | 0 | 0 | 0 | 0 | 0 | 0 | 0 | 0 | 0 | 0 |
|  |  | 0 | 0 | 3 | $0^3$ | 0 | 0 | 3 | 0 | 0 | 0 | 0 | 0 | 0 | 0 | 0 | 0 | 0 | 0 | 0 | 0 | 0 | 0 |
|  |  | 3 | 1 | 3 | 3 | 3 | 3 | 3 | 3 | 3 | 3 | 3 | 3 | 3 | 3 | 3 | 3 | 3 | 3 | 3 | 3 | 3 | 3 |
|  | C-TAB | 0 | 0 | 1 | 0 | 0 | 0 | 2 | 0 | 0 | 0 | 0 | 0 | 0 | 0 | 0 | 0 | 0 | 0 | 0 | 0 | 0 | 0 |
|  | ECTA | 0 | 2 | 3 | 0 | 0 | 0 | 3 | 0 | 0 | 0 | 0 | 0 | 0 | 0 | 0 | 1 | 0 | 0 | 0 | 0 | 0 | 0 |
|  |  | $0^5$ | $2^3$ | 2 | 0 | 0 | 0 | $1^7$ | 0 | 0 | 3 | 2 | 2 | 2 | 3 | 3 | 3 | 3 | 3 | 3 | 3 | 3 | 3 |
|  |  | 2 | 3 | 3 | 3 | 3 | 3 | 3 | 3 | 2 | 2 | 2 | 2 | 3 | 3 | 3 | 3 | 3 | 3 | 3 | 3 | 3 | 3 |
|  | MLLSA | 1 | 3 | 2 | 0 | 0 | 0 | 2 | 1 | 0 | 0 | 0 | 0 | 0 | 0 | 1 | 0 | 0 | 0 | 0 | 0 | 0 | 0 |
|  |  | 1 | 2 | 3 | $0^3$ | 0 | 0 | $1^3$ | $1^5$ | 0 | 0 | 0 | 0 | 0 | 0 | 0 | 0 | 0 | 0 | 0 | 1 | 0 | 0 |
|  |  | 3 | 3 | 3 | 3 | 3 | 3 | 3 | 2 | 3 | 3 | 3 | 3 | 3 | 3 | 3 | 3 | 2 | 2 | 3 | 3 | 3 | 3 |
| Chapter 11 | TSP/DATATRAN | 3 | 2 | 3 | 3 | 2 | 2 | 0 | 2 | 3 | 0 | 0 | 0 | 2 | 2 | 3 | 0 | 0 | 0 | 0 | 2 | 2 | 0 |
|  |  | 2 | 0 | 3 | 3 | 1 | 0 | 0 | 2 | 3 | 1 | 0 | 1 | 1 | 2 | 3 | 0 | 0 | 0 | 0 | 2 | 3 | - |
|  |  | 1 | 1 | 1 | 1 | 1 | 1 | 1 | 1 | 1 | 1 | 1 | 1 | 1 | 1 | 1 | 1 | 1 | 1 | 1 | 1 | 1 | 0 |
|  | B34S | 3 | 3 | 3 | 3 | 3 | 0 | 0 | 0 | 3 | 0 | 0 | 0 | 3 | 2 | 3 | 0 | 0 | 0 | 0 | 2 | 0 | 0 |
|  |  | 2 | 2 | 3 | 3 | 3 | 0 | $0^7$ | 0 | $2^7$ | 1 | 0 | $1^7$ | 3 | 2 | 3 | 0 | 0 | 0 | 0 | $0^5$ | 1 | $1^7$ |
|  |  | 2 | 3 | 3 | 3 | 3 | 3 | 3 | 3 | 3 | 2 | 3 | 3 | 3 | 3 | 3 | 2 | 3 | 3 | 3 | 2 | 3 | 3 |
|  | PACK | 0 | 3 | 3 | 3 | 0 | 0 | 0 | 0 | 3 | 0 | 0 | 0 | 1 | 0 | 2 | 0 | 0 | 0 | 0 | 0 | 0 | 0 |
|  |  | 0 | $1^7$ | $1^3$ | 0 | 0 | 0 | 0 | 0 | 3 | 0 | 0 | 0 | 0 | 0 | 0 | 0 | 0 | 0 | 0 | 0 | 0 | 0 |
|  |  | 2 | 3 | 3 | 2 | 3 | 3 | 3 | 3 | 3 | 3 | 3 | 3 | 2 | 2 | 2 | 3 | 3 | 3 | 3 | 2 | 2 | 2 |
|  | SHAZAM | 2 | 3 | 3 | 3 | 2 | 2 | 0 | 1 | 1 | 0 | 0 | 1 | 2 | 0 | 3 | 0 | 0 | 0 | 0 | 3 | 1 | 0 |
|  |  | $1^5$ | $2^5$ | $2^5$ | 3 | $2^5$ | $1^5$ | $0^5$ | $0^5$ | 0 | $0^5$ | 0 | 1 | $1^5$ | 0 | 3 | 0 | 0 | 0 | 0 | $1^5$ | 0 | 0 |
|  |  | 2 | 2 | 2 | 2 | 2 | 2 | 2 | 2 | 2 | 2 | 2 | 2 | 2 | 2 | 2 | 2 | 2 | 2 | 2 | 2 | 2 | 2 |
|  | TSP | 3 | 2 | 0 | 2 | 0 | 1 | 0 | 0 | 1 | 0 | 0 | 0 | 2 | 0 | 3 | 0 | 0 | 0 | 0 | 0 | 1 | 0 |
|  |  | $1^7$ | $1^3$ | $2^7$ | $2^3$ | $1^3$ | $1^7$ | $0^7$ | $0^7$ | $0^3$ | 0 | 0 | $0^7$ | $1^7$ | 0 | $2^7$ | 0 | 0 | 0 | 0 | 0 | $1^3$ | 0 |
|  |  | 3 | 3 | 3 | 3 | 3 | 3 | 3 | 3 | 3 | 3 | 3 | 3 | 3 | 3 | 3 | 3 | 3 | 3 | 3 | 3 | 3 | 3 |
|  | QUAIL | 3 | 2 | 3 | 1 | 0 | 0 | 1 | 0 | 1 | 0 | 0 | 0 | 0 | 0 | 3 | 0 | 0 | 0 | 0 | 0 | 1 | 0 |
|  | KEIS/ORACLE | 2 | 0 | 2 | 2 | 0 | 0 | 0 | 0 | 1 | 0 | 0 | 0 | 0 | 0 | 2 | 0 | 0 | 0 | 0 | 1 | 0 | 0 |
| Chapter 12 | DATAPAC | 1 | 3 | 2 | 1 | 1 | 1 | 0 | 1 | 2 | 1 | 2 | 1 | 1 | 1 | 0 | 0 | 0 | 0 | 0 | 3 | 1 | 0 |
|  | IMSL LIBRARY | 2 | 3 | 3 | 3 | 3 | 3 | 3 | 3 | 3 | 3 | 1 | 2 | 3 | 1 | 3 | 2 | 2 | 2 | 1 | 3 | 3 | 2 |
|  |  | 0 | 3 | 3 | 2 | 2 | 3 | $1^5$ | 2 | $2^5$ | $2^5$ | 0 | - | 3 | 2 | - | 1 | - | - | - | 3 | 3 | $1^5$ |
|  |  | 2 | 2 | 2 | 2 | 2 | 1 | 2 | 1 | 2 | 2 | 1 | 0 | 1 | 0 | 0 | 1 | 0 | 0 | 0 | 2 | 2 | 2 |
|  | REPOMAT | 3 | 3 | 2 | 2 | 2 | 0 | 0 | 0 | 0 | 0 | 1 | 1 | 0 | 3 | 0 | 0 | 0 | 0 | 0 | 2 | 2 | 1 |
|  |  | 3 | 0 | 3 | $2^3$ | 2 | 2 | $1^3$ | $1^5$ | 0 | 0 | 0 | 0 | 1 | $0^5$ | $1^5$ | 0 | 0 | 0 | 0 | $1^5$ | 3 | $0^5$ |
|  |  | 3 | 1 | 3 | 3 | 3 | 3 | 3 | 2 | 2 | 2 | 1 | 1 | 3 | 2 | 2 | 3 | 3 | 3 | 3 | 2 | 3 | 2 |
|  | NMGS2 | 3 | 3 | 0 | 3 | 2 | 0 | 0 | 0 | 0 | 0 | 0 | 0 | 0 | 0 | 0 | 0 | 0 | 0 | 0 | 0 | 3 | 2 |
|  |  | 3 | 3 | 3 | 3 | 2 | 2 | 0 | 0 | 0 | 0 | 2 | - | 0 | 0 | 0 | 0 | 0 | 0 | 0 | 0 | 3 | - |
|  |  | 1 | 1 | 1 | 1 | 1 | 1 | 1 | 1 | 1 | 1 | 1 | 0 | 1 | 1 | 1 | 1 | 1 | 1 | 1 | 1 | 1 | 0 |

## TABLE A.1:  RATINGS OF A STATISTICAL PROGRAM:  PART III cont.

**Usefulness Ratings**
0=low
1=modest
2=moderate
3=high
-=no data

**Tabular Scheme**

| α |
|---|
| β |
| γ |

Note: $2^7$=2.7

α=Developer rating
β=Average user rating
γ=Number of users

| | PORTABILITY | | | | | | EASE OF LEARNING AND USING | | | | | RELIABILITY | | AVERAGE RATINGS FOR ATTRIBUTES — See Section 1.4.3 | | | | | | | | | |
|---|---|---|---|---|---|---|---|---|---|---|---|---|---|---|---|---|---|---|---|---|---|---|---|
| | Availability | Installations | Computer Makes | Mini Version | Core Requirements | Batch/Interactive | Stat. Training | Computer Training | Language Simplicity | Documentation | User Convenience | Maintenance | Tested for Accuracy | $D_f U_f$ Filing | $D_d U_d$ Detecting Errors | $D_t U_t$ Tabulation | $D_v U_v$ Variance Est. | $D_s U_s$ Statistical Anal. | $D_c U_c$ Contingency Tab. | $D_e U_e$ Econometrics | $D_m U_m$ Mathematics | $D_a U_a$ Availability | $D_u U_u$ Usability |
| | 49 | 50 | 51 | 52 | 53 | 54 | 55 | 56 | 57 | 58 | 59 | 60 | 61 | | | | | | | | | | |
| NAG | 2 | 3 | 3 | 3 | 0 | 0 | 1 | 1 | 0 | 2 | 1 | 3 | 2 | $0^1$ | 0 | 0 | 0 | $0^7$ | 0 | 0 | 2 | 2 | 1 |
| | | | | | | | $1^7$ | $1^7$ | $1^5$ | 3 | $2^3$ | 3 | 2 | $0^1$ | 0 | 0 | 0 | $0^7$ | 0 | 0 | 2 | | 1 |
| | | | | | | | 3 | 3 | 2 | 3 | 3 | 3 | 2 | | | | | | | | | | |
| EISPACK | 3 | 3 | 3 | 3 | 3 | 0 | 3 | 1 | 0 | 3 | 3 | 3 | 3 | 0 | 0 | 0 | 0 | $0^3$ | 0 | 0 | 3 | $2^6$ | 2 |
| | | | | | | | $1^3$ | 1 | - | $2^3$ | $1^7$ | $2^7$ | $2^7$ | 0 | 0 | 0 | 0 | 0 | 0 | 0 | 3 | | $1^3$ |
| | | | | | | | 3 | 3 | 0 | 3 | 3 | 3 | 3 | | | | | | | | | | |

REFERENCES

1. Brode, John (1978), "Rights and Responsibilities of Statistical Language
   Users: Language Standards for Statistical Computing," Proceedings of
   Computer Science and Statistics: 11th Annual Symposium on the Inter-
   face, 7-20.
2. Chambers, J. M. (1973), "Linear Regression Computations: Some Numerical
   Statistical Aspects," Bulletin of the International Statistical
   Institute, 45, 4, 255-267.
3. Dreyfus, H. L. (1972), What Computers Can't Do: A Critique of Artificial
   Reason. New York: Harper and Row.
4. Francis, I. (1977), "The Statistical Profession and the Quality of
   Statistical Software," Bulletin of the International Statistical
   Institute, 46, 212-236.
5. Francis, I. (1979), A Comparative Review of Statistical Software. Voorburg:
   The Netherlands: International Association for Statistical Computing.
6. Francis, I., R. Heiberger, and P. Velleman  (1975), "Criteria and Considerations in the Evaluation of Statistical Program Packages," American
   Statistician, 29, 1, 53-56.
7. Francis, I. and J. Sedransk  (1976), "Software Requirements for the
   Analysis of Surveys," Proceedings Ninth International Biometric Conference, 228-253.
8. Francis, I. and J. Sedransk  (1979), "A Comparison of Software for
   Processing and Analyzing Surveys," Bulletin of the International Statistical Institute, 48.
9. Hartigan, J. (1972), "Direct Clustering of a Data Matrix," Journal of
   American Statistical Association, 67, 123-129.
10. Hartigan, J. (1975), Clustering Algorithms, New York: John Wiley and Sons.
11. Heiberger, R. M. (1976), "A Conceptualization of Experimental Designs and
    Their Specification and Computation with ANOVA Programs," Proceedings
    of Statistical Computing Section, American Statistical Association, 13-
    14.
12. Kaufmann, A. (1913), Theorie und Methoden der Statistik, Tubingen:
    J.C.B. Mohr.
13. Kohm, R. F., T. R. Ryan, and P. F. Velleman  (1977), "Index of Available
    Statistical Software," Proceedings of Statistical Computing Section,
    American Statistical Association, 283-284.
14. Ling, R. F. (1974), "Comparison of Several Algorithms for Computing Sample
    Means and Variances," Journal of American Statistical Association, 69,
    348, 859-866.
15. Rowe, B. C. and M. Scheer  (1976), Computer Software for Social Data,
    Social Science Research Council, (British).
16. Schucany, W. R., B. S. Shannon and P. D. Minton (1972), "Survey of
    Statistical Packages," Computing Surveys, 4, 2, 65.
17. Searle, S. (1979), "Deciphering the Output of ANOVA Programs for Unequal-
    Subclass-Numbers Data Using Benchmark Data Sets," Proceedings of
    Computer Science and Statistics, 12th Annual Symposium on the Interface.
18. Velleman, P. F., J. R. Seaman, and I. E. Allen. (1977), "Evaluating
    Package Regression Routines," Proceedings of Statistical Computing
    Section, American Statistical Association, 82-83.
19. Wexelblat, R. L. (1978), "What is a Language for Statistical Computing?"
    Proceedings of Computer Science and Statistics: 11th Annual Symposium
    on the Interface, 25-37.
20. Wittgenstein, L. (1958), Philosophical Invesigation, 3rd. Ed. New York:
    Macmillan Company.

INDEX